Principles of Imprecise-Information Processing

Shiyou Lian

Principles of Imprecise-Information Processing

A New Theoretical and Technological System

 Springer

Shiyou Lian
Xi'an Shiyou University
Xi'an
China

ISBN 978-981-10-9383-8   ISBN 978-981-10-1549-6   (eBook)
DOI 10.1007/978-981-10-1549-6

Printed on acid-free paper

This Springer imprint is published by Springer Nature
The registered company is Springer Science+Business Media Singapore Pte Ltd.

Preface

With the development of information and intelligence sciences and technologies as well as the rise in social requirements, imprecise-information processing about flexible linguistic values is becoming more and more important and urgent, and it will play an indispensable role in intelligent systems, especially in the anthropomorphic intelligent systems.

Imprecision, which is different from uncertainty, is another independent attribute of information, and the now so-called fuzziness is actually a kind of imprecision. Therefore, the author proposed explicitly the concepts of imprecise information and imprecise-information processing, and the book, just as the title shows, is a monograph on imprecise-information processing.

Actually, on imprecise-information processing, many scholars have been doing research with some results, among which the fuzzy set theory introduced by American Professor Loft Zadeh is the most famous. In fact, since Zadeh proposed the concept of fuzzy sets in 1965, the fuzzy-information processing technology based on fuzzy set theory has developed rapidly and made some achievements. However, so far, some important theoretical and technical problems in fuzzy-information processing have not been solved very well. For this reason, not a few scholars worked to improve and develop fuzzy set theory, and presented many new ideas, theories, and methods, which all have their respective angles of view and characteristics. But on the whole, people have not yet reached a common view, and the existed problems are neither solved really. Making a general survey of the decades of imprecise-information processing, although people presented many theories and methods, a theoretical and technological system has not yet been formed, that is, widely approved and has solid foundation of mathematics and logic like that for uncertain-information processing. In particular, some scholars still put imprecision of information into the category of uncertainty of information or mix the two together to do research. Therefore, imprecise-information processing is still a significant subject necessitating careful research.

After years of concentrated study, the author discovers that the imprecision of information originates from the phenomenon of "continuous distribution or change" of magnitudes of a feature of things (or in other words, "uniform chain similarity" of things) and the treating way of "flexible clustering" of human brain. Thus, based on this and combined with the ways of human brain dealing with imprecise information in daily language, I have examined and explored the principles and methods of imprecise-information processing in an all-round way. As a result, a series of new theories and methods different from fuzzy technology were obtained, which forms a new theoretical and technological system of imprecise information processing. This book is just a summation of these research results. Of course, viewed from the relationship between flexible sets and fuzzy sets, this book can also be viewed as an "amendment" to fuzzy-information processing technology; however, it does not follow traditional thinking to make modifications and supplementations in the existing framework of fuzzy set theory. Rather, tracing to the source and opening a new path, this book researches and explores the imprecise-information processing with new perspectives and ideas.

As early in the start of the 1990s, while building expert systems, from the doubt of Zadeh's CRI (compositional rule of inference) fuzzy reasoning, the author began to think about the problem of "fuzzy." In the period, I analyzed the objective cause that brings about the "fuzziness" of information and proposed some terminologies, concepts and methods such as "flexible linguistic values," "flexible concepts," "degreed logic,"and "reasoning with degrees". In 2000, a book *Degree theory* was published in which I summed up the research results at that time. After that, I continued to explore in this direction, further examined the formation principles and mathematical models of flexible concepts, and realized that fuzzy set is somewhat too general in describing a "fuzzy concept." Accordingly, I introduced the terminology and concept of "flexible sets," further examined the flexible linguistic values, and then founded the corresponding theories of mathematics and logic and meanwhile also found the geometric models and practical models of flexible concepts, the logical semantics of propositions, and the mathematical essence of flexible linguistic rules. The series of new discoveries and new progress made me more confident and determined to continue the cultivation in this field. During further researches, I realized gradually that the more essential characteristic of those so-called fuzzy (vague) concepts modeled by using fuzzy sets should be "flexible" rather than "fuzzy or vague," and the information containing flexible linguistic values is actually a kind of imprecise information. Therefore, I took specifically "imprecise-information processing" as a direction and objective and carried out an all-round research. In September 2009, the results obtained were gathered as a book and formally published with the name of *Principles of Imprecise- Information Processing* (Chinese version).

After this book was published, I also had some further understanding and thinking. For instance, we can also research approximate reasoning and computing at the level of linguistic functions and can extend flexible linguistic value to more general quantifiable linguistic value. Meantime, some problems and deficiencies in this book are also found such as the exposition about "uniform chain similarity relation" of things, the wording of "real number space," the discrimination between flexible concept and vague (fuzzy) concept, the analyses of inference in truth-degreed logic, and the logical semantics of propositions, which all need improvement, and there are some redundancies in Chap. 12. In particular, the comparison is not made in this book between the principles and methods of approximate reasoning and computation we present and those in fuzzy set theory. In addition, some contents in this book are not so closely related to the theme of this book. Thus, I continued again the work nonstop. I did research further and at the same time also made revisions, corrections, and extensions to the original work: deleted or reduced some contents, extended some contents, changed some formulations and especially added many new contents (such as "flexible linguistic functions" and "quantifiable linguistic values" as well as the logical and mathematical principles of approximate reasoning). Thus, some original chapters and sections were deleted, but some new chapters and sections were added, and most of original chapters and sections were rewritten or adapted; correspondingly, the structure of text was also made a large modification—changed from the original 8 parts and 21 chapters to 9 parts and 26 chapters. Thus, a new work about "Principles of Imprecise-Information Processing," that is, the second edition of original book, has been formed. At the beginning of 2015, the manuscript of the new work had been basically completed, and then, some polishing was done. The new work has two versions: one in Chinese and one in English, the latter being this book.

Compared with the original edition, the new edition made much new progress both in depth and in extent—not only the quantity is increased but also the quality is raised, and the whole theoretical and technological system is more compact and coherent. This book has nine parts. The first part gives an outline of imprecise-information processing; the second part reveals the formation principle of imprecise information and establishes its mathematical models; the third part is the mathematical theory on imprecise information; the fourth part is the logic theory on imprecise information; the fifth part expatiates on the principles and methods of reasoning and computation with imprecise information and knowledge; the sixth part is the application and acquiring techniques of imprecise knowledge; the seventh part is the extension of imprecise information; the eighth part expatiates on and deals with the overlap of and the correlation between imprecision and uncertainty; and finally, the ninth part is further work. In terms of structure, Part I is the introduction, Part II the origin, Parts III and IV the basis, Part V the main body, Part VI the application (interface), Part VII the extension, Part VIII the cross, and Part IX the frontier. Their logical relationships and the hierarchy of this book are shown in the following diagram.

This book researches imprecise-information processing by using mathematical and logical methods, but meanwhile, it also develops the corresponding theories of mathematics and logic. The whole book presents over 100 important concepts, derives over 40 theorems and more than 100 formulas, functions, and rules, and gives over 70 specific methods and algorithms. Besides, there are also brief commentaries of some existing viewpoints and methods (which are mainly of fuzzy set theory) in this book.

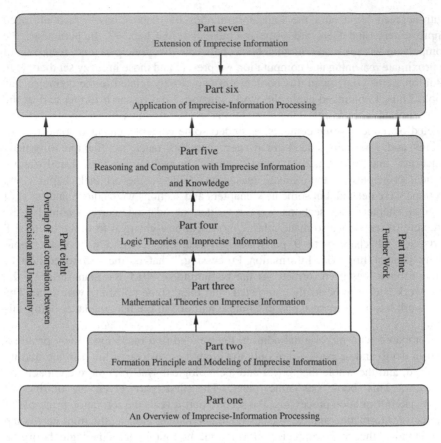

Architecture of *Principles of Imprecise-Information Processing*

This book also has a feature; that is, there are many symmetrical, antithetical, or corresponding concepts and terminologies such as "flexible linguistic value" and "flexible set," "membership function" and "consistency function," "geometric model," and "algebraic model," "combined linguistic value" and "synthetic linguistic value," "form of possession" and "form of membership," "logical composition" and "algebraical composition," "conjunction-type rule" and "disjunction-type rule," "complementary flexible partition" and "exclusive flexible partition," "flexible linguistic value" and "rigid linguistic value," "medium value" and "neutral value," "L-N function" and "N-L function," "certain rule" and "uncertain rule," "natural logical semantics" and "extended logical semantics," "reasoning with truth-degrees" and "reasoning with believability-degrees," "degree-true inference" and "near-true inference," "numerical ××" and "linguistic ××," "conceptual ××" and "practical ××," "×× of single conclusion" and "×× of multiple conclusions," "×× on the same space" and "×× from distinct spaces," "one-dimensional ××" and "multi-dimensional ××," and "typical ××" and "non-typical ××," thus forming many pairs of parallel or complementary theories and methods—they are arranged in a crisscross pattern and together constitute a multidimensional system of theories and technologies.

This book makes an all-round and systematic research of imprecise-information processing, but the focus is on clarifying concepts, straightening out relationships, revealing principles, and presenting methods to lay a theoretical foundation and build a technological platform for further research and application. In fact, on the basis of this book, we can directly develop related applications and also carry out further researches.

Imprecise-information processing is a big subject; in addition, the vision and the level of the author are limited, so the deficiencies and defects in both content and expression in this book are unavoidable although great effort was made. Therefore, the author sincerely invites experts and scholars to grant instructions and the readers to comment and make suggestions!

Xi'an, China   Shiyou Lian
January 2016

Acknowledgements

On the formal publication of this book, I owe a debt of gratitude to my daughter Xiaoya Lian for her great contributions in the translation of the manuscript. Without her help, it is hard to know when this book could ever meet the readers!

I also want to thank my elder daughter's family for their support in spirit!

I would also like to thank all persons who gave me help and support in the completion of this book!

Finally, I still want to thank the people who helped me in the publication of the original work, because without yesterday's original work, there would not be today's new work.

Contents

Part I
Introduction

Chapter 1
Overview of Imprecise-Information Processing

Abstract This chapter introduces firstly what is imprecise information and then examines the origin of imprecise information, thus revealing the formation principle of imprecise information, and then, it discusses the distinction and correlation between imprecision and uncertainty of information, the research issues of imprecise-information processing, and significance of studying imprecise-information processing and the related disciplines and fields; finally, it outlines the work of the book. Besides, a survey of researches on imprecise-information processing is given in the chapter.

Keywords Imprecise information · Flexible linguistic values · Uncertain information · Artificial intelligence

1.1 What Is Imprecise Information?

Imprecise information here refers mainly to the information that is expressed by words with imprecise meanings. For example, "tall" is a word with imprecise meaning in that there is no strict and rigid standard for a certain height to be considered "tall." Therefore, the word "tall" expresses imprecise information [1].

Words with imprecise meanings can be found everywhere in our daily communication and written materials. Here are some examples: "morning" and "evening" characterizing time, "nearby" characterizing location, "far" and "near" characterizing distance, "much" and "little" characterizing quantity, "big" and "small" characterizing volume or space, and "slight," "a little," "very," and "extremely" characterizing strength. Other words such as "fast," "slow," "hot," "cold," "good," "bad," "diligent," "hardworking," "serious," "friendly," "beautiful," "kindhearted," "brave," "ardently love," and "very likely" are all words with imprecise meanings. Thus, it is clear that there is imprecision almost everywhere in our communication (Look, this statement itself contains imprecision: What is "almost"?).

Imprecise information also includes the information expressed by words with precise meanings but which can be replaced by numerical values, because,

© Springer Science+Business Media Singapore 2016
S. Lian, *Principles of Imprecise-Information Processing*,
DOI 10.1007/978-981-10-1549-6_1

compared to numerical values, words appear to be less precise. For example, school records can be represented by "word grades," such as "good" or "excellent," but also can be represented by corresponding numerical scores, such as 85 and 98. Here, the former is not as precise as the latter. Additionally, the occurrence of a random event can be described by the word "likely," but can also be described by a numerical probability; similarly, the former is also not as precise as the latter. Therefore, the information expressed by such words as "good," "excellent," and "likely," is also imprecise.

Note: In addition to the above-mentioned causes, imprecision can also be caused by inappropriate words or inappropriate measuring units (the imprecision caused by measuring units is usually called inexactness). For example, when describing a person's place of residence, the name of country is not as precise as that of the province or city. Also, "ton" is not as precise as "kilogram" and "kilogram" is not as precise as "gram" in describing weights. This book will not cover the study of these two kinds of imprecision.

1.2 Origin of Imprecise Information

As stated above, imprecise information is caused by words with imprecise meanings. Then, why are the meanings of these words not precise? We know that words are actually the linguistic symbolic representations of corresponding concepts in human brain. The reason these words' meanings are imprecise is that the concepts represented by these words have no strict definitions. That is, these concepts' connotations have no rigid standards or conditions and their denotations have no rigid boundaries. In other words, their connotative conditions and denotative boundaries have a certain softness or flexibleness. For example, for the word "tall," heights over 1.75 m are all "tall" to a certain degree, and for "young," ages under 40 years are all "young" to a certain degree. For another example, the boundary between "hot" and "cold" weather is actually a "flexible boundary." That is, "hot" gradually transitions to "cold," and in turn, "cold" also gradually transitions to "hot." Therefore, the concepts expressed by words with imprecise meanings are actually "flexible concepts." That is to say, imprecise information turns out to be caused by flexible concepts in our brain. Then, how are these flexible concepts formed?

We know that everything has some attributes or states, and there are some relationships between things. To facilitate the narration, we call the attributes, states, and relations of things collectively to be the features of things.

Observing and examining the boundless universe we live in, it can be found that for one and the same feature, each relevant object has its specific magnitude and these magnitudes are not exactly the same, but assume continuous distribution or continuous change, thus forming a continuous range. Examples:

- Human heights continuously distribute or change from about 0.3 to 2.5 m, forming a range of [0.3, 2.5], which is a continuous real interval.

- Air temperatures continuously change from about −45 to 45 °C, forming a range of [−45, 45], which is also a continuous real interval.
- Ages of humans continuously distribute or change from about 1 to 120, forming the range of {1, 2, …, 120}, and this is a continuous set of integers.

Note that the "continuous" here includes the "continuous" of real numbers, the "continuous" of rational numbers (i.e., "dense"), and the "consecutive" of integers (i.e., order) and to be the same later.

It can be seen from the above examples that after a certain measure being introduced, the magnitudes of a feature of things are explicitly shown as concrete numbers.

We call the numbers representing magnitudes of a certain feature of things to be the **numerical feature values**, or simply, the **numerical values**, of things. Thus, we will treat the magnitudes and numerical feature values, i.e., numerical values, of the things as synonym later.

It is not hard to see that the continuity of magnitudes of a feature of things makes corresponding things show as the uniform chain similarity relation. Then, facing with one and another things being uniformly chained similar, how should the human brain save and deal with relevant information? Of course, numerical feature values can directly describe things precisely, but if they are used all the time and everywhere, then the human brain would be unable to bear the enormous amount of data and humans would be unable to tolerate the trouble.

It can be seen that in the continuous numerical feature values, the relation between adjacent numbers is the approximation relation. Thus, we can cluster these numerical values according to approximation relation and then express the numbers in one and the same class by using one and the same word. This kind of word summarizing a batch of numerical values is a big-granule value—**linguistic value**. The linguistic values are also a kind of values representing magnitudes of features of things, namely **linguistic feature values**. Clearly, the number of linguistic values of one feature is very finite. Thus, by clustering and summarizing numerical values, we can use finite number of linguistic values to represent an infinite number of numerical values. Thus, the amount of information can be greatly reduced and the complexity lowered.

Actually, according to the law of "quantitative change to qualitative change," there is an obvious difference between things' properties represented by numerical values far apart in range. So, speaking from this point, numerical feature values of things must be clustered and partitioned. Another benefit of clustering and partitioning numerical values and then expressing them in linguistic values is that we can understand and grasp things at a higher level.

However, unfortunately, it is difficult to do the usual natural and objective clustering and partitioning of these continuous numerical values (for detailed analysis, see Sect. 2.1). For this reason, the human brain adopts the clever strategy of flexible clustering and flexible partitioning to obtain corresponding "flexible classes" (for concrete principle and method, see Chaps. 2, 3) and afterward summarize the thing's properties stood for by flexible classes, thus obtaining "flexible concepts"

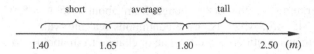

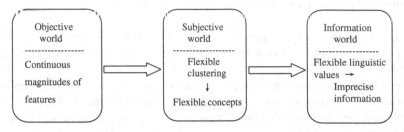

Fig. 1.1 An example of flexible clustering and flexible classes in a range of heights of adults

Fig. 1.2 The diagram of the origin of imprecise information

and "flexible linguistic values" representing flexible concepts. For example, as shown in Fig. 1.1, through flexible partitioning of the range [1.4, 2.5] of heights of adults by flexible clustering, we obtain the corresponding flexible classes, flexible concepts, and flexible linguistic values: "short," "average," and "tall."

From stated above, we see that it is just the phenomenon of "continuous distribution or change" of magnitudes of a feature of things (or in other words, "uniform chain similarity" of things) in the objective world and the treating way of "flexible clustering" of the human brain that result in the flexible concepts in human brain, and then, there occur flexible linguistic values and corresponding imprecise information. Thus, the origin of imprecise information can be diagramed as follows (see Fig. 1.2).

Now, there exists yet another question: When does the human brain flexible treating with respect to the continuous magnitudes of features? In other words, are those flexible concepts in human brain obtained independently by each individual's own flexible clustering of corresponding feature's magnitudes? We will discuss the problem in Sect. 19.1.

There might be readers who think, "Aren't the flexible concepts talked about here same as 'fuzzy concepts' in some other literatures"? Right, the flexible concepts we talk about here are just the fuzzy concepts called in some literatures. Then, why do we call them flexible concepts but not fuzzy concepts? Section 19.3 will give the answer.

1.3   Distinction and Correlation Between Imprecision and Uncertainty of Information

Besides imprecise information, there also exists uncertain information in usual information processing and communication. For example:

It might rain tomorrow.

This sentence carries a piece of uncertain information—it rain tomorrow.

Note that the uncertainty we talk about here refers only to the uncertainty (of information) that is caused from randomness of things or people's lack of knowledge of things. It is not that kind of uncertainty, said in the literature [2], including fuzziness (i.e., imprecision), vagueness, unknownness, non-specificity, strife, discord, conflict, and ignorance. Of course, the uncertainty we talk about here also does not include fallibility, instability, inaccuracy, incompletion, and ambiguity.

1. Distinction between imprecision and uncertainty

Imprecise information is the information that describes the features and relations of things not specifically, strictly, or exactly enough. Uncertain information is the information of which the authenticity cannot be determined, that is, the event, or the properties, relationships, or behaviors of things expressed by which is not certain or not sure.

From the last section, imprecise information originates from the continuous distribution or change of numerical feature values of relevant things (or the uniform chain similarity of things) and the flexible treating mechanism of human brain. Uncertain information originates then from the feature of "partial share" of relevant sets and the relations of "partial correspondence" or "partial inclusion" between relevant sets (see Sect. 25.3).

Although the imprecise information has an objective basis, it is a "man-made" product, so it has a certain subjectivity. The uncertain information is the objective expression of properties or behaviors of things that people can't be sure, but in the description of the degree of uncertainty, there may be subjective factors.

Imprecise information is directly expressed by the relevant statements (of which the uncertainty is shown in the linguistic value(s) of the relevant statements). Uncertainty information, in general, cannot be expressed directly, but it is expressed indirectly by the aid of a main-clause-structured compound sentence (we call it the possibly type modal proposition, see Sect. 25.1). For example, the above uncertain information "It rain tomorrow" is expressed by "It might rain tomorrow," that is, "'It rain tomorrow' is possible."

From the above, we can see that uncertainty and imprecision are two mutually independent attributes of information. Uncertain-information processing solves the possibility problem of the truth or falsity of information, while imprecise-information processing solves the strength problem of the truth or falsity of information.

2. Correlation between imprecision and uncertainty

Now that uncertainty and imprecision are two mutually independent attributes of information, the correlation between them would be not the relation of subordination. But, we find that there are some links between the two of them.

(1) Since different people may define the core and flexible boundary of a flexible concept somewhat differently, uncertainty can be involved when determining the common model (such as the core and support set, or the membership function) of a flexible concept (see Sect. 4.1).

(2) When conjecturing the corresponding numerical value from a flexible linguistic value possessed by an object (i.e., converting a flexible linguistic value into a numerical value), uncertainty will also be encountered (see Sect. 7.3.1).

(3) From some of imprecise information, uncertain information can be drawn, or some uncertain information originates from imprecise information, and imprecise information (processing) and uncertain information (processing) can be translated to each other in some conditions (see Sects. 24.6 and 25.3).

(4) Uncertainty and imprecision of information sometimes occur simultaneously. That is to say, there are both imprecision and uncertainty in one and the same statement. In fact, because the "possible," "probably," and so on themselves are not precise (they are quantifiable rigid linguistic values), (the information expressed by) the main clause of a possibly type modal proposition is imprecise, but the clause of it is uncertain. For example, the main clause of "'It rain tomorrow' is possible" is imprecise, but the clause "It rain tomorrow" of it is uncertain. There is such uncertainty information, it is also imprecise, or there is such imprecise information, and it is also uncertain. For instance, the "It rain heavily tomorrow" in "'It rain heavily tomorrow' is quite possible" is uncertain as well as imprecise, or that it is imprecise as well as uncertain.

Above, we expounded the distinction and correlation between imprecision and uncertainty of information. Actually, imprecise-information processing and uncertain-information processing are both indispensable and important components of artificial intelligence technology. On uncertain-information processing, people have conducted quite deep research and acquired abundant achievements. As a matter of fact, uncertain-information processing already has a solid mathematical basis and a relatively perfect theoretical system. For example, probability theory and mathematical statistics are just special mathematical branches concerned with the processing of uncertain information. By contrast, imprecise-information processing still lacks a solid theoretical basis and the technology is not mature enough. These are the problems that the book is going to solve.

1.4   Research Issues of Imprecise-Information Processing

As a subject, the research issues of imprecise-information processing include basic principles of imprecise information, the theories, technologies, and applications of imprecise-information processing, and the human brain's thinking mechanism concerned with imprecise-information processing.

1. **Basic principles of imprecise information**
   Basic principles of imprecise information involve the following:

   - Objective basis, formation principle, and mathematical models of flexible concepts,
   - Properties, types, relations, operations, and measures of flexible sets,
   - Properties, types, relations, and operations of flexible linguistic values,
   - Properties, types, relations, operations, and measures of propositions containing flexible linguistic values and the corresponding logic and inference,
   - Related theories of quantifiable rigid linguistic values.

2. **Basic technologies of imprecise-information processing**
   Basic technologies of imprecise-information processing include the following:

   - **Techniques of imprecise-knowledge acquisition** It includes artificial acquisition and machine automated acquisition. Artificial acquisition is generally done through such approaches as investigation and statistics to acquire usual commonsense imprecise knowledge. Imprecise knowledge of a professional field should be acquired from domain experts. Machine automated acquisition is to make computers directly induce, discover, and extract imprecise knowledge from relevant data or facts by using means of machine learning.
   - **Techniques of representation, storage, conversion, transformation, and translation of imprecise information and knowledge** Of which, representation and storage include the mathematical models of flexible linguistic values and the representation and storage of propositions, predicate language, rules, functions, frame, and semantic nets, containing flexible linguistic values; conversion includes the interconversion between imprecise information and precise information and the transformation and translation between imprecise information and between the granule sizes of information/knowledge.
   - **Application techniques of imprecise information and knowledge** Which mainly refer to the inference and computation with imprecise information and knowledge, and the approximate reasoning and computation utilizing imprecise-information processing; And they, specifically speaking, include relevant principles and methods, such as inference rules, computation models and algorithms, and so forth.
   - **Machine perception and communication techniques on imprecise information** It includes interconversion between linguistic valued information and numerical valued information, imprecise-information-oriented man–machine interface, machine understanding, and generation of imprecise information.

3. Applications of imprecise-information processing

Application of imprecise-information processing involves the application fields and projects, application ways and methods, and relevant software technologies.

The application of imprecise-information processing is very extensive. The fields and projects that involve obviously imprecise information, such as imprecise-problem solving, classifying, recognition, judging, decision making, natural language processing (including natural language generation and understanding), man machine interfaces, intelligent robots, expert (knowledge) systems, anthropomorphic intelligent systems, and approximate reasoning, can rightfully use imprecise-information processing technology, and some precise problems (e.g., control) can also be indirectly solved by using imprecise-information processing technology. Therefore, in addition that we research how to utilize imprecise-information processing technology to solve the imprecise practical problems and engineering problems, we also need to consider how to introduce the techniques and methods of imprecise-information processing into the precise problem solving, to open up new application areas and projects.

In software technology, what needs studying are data structure, knowledge representation, data/knowledge base structures, relevant algorithms, system architecture, interfaces, and man–machine interfaces, etc., which are suitable for imprecise-information processing. Besides, the combination of imprecise-information processing technology and other existing technologies is also involved.

4. Further research topics

Imprecision exists widely in our daily language, but the human brain can easily grasp it and use it very flexibly. Therefore, in order to further research and apply imprecise-information processing, we should examine thoroughly the human brain's processing mechanism with imprecise information, so we can be enlightened and also borrow ideas from it. This will at least involve the following problems:

- Brain models of flexible concepts;
- Interconversion mechanisms between numerical information and linguistic information in human brain;
- Qualitative thinking mechanism of human brain, that is, taking flexible concepts as an entry point to explore the psychological and physiological models of human brain's clustering and summarizing and qualitative thinking and the relationship between the two;
- Relationships between imprecise-information processing and linguistics, logic, and cognitive science; this subject is also an interface between brain science, psychological science, cognitive science, and intelligence science;
- Principles of imprecise-information processing based on quantum information technology.

1.5   Significance of Studying Imprecise-Information Processing and the Related Disciplines and Fields

As the preface of this book points out, the direct motivation of studying imprecise-information processing is from the intelligentization, especially the anthropomorphic intelligentization, of human society after informatization. From the last section, we can see that the research of imprecise-information processing involves many disciplines and fields, and the imprecise-information processing technology can be applied to many disciplines and fields.

1. **Imprecise-information processing and intelligence science and technology**
   Imprecision exists not only widely in our daily language, but also in our knowledge, especially our experiential knowledge. Therefore, to realize artificial intelligence and develop intelligence science and technology, imprecise-information processing is unavoidable. In fact, at present, imprecise-information processing technology is very important to many fields of artificial intelligence, such as intelligent robots, intelligent Internet/Web, expert (knowledge) systems, pattern recognition, natural language processing, machine learning, knowledge discovery and data mining, machine translation, intelligent control, judging and decision, and intelligent human–computer interface. In the long run, to further develop artificial intelligence, the problem of computer processing human language must be well solved. That is, machines must be capable of perceiving, thinking, and communicating at the level of natural language just like humans. Otherwise, artificial intelligence will always remain at the level of implementation and realization of human intelligence on a machine. That is to say, machines must have the anthropomorphic ability of imprecise-information processing to realize anthropomorphic intelligent systems. Besides, speaking from exploring the mystery of human intelligence, imprecise-information processing is also an important problem. Therefore, the development of intelligence science and technology is bound to face imprecise-information processing, so we must study the imprecise-information processing. Moreover, with the more-and-more thorough research, and wider-and-wider application of intelligence science and technology, imprecise-information processing will appear even more important and more pressing. Therefore, imprecise-information processing is an indispensable and important component of intelligence science and technology.
2. **Imprecise-information processing and information science and technology**
   Just as the name suggests, imprecise-information processing is certainly closely related to information science and technology and should belong to the category of information science and technology. To be more specific, imprecise-information processing should belong to the category of what is now called "intelligence information processing." To be even more accurate, it should be within the research field of "content-based information processing." We know that information science and technology is a big subject and that traditional information processing is mainly processing about the information form, such as information

representation, storage, processing, transformation, transmission, and retrieving. In this respect, a lot of related theories have been established and many techniques developed. However, the research on content-based information processing is still relatively weak. But content-based information processing is an inexorable development trend of information science and technology. Imprecision is concerned with the content of information. Therefore, imprecise-information processing is a very important research subject in content based information processing and a very important research field and development field of information science and technology both at present and in the future.

3. **Imprecise-information processing and computing science and technology**
   The objects processed by conventional computing science and technology are numerical values, while the objects processed by imprecise-information processing are the (flexible) linguistic values. The latter is a summarization of the former, and the former is an instance of the latter. Therefore, imprecise-information processing is related to computing science and technology. In fact, by utilizing the reasoning and computing with flexible linguistic values, some numerical computation problems can be solved. For example, translating some complex nonlinear numerical functions or correlations into simple linguistic functions, we then realize the approximate evaluation of the former by the exact or approximate evaluation of the latter, thus increasing new ideas and approaches, and opening up new approximate computation techniques for conventional numerical computation. Additionally, we can also develop the computer languages and related hardware based on the flexible linguistic values to extend the processing capabilities of existing computers.

4. **Imprecise-information processing and logic**
   Logic studies the form and laws of human thinking, of which the basic objects are concepts and judgments (propositions) and the main issue is inference. Traditional logic deals with rigid concepts and rigid propositions, or treats flexible concepts and propositions as rigid ones, which is a kind of coarse-granule logic at the linguistic level. Imprecise-information processing is a kind of information processing originated from, and based on flexible concepts, so it needs a kind of logic based on the flexible concepts as a support. Thus, the logic based on flexible concepts has a natural link and close relationship with imprecise-information processing. This raises a new issue and opens up a new area for logic. Actually, with flexible concepts, there also occur flexible propositions. Thus, it is possible and necessary to found a new logic system. Examining flexible concepts, flexible propositions, and corresponding inference from the angle of logic, we will find a logic basis for imprecise-information processing; thus, we can use logic to guide the research and development of imprecise-information processing. Conversely, introducing flexible concepts into logic will promote the development of logic. Since flexible concepts have mathematical models, logic and inference based on flexible concepts would certainly be related to numerical values and numerical computation. This, in

turn, would add the color of numerical computation to operations and inference of traditional symbolic logic and provide new theoretical and technological support for the deepening and extension of logic.

5. **Imprecise-information processing and linguistics**

The linguistic representations of flexible concepts, namely flexible linguistic values, are the usual adjectives and adverbs. In linguistics, people generally use qualitative method to study these vocabularies, while in imprecise-information processing, since they have mathematical models, the quantitative method can be used. Thus, introducing ideas and methods from imprecise-information processing into linguistics research will help linguistics step up to a new level and further push its development. Conversely, relevant research results from linguistics can also provide bases and instances for imprecise-information processing. In fact, speaking from the angle of linguistics, the above-mentioned content-based information processing is the semantic representation and processing of natural language. It is now still difficult to establish semantic models of natural language in natural language processing and computational linguistics. Imprecise-information processing raises new research issues for linguistics especially computational linguistics; meanwhile, it also provides new ways of thinking and new approaches. Therefore, the combination of the two will have mutual benefits.

6. **Imprecise-information processing and brain and cognitive science**

Flexible-ening (which is similar to softening) information perceived and processing it freely is an intrinsic mechanism of human brain (see Sect. 19.1), which is also a characteristic and advantage of human brain. Then, to further research the imprecise-information processing mechanism of human brain, brain and cognitive science should certainly be involved. The neural mechanism of imprecise-information processing itself is a very important research subject of brain and cognitive science, and the research on imprecise-information processing will open up a new entry point for brain and cognitive science.

7. **Imprecise-information processing and life science**

The imprecise-information processing mechanism and function of human brain should be innate. That is, the mechanism exits already in DNA. Then, in what manner does it exist? Where is the location? What is the coding? These are all research topics of life science. Thus, imprecise-information processing is closely associated with life science. Imprecise-information processing presents new research problems for life science and at the same time introduces a new entry point for the research of life science.

8. **Imprecise-information processing and mathematical science**

Imprecise-information processing, in the final analysis, is to quantitatively process the usual qualitative information with mathematical sciences as a tool. Specifically, it is the establishment of the mathematical models of imprecise information and development of the mathematical methods for corresponding information processing. In fact, the research on imprecise-information processing involves a lot of mathematical fields and knowledge, such as sets, geometry, algebra, logic, function, and probability, and it extends, deepens, and even amends the relevant concepts, methods, and knowledge systems of these mathematical branches. It should also be mentioned that in recent years, the research on quantum information theory and technology has become more and more active. In information representation and processing, quantum information technology has incomparable characteristics and advantages in comparison with traditional technologies. Then, can quantum information technology be used to represent and process imprecise information? This is obviously an issue worth our attention and thought. The author's intuition is that quantum information technology will very likely be a more suitable and effective new technology for imprecise-information processing. In a word, imprecise-information processing is inseparable from mathematical science.

9. **Imprecise-information processing and system science**
   System science researches the characters and states of systems. The characters and states of some complex systems such as social systems, economic systems, ecosystems, and information systems on the Internet are usually difficult to describe exactly using traditional mathematical models, but they can be described using the models with flexible linguistic values (such as flexible rules or flexible linguistic functions). Although flexible linguistic values are of big granule and not precise enough, the system models with flexible linguistic values are a kind of higher-level general expression of characters and states of systems, which are conducive to us to understand and grasp a complex or large system. Therefore, the description of the characters and states of a system using flexible linguistic values is significant and even necessary. Thus, imprecise-information processing is also related to system science. In fact, the flexible rule and flexible linguistic function discovery techniques in imprecise-information processing can come in handy in Web mining, which is a research hot spot of data mining.

1.6   A Survey of Research on Imprecise-Information Processing

People have long been aware of imprecision in daily language and have put some thought and study into it. As a matter of fact, as early in the 1930s, Polish logician and philosopher Jan Lukasiewiczj studied the mathematical representations of flexible concepts such as "tall," "old," and "hot." He extended the range of logical truth values to all numbers between 0 and 1. But he used numbers in interval [0, 1] to represent the possibility of statements being true [3]. Thereafter, in 1937, philosopher Max Black published a paper titled "Vagueness: an exercise in logical analysis." The paper analyzed the gradual change phenomenon of things and proposed the idea of using numbers to represent degrees. However, he believed that vagueness is a probability problem. The paper (the appendix) defined the first simple fuzzy set and outlined the basic ideas of fuzzy set operations [3].

In 1965, American professor Lotfi Zadeh extended the definition of traditional sets and proposed the concepts of fuzzy set [4], membership grade, membership function, etc., and took fuzzy sets as the mathematical models of flexible concepts. Soon afterward, Professor Zadeh and some other scholars together presented and developed a series of theories, techniques, and methods based on fuzzy sets, such as fuzzy logic and fuzzy reasoning, thus forming the now called fuzzy set theory or fuzzy technology.

In 1975, Ebrahim Mamdani, professor at London University, was the first to apply fuzzy reasoning to the control of a steam engine and boiler combination, putting fuzzy set theory into engineering application [3]. After that, fuzzy control grew vigorously. Particularly in the early 1990s, Japanese engineers applied fuzzy techniques to control electrical home appliances and had great success, thus drawing a huge response and much attention from around the world. After this, Europe and America all entered into this field competitively.

Soon after, there also appeared the techniques of fuzzy pattern recognition, fuzzy judging, fuzzy decision making, etc., and the concepts of fuzzy measure, fuzzy probability, fuzzy integral, fuzzy entropy, etc., also appeared. In 1978, Zadeh proposed the possibility theory.

Although fuzzy technology got some results in practice, some important theoretical and technological problems still have not been well solved thus far, such as the shape of the membership function of a fuzzy set, the objective basis and the logic theoretical basis of the definitions of fuzzy logic operators, and the logical foundation of fuzzy reasoning. These problems directly influence the effects and efficiency of fuzzy technology.

Fuzzy reasoning is a kind of approximate reasoning; it is a key technique in fuzzy technology and also a basic issue of imprecise-information processing. A basic method of fuzzy reasoning is the CRI (compositional rule of inference) proposed by Zadeh. This method can be divided roughly into the following steps: design and select membership functions, convert fuzzy rule (e.g., $A \rightarrow B$) into fuzzy relation (e.g., R) and composition of relations, and converge multiple inference outcomes (if occurring). In addition, for fuzzy control, there are two steps of fuzzification and defuzzification. For each step, people again presented various specific methods. Of them, for converting a fuzzy rule into a fuzzy relation, more than ten "implication operators" were presented, such as the Zadeh operator, Mamdani operator, Goguen operator, and T norm operator. Each of these methods of fuzzy reasoning has its distinguishing feature and shows the unique intelligence and talent of its creators. However, the problems of the general principle and logic basis of fuzzy reasoning still have not been truly or completely solved so far. In fact, the mathematical and logic theories based on fuzzy sets, that is, fuzzy mathematics and fuzzy logic, do not provide much theoretical support for fuzzy reasoning. In addition, we see that nearly all the good applications of fuzzy reasoning in engineering benefited from the introduction of machine learning mechanisms or other mathematical principles. But the result of doing so is that except for the name of "membership function," the so-called fuzzy logic system has almost nothing to do with fuzzy logic in the real sense. Just as the book *Neuro-Fuzzy and Soft Computing* (foreword by Zadeh, authors are his students) said, neural–fuzzy systems encountered "the dilemma between interpretability and precision" [5]. It is no wonder that C. Elkan gave a report "the paradoxical success of fuzzy logic" [6, 7] in the 11th American Association for AI National Conference. Although more than 10 experts rebutted Elkan's opinion later, it is an indisputable fact that logical and mathematical basis of fuzzy reasoning has not yet been really solved up to now.

In imprecise-information processing, in addition to using the fuzzy set, people also proposed the flou set, vague set, and many other sets that extend fuzzy set, such as 2-type fuzzy set, interval-valued fuzzy set, L-fuzzy set, and intuition fuzzy set. There are many scholars devoting themselves to the improvement and development of fuzzy set theory, and many new ideas, theories, and methods have been presented; for example, Chinese scholars presented "cloud model" [8], "interpolation mechanism of fuzzy control" [9], "3I inference algorithm" [10], "universal logics" [11], "weighted fuzzy logic" [12], and "new fuzzy set theory" [13]. In 1996, Zadeh proposed the research direction of computing with words (CW) [2, 14] on the basis of fuzzy set theory. He expresses approximately the relation between computing with words and fuzzy logic as follows: fuzzy logic = computing with words. These new and improved theories and methods all have their own strong points, viewing angles, and features, but on the whole, people have not yet reached a common view, the mathematical and logical basic problems in imprecise-information processing have not be solved, and the scientificity and validity of relevant techniques and methods lack theoretical support. In practice, it is still "a hundred flowers blossoming and a hundred schools of thought contending." In particular, so far some scholars still put imprecision of information into uncertainty of information or mix the two together to do research.

Here, we also need to mention that after the appearance of fuzzy sets, some mathematicians then began to devote themselves to the study of mathematical theories based on fuzzy sets, which were quite hot for a time. But it seems that these researches and achievements have no relation to imprecise-information processing. Besides, people also combine fuzzy set theory with other intelligence technologies. For example, there appear new techniques of fuzzy–neural networks and neural–fuzzy systems on combining fuzzy sets with artificial neural networks. There also appears fuzzy support vector machine on combining fuzzy sets with support vector machine. And Zadeh further combined fuzzy logic, neural computing, probabilistic reasoning, genetic algorithm, chaotic system, and so on and collectively called them to be "soft computing" [5].

In the early 1990s, the author began to think about the problem of "fuzzy" from the doubt of the CRI fuzzy reasoning of Zadeh. However, the author did not make modifications and supplementations in the existing framework of fuzzy set theory by following traditional thinking. Rather, with a new perspective and ideas, the author traces the origin and researches imprecise-information processing in all around, from the formation of imprecise information to modeling of it, from the representations of imprecise information to the conversion and transformation of it, from related mathematical theories to logical theories, from (approximate) reasoning with imprecise information and knowledge to (approximate) computation with imprecise information and knowledge, from imprecise-problem solving to imprecise-knowledge discovery, from flexible linguistic values to quantifiable rigid linguistic values, from pure imprecise-information processing to the overlapping of imprecision and uncertainty, from the principles of imprecise-information processing to the applications and methods of it, etc. After years of unremitting exploring, a series of new theories and methods different from fuzzy technology were obtained; they form a new theoretical and technological system of imprecise-information processing. The book is just a summation of these research results.

1.7 Work of the Book

Starting from its objective basis, the book explores and reveals the cause and principle of forming imprecise information. Then, on the basis of which the book establishes the related mathematical models, it further discusses and reveals the principles and methods of imprecise-information processing in an all-around way, thus establishing a new theoretical and technological system of imprecise-information processing. Specifically, the book mainly does the following:

(1) Examines the characteristics of imprecise information, distinguishes the imprecision of information from the uncertainty of information, treats explicitly imprecise-information processing as an independent research field, discriminates between vagueness (fuzziness) and flexibleness of concepts, proposes the terminologies of "flexible concepts" and "flexible linguistic values," and rectifies the so-called vague (fuzzy) concepts as flexible concepts.

(2) Examines the objective basis of flexible concepts and the cause of flexible concepts, reveals the formation principles of flexible concepts and flexible linguistic values, and presents their general mathematical models and modeling methods.

(3) Proposes the concepts of flexible sets and flexible relations and founds relevant theories and methods.

(4) Examines flexible linguistic values and relevant topics in an all-around way, obtains some important results, and founds relevant theories and methods.

(5) Proposes the concepts of flexible linguistic function and correlation and the concepts of flexible number and flexible function, discusses relevant topics, and founds corresponding theories and methods.

(6) Introduces truth-degrees, founds the basic theory of truth-degreed logic, and finds and presents the principles and methods of corresponding inference.

(7) Introduces flexible linguistic truth values, founds the basic theory of flexible linguistic truth-valued logic, and presents the principles and methods of the corresponding inference.

(8) Finds and proposes the terminology and principle of logical semantics of propositions and establishes the computation models of truth values of basic compound propositions in two-valued logic and truth-degreed logic on the basis. And finds and proposes the concepts of relatively negative-type logic and relatively opposite-type logic.

(9) Examines flexible rules and relevant topics in an all-around way, obtaining a series of important results.

(10) Introduces the adjoint functions of a flexible rule and gives some acquiring methods and reference models.

(11) Studies the reasoning and computation with flexible rules, clarifies their logical and mathematical principles, and gives a series of reasoning and computation approaches.

(12) Studies the approximate evaluation of flexible linguistic functions, reveals its basic principles, and presents some approaches and ideas.

(13) Summarizes and rounds up the practical problems involving imprecise-information processing and presents the corresponding solving techniques and methods.

(14) Explores imprecise-knowledge discovery and presents some methods and ideas.

(15) Introduces several measures to sets and flexible sets and founds relevant theories.

(16) Discusses relevant theories of quantifiable rigid linguistic values.

(17) Talks briefly about the methodology of imprecise-information processing and discusses several application topics.

(18) Founds the probability theory of random flexible events.

(19) Founds the believability-degree theory of flexible propositions and presents the corresponding principle and method of reasoning with believability-degrees.

(20) Analyzes the origin of the uncertain information with a mathematical view and then reveals the correlation between uncertain information (processing) and imprecise information (processing).

The above (1) and (2) are the basic principles of imprecise information; (3), (4), and (5) are the mathematical basic theory on imprecise information; (6), (7), and (8) are the logic basic theory on imprecise information; (9), (10), (11), and (12) are principles and techniques of inference and computation with imprecise information; (13) is the application techniques of imprecise-information processing; (14) is the acquiring techniques of imprecise knowledge; (15), (16), and (17) extend the extent of imprecise-information processing and expatiate the basic methods and techniques of imprecise-information processing; (18), (19), and (20) are the overlapping theories and techniques of imprecise-information processing and uncertain-information processing and also reveals and clears up the connections and relations between the two. Additionally, the book also presents further research directions and topics, and briefly expatiate the issues, approaches, and ideas.

References

1. Lian S (2009) Principles of imprecise-information processing. Science Press, Beijing
2. Ross TJ (2010) Fuzzy logic with engineering applications, 3rd edn. Wiley, New York
3. Negnevitsky M (2002) Artificial intelligence: a guide to intelligent systems, 2nd edn. Pearson Education Limited, London
4. Zadeh LA (1965) Fuzzy sets. Inf Control 8:338–353
5. Jang J-SR, Sun C-T, Mizutani E (1997) Neuro-fuzzy and soft computing. Prentice Hall, Upper Saddle River, pp 342–345, 382–385
6. Elkan C (1994) The paradoxical success of fuzzy logic. IEEE Expert 9:3–8
7. Elkan C (1994) The paradoxical controversy over success of fuzzy logic. IEEE Expert 9:47–49
8. Deyi L, Haijun M, Xuemei S (1995) Membership cloud and membership cloud generator. J Comput Res Dev 32(6):16–21
9. Lee H (1998) The interpolation mechanism of fuzzy control. Sci China (Series E) 28 (3):259–267
10. Wang G (1999) Triple-I algorithm of fuzzy inference. Sci China (Series E) 29(1):43–53
11. He H, Wang H, Liu Y, Wang Y, Du Y (2001) Universal logics principles. Science Press, Beijing
12. He X (1989) Weighted fuzzy logic and its widespread use. Chin J Comput 12(6):458–464
13. Gao Q (2006) New fuzzy set theory basics. China Machine Press, Beijing
14. Zadeh LA (1996) Fuzzy logic = computing with words. IEEE Trans Fuzzy Syst 4(2):103–111

Part II
Formation of Flexible Linguistic Values and Their Mathematical Models: Formation Principles and Modeling of Imprecise Information

Chapter 2
Flexible Concepts and Flexible Linguistic Values and Their Mathematical Models

Abstract This chapter takes real interval $[a, b]$ as a general range of numerical feature values and uses flexible clustering to obtain the corresponding flexible classes and flexible concepts (flexible linguistic values), thus simulating and revealing the objective basis, formation principle, and cause of flexible concepts, and then establishing the mathematical models of flexible concepts, and deriving their general expressions. Besides, it distinguishes between the flexible attributive concept and the flexible entity concept and discusses pseudo-flexible linguistic values.

Keywords Flexible concepts · Membership function · Consistency function

In the last chapter, we introduced and preliminarily discussed flexible concepts and flexible linguistic values. In this chapter, we further analyze concretely the clustering and partitioning of a range of numerical feature values of things, thus revealing the objective basis, formation principle and mathematical essence of flexible concepts and flexible linguistic values, and then establishing the mathematical models of them.

2.1 Flexible Clustering in Range of Numerical Values and Corresponding Flexible Concepts and Flexible Linguistic Values

Let U be a range of numerical values of a feature of certain class of things. As stated in Sect. 1.2, in order to reduce the amount of information, lower complicatedness, and understand and grasp things at a higher level, the numbers in U must be clustered and partitioned. In the following, we consider the corresponding method of clustering.

As we know, the usual methods of clustering can generally be separated into two types: dividing by a threshold and clustering with centers. The basic technique of

© Springer Science+Business Media Singapore 2016
S. Lian, *Principles of Imprecise-Information Processing*,
DOI 10.1007/978-981-10-1549-6_2

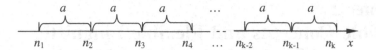

Fig. 2.1 An illustration of equidistant distribution and uniform chain approximation of integers
Here n_1, n_2, ..., n_k are a row of consecutive integers, and a is the distance or approximation-degree
between two adjacent integers

the former is the following: dividing a set awaiting partition according to a set
threshold and based on the similarity measurements between every two points to
realize the partition of the set; and the basic technique of the latter is the following:
according to the similarity measurements of every class center to all other points to
search the optimum partition of a set awaiting partition by employing certain
algorithm under the constraint of clustering criterion.

We also know, the continuum of real numbers means there is no interval
between numbers and their number is infinite. Therefore, for the range of numerical
values, U, consisted of real numbers, the approximation relations between every
two numbers cannot be examined; thus, the dividing by a threshold method cannot
be used to do corresponding clustering. And the continuity of integers is a kind of
succession, which is actually a kind of equidistant distribution. Between integers
distributed equidistantly is really a kind of uniform chain approximation relation;
that is, the degree of approximation between any two adjacent integers is the same
everywhere (as shown in Fig. 2.1). Thus, for the range of numerical values, U,
consisted of integers, doing corresponding clustering by employing dividing by a
threshold would only have two results: One number is a class or all numbers are a
class. Obviously, such clusters are pointless. The following relational matrix is just
a simple example:

	1	2	3	4	5
1	1	a	b	c	d
2	a	1	a	b	c
3	b	a	1	a	b
4	c	b	a	1	a
5	d	c	b	a	1

where set $\{1, 2, 3, 4, 5\}$ of integers is a range of numerical feature values of a class
of things and a, b, c, and d $(0 \leq d < c < b < a < 1)$ are respectively the degrees of
approximation between corresponding two adjacent numbers. To be sure, for a
finite U, the transitive closure $t(R)$ of approximation relation R can also be used to
realize the partition of the set, but since the process of finding a transitive closure is
non-identical transformation, the partition thus obtained whether or not in line with
the actual cannot be guaranteed theoretically.

For the continuous numerical feature values, clustering with centers method may
be used. But also because of continuousness, clustering with centers would make

the two numbers which are originally very close or even adjacent be put into different classes. This is also obviously not reasonable.

Now, on the one hand, it is necessary to do clustering and partition of the continuous numerical feature values; on the other hand, it is difficult to do the usual objective and natural clustering and partition. What should we do?

Analyzing the usual methods of clustering, it can be seen that the characteristics of them are all rigid dividing; that is, an object either belongs completely to a class or does not belong completely to the class. But the degree of approximation between numbers is decreasing progressively with the progressive increase of the distance between two numbers. Therefore, a compromise is that we still use the idea of clustering with centers, but do not draw a line clearly and rigidly between two adjacent classes; rather, we set a gradual change transition region as a "boundary" between them.

Let $U = [a, b]$ $(a, b \in \mathbf{R}$ (real number field)) be a range of numerical values. We mark out some sub-regions $C_1, C_2, \ldots, C_n$ (as shown in Fig. 2.2) with appropriate widths and appropriate intervals. Since there is some interval, the properties represented by various sub-regions (i.e., the summarization of properties represented respectively by numbers in each sub-region) would have obvious differences. Thus, $C_1, C_2, \ldots, C_n$ can be separately treated as the center (region) of classes $A_1, A_2, \ldots, A_n$, and the properties represented by them are respectively treated as the properties stood for by the corresponding classes. Then, the numbers forming the center of a class all have completely the property of the corresponding class, so they are core members of the corresponding class. Since the properties represented by mutually approximate numbers should also be approximate, the numbers outside the center of a class also have the property of the class to some degree. And since the approximation relation is transmitted decreasingly, these degrees will decrease progressively with the progressive increase of the distance between a number and the center of a class. Thus, the numbers in interval B_i $(i = 1, 2, \ldots, n - 1)$ also have the property of A_i, but the farther they are from center C_i, the lower will be their degrees of having the property of A_i. Similarly, the numbers in interval B_i have also the property of A_{i+1}, and the farther they are from center C_{i+1}, the lower will be their degrees of having the property of A_{i+1}. And for one and the same number $x \in B_i$, if its degree of having the property of A_i is high, then its degree of having the property of A_{i+1} is low, and vice versa. Thus, the numbers in interval B_i can be treated respectively as the peripheral members of the classes A_i and A_{i+1}. Then, core members and peripheral members together can form a class about the property represented by center C_i and gathered according to approximation relation, in which the core members form the center of the class and the peripheral members form the boundary of the class. Thus, sub-regions $C_1, C_2, \ldots, C_n$ and $B_1, B_2, \ldots, B_{n-1}$ form classes $A_1, A_2, \ldots, A_n$ (as shown in Fig. 2.2), where $A_1 = C_1 + B_1$, $A_2 = B_1 + C_2 + B_2, A_3 = B_2 + C_3 + B_3, \ldots, A_n = B_{n-1} + C_n$.

It can be seen that interval B_i is the boundary of classes A_i and A_{i+1} separately, and at the same time, it is also the demarcation between the two classes. In the latter case, the interval B_i seems to be the intersection of A_i and A_{i+1}, but not the intersection in the usual sense. In fact, every number in B_i neither entirely belongs to

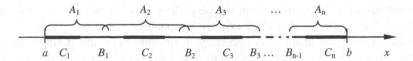

Fig. 2.2 An illustration of clustering and partitioning of continuously distributed numerical feature values, where the parts shown by the thickest line segments are separately the centers of each class and the parts shown by the comparatively thick line segments are the boundaries between these classes

Fig. 2.3 An example of single class

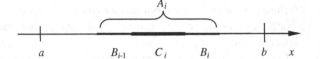

class A_i nor entirely belongs to class A_{i+1}, but belongs to A_i to a certain degree and also belongs to A_{i+1} to a certain degree, and for one and the same $x \in B_i$, if the degree of it belonging to A_i is high, then the degree of it belonging to A_{i+1} is low, and vice versa. That is to say, the numerical objects in region B_i have double identity of "being this and also being that," so the demarcation is also an intermediary transition region between the two classes, and the line of demarcation of two adjacent classes embodies implicitly on the pairs of complementary degrees of membership of members in the region.

Viewed singly, a prototype of the classes stated above is shown in Fig. 2.3. As you can see, the characteristic of this kind of class is that core members completely have the property of the class while peripheral members have the property of the class in some degree (accurately speaking, these degrees decrease progressively from inside to outside). Observed from membership relation, core members completely belong to the class while peripheral members belong to the class to some degree (these degrees also decrease progressively from inside to outside). Thus, the boundary formed by peripheral members is a smooth transition region from the members of the class that completely have the property of the class to those members that do not completely have the property of the class (or in other words, from the members of the class that completely belong to the class to those members that do not completely belong to the class).

To sum up, the boundaries of classes obtained above are not like the usual rigid boundaries, but have a kind of "flexibleness"; that is, they are "flexible boundaries." Thus, this kind of class is also a "flexible class."

Since flexible classes $A_1, A_2, ..., A_n$ have covered the whole interval $[a, b]$, there are no usual intersections between them. Therefore, the group of the flexible classes constitutes an unusual partition of $[a, b]$. Considering $A_1, A_2, ..., A_n$ are flexible classes, this kind of partition is a "flexible partition."

Thus, we have solved the problem of the clustering and partition of range $[a, b]$ of numerical feature values. It can be seen that we realize the clustering and

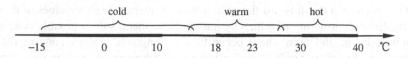

Fig. 2.4 Examples of flexible classes in temperature range [−15, 40]

partition by actually adopting a technique of "flexible dividing" according to the continuousness of numerical feature values. In view of the characteristic of this clustering method, we call it the "flexible clustering."

Using flexible clustering, for example, to do partition of the range [−15, 40] of temperature, we can obtain following flexible classes (as shown in Fig. 2.4).

It can be seen that the flexible clustering is not like the usual rigid clustering to assign an object entirely to a class; rather, it assigns an object "in a certain degree," or in other words "partly," to a class, and meanwhile "partly" to another class. In other words, flexible clustering is not like rigid clustering the boundary being at the outside of points, that is, between points; rather, the boundary is at the inside of points.

In the above, we introduce flexible clustering by taking continuous range [a, b] of real numerical values as an example. It can be seen that the kind of clustering method is also applicable to range $\{n_1, n_2, ..., n_n\}$ of consecutive integers. Besides, the central points of center regions of all classes were not considered in the flexible clustering above. Then, if needed, we can first set the central points and then according to the central points set center regions and boundary regions.

Now, range U of numerical feature values is flexibly partitioned by flexible clustering into flexible classes: $A_1, A_2,..., A_n$. Then, taking these flexible classes as denotations separately, one and another corresponding concepts are obtained. From the above-stated, a flexible class from flexible clustering stands for on the whole a corresponding property of things, so the flexible classes $A_1, A_2,..., A_n$ stand for properties $A_1', A_2',..., A_n'$ with obvious difference (e.g., "cold," "warm," and "hot" in Fig. 2.4). Thus, the concepts stood for by flexible classes $A_1, A_2,..., A_n$ are the attributive concepts. Since the denotations are flexible classes, these attributive concepts are flexible concepts. Thus, the word that denotes a flexible concept, that is, the label of corresponding flexible class, is just a flexible linguistic value.

Of course, considering from angle of feature value, summarizing separately the numerical values in flexible classes $A_1, A_2,..., A_n$, the corresponding one and another flexible linguistic values can also be resulted.

Thus, starting from the continuous numerical feature values of things, we have obtained logically the corresponding flexible concepts and flexible linguistic values through flexible clustering. Examining such concepts as "tall," "big," "many," and "quick" in our brain, obviously they also have the same characteristic as that of the flexible concepts obtained in the above. Therefore, we believe that this type of flexible concept in human brain is also thus formed at the numerical level, or in other words in the mathematical sense.

It has been seen that it is just the continuity of numerical feature values of things and the flexible clustering of which by human brain that result in flexible concepts in human brain, and then, there occur flexible linguistic values and imprecise information. In other words, the continuous distribution or continuous change phenomenon of numerical feature values of things is the objective basis of the flexible concepts, flexible linguistic values, and imprecise information, and the flexible clustering of continuous numerical feature values by human brain is the formation principle of them, and that the rigid clustering cannot be objectively done is the cause resulting in them.

2.2 Denotative Model of a Flexible Concept

The formation principle of flexible concepts shows that a flexible class just is the mathematical essence of the corresponding flexible concept. Yet how should we represent a flexible class?

2.2.1 Core and Support Set

We have already known that one flexible class C contains two types of members: core members and peripheral members. We call the set consisting of core members the core of a flexible class, and denote it as core(C); and we call the set consisting of core members and peripheral members together to be the support set of a flexible class, and denote it as supp(C). Thus, the boundary of flexible class C can also be represented by difference supp(C) − core(C), which can be denoted by boun(C) [1].

Obviously, a flexible class C is completely determined by its core and support set. Therefore, core core(C) and support set supp(C) just form a rough denotative mathematical model of a flexible concept. That is to say, flexible class C can be represented simply as core(C) and supp(C).

2.2.2 Membership Function

First, we introduce the measures of "difference-degree," "sameness-degree," etc.

We know that distance is a measure of the difference between two points in a space. Two points with a distance of 0 are one and the same point, and they would not be the same point if the distance is not 0. However, with the distance decreasing, the difference between two points becomes smaller and smaller; that is, they become closer and closer to be the same. Thus, two points with a nonzero distance can be treated as the same with a degree. Intuitively, the same with a degree is the partial same, while the usual same refers to the complete same.

Although distance can be used to describe the extent of difference between objects, using it to describe the extent of identity degree of objects is difficult. And what the distance reflects is the absolute quantity of difference between objects, which is related to the dimension and measuring unit used, so the comparability between distances is poor. For this reason, we introduce a kind of relative quantity of difference—difference-degree. Since "difference" means "not the same," the degree of difference and degree of sameness should be complementary; that is, the sum of the two is 1. Thus, from difference-degree, another measure—sameness-degree—can be derived.

Definition 2.1 Let $U = [a, b]$ be a range of numerical feature values. For $\forall x, y \in U$, set

$$d(x, y) = |x - y| \tag{2.1}$$

to be called the distance between x and y; take

$$r = b - a \tag{2.2}$$

as the reference distance; set

$$D(x, y) = \frac{d(x, y)}{r} \tag{2.3}$$

to be called the degree of difference, simply written as **difference-degree**, between x and y; set

$$s(x, y) = 1 - D(x, y) = 1 - \frac{d(x, y)}{r} \tag{2.4}$$

to be called the degree of sameness, simply written as **sameness-degree**, between x and y.

From the definition, it can be seen that the sameness-degree is completely determined by the distance, and its range of values is $[0, 1]$.

With the sameness-degree, the relation between objects can be more precisely described. As a matter of fact, the higher the sameness-degree between two two objects is, the closer they are to be completely the same, while similarity is just sameness to some degree, and the higher the sameness-degree is, the higher is the degree of similarity. Therefore, the sameness-degree can be used to portray the similarity relation between objects; or not strictly, the sameness-degree can be treated as the degree of similarity or approximation (for strict definitions of "similarity," see Sect. 3.8.2).

Since peripheral members only belong to flexible class C partly or to some degree, portraying a flexible class by core(C) and supp(C) would appear somewhat rough. Observe that core members belong to class C completely while peripheral members belong to class C partly and that "completely belong to" can be viewed as a particular case of "partly belong to." So we can use a certain kind of measure to describe the degree of point x belonging to flexible class C. We call this measure the degree of point x belonging to flexible class C, simply written as **membership-degree** of x to C, and denote it as $m_C(x)$. We use $m_C(x) = 1$ to indicate point x belongs to C completely and use $m_C(x) = 0$ to indicate point x does not belong to C completely; generally, $m_c(x) = s$ $(s \in [0, 1])$ indicates point x belong to C with membership-degree s.

It can be seen that according to the membership relationship, the membership-degrees of the core members of flexible class C should all be 1, and the membership-degrees of boundary members should be greater than 0 and less than 1, while the membership-degrees of other points in $[a, b]$ to flexible class C are all 0. Thus, a flexible class C can just be represented by a set $\{(x, m_c(x)) | x \in [a, b]\}$ of members attached with membership-degrees. It can be seen that this set also describes a function $f(x) = m_c(x)$ on $[a, b]$. Thus, a flexible class C in $[a, b]$ also determines a function $m_C(x)$ on $[a, b]$. Conversely, function $m_C(x)$ also completely determines flexible class C. We also follow Zadeh to call this function $m_C(x)$ the **membership function** of flexible class C. Thus, membership function $m_C(x)$ is just another kind of mathematical representation of flexible class C.

From the relation between flexible class and flexible concept, a membership function is also a kind of mathematical model of a flexible concept. However, if the mathematical model of a flexible concept only stays on the abstract conception of membership function, it would be not enough for the modeling of a flexible concept. In other words, for a flexible concept, we should study further the concrete form of its membership function.

Definition 2.2 Let C be a flexible class in range $[a, b]$. Set $\{x | x \in U, m_c(x) = 1\}$ is called the core of C and denoted by $\text{core}(C)$; its infimum $\inf(\text{core}(C))$ and supremum $\sup(\text{core}(C))$ are called separately the negative core-boundary point and the positive core-boundary point of flexible class C and denoted by c_C^- and c_C^+; set $\{x | x \in U, 0.5 < m_c(x) \leq 1\}$ is called the extended core of flexible class C and denoted by $\text{core}(C)^+$; set $\{x | x \in U, m_c(x) > 0\}$ is called the support set of flexible class C and denoted by $\text{supp}(C)$; its infimum $\inf(\text{supp}(C))$ and supremum $\sup(\text{supp}(C))$ are called separately the negative critical point and positive critical point of C and denoted by s_C^- and s_C^+; and set $\{x | x \in U, 0 < m_c(x) < 1\}$ is called the boundary of flexible class C and denoted by $\text{boun}(C)$, and the middle points of boundary $\text{boun}(C)$, namely $\frac{s_C^- + c_C^-}{2}$ and $\frac{s_C^+ + c_C^+}{2}$, are called separately the negative median point and positive median point of flexible class C and denoted by m_C^- and m_C^+.

Although Definition 2.2 defines the negative and positive critical points of a flexible class, it can be seen from Fig. 2.2 that only flexible classes located inside space $[a, b]$ have double flexible boundaries, while those located at the boundary of $[a, b]$ have only one flexible boundary.

Since the membership-degrees of the core members of a flexible class are 1 and those of the peripheral members are between 0 and 1, the middle of the graph of membership function $m_C(x)$ must be flat. And it is also known from the relation between the degree of similarity and the distance that the closer the peripheral members are to core-boundary points c_C^- or c_C^+, the closer their membership-degrees will be to 1, and the farther the boundary members are from core-boundary points c_C^- or c_C^+, the closer their membership-degrees will be to 0. Therefore, the part of membership function $m_C(x)$, which located at the boundary of flexible class C, should be monotonic.

Based on the above analysis, let C be a flexible class in range $[a, b]$, $\text{core}(C) = [c_C^-, c_C^+] \subset [a, b]$, and $\text{supp}(C) = (s_C^-, s_C^+) \subset [a, b]$, because when point x changes from s_C^- to c_C^- or from s_C^+ to c_C^+, the membership-degree $m_C(x)$ increases from 0 to 1, so we take

$$r^- = c_C^- - s_C^-, \quad r^+ = s_C^+ - c_C^+$$

as two reference distances. Thus, sameness-degrees

$$s(x, c_C^-) = 1 - \frac{d(x, c_c^-)}{r^-} = 1 - \frac{x - c_C^-}{c_C^- - s_C^-} = \frac{x - s_C^-}{c_C^- - s_C^-} \quad (x \le c_C^-)$$

$$s(x, c_C^+) = 1 - \frac{d(x, c_c^+)}{r^+} = 1 - \frac{c_C^+ - x}{s_C^+ - c_C^+} = \frac{s_C^+ - x}{s_C^+ - c_C^+} \quad (x \ge c_C^+)$$

With sameness-degree $s(x, c_C^-)$, then for $\forall x \in (s_C^-, c_C^-)$, its membership-degree $m_C(x)$ should be the product of membership-degree $m_C(c_C^-)$ of c_C^- and sameness-degree $s(x, c_C^-)$, and while $m_C(c_C^-) = 1$, then

$$m_C(x) = m_C(c_C^-) \cdot s(x, c_C^-) = 1 \cdot s(x, c_C^-) = s(x, c_C^-) = \frac{x - s_C^-}{c_C^- - s_C^-}$$

Similarly, for $\forall x \in (c_C^+, s_C^+)$, we have

$$m_C(x) = m_C(s_C^+) \cdot s(x, s_C^+) = 1 \cdot s(x, c_C^+) = s(x, c_C^+) = \frac{s_C^+ - x}{s_C^+ - c_C^+}$$

In consideration of $\forall x \in [c_C^-, c_C^+]$, $m_C(x) = 1$, and $\forall x \in [a, s_C^-]$ and $\forall x \in [s_C^+, b]$, $m_C(x) = 0$, thus we have

$$m_C(x) = \begin{cases} 0, & a \le x \le s_C^- \\ \frac{x - s_C^-}{c_C^- - s_C^-}, & s_C^- < x < c_C^- \\ 1, & c_C^- \le x \le c_C^+ \\ \frac{s_C^+ - x}{s_C^+ - c_C^+}, & c_C^+ < x < s_C^+ \\ 0, & x' \le x \le b \end{cases} \qquad (2.5)$$

It can be seen that this is a trapezoidal function (its graph is shown in Fig. 2.5).

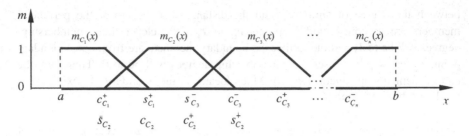

Fig. 2.5 An illustration of the membership functions of flexible concepts

The kind of trapezoidal functions is the membership function of flexible class C with two flexible boundaries. Then, the membership function of the flexible class C with only one flexible boundary (e.g., the flexible classes at the left and right ends in Fig. 2.5 only have one flexible boundary) is a semi-trapezoidal function. It can be seen that the membership function of the flexible class whose flexible boundary is at the positive direction of coordinate axis is

$$m_C(x) = \begin{cases} 1, & a \le x \le c_C^+ \\ \frac{s_C^+ - x}{s_C^+ - c_C^+}, & c_C^+ \le x \le s_C^+ \\ 0, & s_C^+ < x \le b \end{cases} \qquad (2.6)$$

and that the flexible class whose flexible boundary is at the negative direction of coordinate axis is

$$m_C(x) = \begin{cases} 0, & a \le x < s_C^- \\ \frac{x - s_C^-}{c_C^- - s_C^-}, & s_C^- \le x \le c_C^- \\ 1, & c_C^- \le x \le b \end{cases} \qquad (2.7)$$

Thus, from the generality of C, the above Eqs. (2.5), (2.6), and (2.7) are just the general expressions of the membership function of flexible classes in $[a, b]$. It can be seen that the range of values of these membership functions is $[0, 1]$, which means that the degree of a point x belonging to class C can only be 0 or 1 or a number between 0 and 1.

So far, we have derived the general expression, namely trapezoidal function (including semi-trapezoidal function), of the membership functions based on the formation principle of the flexible concepts and the sameness-degree.

For the trapezoidal functions, a kind of piecewise linear function, some readers may have doubt: Could the mathematical model of the flexible concept be so simple? Here, we want to remind our readers: This trapezoidal membership function of a flexible concept is derived on the basis of the continuous distribution of numerical feature values of things, while though the continuous distribution of numerical feature values is a reflection of the gradual change of things, it is not the dynamic process of things gradually changing, but rather a static outcome. That is to say, range [a, b] is really the range of values formed by the gradual change of things, rather than the "domain of definition" describing gradual change of things. If it was the latter, then membership functions are certainly not all linear. Because the process of the gradual change of things is not always uniform and linear, rather, they are more usually nonlinear. For example, suppose on a certain day, the temperature changes from −10 to 10 °C. This change process is generally non-uniform and nonlinear, but the temperature range [−10, 10] formed by this change process is continuous. And the membership functions of the flexible concepts about temperature (e.g., cold and warm) are just defined on such kind of continuous range.

It also needs to be noted that such piecewise membership function is just for flexible concepts on the one-dimensional range of numerical feature values. And in the next chapter, we will see that the membership functions of some flexible concepts on multidimensional measurement space are nonlinear. Further, in Chap. 4 we will see that the extended membership functions of many flexible concepts are also nonlinear.

Actually, this kind of trapezoidal functions is consistent with the understanding and application of human brain for corresponding flexible concepts. In fact, human brain, in general, gives equal treatment to all core members of a flexible concept, but for peripheral members, it then uses degree adverbs such as "comparatively" and "somewhat" to modify. For example, in the sense of "young," people between ages 18 and 25 are generally not discriminated. The reason is human brain treats people in this age group as core members of the flexible concept "young." Of course, in human brain there is no recognized and rigid agreement for the core members and peripheral members of a flexible concept, but every person has his or her own default mental scale.

From the above-stated, we see that the above-given membership functions have objective basis, coincide with human brain's mental reality, and have mathematical basis. Now that a flexible class can be determined by its membership function, and the membership function of a flexible class is a trapezoidal function, then we can give a precise definition for a flexible class.

Definition 2.3 Let $f(x)$ be a trapezoidal function with range [0, 1] on range [a, b] of numerical feature values; then, function $f(x)$ determines a flexible class C in [a, b], and function $f(x)$ is called the membership function of flexible class C and denoted by $m_C(x)$, for $\forall x \in U$, $m_C(x)$ is the membership-degree of x to C.

From Definition 2.3 and Eqs. (2.5), (2.6), and (2.7), for any flexible concept on [a, b], as long as the core-boundary points c_C^- and c_C^+ and critical points s_C^- and s_C^+ of its denotative flexible class are given, then its core and support set can be determined, and expression of its membership function can be written out. Therefore, a membership function can be written as the following parameter form:

$$m_C(x; s_C^-, c_C^-, c_C^+, s_C^+) \tag{2.8}$$

Clearly, the above general expression of the membership functions on range $[a, b]$ of real numerical values is also applicable to flexible concepts on range $\{n_1, n_2, \ldots, n_n\}$ of consecutive integers.

It can be seen that a membership function actually portrays the relationship between the points in a space and the denotation of a flexible concept on the space, so a membership function is actually the denotative mathematical model of a flexible concept, or **denotative model** for short.

2.3   Connotative Model of a Flexible Concept—
Consistency Function

We have known that a flexible linguistic value is a general designation of all numerical values in the corresponding flexible class. The purpose of linguistic values is to enlarge the granularity of information and reduce the amount of information. However, those number objects in the denotation of a flexible concept are objectively not the same but only approximate, so the strengths of objects' features characterized by these numerical values are not the same. So the contributions of these numerical values to the corresponding flexible linguistic value are not the same. Therefore, a kind of measure is needed so as to portray the degree of a numerical value consisting with a corresponding flexible linguistic value, or in other words, the degree of it according with or supporting to the corresponding flexible linguistic value. We call this measure to be the degree of a numerical value consisting with a flexible linguistic value, simply written as **consistency-degree** of x with A.

Definition 2.4 Let A be a flexible concept on range $U = [a, b]$; if the membership function of corresponding flexible class is a semi-trapezoidal increasing function, then the flexible linguistic value A is called an increasing linguistic value; if the membership function is a semi-trapezoidal decreasing function, then the A is called a decreasing linguistic value; and if the membership function is a trapezoidal function, then A is called a convex linguistic value.

In the following, we first analyze the consistency-degree of numerical value x with increasing linguistic value A.

Let A be an increasing linguistic value on range $U = [a, b]$. Observe that the membership-degree of the critical point s_C^- of corresponding flexible class C to C is 0, which shows that s_C^- contributes nothing to the property represented by A, so the consistency-degree of it with A should be 0; and from s_C^- to right, with the increase of distance from s_C^-, the consistency-degree of x with the property represented by A becomes higher and higher, and the support becomes stronger and stronger; and from s_C^- to the left, with the increase of distance from s_C^-, the difference between x and the property represented by A is larger and larger. Therefore, critical point s_C^-

is tantamount to the origin of coordinates of linguistic value A. Thus, the difference of point x in U and critical point s_C^-

$$x - s_C^- = x' \tag{2.9}$$

is also a kind of magnitude of x relative to linguistic value A. In fact, Eq. (2.9) is just a translation transformation of coordinate, which transforms measurement x of objects about the original attribute to measurement x' of the objects about its sub-attribute (i.e., is, linguistic value A). Therefore, x' indeed represents the quantity of point x having linguistic value A. But this quantity is only a kind of absolute quantity with poor comparability. Now we consider the core-boundary point c_C^-; obviously, it is a standard object of A and the membership-degree of it to A is 1, so the consistency-degree of it with A should also be 1. Thus, we take the difference between core-boundary point c_C^- and critical point s_C^- (i.e., the measurement of point c_C^- relative to A)

$$x_1 = c_C^- - s_C^- \tag{2.10}$$

as an unit quantity of x', now, set

$$g(x) = \frac{x'}{x_1} = \frac{x - s_C^-}{c_C^- - s_C^-} \tag{2.11}$$

It can be seen that what $g(x)$ represents is a relative quantity of number x with linguistic value A. Obviously, $g(s_C^-) = 0$ and $g(c_C^-) = 1$. And it can be verified that $g(x)$ also satisfies the following properties:

For $\forall x$ and $y \in U$,

(1)   When $x \neq y$, then $g(x) \neq g(y)$;
(2)   When $x < y$, then $g(x) < g(y)$;
(3)   When $x < s_C^-$, $g(x) < 0$; when $x > c_C^-$, $g(x) > 1$.

This shows that $g(x)$ not only reflects the interrelation between all the numbers in range U and linguistic value A, but it also maintains the original order relation among all the numbers. Property (3), in particular, reflects the objectivity of $g(x)$. Therefore, $g(x)$ can be taken as a measure of numbers x with linguistic value A—consistency-degree.

Thus, the consistency-degrees of all points in U with flexible linguistic value A form a function $g(x)$ $(x \in [a, b])$ on range U. We call such a function the consistency function of flexible linguistic value (or flexible concept) A, and denote it $c_A(x)$.

On the basis of the above analysis, we give the following definition.

Definition 2.5 Let A be an increasing linguistic value on range $U = [a, b]$, s_C^- be the critical point of corresponding flexible class C, and c_C^- be the core-boundary point of C, then

$$c_A(x) = \frac{x - s_C^-}{c_C^- - s_C^-} \quad (x \in U) \tag{2.12}$$

is called the consistency function of this linguistic value A. For $\forall x \in U$, $c_A(x)$ is the consistency-degree of x with A.

Similar analyzing of decreasing linguistic value can also be done. Yet from the symmetric relation between it and increasing linguistic value, we directly give the definition.

Definition 2.6 Let A be a decreasing linguistic value on range $U = [a, b]$, s_C^+ be the critical point of corresponding flexible class C, and c_C^+ be the core-boundary point of C, then

$$c_A(x) = \frac{s_C^+ - x}{s_C^+ + c_C^+} \quad (x \in U) \tag{2.13}$$

is called the consistency function of this linguistic value A. For $\forall x \in U$, $c_A(x)$ is the consistency-degree of x with A.

Since convex linguistic value has characteristics of both increasing linguistic value and decreasing linguistic value, the expression of its consistency function should be the combination of the above two consistency functions. But, the intersection point of these two function curves needs to be determined.

Let the intersection point of these two function curves be (x^*, y^*), it is easy to obtain that

$$x^* = \frac{s_C^+ c_C^- - s_C^- c_C^+}{(s_C^+ - s_C^-) - (c_C^+ - c_C^-)}, \quad y^* = \frac{s_C^+ - s_C^-}{(s_C^+ - s_C^-) - (c_C^+ - c_C^-)}$$

Thus, we have the definition.

Definition 2.7 Let A be a convex linguistic value on range $U = [a, b]$, s_C^- and s_C^+ be critical points of corresponding flexible class C, and c_C^- and c_C^+ be the core-boundary points of C, then

$$c_A(x) = \begin{cases} \frac{x - s_C^-}{c_C^- - s_C^-}, & a \le x \le x^* \\ \frac{s_C^+ - x}{s_C^+ - c_C^+}, & x^* \le x \le b \end{cases} \tag{2.14}$$

is called the consistency function of this linguistic value A. For $\forall x \in U$, $c_A(x)$ is the consistency-degree of x with A.

Actually, a convex linguistic value can also be viewed as the conjunction of an increasing linguistic value and a decreasing linguistic value. So its consistency function can also be expressed as

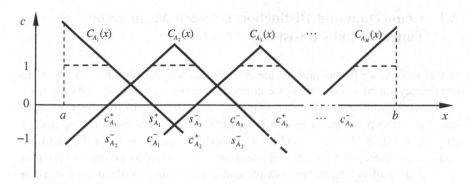

Fig. 2.6 An illustration of the consistency functions of flexible linguistic values

$$c_A(x) = \min\left\{\frac{x - s_C^-}{c_C^- - s_C^-}, \frac{s_C^+ - x}{s_C^+ - c_C^+}\right\} \quad (x \in U) \tag{2.15}$$

It can be seen from the above-stated three functional expressions that the consistency function of a convex linguistic value is a triangular function and that the consistency functions of an increasing linguistic value and a decreasing linguistic value are linear functions. We may as well call the latter two the semi-triangular functions. The graphs of these three functions are shown in Fig. 2.6. It can also be seen that the range of a consistency function is interval $[\alpha, \beta]$ $(\alpha \le 0, 1 \le \beta)$.

From the graphs of the functions, it can be visually seen that a consistency function indeed expresses the correlation between the thing's property represented by numerical values and that represented by the corresponding linguistic value, and reflects the distribution of the essential attribute of a flexible concept on the measurement space. Or in other words, the essential attribute of a flexible concept is just fully reflected and completely expressed by the consistency-degree that every number in the range of numerical values is with the corresponding linguistic value. Therefore, speaking in this sense, the consistency function of a flexible linguistic value can also be viewed as a kind of connotative mathematical model of the corresponding flexible concept, or **connotative model** for short.

From the general expression of consistency functions above, for any flexible concept, only its critical points s_C^- and s_C^+, its core-boundary points c_C^- and c_C^+, and peak value point ξ_C are needed to be given, and the specific expression of consistency function can be written. Thus, a consistency function can be written as the following parametric form:

$$c_A(x; s_C^-, c_C^-, \xi_C, c_C^+, s_C^+) \tag{2.16}$$

Clearly, the above general expression of consistency functions on range $[a, b]$ of real numerical values is also applicable to the flexible linguistic values on range $\{n_1, n_2, \ldots, n_n\}$ of consecutive integer values.

2.4   Connection and Distinction Between Membership Function and Consistency Function

In this section, we further analyze the characteristics of membership functions and consistency functions as well as the connection and distinction between the two.

Comparing the expressions and graphs of the two kinds of functions, it can be seen that the expressions of them are both based on the critical points s_C^- and s_C^+ and core-boundary points c_C^- and c_C^+ of a flexible concept and that the two functions are completely the same for critical points, core-boundary points, and points in the boundary of corresponding flexible class. Therefore, if only one of $m_C(x)$ or $c_A(x)$ is known, the other can be obtained. Such is just the connection between membership functions and consistency functions.

However, the two also have the following important distinctions:

(1) **Functional differences**

As viewed from functions, a membership function describes the denotation of a flexible concept and reflects the clustering and summarization of the properties of objects, whereas a consistency function describes the connotation of a flexible concept and reflects the distribution and detailing of the properties of objects.

The biggest characteristic of the membership function is the following: mapping all the objects inside the core of a flexible concept into 1 and mapping all the objects outside the support set into 0, which play the role of classification and summarization, meanwhile, which is also a bridge of the conversion of information granularity from fine to coarse and the conversion of information description from quantitative to qualitative.

In fact, the denotations of flexible concepts are a kind of flexible class. Therefore, just like usual rigid classes, human brain sets up these flexible classes also in order to enlarge granularity of information, reduce amount of information, and simplify representation and processing of information. Therefore, the number of elements in the core of a flexible class should generally be greater than 1 and the membership-degrees of the elements in the core should all equal to 1; that is, elements in the core are not discriminated anymore; meantime, the boundary must be non-empty and the membership-degrees of elements in it should be between 0 and 1. Besides, the membership-degrees of elements outside the support set should all be equal to 0; that is, these elements are also not discriminated anymore. Only thus the effect of classification and summarization could be achieved. And only thus would it coincide with the original intention of human brain to set flexible concepts. The membership function is just a mathematical realization of this kind of functions of clustering and summarization.

The characteristics of consistency functions are distinguishing between various number objects (including the number objects in the core of a flexible concept and those outside the support set) according to their original approximation relation; that is, the consistency-degrees of different numbers are not the same. Thus, the degree of a number object x having property A can not only be numbers 0, 1, and numbers between 0 and 1, but it also can be greater than 1 or less than 0. Therefore, the consistency function is also the degree distribution function of the connotation of a flexible concept.

Just because of the above differences, a membership function is a trapezoidal function and can only be a trapezoidal function, and a consistency function is a triangular function and can only be a triangular function. The range of a membership function is [0, 1] and must be [0, 1], while the range of a consistency function is $[\alpha, \beta]$ $(\alpha \leq 0, 1 \leq \beta)$ and must be $[\alpha, \beta]$.

(2) **Differences in the natures**

Viewed from the nature, membership functions show a kind of subjective classification by human (brain) of continuously distributed numerical feature values of things, while consistency functions reflect the objective relations between the numerical values that characterize feature of things and the corresponding linguistic values.

(3) **Differences in application**

Viewed from the angle of application, membership functions place emphasis on classification and summarization, which solve the problems of "what is," so which can be used to solve such problems as classification, recognition, diagnosis, and prediction; and consistency functions facilitate detailing and accuracy, which solve the problem of "how," and so can be used to solve such problems as judgment, decision, control, and planning.

In a word, the membership function and the consistency function both have characteristics of their own and the two have both connection and distinction, and they complement each other and form a complete representation of a flexible concept, which supplement each other in application and are both indispensable.

2.5 Flexible Entity Concepts and Their Mathematical Models

Strictly speaking, the flexible concepts stated above are all flexible attributive concepts. Besides, there are flexible entity concepts in our brains. For example, "good student" is a flexible entity concept. Then, what are the formation principle and the mathematical models of flexible entity concepts? It can be seen that the flexible entity concepts are closely linked with flexible attributive concepts. In fact, in macro, a flexible entity concept is an entity concept modified by flexible attributive concept, or in other words, it is a compound flexible concept combined by a flexible attributive concept and an entity concept. In micro, the numbers in the numerical valued range that a flexible attributive concept is on and the subsets (equivalence classes) consisting of entity objects which take these numbers as respective numerical feature value in corresponding entity objected set are one-to-one correspondence. Thus, in the sense of corresponding feature, the former represents the latter. Thus, with a flexible attributive concept on numerical valued range U being formed, corresponding flexible entity concept is also formed on corresponding entity objected set S, and the mathematical model of the flexible entity concept depends on or is reduced to those of corresponding flexible attributive concept. Thus, the former can be obtained from the latter. For example, for students, let "good" be a flexible attributive concept on the range [0, 100] of scores of their synthetic evaluation. Suppose the support set and core of "good" are supp(good) = (80, 100] and core(good) = [90, 100]; then, the corresponding membership function is

$$m_{\text{good}}(x) = \begin{cases} 0, & 0 \le x \le 80 \\ \frac{x-80}{10}, & 80 < x < 90 \\ 1, & 90 \le x \le 100 \end{cases} \quad (x \text{ is a score of synthetic evaluation})$$

and the consistency function is

$$c_{\text{good}}(x) = \frac{x-80}{10}, \quad 0 \le x \le 100$$

Thus, the membership function of "good student" is

$$m_{\text{good student}}(s_x) = m_{\text{good}}(x)$$

and the consistency function can be

$$c_{\text{good student}}(s_x) = c_{\text{good}}(x)$$

And the corresponding flexible class of "good student," the flexible entity concept, is a flexible class in universe {all students}, its support set, and core are

$$\text{supp(good student)} = \left\{ s_x | s_x \in \{\text{all students}\}, 0 < m_{\text{good student}}(s_x) \le 1 \right\}$$
$$\text{core(good student)} = \left\{ s_x | s_x \in \{\text{all students}\}, m_{\text{goodstudent}}(s_x) = 1 \right\}$$

Generally, let AE be a flexible entity concept combined by flexible attributive concept A and entity concept E; then, its mathematical model is as follows:

$$\text{supp}(AE) = \{e | e \in E, 0 < m_{ES}(e) \le 1\} \tag{2.17}$$

$$\text{core}(AE) = \{e | e \in E, m_{AE}(e) = 1\} \tag{2.18}$$

$$m_{AE}(e_x) = m_A(x) \tag{2.19}$$

$$c_{AE}(e_x) = c_A(x) \qquad\qquad (2.20)$$

where x is numerical feature value of an entity object and e_x is the entity objects whose numerical feature values is x.

2.6 About Pseudo-Flexible Linguistic Values

In the above, we reveal the formation principle of flexible linguistic values and present their mathematical models; however, in daily language and information exchange, people employ also flexible linguistic values to describe two, three or a handful of discrete numerical values. For example, for 3 and 5, the two numbers, speaking relatively, people would say that 3 (is) "little" while 5 (is) "much" (if the two numbers characterize the quantities of objects), or say that 3 (is) "small" while 5 (is) "big" (if the two numbers characterize the volumes of objects); and then, if there is also a 6 here, people would say then that 6 (is) "more" or "bigger." It can be seen that people usually use two opposite flexible linguistic values to describe the comparison between two numerical values, whereas use the flexible linguistic values having progressive relationship to describe the comparison between more than two numerical values. However, note that the flexible linguistic values used in this situation are not formed by clustering and summarization, merely relative. In fact, they are a judgment of corresponding discrete numerical values. Of course, this can also be regarded as a partition of the set of corresponding numerical values, but which is flexible partition not rigid partition, and each flexible linguistic value describes or represents only one numerical value. In a word, these relative flexible linguistic values are not real flexible linguistic values we discussed previously, which is really a kind of **pseudo flexible linguistic value**.

Pseudo-flexible linguistic values frequently occur in our daily language and information exchange, and the most used are those opposite flexible linguistic values. Besides the examples above, such as "high," "low," "fast," "slow," "hot," "cold," "good," "bad," "young," and "old" are all frequently used pseudo-flexible linguistic values. They will occur when we describe relative feelings. For instance, in severe cold, people would feel "warm" when the temperature rises a little; here, "warm" is just a pseudo-flexible linguistic value that describes the corresponding temperature relative to (that temperature) severe cold. Similarly, in intense heat, people would feel "cool" when the temperature drops slightly; here, "cool" is just a pseudo-flexible linguistic value that describes the corresponding temperature relative to (that temperature) intense heat.

Pseudo-flexible linguistic values are literally no different from real flexible linguistic values. In natural language processing, whether a flexible linguistic value is real or pseudo can be discriminated according to the context. Real flexible linguistic values, as stated in previous sections, result from flexible clustering and flexible partition, which represent one and another relatively fixed continuous sets of numerical values, and which are the flexible linguistic values on the corresponding ranges of numerical values and also the flexible linguistic values often existing in human brains; but pseudo-flexible linguistic values are provisionally generated, which have only relative significance but have no connotation and denotation, and one and the same pseudo-flexible linguistic value can represent different quantities in different contexts. For example, we can still use "little" and "much" or "small" and "large" to describe 30 and 50.

Although pseudo-flexible linguistic values occur frequently in our information exchange, since they are not real flexible linguistic values summarizing a batch of numerical values, in imprecise-information processing the kind of flexible linguistic value is not involved actually. In view of this, we will not discuss the kind of pseudo-flexible linguistic value in this book.

2.7 Summary

In this chapter, we took real interval $[a, b]$ as a general range of numerical feature values and used flexible clustering to obtain the corresponding flexible classes and flexible concepts (flexible linguistic values), thus simulating and revealing the objective basis, formation principle, and cause of flexible concepts, and then we established the mathematical models of flexible concepts and derived their general expressions. Besides, we distinguished between the flexible attributive concept and the flexible entity concept and discussed pseudo-flexible linguistic values.

The main points of the chapter are as follows:

- The phenomenon of "continuous distribution or change" of magnitudes of a feature, i.e., numerical feature values, of things and the treating way of "flexible clustering" of human brain result in flexible concepts (flexible linguistic values) in human brain. In other words, the continuous distribution or change of magnitudes of a feature of things is the objective basis of flexible concepts, and the flexible clustering of continuous magnitudes of a feature by human brain is the formation principle of flexible concepts, and that rigid clustering can not be done objectively is the cause resulting in flexible concepts.
- A flexible concept can have two kinds of mathematical models: denotative model and connotative model, the former being core + support set and membership function, and the latter consistency function; they constitute the complete representation of a flexible concept.
- The membership functions of flexible concepts on numerical ranges are trapezoidal or semi-trapezoidal functions, and the consistency functions are triangular or semi-triangular functions.
- The flexible concepts can be classified into flexible attributive concepts and flexible entity concepts, but the former is the abstract of the latter and the latter can be reduced to or is dependent on the former.

Reference

1. Lian S (2009) Principles of imprecise-information processing. Science Press, Beijing

Chapter 3
Multidimensional Flexible Concepts and Flexible Linguistic Values and Their Mathematical Models

Abstract This chapter further considers the flexible clustering and flexible classes in measurement spaces and reveals the formation principles of multidimensional flexible concepts and flexible linguistic values and establishes their mathematical models, especially presenting the universal mathematical models of flexible properties (concepts) and flexible relations (concepts).

Keywords Multidimensional flexible concepts and flexible linguistic values · Flexible attributive concepts · Flexible properties (concepts) · Flexible relations (concepts)

In the last chapter, we did flexible clustering of numbers in a range of numerical feature values of things and obtained the corresponding flexible concepts and flexible linguistic values. However, the range of numerical feature values is one-dimensional, so the flexible concepts on it are only the flexible concepts about single feature of things. We call the flexible concepts to be "one-dimensional" flexible concepts. Besides this kind of one-dimensional flexible concepts, there are also the flexible concepts about multiple features of things in our brains. These flexible concepts are on multidimensional ranges of numerical values, which can be called the "multidimensional" flexible concepts. For example, "nearby circle O" is just a two-dimensional flexible concept on a two-dimensional range of numerical values. In addition, those flexible concepts about certain relations between things can also be seen as multidimensional. Therefore, in this chapter, we will examine further multidimensional flexible concepts and flexible linguistic values and their mathematical models.

3.1 Measurement Space and Corresponding Flexible Clustering

A multidimensional region formed by the numerical values of multiple features of things is just the Cartesian product of the corresponding multiple ranges of numerical feature values. Therefore, this kind of multidimensional region is a kind of "space." Also, considering that the components of vectors wherein, namely numerical feature values, are the values of corresponding measures, so we call this kind of space the "measurement space."

Definition 3.1 Let $\mathscr{F}_1, \mathscr{F}_2, \ldots, \mathscr{F}_n$ be n features of a certain type of things, and let $U_1, U_2, \ldots, U_n$ be successively their ranges of numerical values. We call the Cartesian product $U_1 \times U_2 \times \cdots \times U_n = U$ to be the measurement space of the type of things.

Example 3.1 Suppose the range of human's heights is [0.5, 2.5] and the range of human's weights is [1, 120], then [0.5, 2.5] $\times$ [1, 120] is a measurement space of human being.

From the definition and example, we can see that the points in a measurement space are also continuous.

Now, speaking in terms of measurement space, a range of numerical feature values is just a one-dimensional measurement space.

Note: The "space" here only refers to a kind of region but not involving the operations on it, so the measurement space is not an n-dimensional vector space in the strict sense, even though it consists of n-dimensional vectors.

Let $U = U_1 \times U_2 \times \cdots \times U_n$ ($U_i = [a_i, b_i]$, $i = 1, 2, \ldots, n$; $n > 1$) be a measurement space.

Like the case of range $[a, b]$ of numerical feature values, in order to reduce the amount of information, lower complicatedness and understand and grasp things at a higher level, clustering and partitioning of points in space U *also need to be done*, and on the other hand, the points in U are also continuous. Therefore, the clustering and partitioning of the points in space U also have to use the flexible clustering and flexible partitioning.

Since points in measurement space U have multiple coordinates, they can be continuous in multiple directions of coordinates and the multiple coordinates can be combined to more or even infinite number of directions. Taking the two-dimensional space $U_1 \times U_2$ as an instance, it can be seen that points in multidimensional space U actually can be continuous in infinite directions, which means that for multidimensional measurement space U, there can be multiple methods of flexible clustering, such as bar flexible clustering, square flexible clustering, circle flexible clustering, and even irregular flexible clustering, so there can be multiple kinds of shapes of flexible classes (of course, we only need to discuss those meaningful flexible clustering and flexible classes). However, we can imagine visually that only bar flexible clustering and square flexible clustering, as well as the flexible clustering based on straight lines and planes, can realize the flexible partition of the measurement space U.

Next, we give some related concepts of a flexible class in the n-dimensional measurement space.

Definition 3.2 Let C be a flexible class in an n-dimensional measurement space U. Set $\{x \mid x \in U, m_C(x) = 1\}$ is called the core of C, denoted core(C); set $\{x \mid x \in U, 0.5 < m_C(x) \leq 1\}$ is called the extended core of flexible class C, denoted core(C)$^+$; set $\{x \mid x \in U, m_C(x) > 0\}$ is called the support set of flexible class C, denoted supp(C); set $\{x \mid x \in U, 0 < m_C(x) < 1\}$ is called the boundary of flexible class C, denoted boun(C). The boundary point, boundary line, or boundary plane of the support set are called the critical point, critical line, and critical plane of flexible class C, separately; the boundary point, boundary line, or boundary plane of the core are called the core–boundary point, core–boundary line, and core–boundary plane of flexible class C, separately; the middle point, middle line, or middle plane of the boundary (region) are called the median point, median line, and median plane of flexible class C, separately.

Definition 3.3 Let $U \subset \mathbf{R}^n$ be an n-dimensional measurement space ($n \geq 1$). For $\forall$ $x = (x_1, x_2, \ldots, x_n)$ and $y = (y_1, y_2, \ldots, y_n) \in U$; set

$$d(x,y) = \sqrt{\sum_{i=1}^{n} (x_i - y_i)^2} \tag{3.1}$$

is called the distance between x and y; take

$$r = \max_{x,y \in U} d(x,y) \tag{3.2}$$

as the reference distance; set

$$D(x,y) = \frac{d(x,y)}{r} \tag{3.3}$$

is called the degree of difference, simply written as difference-degree, between x and y; set

$$s(x,y) = 1 - D(x,y) = 1 - \frac{d(x,y)}{r} \tag{3.4}$$

is called the degree of sameness, simply written as sameness-degree, between x and y.

3.2 Flexible Clustering with Respect to Partial Coordinate Components

Firstly, let us consider flexible clustering with respect to one coordinate component in two-dimensional and three-dimensional measurement spaces.

Let $U = U \times V \subset \mathbf{R}^2$ be a two-dimensional measurement space, where $U = [a, b]$ and $V = [c, d]$, and then, U is a rectangle region on a two-dimensional plane. As shown in Fig. 3.1, in $U \times V$, a flexible class A is resulted by flexible clustering merely with respect to the x coordinate of points, and a flexible class B is resulted by flexible clustering merely with respect to the y coordinate of points. The dark gray part in the figure is the core of the corresponding flexible class, and the light gray part is the flexible boundary [1].

It can be seen that the boundaries of the core and support set of a flexible class in a two-dimensional space are no longer points but lines, that is, core–boundary line and critical line. Just like a one-dimensional flexible class, these two two-dimensional flexible classes stand for two flexible concepts on a two-dimensional measurement space U. Since flexible class A is resulted by clustering according to the approximation relation between the x coordinates, for any $(x, y) \in U \times V$, the membership-degree $m_A(x, y)$ is only related to x. Thus, we have

$$m_A(x,y) = m_A(x) \quad (x \in U, y \in V) \tag{3.5}$$

Similarly, we have

$$m_B(x,y) = m_B(y) \quad (x \in U, y \in V) \tag{3.6}$$

whose graphs are shown in Fig. 3.2. They are truncated ridged surface.

Correspondingly, the consistency functions of flexible linguistic values A and B are

$$c_A(x,y) = c_A(x) \quad (x \in U, y \in V) \tag{3.7}$$

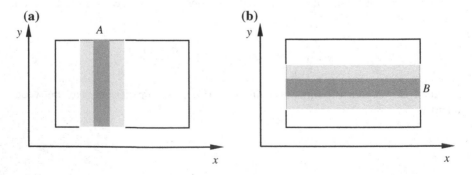

Fig. 3.1 Examples of two-dimensional flexible classes with respect to one coordinate component

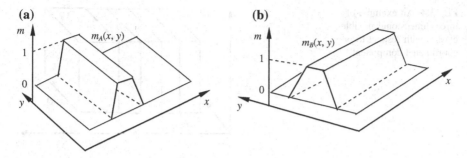

Fig. 3.2 Examples of the graphs of membership functions of flexible classes in a two-dimensional space

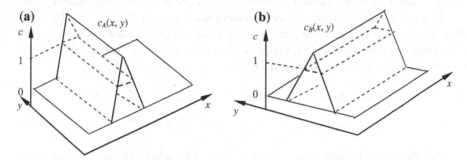

Fig. 3.3 Examples of the graphs of consistency functions of flexible linguistic values on a two-dimensional space

$$c_B(x, y) = c_B(y) \quad (x \in U, y \in V) \tag{3.8}$$

whose graphs are shown in Fig. 3.3. They are ridged surface.

What Fig. 3.4 shows is an example of flexible classes with respect to one coordinate component in a three-dimensional measurement space. Flexible class A stands for a flexible concept on the corresponding three-dimensional measurement space. Because flexible class A is obtained from flexible clustering only with respect to the approximation relation between coordinate components x_s, its membership function and consistency function are

$$m_A(x, y, z) = m_A(x) \quad (x \in U, y \in V, z \in W) \tag{3.9}$$

$$c_A(x, y, z) = c_A(x) \quad (x \in U, y \in V, z \in W) \tag{3.10}$$

It can be seen that this kind of flexible clustering in two-dimensional and three-dimensional measurement spaces and their flexible classes can be completely generalized to n-dimensional measurement space. In fact, the above two-dimensional flexible classes A and B are tantamount to the extension of

Fig. 3.4 An example of
three-dimensional flexible
classes with respect to one
coordinated component

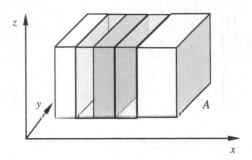

one-dimensional flexible classes A and B on a two-dimensional space. Conversely, the original one-dimensional flexible classes A and B are then tantamount to the projections of two-dimensional flexible classes A and B on one-dimensional space.

Generally, for n-dimensional measurement space $U = U_1 \times U_2 \times \cdots \times U_n \subset \mathbf{R}^n$, the membership function and consistency function of the flexible class A obtained by doing flexible cluster in U with respect to coordinate component x_k, which stands for flexible concept A on U, separately are

$$m_A(x_1, x_2, \ldots, x_n) = m_A(x_k), \quad x_j \in U_j, \; j = 1, 2, \ldots, n \qquad (3.11)$$

$$c_A(x_1, x_2, \ldots, x_n) = c_A(x_k), \quad x_j \in U_j, \; j = 1, 2, \ldots, n \qquad (3.12)$$

In a three-dimensional space, we could also do flexible clustering with respect to two coordinate components simultaneously. For example, Fig. 3.5 just shows an example of a flexible class obtained by flexible clustering with respect to two coordinate components x and y simultaneously in a three-dimensional measurement space, where (a) is a clustering without a center and (b) is a clustering with a center. It can be seen that flexible class A here is only related to two coordinates x and y of a point but not related to coordinate z, so whose membership function and the consistency function of the corresponding flexible linguistic value should be

$$m_A(x, y, z) = m_A(x, y) \quad (x \in U, y \in V, z \in W) \qquad (3.13)$$

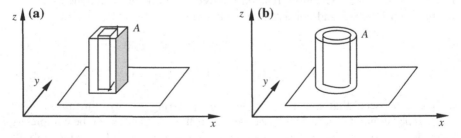

Fig. 3.5 Examples of flexible classes with respect to two coordinates x and y in a three-dimensional space

$$c_A(x, y, z) = c_A(x, y) \quad (x \in U, y \in V, z \in W) \tag{3.14}$$

From this, it is not hard for us to derive the membership function of an n-dimensional flexible class obtained by clustering with respect to k $(1 < k < n)$ coordinate components and the consistency function of the corresponding linguistic value.

Note that although the above n-dimensional flexible classes $(n > 1)$ and the flexible concepts and flexible linguistic values stood for by them are obtained with respect to one or multiple coordinate components of points, in appellation, we still say that n-dimensional point $(x_1, x_2, ..., x_n)$ has flexible linguistic value A.

3.3 Square Flexible Clustering and Circular Flexible Clustering

In the following, we do flexible clustering of points in a space with respect to all coordinate components of a point, that is, with respect to a whole point, and then get corresponding flexible classes and flexible concepts. This section discusses square flexible clustering and circular flexible clustering.

3.3.1 Square Flexible Clustering and Flexible Squares

Square flexible clustering is the flexible clustering taking a square (including also cubical and hypercubical) region as center.

Let $U = U \times V \subset \mathbf{R}^2$ be a two-dimensional measurement space, where $U = [a, b]$ and $V = [c, d]$. As shown in Fig. 3.6, draw two square regions with proper size in U, of them one contains another and each side of them parallel to the corresponding side of space U, and then take smaller square region (the part of dark gray in the figure) as core and bigger square region as support set, forming then a square flexible class C in space U. Unlike one-dimensional flexible classes, the

Fig. 3.6 Examples of square flexible clustering and flexible square in a two-dimensional space

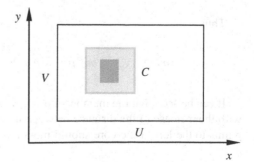

two-dimensional flexible class C has 4 critical lines and 4 core–boundary lines. Viewed from shape, this kind of flexible classes is just a **flexible square**. The flexible square C stands for a flexible concept and flexible linguistic value on a two-dimensional space U.

Obviously, the square flexible clustering in a two-dimensional measurement space can also be generalized to general n-dimensional measurement space $U = U_1 \times U_2 \times \cdots \times U_n$ $(U_i = [a_i, b_i], i = 1, 2, \ldots, n)$ and to obtain an n dimensional flexible square.

Actually, the flexible squares in multidimensional spaces are also a kind of generalizations of the flexible intervals in one-dimensional spaces, but a flexible square can also be viewed as the intersection of mutually orthogonal flexible classes in a multidimensional space. For instance, the two-dimensional flexible square C in Fig. 3.6 can also be viewed as the intersection of two orthogonal bar flexible classes whose shapes are similar to A and B in Fig. 3.1.

Next, we consider then the membership function of a flexible square and corresponding consistency function. It can be seen that for point (x, y) in core of flexible square C, the membership-degree $m_C(x, y)$ should certainly be 1, and for points (x, y) on or outside critical line of the support set, the membership-degrees $m_C(x, y)$ should be 0. In the following, we consider the membership-degrees of points in the flexible boundary around the core for flexible square C. Let point $p = (x^*, y^*)$ be a point to the left of the core in the support set. It can be seen that the distance from point p to the left core–boundary line of flexible square C is

$$d(p, p') = \sqrt{(x' - x^*)^2 + (y^* - y^*)^2} = x' - x^*$$

Here, $p' = (x' - y^*)$ is the foot of perpendicular of point p to left core–boundary line. Since in the direction of x, points change from (x', y) to (x'', y) (the latter is the corresponding points on the left critical line), their membership-degrees change from 0 to 1, for Definition 3.3, we take $r = c_{C_x}^- - s_{C_x}^-$ ($c_{C_x}^-$ and $s_{C_x}^-$ are separately the negative core–boundary point and negative critical point of flexible square C in the direction of x) as the reference distance. Therefore, we have

$$s(P, P') = 1 - \frac{d(P, P')}{r} = 1 - \frac{x' - x^*}{c_{C_x}^- - s_{C_x}^-} = \frac{x^* - x'}{c_{C_x}^- - s_{C_x}^-}$$

Thus,

$$m_C(x^*, y^*) = m_C(p') \cdot s(p, p') = 1 \cdot s(p, p') = \frac{x^* - x'}{c_{C_x}^- - s_{C_x}^-}$$

It can be seen that the membership-degree of point p should be between 0 and 1; with the increase of the distance between it and the core, the membership-degree of points to the left of the core should monotonically decreasing from 1 to 0. And by

$x' = c_{C_x}^-$ and the point (x^*, y^*) is arbitrary, so for points at the left of the core of C, we have

$$m_C(x, y) = \frac{x - c_{C_x}^-}{c_{C_x}^- - s_{C_x}^-}$$

It can be seen from this equation that in the direction of x, for point (x, y) in the negative- or positive-side boundary of core, the membership-degree $m_C(x, y)$ is merely related to x. Similarly, in the direction of y, for point (x, y) in the negative- or positive-side boundary of core, the membership-degree $m_C(x, y)$ is merely related to y. Thus, viewed from x direction and y direction, the shape of membership function $m_C(x, y)$ should all be trapezoidal. So it can be imagined that the shape of the membership function of flexible square C should be a prismoid (also called truncated square cone) surface as shown in Fig. 3.7a (of course, this prismoid is the geometry of the membership function of flexible classes on the non-edge part of space U. For those flexible classes at the edge of space U, the graphs of their membership functions are then not of standard prismoid but "semi-prismoid").

Overlooking the prismoid in Fig. 3.7a, we obtain Fig. 3.7b. It can be seen that the support set of flexible square C is actually divided into 5 small regions by the projections of 8 edges of the prismoid. Numbering these small regions, then region a_1 is the core of flexible class C, and the others are all the boundaries of C.

From the above analysis and Figs. 3.7a, b, it is not hard to see that for $\forall (x, y) \in a_1$, $m_C(x, y) = 1$; for $\forall (x, y) \in a_3$, $m_C(x, y) = \frac{x - s_{C_x}^-}{c_{C_x}^- - s_{C_x}^-}$; for $\forall (x, y) \in a_2$, $m_C(x, y) = \frac{s_{C_y}^+ - y}{s_{C_y}^+ - c_{C_y}^+}$; for $\forall (x, y) \ (x, y) \in a_4$, $m_C(x, y) = \frac{s_{C_x}^+ - x}{s_{C_x}^+ - c_{C_x}^+}$; for $\forall (x, y) \in a_5$, $m_C(x, y) = \frac{y - s_{C_y}^-}{c_{C_y}^- - s_{C_y}^-}$. Therefore, to sum up, we have

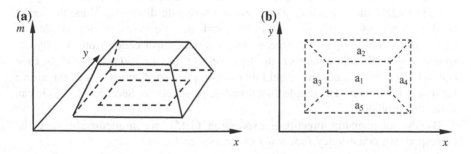

Fig. 3.7 An example of a two-dimensional flexible square and the graph of its membership function

$$m_C(x,y) = \begin{cases} \dfrac{x - s_{C_x}^-}{c_{C_x}^- - s_{C_x}^-}, & (x,y) \in a_3 \\[2ex] \dfrac{s_{C_x}^+ - x}{s_{C_x}^+ - c_{C_x}^+}, & (x,y) \in a_4 \\[2ex] 1, & (x,y) \in a_1 \\[2ex] \dfrac{y - s_{C_y}^-}{c_{C_y}^- - s_{C_y}^-}, & (x,y) \in a_5 \\[2ex] \dfrac{s_{C_y}^+ - y}{s_{C_y}^+ - c_{C_y}^+}, & (x,y) \in a_2 \\[2ex] 0, & (x,y) \notin a_1 \cup a_2 \cup a_3 \cup a_4 \cup a_5 \end{cases} \qquad (3.15)$$

where $s_{C_x}^-$ and $s_{C_x}^+$ are the negative and positive critical points of flexible square C about x, and $c_{C_x}^-$ and $c_{C_x}^+$ are the negative and positive core–boundary points of C about x; $s_{C_y}^-$ and $s_{C_y}^+$ are the negative and positive critical points of C about y, and $c_{C_y}^-$ and $c_{C_y}^+$ are the negative and positive core–boundary points of C about y.

Equation (3.15) is the general expression of the membership functions of flexible squares based on square flexible clustering on a two-dimensional measurement space U. The geometry of this function is prismoid surface (also called truncated square cone). Of course, Eq. (3.15) is the membership function of flexible squares at the non-edge part of space U. For those flexible squares at the edge of space U, since the graph of their membership functions is not a standard prismoid but a semi-prismoid, the expressions of their membership functions should be somewhat different from the expression in Eq. (3.15).

From the membership functions of two-dimensional flexible squares, it is not hard to derive that the membership functions of the multidimensional flexible squares of over 3 dimensions should have the same characteristics, that is, they are all trapezoidal functions viewed from every coordinate direction. Thus, the membership functions of these multidimensional flexible squares are similar to Eq. (3.15), whose graphs are hyperprismoid surface (also called truncated hypersquare cone). But the membership functions of the multidimensional flexible squares of over 3 dimensions would be too complicated if written as in the above form. Luckily, there are simple functional expressions in Sect. 5.5.2 which can solve this problem.

By the membership functional expression (3.15), we immediately have the corresponding consistency functional expression as follows:

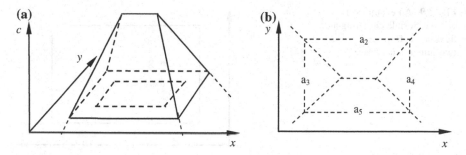

Fig. 3.8 An example of the graph of the corresponding consistency function of a two-dimensional flexible square

$$c_C(x,y) = \begin{cases} \dfrac{x - s_{C_x}^-}{c_{C_x}^- - s_{C_x}^-}, & (x,y) \in a_3 \\[2ex] \dfrac{s_{C_x}^+ - x}{s_{C_x}^+ - c_{C_x}^+}, & (x,y) \in a_4 \\[2ex] \dfrac{y - s_{C_y}^-}{c_{C_y}^- - s_{C_y}^-}, & (x,y) \in a_5 \\[2ex] \dfrac{s_{C_y}^+ - y}{s_{C_y}^+ - c_{C_y}^+}, & (x,y) \in a_2 \end{cases} \tag{3.16}$$

Here, regions a_2, a_3, a_4, and a_5 are shown in Fig. 3.8b, and the graph of the function is shown in Fig. 3.8a, whose geometry is a wedge surface. Similarly, the kind of consistency functions has also a simple general expression (reader can see the Eq. (6.10) in Sect. 6.4.1).

3.3.2 Circular Flexible Clustering and Flexible Circles

Circular flexible clustering is the flexible clustering taking a circular (including also spherical and hyperspherical) region as center. We still take a two-dimensional space $U = U \times V$ as an instance. As shown in Fig. 3.9, drawing two concentric circular regions of proper radius in U, then taking smaller circular region (the part of dark gray in the figure) as core, bigger circular region as support set, thus forms a circular flexible class C in space U. Viewed from shape, this kind of flexible class is just a **flexible circle**. The flexible circle C stands for a flexible concept and flexible linguistic value on a two-dimensional space U.

Similarly, the circular flexible clustering in a two-dimensional measurement space can also be generalized to general n-dimensional measurement space $U = U_1 \times U_2 \times \cdots \times U_n$ ($U_i = [a_i, b_i]$, $i = 1, 2, \ldots, n$) and to obtain an n-dimensional flexible circle—including flexible circle, flexible sphere, and flexible hypersphere, especially also including flexible interval.

Fig. 3.9 An example of circular flexible clustering and flexible circles in a two-dimensional space

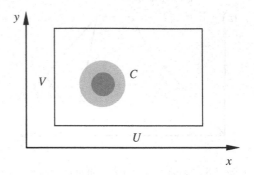

Fig. 3.10 An illustration of the principle of deriving membership function $m_C(P)$

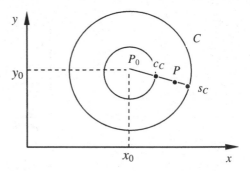

Next, we consider then the membership function of a flexible circle and corresponding consistency function.

It is shown in Fig. 3.9 that for any point $P \in U$, if $P \in \text{core}(C)$, then $m_C(P) = 1$; if $P \notin \text{supp}(C)$, the $m_C(P) = 0$. Next, let us consider how to compute the membership-degree of point $P \in \text{boun}(C)$.

As shown in Fig. 3.10, draw a straight line through the circle's center P_0 and point P, which intersects separately at P_c and P_s with the core–boundary line and critical line of flexible circle C, and then, intersection points P_c and P_s are the core–boundary point and critical point of flexible circle C that correspond to P. Therefore, denote P_c and P_s separately as c_C and s_C, and take $d(s_C, c_C)$ as the reference distance; then,

$$s(P, c_C) = 1 - D(P, c_C)$$

$$= 1 - \frac{d(P, c_C)}{d(s_C, c_C)} = \frac{d(P, s_C)}{d(s_C, c_C)}$$

Thus, the membership-degree of point P for flexible circle C

$$
\begin{aligned}
m_C(P) &= m_C(c_C) \cdot s(P, c_C) \\
&= 1 \cdot s(P, c_C) = s(P, c_C) \\
&= \frac{d(P, s_C)}{d(s_C, c_C)}
\end{aligned}
$$

Namely,

$$
m_C(P) = \frac{d(P, s_C)}{d(s_C, c_C)}
$$

Thus, to sum up the above analysis, we have

$$
m_C(P) = \begin{cases} 1, & P \in \mathrm{core}(C) \\ \dfrac{d(P, s_C)}{d(s_C, c_C)}, & P \in \mathrm{boun}(C) \\ 0, & P \notin \mathrm{supp}(C) \end{cases} \tag{3.17}
$$

From the point P being arbitrary, so what Eq. (3.17) shows is the membership function of flexible circle C. But this functional expression is an expression about (whole) point P, and we may as well call it the **point-level membership function** of flexible circle C.

Let $P = (x, y)$, $c_C = (x_c, y_c)$, and $s_C = (x_s, y_s)$, and then

$$
d(P, s_C) = \sqrt{(x - x_s)^2 + (y - y_s)^2}
$$
$$
d(s_C, c_C) = \sqrt{(x_c - x_s)^2 + (y_c - y_s)^2}
$$

Thus,

$$
m_C(x, y) = \begin{cases} 1, & (x, y) \in \mathrm{core}(C) \\ \dfrac{\sqrt{(x-x_s)^2 + (y-y_s)^2}}{\sqrt{(x_c-x_s)^2 + (y_c-y_s)^2}} & (x, y) \in \mathrm{boun}(C) \\ 0, & (x, y) \notin \mathrm{supp}(C) \end{cases} \tag{3.18}
$$

This functional expression is an expression about the coordinates of point P, and we may as well call it the **coordinate-level membership function** of flexible circle C.

From the above functional expression, it can be seen that for any point $P \in U$, as long as the corresponding core–boundary point c_C and critical point s_C are known, the membership-degree of it for flexible circle C can be found. However, when finding a membership-degree, every time we have to find the corresponding core–boundary point c_C and critical point s_C firstly, which is apparently rather cumbersome. Considering that the center point of a flexible circle and its core radius and support set radius are fixed and known, and we find that the center and the two radii are related to two distances in the functional expression, so these relations can be utilized to transform the above membership functional expression.

Let r_c and r_s be separately radius of the core (i.e., inner circle) and support set (i.e., the excircle) of C, and it is not hard to see that $r_s - r_c = d(s_C, c_C)$ and $r_s - d(P, P_0) = d(P, s_C)$. Consequently, the above Eqs. (3.16) and (3.17) are transformed to

$$m_C(P) = \begin{cases} 1, & P \in \text{core}(C) \\ \frac{r_s - d(P,P_0)}{r_s - r_c}, & P \in \text{boun}(C) \\ 0, & P \notin \text{supp}(C) \end{cases} \qquad (3.19)$$

and

$$m_C(x,y) = \begin{cases} 1, & \sqrt{(x-x_0)^2 + (y-y_0)^2} \le r_c \\ \frac{r_s - \sqrt{(x-x_0)^2+(y-y_0)^2}}{r_s - r_c}, & r_c < \sqrt{(x-x_0)^2 + (y-y_0)^2} < r_s \\ 0, & r_s \le \sqrt{(x-x_0)^2 + (y-y_0)^2} \end{cases} \qquad (3.20)$$

Here, r_1 and r_2, and x_0 and y_0 are all known constants, and the evaluation of a function is much simplified. Therefore, these two membership functional expressions are more practical.

In the following, we then consider the consistency function. Since c_C and s_C are separately the core–boundary point and critical point of flexible class C that correspond to P, we set consistency-degree $c_C(s_C) = 0$ and $c_C(c_C) = 1$. Thus, distance d (s_C, c_C) can be treated as a unit quantity, further to determine the consistency-degree of any vector in space U with flexible linguistic value C. From Fig. 3.10, it can be seen that for any vector $P \in U$, if $P \in \text{supp}(C)$, then $c_C(P) = \frac{d(P,s_C)}{d(s_C,c_C)}$; if $P(x, y) \notin$ supp(C), then $c_C(P) = -\frac{d(P,s_C)}{d(s_C,c_C)}$. Thus, to sum up, we have

$$c_C(P) = \begin{cases} \frac{d(P,s_C)}{d(s_C,c_C)}, & P \in \text{supp}(C) \\ -\frac{d(P,s_C)}{d(s_C,c_C)}, & P \notin \text{supp}(C) \end{cases} \qquad (3.21)$$

From the vector P being arbitrary, so Eq. (3.21) is the consistency function of flexible linguistic value C. We may as well call this kind of consistency function about vector P to be the **vector-level consistency function** of flexible linguistic value C.

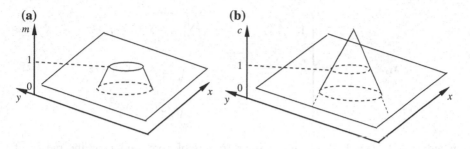

Fig. 3.11 Examples of the graphs of membership function of flexible circle and corresponding consistency function

From this function, we can further have

$$
c_C(x,y) = \begin{cases} \dfrac{\sqrt{(x-x_s)^2+(y-y_s)^2}}{\sqrt{(x_c-x_s)^2+(y_c-y_s)^2}}, & (x,y) \in \mathrm{supp}(C) \\[4mm] -\dfrac{\sqrt{(x-x_s)^2+(y-y_s)^2}}{\sqrt{(x_c-x_s)^2+(y_c-y_s)^2}}, & (x,y) \notin \mathrm{supp}(C) \end{cases} \tag{3.22}
$$

This is the **component-level consistency function** of flexible linguistic value C.

Likewise, the computation is somewhat cumbersome in application of two functions, so we also transform them into the expressions about the center of circle and the radius of core and radius of support set:

$$
c_C(P) = \frac{r_s - d(P,P_0)}{r_s - r_c}, \quad P \in U \times V \tag{3.23}
$$

$$
c_C(x,y) = \frac{r_s - \sqrt{(x-x_0)^2+(y-y_0)^2}}{r_s - r_c}, \quad (x,y) \in U \times V \tag{3.24}
$$

The graph of function $m_C(x,y)$ is shown in Fig. 3.11a, whose geometry is a round platform (also called truncated circular cone) surface. The graph of $c_C(x,y)$ is shown in Fig. 3.11b, whose geometry is a circular cone surface.

It is conceivable that generalizing the above flexible circle on a two-dimensional space to a three-dimensional measurement space and multidimensional measurement spaces of over three dimensions, their geometries would be a **flexible sphere** and a **flexible hypersphere**, and their membership function of coordinate level is then

$$
m_C(x_1, x_2, \ldots, x_n) = \begin{cases} 1, & \sqrt{\sum_{i=1}^{n}(x_i - x_{i_0})^2} \leq r_1 \\[2mm] \frac{r_2 - \sqrt{\sum_{i=1}^{n}(x_i - x_{i_0})^2}}{r_2 - r_1}, & r_1 < \sqrt{\sum_{i=1}^{n}(x_i - x_{i_0})^2} < r_2 \\[2mm] 0, & r_2 \leq \sqrt{\sum_{i=1}^{n}(x_i - x_{i_0})^2} \end{cases} \quad (3.25)
$$

where $x_i \in U_i$ ($i = 1, 2, \ldots, n$), r_1 is the radius of core, and r_2 is the radius of support set. The graph of function $m_C(x_1, x_2, \ldots, x_n)$ is an hyper-round platform (also called hypertruncated circular cone) surface.

Correspondingly, the component-level consistency function of flexible linguistic value C on $n(n > 2)$-dimensional measurement space is

$$
c_C(x_1, x_2, \ldots, x_n) = \frac{r_2 - \sqrt{\sum_{i=1}^{n}(x_i - x_{i_0})^2}}{r_2 - r_1}, \quad (x_1, x_2, \ldots, x_n) \in U \quad (3.26)
$$

whose graph is a hypercircular cone surface.

3.4   Datum-Based Flexible Clustering, "About ××" and "Near ××"

The flexible clustering above are all by determining directly core and support set to obtain a corresponding flexible class. However, some classes are about some special points or point sets in space. For example, the two flexible classes that stand for "about point P" and "near point P" are just for the specific point P. It can be seen that to obtain the classes about specific point or point set, we need to determine the point or point set focused firstly and then to determine corresponding core and support set when doing clustering. Thus, the points or point sets focused are a kind of datum of corresponding flexible clustering. Thus, in order to distinguish, we call the flexible clustering with a datum to be the **datum-based flexible clustering**.

In the following, we introduce separately the datum-based flexible clustering that takes, respectively, "point," "line," and "plane" as a datum.

3.4.1   Point-Based Flexible Clustering and Flexible Points

Point-based flexible clustering is the square flexible clustering or circular flexible clustering that takes a point in a space as the center point.

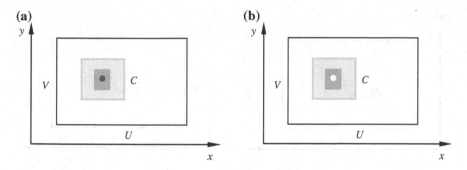

Fig. 3.12 An example of point-based square flexible clustering in a two-dimensional measurement space

Point-based square flexible clustering is as follows: Take firstly a point $P_0(x_0, y_0)$ in space U (we still take the two-dimensional space $U = U \times V$ as instance) as the center point, draw two concentric squares with appropriate size, and then take smaller square region as core and bigger square region as support set, forming a square flexible class C (as shown in Fig. 3.12a), that is, a flexible square in the space. Of course, it stands also for a flexible concept and flexible linguistic value on the space. Because the flexible square is determined actually by two coordinates x_0 and y_0 of its center point P_0, its geometric interpretation is "about x_0 and about y_0." But if the center point P_0 is not included, then the geometric interpretation of the flexible square—strictly speaking, should be a hollow flexible square (as shown in Fig. 3.12b)— is "close to x_0 and close to y_0" or "near x_0 and near y_0." As for the membership function of the flexible class C and the corresponding consistency function, do not hard to see, are actually the same as the previous Eqs. (3.15) and (3.16).

Generally, the point-based square flexible clustering in an n-dimensional measurement space is also analogous, that is, the corresponding flexible class is a solid or hollow flexible "square."

Point-based circular flexible clustering is as follows: Take firstly a point $P_0(x_0, y_0)$ in space U as the center point, draw two concentric circles with appropriate radius, and then take smaller circle region as core and bigger circle region as support set, forming a circular flexible class C (as shown in Fig. 3.13a), that is, a flexible circle in the space. Of course, it stands also for a flexible concept and flexible linguistic value on the space. Because the flexible circle is determined by its center point P_0, we call the kind of flexible class obtained by point-based flexible clustering a **flexible point**, and its geometric interpretation is "about P_0." Similarly, if the center point P_0 is not included, then the geometric interpretation of the flexible circle—strictly speaking, should be a hollow flexible circle (as shown in Fig. 3.13b)— is "close to P_0" or "near P_0." As for the membership function of the flexible class C and the corresponding consistency function, do not hard to see, are actually the same as the previous Eqs. (3.17)–(3.24).

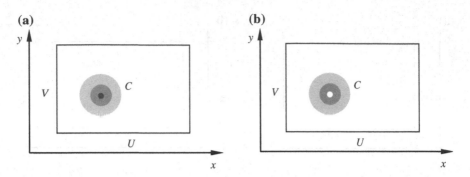

Fig. 3.13 An example of point-based circular flexible clustering in a two-dimensional measurement space

Generally, the point-based circular flexible clustering in an n-dimensional measurement space is also analogous, that is, the corresponding flexible class is a solid or hollow flexible sphere.

From the flexible classes in the above Figs. 3.12b and 3.13b and their geometric interpretations, we see "near x_0 and near y_0" refers to a hollow flexible square with point (x_0, y_0) as center point, while "near (x_0, y_0)" refers to a hollow flexible circle with point (x_0, y_0) as center point. That is to say, the meanings of "near (x_0, y_0)" and "near x_0 and near y_0" are not same actually. Given extended application, the meanings of "near $(x_1, x_2, ..., x_n)$" and "near x_1 and near x_2 and...and near x_n" are also not the same.

Similarly, the geometric interpretations of the flexible classes in Figs. 3.12a and 3.13a above show that the meanings of "about (x_0, y_0)" and "about x_0 and about y_0" are also not the same. In the same way, the meanings of "about $(x_1, x_2, ..., x_n)$" and "about x_1 and about x_2 and...and about x_n" are also not the same.

Actually, extending point P_0 into a square region R_s, then the analogous square-based flexible clustering can also be doing and obtaining a corresponding flexible class. Further, if the flexible class contains the square region R_s as datum (as shown in Fig. 3.14a), then the geometric interpretation of the corresponding flexible class is "about R_s"; otherwise, the corresponding flexible class is a semi-flexible square frame (as shown in Fig. 3.14b), and its geometric interpretation is then "near R_s."

Similarly, extending point P_0 into a circular region R_c, then the analogous circle-based flexible clustering can also be doing and obtaining a corresponding flexible class. Further, if the flexible class contains the circular region R_c as datum (as shown in Fig. 3.15a), then the geometric interpretation of the corresponding flexible class is "about R_c"; otherwise, the corresponding flexible class is a semi-flexible circular ring (as shown in Fig. 3.15b), and its geometric interpretation is then "close to Rc" or "near R_c."

From Figs. 3.14a and 3.15a, we can see that the core of "about R_s" contains the square region R_s as datum, and the core of "about R_c" contains the circular region R_c

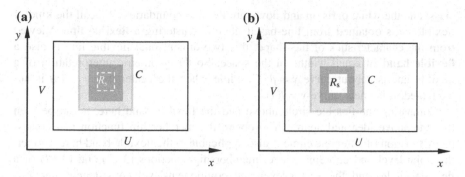

Fig. 3.14 An example of square-based flexible clustering in a two-dimensional measurement space

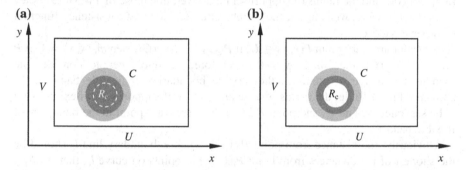

Fig. 3.15 An example of circle-based flexible clustering in a two-dimensional measurement space

as datum. Then, speaking conversely, a flexible square B in measurement space can also be viewed as the flexible class of a certain "about R_s" and $R_s \subset \text{core}(B)$; similarly, a flexible circle C in measurement space can also be viewed as the flexible class of a certain "about R_c" and $R_c \subset \text{core}(C)$. For the one-dimensional measurement space, it is that a flexible interval can be viewed as a flexible class whose core contains a certain interval $[a, b]$, that is, the flexible interval can be called "about $[a, b]$"; conversely, "about $[a, b]$" is also a flexible interval whose core contains $[a, b]$.

3.4.2 Line-Based Flexible Clustering and Flexible Lines

Line-based flexible clustering is a kind of datum-based flexible clustering taking a curve in a space as the center line. For example, doing line-based flexible clustering with curve $y = f(x)$ in a two-dimensional space as the center line, we get a flexible class as shown in Fig. 3.16. The gray part in the figure is the core of the flexible class, and the white parts up and down are flexible boundaries. We call the kind of flexible class obtained from line-based flexible clustering a **flexible line**. Viewed from the characteristics of the shape, this two-dimensional flexible line is also a **flexible band** of equal widths in the space. So the geometric interpretation of a flexible line is "about curve $y = f(x)$"; while when the center line $y = f(x)$ is not contained, it is "near curve $y = f(x)$."

Comparing the flexible circle above and the flexible band here, it can be seen that the above idea and approach to obtain the membership function and consistency function of a flexible circle is also applicable to the flexible band here, further, the point-level and coordinate-level membership functions (3.17) and (3.19) of a flexible circle, and the vector-level and component-level consistency functions (3.18) and (3.20) are also applicable to flexible band C here. However, the problem is for a point $P(x, y) \in U$, and how are the corresponding core–boundary point c_C and critical point s_C to be determined? Or can the functional expressions about the radius of core and the radius of support set be derived like those of a flexible circle?

In the following, we consider the membership function and consistency function of flexible band C.

Since for arbitrary point $P(x, y) \in U$, if $P(x, y) \in \text{core}(C)$, then $m_C(x, y) = 1$; if $P(x, y) \notin \text{supp}(C)$, then $m_C(x, y) = 0$. Therefore, we only need to consider the membership-degree of points in the flexible boundaries of flexible band C. As shown in Fig. 3.16, let the radius of core and radius of support set of flexible band C be separately r_1 and r_2, and point $P_i(x_i, y_i)$ be a arbitrary point in the boundary of flexible band C.

Obviously, the distance from point $P_i(x_i, y_i)$ to core–boundary line l_c should be the shortest of the distances from point $P_i(x_i, y_i)$ to points on curve l_c, that is, $d(P_i, l_c) = \min_{P_j \in l_c} d(P_i, P_j)$. Then, how is the point on curve l_c closest to point $P_i(x_i, y_i)$ to be determined? Suppose there is also a curve $y = f_1(x)$ through point $P_i(x_i, y_i)$ that is parallel to center line $y = f(x)$ (as shown by the broken line in the figure). And the slope of tangent of this curve at point $P_i(x_i, y_i)$ is $f'(x_i)$, and thus, tangent l of curve $y = f_1(x)$ through point $P_i(x_i, y_i)$ is $y - y_i = f'(x_i)(x - x_i)$ (as shown in Fig. 3.16).

From this, the straight line l' through point $P_i(x_i, y_i)$ and perpendicular to tangent line l is $y - y_i = \text{tg}(\text{arctg}(f'(x_i)) + \frac{\pi}{2})(x - x_i)$. With this straight line l', we can then

Fig. 3.16 Illustration of flexible line and its membership function deriving process

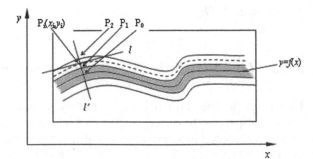

find separately the intersection points of it and the core–boundary line and critical line. From the "parallelism" of the three of the center line, core–boundary line, and critical line, it is known that the intersection points of straight line l' and the core–boundary line and critical line are separately the core–boundary points c_C and critical point s_C of flexible band C that is correlative with point $P_i(x_i, y_i)$. Thus, from the point $P_i(x_i, y_i)$ being arbitrary, we can obtain the point-level and coordinate-level membership functions and vector-level and component-level consistency functions of flexible band C [the expressions are the same as shown in Eqs. (3.17), (3.18), (3.21), and (3.22)].

However, the precondition of this method is that the equations of the core–boundary line and critical line must be known and that cumbersome computation would be met when finding function value. Next, we consider whether the radius of core and radius of support set of flexible band C can be used to derive the simpler expressions of membership function and consistency function.

Solving equations set

$$\begin{cases} y = f(x) \\ y - y_i = \text{tg}(\text{arctg}(f'(x_i)) + \frac{\pi}{2})(x - x_i) \end{cases}$$

the intersection point $P_0(x_0, y_0)$ of straight line l' and center line $y = f(x)$ can be obtained. Then, the distance from $P_i(x_i, y_i)$ to center line $y = f(x)$ is

$$d(P_i, P_0) = \sqrt{(x_i - x_0)^2 + (y_i - y_0)^2}$$

Thus, the distance between $P_i(x_i, y_i)$ and negative core–boundary line l_c is

$$d(P_i, l_c) = \sqrt{(x_i - x_0)^2 + (y_i - y_0)^2} - r_1$$

whereas $r_s - r_c = d(s_C, c_C)$ is just the reference distance, so the closeness of $P_i(x_i, y_i)$ to core–boundary line l_c is

$$\begin{aligned} s(P_i, l_c) &= 1 - \frac{d(P_i, l_c^-)}{r} \\ &= 1 - \frac{\sqrt{(x_i - x_0)^2 + (y_i - y_0)^2} - r_1}{r_2 - r_1} \\ &= \frac{r_2 - \sqrt{(x_i - x_0)^2 + (y_i - y_0)^2}}{r_2 - r_1} \end{aligned}$$

Further, we have

$$m_C(x_i, y_i) = 1 \cdot s(P_i, l_c^-) = \frac{r_s - \sqrt{(x_i - x_0)^2 + (y_i - y_0)^2}}{r_s - r_c}$$

Thus, from the point $P_i(x_i, y_i)$ being arbitrary, we have

$$m_C(x, y) = \begin{cases} 1, & \sqrt{(x - x_0)^2 + (y - y_0)^2} \leq r_c \\ \frac{r_s - \sqrt{(x-x_0)^2 + (y-y_0)^2}}{r_s - r_c}, & r_c < \sqrt{(x - x_0)^2 + (y - y_0)^2} < r_s \\ 0, & r_s \leq \sqrt{(x - x_0)^2 + (y - y_0)^2} \end{cases} \quad (3.27)$$

where r_c and r_s are separately the radius of core and radius of support set of flexible band C, and x_0 and y_0 are the coordinates of a point that correspond to point $P(x, y)$ on center line $y = f(x)$. Actually, point (x_0, y_0) is tantamount to the "circle's center" that point (x, y) corresponds to.

This is the coordinate-level membership function of flexible band C based on the radius of core and radius of support set. Here, r_c and r_s are known, but point (x_0, y_0) needs to be found with point (x, y) for the occasion. It is not hard to imagine that the graph of the membership function of this kind of flexible band is truncated ridge-shaped.

In the same way, we can obtain the component-level consistency function of the flexible linguistic value C that corresponds to flexible band C as follows:

$$c_C(x, y) = \frac{r_s - \sqrt{(x - x_0)^2 + (y - y_0)^2}}{r_s - r_c}, \quad (x, y) \in U \times V \quad (3.28)$$

Its graph is ridge-shaped.

It should be noted that viewed from the form, here expressions in Eqs. (3.27) and (3.28) seem to be no different from the previous expressions in Eqs. (3.20) and (3.24). But there (x_0, y_0) is the center of a flexible circle, so it is unique and changeless, that is, a constant, while here (x_0, y_0) is a point on the center line $y = f(x)$ of a flexible band. From the above equations' set, it can be seen that the solutions x_0 and y_0 are actually also functions of x_i and y_i, respectively, while x_i and y_i are just x and y in Eqs. (3.27) and (3.28), so for the specific center line $y = f(x)$, there would not appear x_0 and y_0 in Eqs. (3.27) and (3.28), but the corresponding functions $g_{x0}(x, y)$ and $g_{y0}(x, y)$.

In the above, we studied line-based flexible clustering and its flexible classes in a two-dimensional space, and the line-based flexible clustering of over three dimensions should be analogous. It can be imaged that a three-dimensional flexible line would be a "flexible rope" with a center line as the axis. The flexible rope in a three-dimensional space can also be viewed as the trace formed by the flexible sphere that moves along the center line of the flexible rope with a point on the center line as the center. On the basis of that, the membership function and consistency function of a flexible rope can be constructed.

Let the equation of curve l in a three-dimensional space be

$$\begin{cases} x = x(t) \\ y = y(t) \\ z = z(t) \end{cases}$$

where t is the parameter. Thus, the coordinate-level membership function and component-level consistency function of flexible line C with curve l as the center line are

$$m_C(x, y, z) = \begin{cases} 1, & d \leq r_1 \\ \frac{r_2 - \sqrt{(x-x(t_0))^2 + (y-y(t_0))^2 + (z-z(t_0))^2}}{r_2 - r_1}, & r_1 < d < r_2 \\ 0, & r_2 \leq d \end{cases} \quad (3.29)$$

$$c_C(x, y, z) = \frac{r_2 - \sqrt{(x - x(t_0))^2 + (y - y(t_0))^2 + (z - z(t_0))^2}}{r_2 - r_1}, \quad (x, y, z) \in U \tag{3.30}$$

where $x(t_0)$, $y(t_0)$, and $z(t_0)$ are coordinates of "sphere's center" point that corresponds to point $P(x, y, z)$ on center line l, and r_1 and r_2 are separately the radius of core and radius of support set of flexible line C, and $d = \sqrt{(x - x(t_0))^2 + (y - y(t_0))^2 + (z - z(t_0))^2}$.

Generalizing the three-dimensional flexible line, we have the following coordinate-level membership function and component-level consistency function of flexible line C in an n-dimensional space:

$$m_C(x_1, x_2, \ldots, x_n) = \begin{cases} 1, & \sqrt{\sum_{i=1}^{n} (x_i - x_l(t_0))^2} \leq r_1 \\ \frac{r_2 - \sqrt{\sum_{i=1}^{n} (x_i - x_l(t_0))^2}}{r_2 - r_1}, & r_1 < \sqrt{\sum_{i=1}^{n} (x_i - x_l(t_0))^2} < r_2 \quad (3.31) \\ 0, & r_2 \leq \sqrt{\sum_{i=1}^{n} (x_i - x_l(t_0))^2} \end{cases}$$

$$c_C(x_1, x_2, \ldots, x_n) = \frac{r_s - \sqrt{\sum_{i=1}^{n} (x_i - x_l(t_0))^2}}{r_s - r_c}, \quad (x_1, x_2, \ldots, x_n) \in U \tag{3.32}$$

Here, $x_i(t_0)$ ($i = 1, 2, \ldots, n$) is the coordinate of "sphere's center" point that corresponds to point $P(x_1, x_2, \ldots, x_n)$ on center line l, and r_1 and r_2 are separately the radius of core and radius of support set of flexible line C.

3.4.3 Plane-Based Flexible Clustering and Flexible Planes

Analogous to line-based flexible clustering, we can also take a curved surface as the center plane to do datum-based flexible clustering for points in a space, and this is what we call plane-based flexible clustering. For example, taking curved surface $z = f(x, y)$ in a three-dimensional space as the center plane to do plane-based flexible clustering, we obtain a flexible class as shown in Fig. 3.17. The gray part in the figure is the core of the flexible class, and the white parts up and down are the flexible boundaries. We call this kind of flexible class obtained from plane-based clustering a **flexible plane**. Viewed from the characteristics of the shape, a three-dimensional flexible plane is also a **flexible plate** in the space. Therefore, the geometric interpretation of a flexible plane is "about curved surface P"; while when the center curved surface $z = f(x, y)$ is not contained, it is "near curved surface P."

For the kind of flexible class of flexible plate, we can use the above idea and methods obtaining the membership function and consistency function of a flexible circle and a flexible band to obtain its membership function and consistency function. And it is not hard to see that the expressions of the point-level and coordinate-level membership functions and vector-level and component-level consistency functions of a flexible plate are still the previous expressions in Eqs. (3.17) and (3.18) and Eqs. (3.21) and (3.22). In the following, we consider whether a flexible plate has a membership function and consistency function based on the radius of core and radius of support set.

It can be seen that the projections of a flexible plate in a three-dimensional space on plane x–z or plane y–z are also flexible bands. Therefore, from the above membership function and consistency function of a flexible band, we can immediately get the membership function and consistency function of flexible plate C as follows:

$$m_C(x, y, z) = \begin{cases} 1, & d \leq r_1 \\ \frac{r_2 - \sqrt{(x-x_0)^2 + (y-y_0)^2 + (z-z_0)^2}}{r_2 - r_1}, & r_1 < d < r_2 \\ 0, & r_2 \leq d \end{cases} \qquad (3.33)$$

Fig. 3.17 An example of plane-based flexible clustering and flexible planes

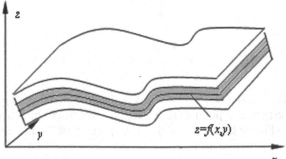

$z = f(x, y)$

$$c_C(x,y,z) = \frac{r_2 - \sqrt{(x-x_0)^2 + (y-y_0)^2 + (z-z_0)^2}}{r_2 - r_1}, \quad (x,y,z) \in U \qquad (3.34)$$

Here, r_1 and r_2 are separately the radius of core and radius of support set of a flexible plate, x_0, y_0, and z_0 are the coordinates of the "sphere's center" point that corresponds to point $P(x, \ y, \ z)$ in center plane $z = f(x, \ y)$, and $d = \sqrt{(x-x(t_0))^2 + (y-y(t_0))^2 + (z-z(t_0))^2}$.

More generally, the membership function and consistency function of flexible hyperplane C on an $n(n > 3)$-dimensional space are

$$m_C(x_1, x_2, \ldots, x_n) = \begin{cases} 1, & \sqrt{\sum_{i=1}^{n}(x_i - x_{i_0}))^2} \leq r_c \\ \frac{r_s - \sqrt{\sum_{i=1}^{n}(x_i - x_{i_0})^2}}{r_s - r_c}, & r_c < \sqrt{\sum_{i=1}^{n}(x_i - x_{i_0}))^2} < r_s \\ 0, & r_s \leq \sqrt{\sum_{i=1}^{n}(x_i - x_{i_0}))^2} \end{cases} \qquad (3.35)$$

$$c_C(x_1, x_2, \ldots, x_n) = \frac{r_s - \sqrt{\sum_{i=1}^{n}(x_i - x_{i_0})^2}}{r_s - r_c}, \quad (x_1, x_2, \ldots, x_n) \in U \qquad (3.36)$$

Here, r_c and r_s are separately the radius of core and radius of support set of a flexible hyperplane, and $x_{1_0}, x_{2_0}, \ldots, x_{n_0}$ are the coordinates of the "center of sphere" point that corresponds to point $P(x_1, x_2, \ldots, x_n)$ in center plane $F(x_1, x_2, \ldots, x_n) = 0$.

3.5   Universal Mathematical Models of Flexible Properties (Concepts)

A flexible class in a multidimensional measurement space which is formed by flexible clustering with respect to points (or their partial coordinate components) stands for a corresponding attributive concept. In consideration of its denotation being flexible classes in a multidimensional space, so we call this kind of flexible concepts to be the multidimensional flexible attributive concept. But, up to now, the flexible attributive concepts we talk refer actually to "flexible properties" things have.

Examining the various flexible classes and their membership functions and consistency functions in the above sections, it is not hard to see that the mathematical models of multidimensional flexible properties (concepts) can be unified.

In fact, previous point-level membership function (3.17), coordinate-level membership function (3.18), vector-level consistency function (3.21), and component-level consistency function (3.22) are also applicable to any flexible property (concept) on multidimensional spaces.

Further, it can be seen that the mathematical models of multidimensional flexible properties (concepts) and those of one-dimensional flexible properties (concepts) in the last chapter can also be unified. In fact, the idea and method obtaining the membership function and consistency function of a multidimensional flexible property (concept) are the generalization of those of one-dimensional flexible properties (concepts); conversely, applying the point-level membership function and vector-level consistency function of a multidimensional flexible property (concept) here to a one-dimensional flexible property (concept), the membership function and consistency function obtained are just the general expressions (2.5) and (2.14) given in Chap. 2. Only, since a one-dimensional space only has two directions, for any point $P \in [a, b]$ in the same direction, its corresponding core–boundary point c_C and critical point s_C are both changeless. Therefore, there c_C and s_C are constants. And since a point in a one-dimensional space and its coordinates are the same number, expression (2.5) is both a point-level membership function and a coordinate-level membership function, and expression (2.14) is both a vector-level consistency function and a component-level consistency function.

Thus, the following functions

$$m_C(P) = \begin{cases} 1, & P \in \text{core}(C) \\ \frac{d(P,s_C)}{d(s_C,c_C)}, & P \in \text{boun}(C) \\ 0, & P \notin \text{supp}(C) \end{cases} \tag{3.37}$$

$$c_C(P) = \begin{cases} \frac{d(P,s_C)}{d(s_C,c_C)}, & P \in \text{supp}(C) \\ -\frac{d(P,s_C)}{d(s_C,c_C)}, & P \notin \text{supp}(C) \end{cases} \tag{3.38}$$

are separately the general expressions of point-level membership functions and vector-level consistency functions of flexible properties (concepts) in an $n(n \geq 1)$-dimensional measurement space, where P is a point variable, and c_C and s_C are separately the core–boundary point and critical point for point P in the corresponding flexible class C. Functions

$$m_C(x_1,x_2,\ldots,x_n) = \begin{cases} 1, & (x_1,x_2,\ldots,x_n) \in \text{core}(C) \\ \frac{\sqrt{\sum_{i=1}^n (x_i-x_{is})^2}}{\sqrt{\sum_{i=1}^n (x_{ic}-x_{is})^2}}, & (x_1,x_2,\ldots,x_n) \in \text{boun}(C) \\ 0, & (x_1,x_2,\ldots,x_n) \notin \text{supp}(C) \end{cases} \tag{3.39}$$

$$c_C(x_1, x_2, \ldots, x_n) = \begin{cases} \dfrac{\sqrt{\sum_{i=1}^{n} (x_i - x_{i_s})^2}}{\sqrt{\sum_{i=1}^{n} (x_{i_c} - x_{i_s})^2}}, & (x_1, x_2, \ldots, x_n) \in \mathrm{supp}(C) \\[4ex] -\dfrac{\sqrt{\sum_{i=1}^{n} (x_i - x_{i_s})^2}}{\sqrt{\sum_{i=1}^{n} (x_{i_c} - x_{i_s})^2}}, & (x_1, x_2, \ldots, x_n) \notin \mathrm{supp}(C) \end{cases} \tag{3.40}$$

are separately the general expressions of coordinate-level membership functions and component-level consistency functions of flexible properties (concepts) in an n ($n \geq 1$)-dimensional measurement space, where x_i is a point coordinate variable, and x_{i_c} and x_{i_s} ($i = 1, 2, \ldots, n$) are separately the coordinates of c_C and s_C.

For a flexible class with center point, the general expressions of the point-level membership function and vector-level consistency function can also be

$$m_C(P) = \begin{cases} 1, & P \in \mathrm{core}(C) \\ \dfrac{r_s - d(P, P_0)}{r_s - r_c}, & P \in \mathrm{boun}(C) \\ 0, & P \notin \mathrm{supp}(C) \end{cases} \tag{3.41}$$

$$c_C(P) = \frac{r_s - d(P, P_0)}{r_s - r_c}, \quad P \in U \tag{3.42}$$

The general expressions of the coordinate-level membership function and component-level consistency function can also be

$$m_C(x_1, x_2, \ldots, x_n) = \begin{cases} 1, & \sqrt{\sum_{i=1}^{n} (x_i - x_{i_0})^2} \leq r_c \\[3ex] \dfrac{r_s - \sqrt{\sum_{i=1}^{n} (x_i - x_{i_0})^2}}{r_s - r_c}, & r_c < \sqrt{\sum_{i=1}^{n} (x_i - x_{i_0})^2} < r_s \\[3ex] 0, & r_s \leq \sqrt{\sum_{i=1}^{n} (x_i - x_{i_0})^2} \end{cases} \tag{3.43}$$

$$c_C(x_1, x_2, \ldots, x_n) = \frac{r_s - \sqrt{\sum_{i=1}^{n} (x_i - x_{i_0})^2}}{r_s - r_c}, \quad (x_1, x_2, \ldots, x_n) \in U \tag{3.44}$$

Here, P_0 is the center point of flexible class C that corresponds to point variable P, x_{i_0} ($i = 1, 2, \ldots, n$) is its coordinate, and r_c and r_s are separately the radius of core and radius of support set of flexible class C.

3.6 Flexible Relations (Concepts) and Their Mathematical Models

Besides flexible properties, there are also "flexible relations" in flexible attributive concepts (or, in human brain), that is, the flexible concepts on relations between things, such as "similar," "analogous," "approximate," "approximately equal to," "far greater than," and "good friend" that are all flexible relations (concepts). In this section, we will discuss the formation principle and mathematical models of flexible relations (concepts).

3.6.1 Flexible Clustering in a Product Space

A flexible relation (concept) can be formed by using the method of flexible clustering in measurement space U of objects, but in the space U, it is difficult or even impossible to directly realize flexible clustering based on the relation of objects. However, we know that a subset of Cartesian product $S_1 \times S_2 \times \cdots \times S_n$ stands for an n-ary relation. Then, analogously, a flexible subset of product space U^n can stand for an n-ary flexible relation. That is to say, doing flexible clustering in product space U^n, we can just obtain the flexible classes standing for the flexible relations between points in U. Then, generally, to do flexible clustering in product space $U_1 \times U_2 \times \cdots \times U_n$ (U_i is a k_i ($k_i \geq 1$)-dimensional measurement space, $i = 1, 2, \ldots, n$), we can obtain more general flexible classes representing flexible relations.

Next, we consider the problem of flexible clustering in a product space.

It can be seen that when U is a one-dimensional space, it is very easy to do flexible clustering in product space U^n. As a matter of fact, here n-dimensional point $(x_1, x_2, \ldots, x_n) \in U^n$ is also a group of n-ary one-dimensional points, and $x_1, x_2, \ldots, x_n$ are not coordinates but points in one-dimensional space U. Thus, any flexible class in product space U^n can be seen as a flexible class representing a certain flexible relation. As, thus, any multidimensional flexible class representing a flexible property can also be viewed as a flexible class representing a certain flexible relation (of course, for one and the same multidimensional flexible class, the flexible property and the flexible relation represented by which are two different flexible concepts, and the names of the two are not the same either). That means to do flexible clustering arbitrarily in an $n(n > 1)$-dimensional product space U^n, and the flexible classes obtained are all seen as a flexible class representing a certain flexible relation (of course, which can also be seen as a flexible class representing a certain flexible property).

However, when U is a multidimensional space, it is difficult to do flexible clustering in product space U^n. For example, when U is a two-dimensional space, how is flexible clustering to be conducted in $U \times U$? Even the shapes of the flexible classes therein are all hard to image. As to further doing flexible clustering in more general product space $U_1 \times U_2 \times \cdots \times U_n$, it is even more difficult.

Since the flexible clustering based on relation focuses on a certain relation between points but not the property of points, it is not as direct and visual as flexible clustering based on property. Therefore, though theoretically speaking, to do flexible clustering in a product space, we can obtain corresponding flexible relations (concepts), and how it is to be done is hard to formulate.

There is another problem, that is, although in an $n(n > 1)$-dimensional product space U^n, flexible clustering can be done arbitrarily; what flexible relations (concepts) do flexible classes thus obtained stand for? Obviously, most are hard to formulate. And for those flexible classes with practical meanings, how the membership functions are to be obtained is also a problem.

Actually, just as a flexible property (concept) is the result of human brain clustering and summarizing related properties that are continuously distributed, a flexible relation (concept) is the result of human brain clustering and summarizing related relations that are continuously distributed. For this reason, we need to look for a kind of space that can directly characterize the continuous distribution of the relation between objects.

3.6.2 Space Transformation and the Formation of a Flexible Relation (Concept)

In the following, we take "similar" relation as an example to further analyze and discuss the formation principle and mathematical models of a flexible relation (concept).

Analyzing carefully the semantics of "similar," it can be seen that "similar" is actually reaching or exceeding "same" to a certain degree. Therefore, we can take sameness-degree as a measure of the strengths of sameness between objects. Thus, "similar" is a flexible concept on the range of sameness-degrees. From Sect. 2.1.2, it is known that the range of sameness-degrees is [0, 1], so on which flexible clustering can be done, and further, a flexible class standing for "similar" can be obtained and so can its membership function and consistency function.

Let $s \in [0, 1]$ be a sameness-degree and $w_0, w_1 \in [0, 1]$ be separately the critical point and core–boundary point of "similar," and then,

$$m_{\text{similar}}(s) = \begin{cases} 1, & w_1 \leq s \leq 1 \\ \frac{s - w_0}{w_1 - w_0}, & w_0 < s < w_1 \\ 0, & 0 \leq s \leq w_0 \end{cases} \tag{3.45}$$

$$c_{\text{similar}}(s) = \frac{s - w_0}{w_1 - w_0}, \quad s \in [0, 1] \tag{3.46}$$

are separately the membership function and consistency function of "similar" on the range [0, 1] of sameness-degrees (whose graphs are shown in Fig. 3.18).

Fig. 3.18 An illustration of the membership function and consistency function of flexible relation (concept) "similar"

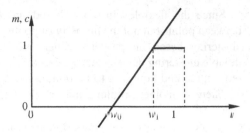

Thus, we obtain the flexible relation (concept) of "similar" by flexible clustering in one-dimensional measurement space [0, 1] and establish its mathematical models. However, that is only a conceptual mathematical model, and we still need to further derive the mathematical models of "similar" in the practical measurement space.

Let U be an n-dimensional measurement space, $x = (x_1, x_2, \ldots, x_n)$, and $y = (y_1, y_2, \ldots, y_n) \in U$, and by the definition and computation formula of sameness-degree given in Sect. 2.1.2, we have

$$s(x, y) = 1 - \frac{d(x, y)}{r}$$

where r is a reference distance, generally taking the maximum value of the distances between points within a certain scope, here taking $r = \max\limits_{x, y \in U} d(x, y)$, whereas distance $d(x, y) = \sqrt{\sum_{i=1}^{n} (x_i - y_i)^2}$ is the Euclidean distance. Thus, substituting $s = s(x, y)$ into expression $\frac{s - w_0}{w_1 - w_0}$, we have

$$\frac{s(x, y) - w_0}{w_1 - w_0} = \frac{\left(1 - \frac{d(x, y)}{r}\right) - w_0}{w_1 - w_0}$$

Thus,

$$m_{\text{similar}}(x, y) = \begin{cases} 1, & w_1 \leq 1 - \frac{d(x, y)}{r} \leq 1 \\ \frac{\left(1 - \frac{d(x, y)}{r}\right) - w_0}{w_1 - w_0}, & w_0 < 1 - \frac{d(x, y)}{r} < w_1 \\ 0, & 0 \leq 1 - \frac{d(x, y)}{r} \leq w_0 \end{cases} \qquad (3.47)$$

$$c_{\text{similar}}(x, y) = \frac{\left(1 - \frac{d(x, y)}{r}\right) - w_0}{w_1 - w_0}, \quad 1 - \frac{d(x, y)}{r} \in [0, 1] \qquad (3.48)$$

They are the point-level membership function and vector-level consistency function of "similar" on product space $U \times U$.

Further, substituting $x = (x_1, x_2, \ldots, x_n)$ and $y = (y_1, y_2, \ldots, y_n)$ into the above expressions, we have

$$m_{\text{similar}}(x_1, x_2, \ldots, x_n, y_1, y_2, \ldots, y_n)$$

$$= \begin{cases} 1, & w_1 \leq 1 - \dfrac{\sqrt{\sum_{i=1}^{n}(x_i - y_i)^2}}{r} \leq 1 \\[3mm] \dfrac{\left(1 - \dfrac{\sqrt{\sum_{i=1}^{n}(x_i - y_i)^2}}{r}\right) - w_0}{w_1 - w_0}, & w_0 < 1 - \dfrac{\sqrt{\sum_{i=1}^{n}(x_i - y_i)^2}}{r} < w_1 \\[3mm] 0, & 0 \leq 1 - \dfrac{\sqrt{\sum_{i=1}^{n}(x_i - y_i)^2}}{r} \leq w_0 \end{cases} \tag{3.49}$$

$$c_{\text{similar}}(x_1, x_2, \ldots, x_n, y_1, y_2, \ldots, y_n)$$

$$= \dfrac{\left(1 - \dfrac{\sqrt{\sum_{i=1}^{n}(x_i - y_i)^2}}{r}\right) - w_0}{w_1 - w_0}, \quad 1 - \dfrac{\sqrt{\sum_{i=1}^{n}(x_i - y_i)^2}}{r} \in [0, 1] \tag{3.50}$$

They are the coordinate-level membership function and component-level consistency function of "similar" on product space $U \times U$.

In particular, when $U = [a, b]$, the $r = b - a$ and the $d(x, y) = |x - y|$, and thus,

$$m_{\text{similar}}(x, y) = \begin{cases} 1, & w_1 \leq 1 - \frac{|x-y|}{b-a} \leq 1 \\[2mm] \dfrac{\left(1 - \frac{|x-y|}{b-a}\right) - w_0}{w_1 - w_0}, & w_0 < 1 - \frac{|x-y|}{b-a} < w_1 \\[2mm] 0, & 0 \leq 1 - \frac{|x-y|}{b-a} \leq w_0 \end{cases} \tag{3.51}$$

$$c_{\text{similar}}(x, y) = \dfrac{\left(1 - \frac{|x-y|}{b-a}\right) - w_0}{w_1 - w_0}, \quad 1 - \dfrac{|x - y|}{b - a} \in [0, 1] \tag{3.52}$$

These two functions are separately a point-level membership function and a vector-level consistency function and also a coordinate-level membership function and component-level consistency function separately.

We notice that $m_{\text{similar}}(x_1, x_2, \ldots, x_n, y_1, y_2, \ldots, y_n)$ and $c_{\text{similar}}(x_1, x_2, \ldots, x_n, y_1, y_2, \ldots, y_n)$ are functions on a multidimensional space. So the variable substitutions above makes the above flexible linguistic value "similar" on one-dimensional space $[0, 1]$ extended to multidimensional space $U \times U$, thus, the flexible relation (concept) of "similar" also becomes a flexible concept on multidimensional space; moreover, the flexible class that "similar" denotes is a flexible subset R of product space $U \times U$.

Now, we see that relation "similar" is originally between two objects, but which is hard to be obtained directly by flexible clustering in measurement space U of objects or product space $U \times U$. However, we transform every ordered pair $(x, y) \in U \times U$ to a real number $s = s(x, y)$ through the measure of sameness-degree, thus obtaining one-dimensional space $[0, 1]$ from product space $U \times U$; then, we obtain flexible concept "similar" on [0, 1] by flexible clustering and establish its mathematical models; next, we extend the mathematical models of "similar" on one-dimensional space [0, 1] to product space $U \times U$ through sameness-degree (function) $s = s(x, y)$, which is tantamount to transforming back "similar" from one-dimensional space [0, 1] to product space $U \times U$, making it into a flexible concept on the latter.

It can be seen that in the formation process of the flexible concept "similar," sameness-degree $s = s(x, y)$ plays a key role of a "bridge." Actually, sameness-degree $s = s(x, y)$ is also a mapping or transformation. It is just this mapping that transforms product space $U \times U$ into one-dimensional space [0, 1] that makes this flexible concept of "similar" to be formed and then also makes it returned to product space $U \times U$ and original space U.

Actually, to put it another way, "similar" is also "slightly different." Thus, "similar" can also be defined on difference-degree range [0, 1], or even defined directly on distance range [0, b]. That is to say, we can use multiple kinds of measures and transformations to establish the mathematical models of one and the same flexible relation (concept). Of course, these expression forms of mathematical models in different measures and transformations are not the same.

From the formation process of the flexible concept of "similar," we see that the formation of a flexible relation (concept) is tantamount to a process of going to a one-dimensional measurement space from a product space and then returning back to the product space from the one-dimensional measurement space. Therefore, the formation principles of a flexible relation (concept) can be shown in Fig. 3.19. In the figure, U_i is a k_i ($k_i \geq 1$)-dimensional measurement space, $P_i \in U_i$ ($i = 1, 2, \ldots, n$), and $v = \varphi (P_1, P_2, \ldots, P_n)$ are a certain measure about points $P_1, P_2, \ldots, P_n$, that is, a certain kind of transformation from product space $U_1 \times U_2 \times \cdots \times U_n$ to one-dimensional space [a, b]; R is the corresponding flexible relation (concept).

3.6.3   Universal Mathematical Models of Flexible Relations (Concepts)

From the formation principle of flexible relations (concepts), we obtain the universal mathematical models of them.

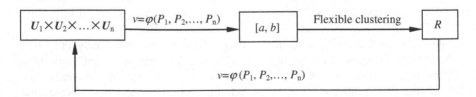

Fig. 3.19 Diagram of the formation principle of a flexible relation (concept)

1. **General expressions of the membership function and consistency function of binary flexible relations**

$$m_R(P_1, P_2) = \begin{cases} 0, & a \le \varphi(P_1, P_2) \le s_R^- \\ \frac{\varphi(P_1,P_2)-s_R^-}{c_R^- - s_R^-}, & s_R^- < \varphi(P_1,P_2) < c_R^- \\ 1, & c_R^- \le \varphi(P_1, P_2) \le c_R^+ \\ \frac{s_R^+ - \varphi(P_1,P_2)}{s_R^+ - c_R^+}, & c_R^+ < \varphi(P_1, P_2) < s_R^+ \\ 0, & s_R^+ \le \psi(P_1, P_2) \le b \end{cases} \qquad (3.53)$$

$$m_R(x,y) = \begin{cases} 0, & a \le \varphi(x,y) \le s_R^- \\ \frac{\varphi(x,y)-s_R^-}{c_R^- - s_R^-}, & s_R^- < \varphi(x,y) \le c_R^- \\ 1, & c_R^- \le \varphi(x,y) \le c_R^+ \\ \frac{s_R^+ - \varphi(x,y)}{s_R^+ - c_R^+}, & c_R^+ < \varphi(x,y) \le s_R^+ \\ 0, & s_R^+ \le \varphi(x,y) \le b \end{cases} \qquad (3.54)$$

$$c_R(P_1, P_2) = \min\left\{ \frac{\varphi(P_1, P_2) - s_R^-}{c_R^- - s_R^-}, \frac{s_R^+ - \varphi(P_1,P_2)}{s_R^+ - c_R^+} \right\}, \quad \varphi(P_1, P_2) \in [a, b] \qquad (3.55)$$

$$c_R(x,y) = \min\left\{ \frac{\varphi(x, y) - s_R^-}{c_R^- - s_R^-}, \frac{s_R^+ - \varphi(x,y)}{s_R^+ - c_R^+} \right\}, \quad \varphi(x,y) \in [a, b] \qquad (3.56)$$

where $P_1 \in U_1, P_2 \in U_2, (P_1, P_2) \in U_1 \times U_2$, and $\varphi(P_1, P_2)$ is a measure about point variables P_1 and P_2; c_R^-, c_R^+ and s_R^-, s_R^+ are separately the core–boundary points and critical points of flexible relation R on range $[a, b]$ of measures, and x and y are separately the coordinate variables of P_1 and P_2.

2. **General expressions of the membership function and consistency function of n-ary flexible relations**

$$m_R(P_1, \ldots, P_n) = \begin{cases} 0, & a \le \varphi(P_1, \ldots, P_n) \le s_R^- \\ \frac{\varphi(P_1,\ldots,P_n)-s_R^-}{c_R^- - s_R^-}, & s_R^- < \varphi(P_1, \ldots, P_n) < c_R^- \\ 1, & c_R^- \le \varphi(P_1, \ldots, P_n) \le c_R^+ \\ \frac{s_R^+ - \varphi(P_1,\ldots,P_n)}{s_R^+ - c_R^+}, & c_R^+ < \varphi(P_1, \ldots, P_n) < s_R^+ \\ 0, & s_R^+ \le \varphi(P_1, \ldots, P_n) \le b \end{cases} \qquad (3.57)$$

$$
m_R(x_1,\ldots,x_n) = \begin{cases} 0, & a \le \varphi(x_1,\ldots,x_n) \le s_R^- \\ \dfrac{\varphi(x_1,\ldots,x_n)-s_R^-}{c_R^- - s_R^-}, & s_R^- < \varphi(x_1,\ldots,x_n) \le c_R^- \\ 1, & c_R^- \le \varphi(x_1,\ldots,x_n) \le c_R^+ \\ \dfrac{s_R^+ - \varphi(x_1,\ldots,x_n)}{s_R^+ - c_R^+} & c_R^+ < \varphi(x_1,\ldots,x_n) \le s_R^+ \\ 0, & s_R^+ \le \varphi(x_1,\ldots,x_n) \le b \end{cases} \tag{3.58}
$$

$$
c_R(P_1,\ldots,P_n) = \min\left\{ \frac{\varphi(P_1,\ldots,P_n)-s_R^-}{c_R^- - s_R^-}, \frac{s_R^+ - \varphi(P_1,\ldots,P_n)}{s_R^+ - c_R^+} \right\}, \\ \varphi(P_1,\ldots,P_n) \in [a,b] \tag{3.59}
$$

$$
c_R(x_1,\ldots,x_n) = \min\left\{ \frac{\varphi(x_1,\ldots,x_n)-s_R^-}{c_R^- - s_R^-}, \frac{s_R^+ - \varphi(x_1,\ldots,x_n)}{s_R^+ - c_R^+} \right\}, \\ \varphi(x_1,\ldots,x_n) \in [a,b] \tag{3.60}
$$

where $P_i \in U_i$ ($i = 1, 2, \ldots, n$), $(P_1, \ldots, P_n) \in U_1 \times U_2 \times \cdots \times U_n$, and $\varphi(P_1, \ldots, P_n)$ is a measure about point variables $P_1, \ldots, P_n$; c_R^-, c_R^+ and s_R^-, s_R^+ are separately the core–boundary points and critical points of flexible relation R on range $[a, b]$ of measures, and $x_1, \ldots, x_n$ are separately the coordinate variables of $P_1, \ldots, P_n$.

3.7 Summary

In this chapter, we further considered the flexible clustering and flexible classes in measurement spaces and revealed the formation principles of multidimensional flexible concepts and flexible linguistic values and established their mathematical models. In particular, we presented the universal mathematical models of flexible properties (concepts) and flexible relations (concepts).

The main points and results of the chapter are as follows:

- There are various ways of flexible clustering in a measurement space, such as flexible clustering with respect to the partial coordinate components of a point and flexible clustering with respect to whole point, and the latter can be further classified as square flexible clustering, circular flexible clustering, point-based flexible clustering, line-based flexible clustering, plane-based flexible clustering, etc. The flexible classes obtained are flexible squares, flexible circles, flexible points, flexible lines, flexible planes, etc., whose geometrical shapes are separately flexible squares, flexible circles, flexible spheres, flexible bands, flexible ropes, flexible plates, etc. These flexible classes stand for various kinds of multidimensional flexible concepts. Since these flexible geometric bodies as flexible classes are all formed by the overlapping of a pair of corresponding core and support set, the core and support set together can be viewed as the geometric model of a flexible concept (correspondingly, the membership function and consistency function can be said to be the algebraic model of a flexible concept).

- Flexible attributive concepts obtained by flexible clustering with respect to points are generally flexible properties (concepts). The flexible properties (concepts) on an $n(n \geq 1)$-dimensional space have universal mathematical models.
- Except for special cases, a flexible relation (concept) is hard to be directly obtained from flexible clustering in the corresponding product measurement space, so we need to employ the method of space transformation to indirectly obtain its mathematical model. The flexible relations (concepts) have also universal mathematical models.

Reference

1. Lian S (2009) Principles of imprecise-information processing. Science Press, Beijing

Chapter 4
Modeling of Flexible Concepts

Abstract This chapter discusses the methods of the determination and acquisition of the membership functions and consistency functions of known flexible concepts and discusses the dynamics and polymorphism of mathematical models of a flexible concept.

Keywords Flexible concepts · Mathematical models · Membership function · Consistency function

In Chaps. 2 and 3 we revealed the formation principle of flexible concepts and presented their general mathematical models. In this chapter, we further discuss the modeling methods of flexible concepts, that is, how the core and support set as well as the membership function and consistency function of a given or known flexible concept can be determined and acquired. For convenience of stating, we shorten the membership function and consistency function as membership-consistency functions in what follows.

4.1 Determination of Measurement Space and Directly Modeling

We know that the determination of the core and support set as well as the membership-consistency functions of a flexible concept actually also is reduced to the determination of its core–boundary points (lines and planes) and critical points (lines and planes). While to determine these parameters, we need to determine the measurement space of corresponding feature firstly.

Any feature of objects has either numerical values or linguistic values, or both. The numerical values are a certain kind of measurement, and non-symbolic linguistic values are then the summarization of a batch of numerical values. If a feature of objects has already a measure, then the range of values of the measure is the range of numerical values of this feature. If a feature has not a measure (for

© Springer Science+Business Media Singapore 2016
S. Lian, *Principles of Imprecise-Information Processing*,
DOI 10.1007/978-981-10-1549-6_4

instance, the feature of "attitude toward learning" can have linguistic values: "conscientious," "not conscientious," etc., but generally it has no a measure), then we can define its measure by the characteristics of the feature or the semantics of corresponding linguistic value. For instance, those of the previous sameness-degree and difference-degree are just done in this way. Of course, there exist some features whose objective measure may be difficult to find. For these features, we can use the method of marking subjectively to acquire their numerical values. Given a measure, the measurement space of corresponding feature also be got, and then, we can define corresponding flexible linguistic values on it.

As to how to determine the core–boundary points (lines and planes) and critical points (lines and planes) of a flexible concept, we give some methods in the following for Ref. [1].

1. **"Personal preference" method**
   The so-called personal preference is to give the core–boundary points (lines and planes) and critical points (lines and planes) of a corresponding flexible concept according to one's personal subjective understanding. Of course, for the flexible concept with a center point (lines and planes), if the center of core, the radius of core and the radius of support set can be given, then the corresponding core–boundary points (lines and planes) and critical points (lines and planes) can also be derived. Personal preference method is suitable for the modeling of the related flexible concepts in the specified fields. The parameters of the core–boundary points (lines and planes) and critical points (lines and planes) can be directly given by domain experts from their knowledge and experience.

2. **"Statistics from a group" method**
   The so-called statistics from a group is to collect a certain amount of "public opinion" data by consulting in a certain number and part of the population, then using mathematical statistical method to determine the parameters of core–boundary points (lines and planes) and critical points (lines and planes). For example, the values that mostly frequently occur or the mathematical expectation of the parameter variable, that is, the mean value, can be used as the value of the corresponding parameter. The method of statistics from a group is suitable for the modeling of the ordinary flexible concepts in daily language. Therefore, this method can be used in natural language processing to determine those cores and support sets as well as membership-consistency functions of the related flexible concepts.

3. **"Derivation with instances" method**
   This method is to derive the membership-consistency functions of a flexible concept from several instances of this flexible concept, for example, according to the heights of a class of students and the corresponding membership-degrees to "tall" to derive the membership function of "tall." In this method, we can use function fitting or piecewise function fitting, and solving the corresponding simultaneous equation to obtain core–boundary points and critical points.

4. **"Generation by translating" method**

The so-called generation by translating is using the related parameters of the known linguistic values and the relation between the known linguistic values and the targeted linguistic values to derive the related parameters of the targeted linguistic values through translation transformation and then obtain the corresponding membership-consistency functions, or directly doing translation transformation of the original linguistic values to derive the membership-consistency functions of the targeted linguistic values. This method is generally fit for the modeling of superposed linguistic values (which is to do so in Sect. 7.1), but it can also be conversely used, that is, from the membership-consistency functions of a superposed linguistic value to derive the membership-consistency functions of the original linguistic value through translation transformation.

Finally, it should be noted that although speaking for single flexible concept, its core–boundary points (lines and planes) and critical points (lines and planes) need to determined, for a group of flexible concepts $A_1, A_2, \ldots, A_n$ of the corresponding partition of a space, only the core–boundary points (lines and planes) of all flexible concepts need to be determined. Because in this case, the negative core–boundary point $c_{A_i}^-$ of flexible concept A_i is just the positive critical point $s_{A_{i-1}}^+$ of flexible concept A_{i-1}, and the positive core–boundary point $c_{A_i}^+$ of flexible concept A_i is just the negative critical point $s_{A_{i+1}}^-$ of flexible concept A_{i+1}.

4.2 Space-Transforming Method

4.2.1 Space-Transforming Method for the Modeling of a Flexible Relation (Concept)

Actually, the formation principle of flexible relations (concepts) in Sect. 3.6 also gave a general method for acquiring mathematical models of flexible relations (concepts), that is, first utilize the expression of definition of corresponding measure to transform a flexible relation on a multidimensional product space into a flexible concept on a one-dimensional measurement space (i.e., a range of numerical feature values) and modeling for it, then transform conversely the mathematical models obtained back to the original multidimensional space. In consideration of the characteristic, we call this method of modeling of flexible relations (concepts) to be the space-transforming method, the concrete steps of which are as follows:

① Select or define an appropriate measure $v = \varphi(P_1, P_2, \ldots, P_n)$ according to the semantics of relation R $(P_i \in U_i$ is a point variable, U_i is a k_i $(k_i \geq 1)$-dimensional space, $i = 1, 2, \ldots, n)$, take $v = \varphi(P_1, P_2, \ldots, P_n)$ as a function on product space $U_1 \times U_2 \times \cdots \times U_n$, determine its range U and treat U as the measurement space that R belongs to.

② Define flexible concept R on U by flexible clustering and obtain the corresponding membership-consistency function $m_R(v)$ and $c_R(v)$.

③ Substitute $v = \varphi(x_1, x_2, \ldots, x_n)$ (x_i is the coordinate variable of P_i, $i = 1, 2, \ldots, n$) into $m_R(v)$ and $c_R(v)$ as well as the corresponding core–boundary points and critical points, then obtaining immediately the membership-consistency function $m_R(x_1, x_2, \ldots, x_n)$ and $c_R(x_1, x_2, \ldots, x_n)$ as well as the corresponding core and support set of R on product space $U_1 \times U_2 \times \cdots \times U_n$.

Example 4.1 "Approximately equal" is a flexible relation (concept) between real numbers. Try to find its membership-consistency function by using space-transforming method.

Solution "Approximately equal" is reaching or exceeding "equal" to a certain degree. Then, how is the degree of equality between two numbers portrayed? We know that if two numbers having same sign are approximately equal then their ratio is close to 1, and vice versa. Thus, $z = \dfrac{n_1}{n_2}$ (n_1 and n_2 have same sign and $|n_1| < |n_2|$) can be treated as a measure that portrays the degree of the equality between two numbers and we may as well call it as equality-degree. Obviously, the range of this equality-degree is interval $(0, 1]$, which can be called the equality-degree range. Then, from the semantics, "approximately equal" can be a flexible concept on range $(0, 1]$ of equality-degrees. Let the membership function of "approximately equal" on $(0, 1]$ be as follows

$$m_{\text{approximately equal}}(z) = \begin{cases} 0, & 0 < z < z_1 \\ \dfrac{z - z_1}{z_2 - z_1}, & z_1 < z < z_2 \\ 1, & z_2 < z < 1 \end{cases} \tag{4.1}$$

where z_1 and z_2 are separately the critical point and core–boundary point of "approximately equal," the graph of function is shown in Fig. 4.1.

The corresponding consistency function is

$$c_{\text{approximately equal}}(z) = \dfrac{z - z_1}{z_2 - z_1}, \quad 0 < z < 1 \tag{4.2}$$

Fig. 4.1 An example of the membership function of "approximately equal" on range of equality-degrees

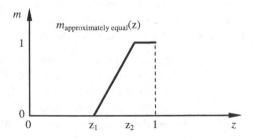

Now, for arbitrary $x, y \in (-\infty, 0)$ or $(0, +\infty)$, when $|x| < |y|$, substitute $z = \frac{x}{y}$ into the expression at right side of Eq. (4.1); then, it follows that

$$m_{\text{approximately equal}}(x, y) = \begin{cases} 0, & 0 < \frac{x}{y} \le z_1 \\ \dfrac{\frac{x}{y} - z_1}{z_2 - z_1}, & z_1 < \frac{x}{y} < z_2 \\ 1, & z_2 \le \frac{x}{y} < 1 \end{cases} \tag{4.3}$$

when $|x| > |y|$, substitute $z = \frac{y}{x}$ into the expression at right side of Eq. (4.1), then it follows that

$$m_{\text{approximately equal}}(x, y) = \begin{cases} 0, & 0 < \frac{y}{x} \le z_1 \\ \dfrac{\frac{y}{x} - z_1}{z_2 - z_1}, & z_1 < \frac{y}{x} < z_2 \\ 1, & z_2 \le \frac{y}{x} < 1 \end{cases} \tag{4.3'}$$

These two functional expressions joined together are just the membership function of "approximately equal" on product space $(-\infty, +\infty) \times (-\infty, +\infty)$ (a part of its graph is shown in Fig. 4.2).

And the corresponding consistency function is

$$c_{\text{approximately equal}}(x, y) = \begin{cases} \dfrac{\frac{x}{y} - z_1}{z_2 - z_1}, & 0 < \frac{x}{y} < 1 \\ \dfrac{\frac{y}{x} - z_1}{z_2 - z_1}, & 0 < \frac{y}{x} < 1 \end{cases} \tag{4.4}$$

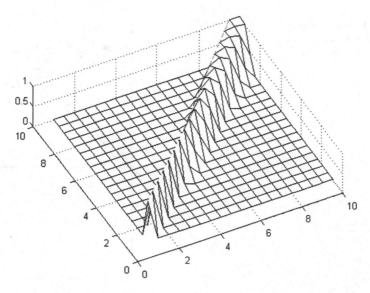

Fig. 4.2 An example of the membership function of "approximately equal" on product space

Similarly, by transformations $z = \frac{x}{y}$ and $z = \frac{y}{x}$ as well as the original critical point z_1 and core–boundary point z_2 of "approximately equal," we can obtain the critical line and core–boundary line of "approximately equal" in two-dimensional space $(-\infty, +\infty) \times (-\infty, +\infty)$ separately as

$$y = \frac{1}{z_1}x, \quad y = z_1 x$$

and

$$y = \frac{1}{z_2}x, \quad y = z_2 x$$

The core and support set enclosed by them are shown in Fig. 4.3.

Of course, we can also put it another way—"approximately equal" is "difference being very small." Thus, the "approximately equal" can also be defined on range $[0, 1]$ of difference-degrees, or even directly be defined on range $[0, b]$ of distances.

Example 4.2 Try to give the membership-consistency function of the "far greater than" relation between two numbers on interval $[a, b]$.

Solution "Far greater than" should be reaching or exceeding "greater than" to a certain degree, so we define measure $z = \frac{x-y}{b-a}$ ($x, y \in [a, b]$, $x \geq y$) as greater than degree. It is not hard to see that the corresponding range of measurements, that is, the range of greater than degree, is $[0, 1]$, while "far greater than" is a flexible concept on the range $[0, 1]$ of greater than degrees. Let its membership function be

Fig. 4.3 Core and support set of "approximately equal" in two-dimensional space

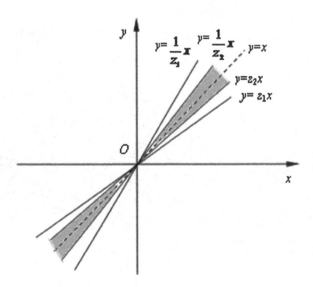

$$m_{\text{far greater than}}(z) = \begin{cases} 0, & 0 \le z \le z_1 \\ \dfrac{z - z_1}{z_2 - z_1}, & z_1 \le z \le z_2 \\ 1, & z_2 \le z \le 1 \end{cases} \qquad (4.5)$$

and the consistency function be

$$c_{\text{far greater than}}(z) = \dfrac{z - z_1}{z_2 - z_1}, \quad z \in [0, 1] \qquad (4.6)$$

where z_1 and z_2 are separately the critical point and core–boundary point of "far greater than."

Substitute $z = \frac{x-y}{b-a}$ into the above two expressions, we have

$$m_{\text{far greater than}}(x, y) = \begin{cases} 0, & 0 \le \frac{x-y}{b-a} \le z_1 \\ \dfrac{\frac{x-y}{b-a} - z_1}{z_2 - z_1}, & z_1 < \frac{x-y}{b-a} < z_2 \\ 1, & z_2 \le \frac{x-y}{b-a} \le 1 \end{cases} \qquad (4.7)$$

$$c_{\text{far greater than}}(x, y) = \dfrac{\frac{x-y}{b-a} - z_1}{z_2 - z_1}, \quad \frac{x-y}{b-a} \in [0, 1] \qquad (4.8)$$

They are the membership function and consistency function of "far greater than" on product space $[a, b] \times [a, b]$ (of course, we can also find the core and support set of "far greater than" in two-dimensional space $[a, b] \times [a, b]$). Using them, we can compute the degree of x far greater than y in interval $[a, b]$.

In the above, we introduced space-transforming method for the modeling of a flexible relation (concept). However, there exist such flexible relations, which cannot be represented into the flexible relations between numerical feature values or feature vectors of related entity objects. For example, "friend relation" is just such a flexible relation. Obviously, we cannot use a certain measurement to stand for a person to define the fried relation between people. Then, how do we establish models for this kind of flexible relations? For this kind of flexible relations, we can use the method similar to space transforming to modeling.

In fact, we can use directly numbers to represent the relation between entity objects, that is, defining a measure of corresponding relations between objects (for example, we can give a mark to the friend relation between two people), then define the flexible relation on the measurement range obtained, and then through "inverse transforming" to obtain the mathematical models of the flexible relation about corresponding entity objects.

Concretely speaking, let R be an n-ary flexible relation between entity objects which cannot be represented by numbers. Then, we define measure $v = \varphi(o_1, o_2, \ldots, o_n)$ ($o_i \in E_i$ is entity object, E_i is a set of entity objects, $i = 1, 2, \ldots, n$), then define flexible relation R on range $[a, b]$ of measure function $v = \varphi(o_1, o_2, \ldots, o_n)$, and obtain the corresponding membership-consistency functions $m_R(v)$ and $c_R(v)$;

from again measure function $v = \varphi(o_1, o_2, \ldots, o_n)$, we obtain membership-consistency of flexible relation R on Cartesian product $\overset{n}{\underset{i=1}{\times}} E_i$:

$$m_R\big((o_1, o_2, \ldots, o_n)_v\big) = m_R(v) \tag{4.9}$$

$$c_R\big((o_1, o_2, \ldots, o_n)_v\big) = c_R(v) \tag{4.10}$$

the corresponding support set and core are

$$\text{supp}(R) = \left\{ o \,\middle|\, o \in \overset{n}{\underset{i=1}{\times}} E_i, 0 < m_R(o) \le 1 \right\} \tag{4.11}$$

$$\text{core}(R) = \left\{ o \,\middle|\, o \in \overset{n}{\underset{i=1}{\times}} E_i, m_R(o) = 1 \right\} \tag{4.12}$$

where $o = (o_1, o_2, \ldots, o_n)$.

Of course, if viewed object group $(o_1, o_2, \ldots, o_n)$ as one object, then relation R is a property of $(o_1, o_2, \ldots, o_n)$. At that time, we can use measurement v in place of $(o_1, o_2, \ldots, o_n)$ and take directly $m_R(v)$ and $c_R(v)$ as membership-consistency functions of R.

4.2.2 Space-Transforming Method for the Modeling of a Multidimensional Flexible Property (Concept)

In Chap. 3, we have discussed the formation principle and mathematical models of multidimensional flexible properties (concepts), and given the general expressions of the membership-consistency functions of flexible properties (concepts). However, we see that if according to the formation principle we modeled for a multidimensional flexible property (concept) using flexible clustering in multidimensional space, then the computation would be tedious and which would be hard to realize with the increase of the dimensions of the space. Even if directly using the general expressions of the membership-consistency functions of multidimensional flexible properties (concepts), there still exist difficult. Because the precondition of using these general expressions is that the corresponding flexible class already is known, but the flexible classes of those usually known flexible properties are often unknown. Then, are there other methods for the modeling of the multidimensional flexible properties (concepts)?

Actually, we find that the above-stated method of space transforming for the modeling of flexible relations (concepts) can also be applied to the modeling of the multidimensional flexible properties (concepts). Next, we take flexible property (concept) "near the point P_0" on multidimensional space U as an example to illustrate the approach in detail.

It can be seen that whether in one-dimensional space or multidimensional space, "near" is always a kind of flexible linguistic value that characterizes distance. Therefore, range $D = (0, b]$ of distances (based on a certain unit of length) is the direct measurement space of flexible concept "near." Thus, we can define the membership-consistency functions of "near" on range D of distances.

Let

$$m_{near}(z) = \begin{cases} 1, & 0 < z \le z_1 \\ \frac{z_2 - z}{z_2 - z_1}, & z_1 < z < z_2 \\ 0, & z_2 \le z \end{cases} \quad (4.13)$$

$$c_{near}(z) = \frac{z_2 - z}{z_2 - z_1}, \quad z \in D \quad (4.14)$$

here, $z \in D$ is distance, z_1 and z_2 are separately the core–boundary point and critical point of "near." The graph of the membership function is shown in Fig. 4.4.

We know that in multidimensional space U, for any point $P(x) \in U$, distance $d(P, P_0) = \sqrt{\sum_{i=1}^{n} (x_i - x_0)^2}$. Thus, set

$$z = d(P, P_0) = \sqrt{\sum_{i=1}^{n} (x_i - x_0)^2} \quad (4.15)$$

Then, substitute this expression into the above expressions (4.9) and (4.10), and then we have

$$m_{near}(x_1, x_2, \ldots, x_n) = \begin{cases} 1, & 0 < \sqrt{\sum_{i=1}^{n} (x_i - x_0)^2} \le z_1 \\ \frac{z_2 - \sqrt{\sum_{i=1}^{n} (x_i - x_0)^2}}{z_2 - z_1}, & z_1 < \sqrt{\sum_{i=1}^{n} (x_i - x_0)^2} < z_2 \\ 0, & z_2 \le \sqrt{\sum_{i=1}^{n} (x_i - x_0)^2} \end{cases} \quad (4.16)$$

Fig. 4.4 Membership function of flexible linguistic value "near" on range of distances

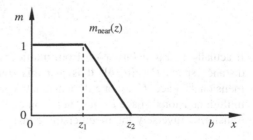

$$c_{near}(x_1, x_2, \ldots, x_n) = \frac{z_2 - \sqrt{\sum_{i=1}^{n}(x_i - x_0)^2}}{z_2 - z_1}, \quad (x_1, x_2, \ldots, x_n) \in U \qquad (4.17)$$

They are the membership-consistency functions of flexible concept "near the point P_0" on multidimensional space U. Besides, we can also consider the core and support set of "near the point P_0" in multidimensional space U.

If the radius of core and radius of support set of "near" are already known, then the membership-consistency functions of "near the point P_0" can also be rewritten as

$$m_{near}(x_1, x_2, \ldots, x_n) = \begin{cases} 1, & 0 < \sqrt{\sum_{i=1}^{n}(x_i - x_0)^2} \leq r_1 \\ \frac{r_2 - \sqrt{\sum_{i=1}^{n}(x_i - x_0)^2}}{r_2 - r_1}, & r_1 < \sqrt{\sum_{i=1}^{n}(x_i - x_0)^2} < r_2 \qquad (4.18) \\ 0, & r_2 \leq \sqrt{\sum_{i=1}^{n}(x_i - x_0)^2} \end{cases}$$

$$c_{near}(x_1, x_2, \ldots, x_n) = \frac{r_2 - \sqrt{\sum_{i=1}^{n}(x_i - x_0)^2}}{r_2 - r_1}, \quad (x_1, x_2, \ldots, x_n) \in U \qquad (4.19)$$

It can be seen that these two functions are the same as the membership-consistency functions of "near the point P_0" in Sect. 3.4.

The above two functional expressions are actually also the common expressions of the membership-consistency functions of "near the point P_0." Core–boundary point and critical point z_1 and z_2, and radius of core and radius of support set r_1 and r_2 in the expressions are all adjustable parameters, and for different practical problems, values of different meanings and sizes should be given to characterize the concept of "near" of different scales. For example, for "near the sun," radii r_1 and r_2 are several tens of thousands of kilometers, for "near the Milky Way Galaxy," r_1 and r_2 are several light years, for "near city A," r_1 and r_2 are several kilometers, and for "near the nucleus," r_1 and r_2 are just several nanometers.

It can be seen that the key of the above-stated modeling method is the equation

$$z = d(P, P_0) = \sqrt{\sum_{i=1}^{n}(x_i - x_0)^2}$$

It actually is a transformation from multidimensional space U to one-dimensional distance space $D = [0, b]$. It is just this transformation that transforms multidimensional space U to one-dimensional space $[0, b]$, thus changing "near" on multidimensional space U into "near" on one-dimensional space $[0, b]$; then, it also transforms inversely the membership-consistency functions of the latter into those of the former. Therefore, the modeling method of multidimensional flexible attributive concept "near the point P_0" is also space-transforming method.

From this example, it can be seen that the space-transforming method can also be used in the modeling of multidimensional flexible properties (concepts). The specific steps are similar with minor differences to the space-transforming method in the modeling of flexible relations (concepts).

Let A be a known multidimensional flexible property (concept), we can obtain its membership-consistency functions by the following steps and method:

① Select or define appropriate measure $u = \psi(P)$ by the semantics of A ($P \in U$ is a point variable, U is an $n(n > 1)$-dimensional space), treat $u = \psi(P)$ as a function on multidimensional space U, and determine the range V of it, and then treat V as the measurement space that A belongs to.

② Define flexible concept A on V by flexible clustering, obtaining the corresponding membership-consistency functions $m_A(u)$ and $c_A(u)$.

③ Substitute $u = \psi(x_1, x_2, \ldots, x_n)$ ($x_1, x_2, \ldots, x_n$ are the coordinate variables of point P) into $m_A(u)$ and $c_A(u)$ as well as the corresponding core–boundary point and critical point, then obtaining immediately membership-consistency functions $m_A(x_1, x_2, \ldots, x_n)$ and $c_A(x_1, x_2, \ldots, x_n)$ as well as the corresponding core and support set of A on multidimensional space U.

Thus, the space-transforming method is also a common method for the modeling of multidimensional flexible concepts.

For the space-transforming method, we need also to make some explanation here.

We know that a one-dimensional point and its coordinate are the same real number, so a flexible concept on one-dimensional measurement space U is also the summarization of a set of numerical values on U. For example, "hot" is the summarization of a set of temperatures, "tall" is the summarization of a set of heights, and "old" is the summarization of a set of ages, etc. Therefore, the membership-consistency functions of flexible concepts on one-dimensional measurement spaces are all the direct functions of corresponding measures. However, a point on a multidimensional space is no more single real number but a vectors (x_1, x_2, \ldots, x_m) consisting of multiple real numbers, so a flexible property (concept) on multidimensional space U is the summarization of properties stood for by a batch of vectors on U rather than the direct summarization of vector's components or point's coordinates $x_1, x_2, \ldots, x_m$, while a flexible relation (concept) is even more the summarization of the relation stood for by a batch of "tuple of vector" on product space $U_1 \times U_2 \times \cdots \times U_n$ rather than the direct summarization of members v_1, $v_2, \ldots, v_n$ of vector tuples. Therefore, nor are the membership-consistency functions of flexible concepts on multidimensional spaces the direct functions of corresponding measures. For instance, flexible concept "near the point P" on a two-dimensional space can be stood for by a two-dimensional flexible class (a flexible circle), but this "near" is for the whole of two-dimensional point (x, y) rather than directly for point coordinates x and y. Therefore, the membership-consistency functions of "near the point P" are not the direct functions of x and y, either. For another instance, the flexible relation (concept) of "approximately equal" can be stood for by a two-dimensional flexible class (a flexible band), but this "approximately equal" is for the relation between x and y in two-dimensional point (x, y) rather than directly for point coordinates x and y. Therefore, the membership-consistency functions of "approximately equal" are not the direct functions of x and y. For this reason, we need to use the space-transforming method to indirectly obtain the membership-consistency functions of flexible concepts on a multidimensional space.

4.3 Variable Substitution and Extended Membership-Consistency Functions

Definition 4.1 Let $m_A(u)$ and $c_A(u)$ be the membership-consistency functions of flexible concept A, and u be the function of variables $x_1, x_2, \ldots, x_n$, namely $u = f(x_1, x_2, \ldots, x_n)$. Then, functions

$$m_A(f(x_1, x_2, \ldots, x_n)) = m_A(x_1, x_2, \ldots, x_n)$$
$$c_A(f(x_1, x_2, \ldots, x_n)) = c_A(x_1, x_2, \ldots, x_n)$$

are called the extended membership-consistency functions of flexible concept A.

Example 4.3 Suppose "excellent" is a flexible linguistic value on the range [0, 100] of learning achievements, and whose membership function is $m_{\text{excellent}}(u)$. Also, it is known that the learning achievement u is the average of exam scores x, y, and z of three courses A, B, and C, namely

$$u = \frac{x + y + z}{3}$$

Then, substitute that expression into $m_{\text{excellent}}(u)$, we have

$$m_{\text{excellent}}(u) = m_{\text{excellent}}\left(\frac{x + y + z}{3}\right) = m_{\text{excellent}}(x, y, z)$$

The function is a function about scores x, y, and z, which is just the extended membership function of "excellence."

Extended membership-consistency functions are needed in some practical problems yet. From the mathematical point of view, the extended membership-consistency function of a flexible concept is obtained by variable substitution $u = f(x_1, x_2, \ldots, x_n)$ from its original membership-consistency functions.

Example 4.4 Find the membership-consistency functions and extended membership-consistency functions of an "approximate right triangle."

Solution We know that the right-angle in a right triangle is certainly a maximum angle. Therefore, an approximate right triangle is the triangle whose max angle is approximate to 90°. So this problem is also to obtain the membership-consistency functions of "a triangle whose max angle is approximate to 90°." Thus, we firstly find the membership-consistency functions of the flexible concept of "approximate to 90°" (simply ap90). It is not hard to see that the measurement range corresponding to ap90 is interval [0, 180], and ap90 is a flexible class on space [0, 180]. Since the max angle of a triangle is certainly not smaller than 60°, so we take 60° and 120°as the critical points and 80°and 100°as the core–boundary points. Thus, the membership-consistency functions of ap90 are

$$
m_{ap90}(\theta) = \begin{cases} 0, & 0 \le \theta \le 60 \\ \frac{\theta-60}{20}, & 60 < \theta < 80 \\ 1, & 80 \le \theta \le 100 \\ \frac{120-\theta}{20}, & 100 < \theta < 120 \\ 0, & 120 \le \theta \le 180 \end{cases} \tag{4.20}
$$

$$
c_{ap90}(\theta) = \begin{cases} \frac{\theta-60}{20}, & \theta \le 90 \\ \frac{120-\theta}{20}, & 90 \le \theta \end{cases} \tag{4.21}
$$

From the above analysis, these two functions are the membership-consistency functions of "approximate right triangle," where $\theta = \max\{x, y, z\}$, and x, y, and z are separately the degrees of three internal angles of a triangle. Then, substitute these expressions into the above expressions (4.16) and (4.17), we have

$$
m_{ap90}(x, y, z) = \begin{cases} 0, & \max\{x, y, z\} \le 60 \\ \frac{\max\{x,y,z\}-60}{20}, & 60 < \max\{x, y, z\} < 80 \\ 1 & 80 \le \max\{x, y, z\} \le 100 \\ \frac{120-\max\{x,y,z\}}{20}, & 100 < \max\{x, y, z\} < 120 \\ 0 & 120 \le \max\{x, y, z\} \end{cases} \tag{4.22}
$$

$$
c_{ap90}(x, y, z) = \begin{cases} \dfrac{\max\{x, y, z\} - 60}{20}, & \max\{x, y, z\} \le 90 \\ \dfrac{120 - \max\{x, y, z\}}{20}, & 90 \le \max\{x, y, z\} \end{cases} \tag{4.23}
$$

The two functions are the extended membership-consistency functions of "approximate right triangle."

Example 4.5 Find the membership-consistency functions and extended membership-consistency functions of an "approximate isosceles triangle."

Solution An approximate isosceles triangle is a triangle in which the lengths of two sides are approximate, that is, the two internal angles are approximate. "Two internal angles are approximate" also means their difference is approximate to 0°. Therefore, we firstly find the membership-consistency functions of "approximate to 0°" (simply ap0). According to the problem, we take [0, 180] as the corresponding measure range. Then, ap0 is a flexible class on space [0, 180]. Suppose its radius of core and radius of support set separately is 2.5° and 5°, then the membership-consistency functions of ap0 are

$$m_{ap0}(\theta) = \begin{cases} 1, & 0 \le \theta \le 2.5 \\ \dfrac{5-\theta}{2.5}, & 2.5 < \theta < 5 \\ 0, & 5 \le \theta \le 180 \end{cases} \tag{4.24}$$

$$c_{ap0}(\theta) = \frac{5-\theta}{2.5}, \quad 0 \le \theta \le 180 \tag{4.25}$$

These are the membership-consistency functions of "approximate isosceles triangle," where $\theta = \min\{|x - y|, |y - z|, |z - x|\}$, and x, y, and z are separately the degrees of three internal angles of a triangle. Then, substitute these expressions into the above expressions (4.24) and (4.25), we have immediately

$$m_{s0}(x, y, z) = \begin{cases} 1, & 0 \le \theta \le 2.5 \\ \dfrac{5 - \min\{|x - y|, |y - z|, |z - x|\}}{2.5}, & 2.5 < \theta < 5 \\ 0, & 5 \le \theta \le 180 \end{cases} \tag{4.26}$$

$$c_{s0}(x, y, z) = \frac{5 - \min\{|x - y|, |y - z|, |z - x|\}}{2.5}, \quad 0 \le \theta \le 180 \tag{4.27}$$

These are the extended membership-consistency functions of "approximate isosceles triangle."

Note that the domain of an extended membership-consistency function cannot certainly be treated as the measurement space that the corresponding flexible concept corresponds to.

4.4 Dynamic and Polymorphism of Mathematical Models of Flexible Concepts

As stated in Sect. 2.2, the cores and support sets of some flexible concepts in human brain can expand or contract dynamically, and the measurement spaces that some flexible concepts belong to can also expand or contract dynamically, while expanding or contracting of a measurement space will also lead to adjusting of the corresponding cores and support sets or redefining of the corresponding flexible linguistic values. On the other head, we also find that with different problems one and the same flexible concept (i.e., one and the same flexible linguistic value) will belong to different range of numerical values, or in other words, different measurement spaces. For example, the "tall" of people, the "tall" of trees, and the "tall" of mountains are all "tall," but the measurement spaces that such three "tall" belong to are not same obviously. For another example, the "quick" of people, the "quick" of cars, and the "quick" of aircrafts are all "quick," but the measurement spaces that such three "quick" belong to are not same, either. There is another situation that some flexible concepts have special denotations except basic denotation. For example, the "old" of ordinary people and the "old" of sportsman, and the "young" of ordinary people and the "young" of scientists are in such situations. That one and the same flexible linguistic value has multiple belonging measurement spaces means that which has multiple cores and support sets and multiple membership-consistency functions. We call the characteristic to be the polymorphism of a flexible concept, i.e., a flexible linguistic value.

Thus, dynamics and polymorphism are problems encountered in the modeling of flexible concepts. Dynamics of a flexible concept is reflected in the mathematical models so that the belonging measurement space of a flexible linguistic value can change, the core and support set; that is, the domains of membership-consistency functions can also change; specifically speaking, it is that the values of related parameters (as s_A^-, c_A^-, c_A^+ and s_A^+) can change, that is, these parameters become variables. If parameters s_A^-, c_A^-, c_A^+ and s_A^+ are also variables, then the corresponding expressions of membership-consistency functions represent separately a cluster of functions. Polymorphism of a flexible concept is reflected in the mathematical models so that one and the same flexible linguistic value corresponds to multiple measurement spaces, multiple cores and support sets, and multiple membership-consistency functions. As for the parameters and domains suitable for dynamic and polymorphism of membership-consistency functions, they need then to be set and determined according to specific problems, where related domain knowledge and common sense will be involved.

4.5 Summary

In this chapter, we discussed the methods of the determination and acquisition of the membership functions and consistency functions of known flexible concepts and discussed the dynamics and polymorphism of mathematical models of a flexible concept. The main points and results of the chapter are as follows:

- The basic method of the modeling of a flexible concept: first, determine the corresponding measure and measurement space according to the feature of things that a target flexible concept belongs to, then, determine the core–boundary points (lines and planes) and critical points (lines and planes) of the corresponding flexible concept, and then give the corresponding core and support set as well as membership-consistency functions.

- According to corresponding practical problem, we can select one of the methods of personal preference, statistics from a group, derivation with instances, and generation by translating to determine core–boundary points (lines and planes) and critical points (lines and planes) of a flexible concept.
- For flexible relations (concepts) and multidimensional flexible properties (concepts), we can also adopt space-transforming method to modeling. That is, first, transform a multidimensional space to a one-dimensional space through a certain measure, then establish mathematical models for the corresponding flexible concept on the one-dimensional space, after then transform it back to the original space. The basic idea of the space-transforming method is "translating problem," which, generally speaking, is necessary for flexible relations (concepts), while for multidimensional flexible properties (concepts) it is then a kind of modeling technique.
- For some flexible concepts, we can firstly find the membership-consistency functions on the corresponding measurement space, then derive the more practical extended membership-consistency functional expressions by using the method of variable substitution.

Reference

1. Lian S (2009) Principles of imprecise-information processing. Science Press, Beijing

Fundamental Theories of Flexible Sets and Flexible Linguistic Values: Mathematical Theories on Imprecise Information

Chapter 5
Flexible Sets and Operations on Flexible Sets

Abstract This chapter founds the fundamental theory of flexible sets. Firstly, it gives the types and definitions of flexible sets and analyzes and expounds the relationships and similarities and differences between flexible set and ordinary (rigid) set, flexible set and fuzzy set, and flexible set and rough set, respectively; then, it defines the operations on flexible sets and the relationships between flexible sets. In addition, the concept of flexible relations is also presented.

Keywords Flexible sets · Fuzzy sets · Rough sets · Flexible relations

In Chaps. 2 and 3, we focused on the formation principles of flexible linguistic values and obtained various flexible classes by flexible clustering in measurement spaces. In fact, the flexible classes are also a kind of subset of measurement space, but they are not usual subsets. In this chapter, we will further examine this special subset from the angle of set theory.

5.1 Types and Definitions of Flexible Sets

We know that a flexible class is the denotative model of a flexible concept. Since flexible concepts can be classified into flexible attribute concept and flexible entity concept, flexible classes can also be classified into two types: One type is the flexible classes corresponding to flexible attribute concepts, each of which consists of numerical feature values or feature vectors, and we call them the flexible class of numerical values; the other type is the flexible classes corresponding to flexible entity concepts, each of which consists of entity objects, and we call them the flexible classes of entities. Examining the two types of flexible classes (see Chaps. 2 and 3), it can be seen that the common characteristics of flexible classes are as follows:

© Springer Science+Business Media Singapore 2016
S. Lian, *Principles of Imprecise-Information Processing*,
DOI 10.1007/978-981-10-1549-6_5

① A flexible class is completely determined by its core and support set;
② The core and boundary of a flexible class are both non-empty;
③ Every element in the core of a flexible class belongs to the flexible class completely, and every element in the boundary belongs to the flexible class partially, or in other words, to a certain degree.

And the flexible classes of numerical values have also the following characteristics:

① The base set that a flexible class of numerical values belongs to is a continuous measurement space, and the flexible class itself is also continuous (here "continuous" includes both the "continuous" of real numbers and the "consecutive" of integers, and for multidimensional measurement space, it is the succession and the equidistant distribution of points; the same hereinafter);
② The membership-degree of a point in the core of a flexible class of numerical values to the flexible class is 1, and the membership-degree of a point in the boundary to the flexible class is equal numerically to the degree of this point closing to the corresponding core-boundary point, that is, the degree of approximation or similarity (strictly, the sameness-degree) between the two.

Besides, a flexible class of entities is always dependent on the corresponding flexible class of numerical values.

On the basis of the types and characteristics of flexible classes, we introduce the terminology and concept of flexible sets.

Definition 5.1 Let U be an n-dimensional measurement space as a universe of discourse, and let S_1 and S_2 be two non-empty subsets of U. If S_1 and S_2 are both continuous and $S_1 \subset S_2$, then S_1 and S_2 determine a flexible subset A of U; we call it the flexible set of numerical values, where S_1 is called the core of A, denoted core (A), each member of which belongs to A completely; S_2 is called the support set of A, denoted supp(A); and $S_2 - S_1$ is called the flexible boundary of A, denoted boun (A), each member of which belongs to A to a certain degree that numerically equals the degree of approximation between the member and the corresponding boundary point of S_1. And set $\{x|x \in U, 0.5 < m_A(x) \leq 1, m_A(x)$ is the membership-degree of x to $A\}$ is called the extended core of flexible set A, denoted core$(A)^+$.

It can be seen that the membership-degrees of the objects in core(A) to A should be all 1, and by the definition (Definition 2.2) of sameness-degree, the membership-degrees of the objects in boun(A) to A should be between 0 and 1. Besides, the objects in $U - S_2$ completely do not belong to A, so their membership-degree to A can only be 0. Thus, flexible set A actually defines a mapping φ from measurement space U to interval $[0, 1]$. Conversely, if there exists a mapping $\varphi: U \to [0, 1]$ satisfying conditions in the above definition, then this mapping also determines a flexible subset A of U. Thus, we have another definition of the flexible set.

Definition 5.1' Let U be an n-dimensional measurement space as a universe of discourse, $\varphi : U \to [0, 1]$ be a mapping from U to interval $[0, 1]$, $S_1 = \{x|x \in$

$U, \varphi(x) = 1\}$ and $S_2 = \{x|x \in U, 0 < \varphi(x) \leq 1\}$. If the following conditions are satisfied:

(1) S_1 and S_2 are both non-empty and continuous;
(2) For $\forall x \in S_2 - S_1$, always $0 < \varphi(x) < 1$, and $\varphi(x)$ is numerically equal to the degree of approximation between x and the corresponding boundary point of S_1;
(3) For $\forall x \in U - S_2$, always $\varphi(x) = 0$;

then set

$$\{(x, \varphi(x))|x \in U, \varphi(x) \in [0, 1]\} = A$$

is called a flexible subset of U, where $\varphi(x)(x \in U)$ is called the membership function of A, denoted $m_A(x)$. For any $x \in U, m_A(x) \in [0, 1]$ is called the degree of x belonging to flexible set A, or simply, the membership-degree of x [1]. Sets S_1 and S_2 are separately called the core and support set of flexible set A, denoted separately core(A) and supp(A), difference $S_2 - S_1$ is called the boundary of flexible set A, denoted boun(A). And set $\{x|x \in U, 0.5 < m_A(x) \leq 1\}$ is called the extended core of flexible set A, denoted core(A)$^+$.

It can be verified that if we take the sameness-degree in Definition 2.1 as the degree of approximation between two points, then by Definition 5.1′, it can still be derived that the membership function of flexible set A in one-dimensional measurement space $[a, b]$ is a trapezoidal function, namely

$$m_A(x) = \begin{cases} 0, & a \leq x \leq s_A^- \\ \frac{x - s_A^-}{c_A^- - s_A^-}, & s_A^- < x < c_A^- \\ 1, & c_A^- \leq x \leq c_A^+ \\ \frac{s_A^+ - x}{s_A^+ - c_A^+}, & c_A^+ < x < s_A^+ \\ 0, & s_A^+ \leq x \leq b \end{cases} \tag{5.1}$$

Definition 5.2 Let E be a set of entities, and E_1 and E_2 be two non-empty subsets of $E, E_1 \subset E_2$, and let $v(o)$ be the feature vector of object o in E, and U be a continuous measurement space consisting of all $v(o)$. If $S_1 = \{v(o)|o \in E_1\}$ and $S_2 = \{v(o)|o \in E_2\}$ can determine a flexible set A in U, then E_1 and E_2 determine a flexible set O in E. We call the flexible set O to be the flexible set of entities, where E_1 is called the core of O, denoted core(O), each member of which belongs to O completely; E_2 is called the support set of O, denoted supp(O); and $E_2 - E_1$ is called the flexible boundary of O, denoted boun(O), each member of which belongs to O to a certain degree that equals the membership-degree of feature vector of the object to flexible set A of numerical values; that is, $m_O(o) = m_A(v(o))$ ($o \in E_2 - E_1, m_O(o)$ is the membership-degree of o to O). And set $\{o|o \in E, 0.5 < m_O(o) \leq 1\}$ is called the extended core of flexible set O, denoted core(O)$^+$.

Definition 5.2′ Let E be a set of entities, and $v(o)$ be the feature vector of object o in E, and let U be a continuous measurement space consisting of all $v(o)$, and S_1

and S_2 be two non-empty subsets of U, and $S_1 \subset S_2$. If S_1 and S_2 can determine a flexible set A in U, then $E_1 = \{o|v(o) \in S_1\}$ and $E_2 = \{o|v(o) \in S_2\}$ determine a flexible set O in E. We call the flexible set O to be the flexible set of entities, where E_1 is the core of O, and E_2 is the support set of O.

Actually, feature vector $v(o)$ also denotes a correspondence relation from set E of entities to measurement space U (generally speaking, it is a many-to-one correspondence). Thus, the Definitions 5.2 and 5.2' above can also be simply stated as the Definition 5.2" below.

Definition 5.2" If $A = v(O)$ is a flexible set of numerical values in U, then $O = v^{-1}(A)$ is a flexible set of entities in corresponding set E of entities (here $v^{-1}(A)$ denotes the inverse correspondence of $v(O)$).

It can be seen that like the relation between flexible entity concepts and flexible attribute concepts, flexible sets of entities are established on the base of flexible sets of numerical values, or in other words, flexible sets of entities are dependent on the flexible sets of numerical values. This is to say, to establish a flexible set of entities, we need to establish the corresponding flexible set of numerical values firstly.

Example 5.1 Let $U = [1, 200]$ be a universe of discourse. We take subsets $[15, 40]$ and $[18, 25] \subset U$ as the support set and core, respectively, then a flexible set of numerical values of U, A, is obtained, whose membership function is

$$m_A(x) = \begin{cases} 0, & x \leq 15 \\ \frac{x-15}{3}, & 15 < x < 18 \\ 1, & 18 \leq x \leq 25 \\ \frac{40-x}{15}, & 25 < x < 40 \\ 0, & 40 \leq x \end{cases}$$

It can be seen that if U is interpreted as the range of ages of mankind, then this flexible set of numerical values, A, can be treated as the denotative model of flexible attribution concept "young". With the flexible set of numerical values, A, the denotative model of flexible entity concept "young people" is just the flexible set, denoted AP, of entities based on A, whose support set and core are

$$\text{supp}(AP) = \{p_x|p_x \in P, 0 < m_{AP}(p_x) \leq 1\}$$
$$\text{core}(AP) = \{p_x|p_x \in P, m_{AP}(p_x) = 1\}$$

and the membership function is

$$m_{AP}(p_x) = m_A(x), \quad x \in U$$

here, p_x denotes a person aged x, and P is the mankind set.

As an extreme case, there may be only one element in the core of a flexible set. We call the flexible set whose core contains only one element to be the **single-point-core flexible set**. Since there is only one element in the core, this element is not only the center of the core, but also the core-boundary point, core-boundary line, or core-boundary plane of this flexible set. The membership function of a single-point-core flexible set in one-dimensional measurement space is a triangle or semi-triangle function. The common expression of the membership function of a single-point-core flexible set with center of core number n is as follows:

$$m_n(x) = \begin{cases} 0, & a \leq x \leq s_n^- \\ \frac{x - s_n^-}{c_n^- - s_n^-}, & s_n^- < x < n \\ 1, & x = n \\ \frac{s_n^+ - x}{s_n^+ - c_n^+}, & n < x < s_n^+ \\ 0, & s_n^+ \leq x \leq b \end{cases} \tag{5.1'}$$

From the definition of a flexible set, it is not hard to see that a flexible set can be viewed as the extension of an ordinary (rigid) set, while an ordinary (rigid) set can then be viewed as the contraction of a flexible set. The relationship between the two is analogous to that between one-dimensional geometric space and two-dimensional geometric space.

With this terminology of flexible set, the denotative representation of a flexible concept can be said to be a flexible set.

5.2 Flexible Sets Versus Fuzzy Sets

We know that the fuzzy set [2] can be defined as follows.

Definition 5.3 Let X be a set as a universe of discourse, and let $\mu: X \rightarrow [0, 1]$ be a mapping from X to interval [0, 1]. Then set

$$A = \{(x, \mu(x)) | x \in X, \mu(x) \in [0, 1]\}$$

to be called a fuzzy subset of X, where $\mu(x)$ $(x \in X)$ is called the membership function (MF) of A. And sets $S_1 = \{x | x \in X, \mu(x) = 1\}$ and $S_2 = \{x | x \in X, 0 < \mu(x) \leq 1\}$ are separately called the core and support (set) of fuzzy set A.

Comparing this Definition 5.3 with the above Definition 5.1', it can be seen that if removing conditions (1)–(3) in Definition 5.1' and changing measurement space U into an ordinary set X, then the definition of flexible set becomes the definition of fuzzy sets. That is to say, the flexible set of numerical values is actually a kind of special fuzzy set. And from Definition 5.2', the flexible set of entities is based on the flexible set of numerical values, and the membership function of a flexible set of numerical values is also tantamount to the membership function of corresponding flexible set of entities. So, a flexible set of entities can also be regarded as being determined by the mapping, that is, membership function, from corresponding universe to interval [0, 1]. Thus, flexible sets of entities are also a kind of special fuzzy sets. Thus, conceptually speaking, flexible sets of numerical values and flexible sets of entities are all special fuzzy sets.

However, our flexible sets discriminate explicitly an entity object from its measurement, and discriminate explicitly between flexible sets of numerical values and flexible sets of entities, and take the former as the basis of the latter, which just coincides with the denotations of flexible attribution concept and flexible entity concept and the relationship between the two; but fuzzy sets do not discriminate explicitly an entity object from its measurement, do not discriminate explicitly between fuzzy sets of numerical values and fuzzy sets of entities either, but only give generally a concept of fuzzy set.

Also, viewed from the connotation, ① the base set of a flexible set of numerical values must be a continuous measurement space, ② the core and support set of a flexible set must be non-empty, ③ the membership function of a one-dimensional flexible set is trapezoidal function, but fuzzy sets have no these requirements, and only a general and not specific membership function (mapping: $X \rightarrow [0, 1]$) being given as the definition of a fuzzy set. As to the shape of the membership functions, due to lack of objective basis, it can merely be decided subjectively.

We now examine, respectively, the relationship between flexible sets and flexible concepts and that between fuzzy set and flexible concept. We know that for any flexible concept, its denotation can be represented as a flexible set; conversely, for any flexible set, we can obtain a flexible concept by taking the flexible set as the denotation. However, the relationship between fuzzy sets and flexible concepts is not such. For instance, although the fuzzy subset in a non-continuous universe can also be treated as the denotation of a certain concept, the concept is not a flexible concept. For another example, a fuzzy set without a core does not stand for any flexible concept.

In fact, when the support set of a discrete fuzzy set is a non-continuous subset in a measurement space, this fuzzy set is only a subset of instances of the corresponding flexible concept. For instance, the fuzzy set "tall men" consisting of students in a class is a subset of instances of flexible concept "tall men."

The reason it is called a subset of instances here is because this kind of fuzzy set is not the denotation of the corresponding flexible concept but only some instances in the denotation. A subset of instances is not a flexible set, but that some subsets of instances may be represented in the form of flexible set, such as $\{(x_1, m(x_1)), (x_2, m(x_2)), \ldots, (x_n, m(x_n))\}$ or $\{(x, m(x)) | x \in U_1 \in U\}$, and they may also be represented in the form of membership function, such as $m_A(x), x \in U_1$ (U_1 is not continuous).

Now we are clear that many discrete fuzzy sets actually are only subsets of instances of the corresponding flexible concepts. In this sense, the flexible sets are also the background sets of this type of fuzzy set.

Having made clear the relationship and difference between flexible sets and fuzzy sets, we see more clearly the problem in using fuzzy sets for the modeling of flexible concepts. Meanwhile, we also see the reasons why the form of the membership function of a fuzzy set is chosen arbitrarily: Besides people's unclear understanding of the objective basis and formation principle of flexible concepts, the definition of the fuzzy set also appears vague and general and not appropriate enough for the expression of the denotation of a flexible concept.

We use flexible sets of numerical values to model flexible attribution concepts and use flexible sets of entities to model flexible entity concepts. However, fuzzy set theory does not discriminate the two and all use fuzzy sets for modeling. By comparison, the maladies of using fuzzy sets to model flexible concepts appear obviously.

5.3 Flexible Sets Versus Rough Sets

In 1982, Polish mathematician Pawlak proposed the theory of rough sets [3]. Rough set is considered an important tool to solve uncertainty problems. Next we will make a comparison between the flexible set and the rough set.

A rough set can be defined as follows:

Definition 5.4 Let U be a set as a universe of discourse, and let R be an equivalence relation on U. $X \subseteq U$ cannot be exactly represented by a union of some equivalence classes $[x]_R$ ($x \in U$) in quotient set U/R, but can be roughly represented by two such unions approximating to X. Then, X is called a rough subset of U, or simply, a rough set. These two sets that jointly describe X are separately called the upper approximation and lower approximation of rough set X, and denoted separately $R^-(X)$ and $R\_(X)$, namely

$$R^-(X) = \bigcup_i \{Y_i | Y_i \in U/R, Y_i \cap X \neq \Phi\}$$

$$R\_(X) = \bigcup_j \{Y_j | Y_j \in U/R, Y_j \subseteq X\}$$

difference

$$BN_R(X) = R^-(X) - R\_(X)$$

is called the boundary of X.

Note: In rough set theory, what plays the role of equivalence relation R in Definition 5.4 is the so-called indiscernibility relation. But the indiscernibility relation is also an equivalence relation on U. Therefore, here we directly use the equivalence relation to define a rough set.

It can be seen that upper approximation $R^-(X)$ is the smallest set that contains X in U formed by certain equivalence classes $[x]_R$, and the lower approximation $R\_(X)$ is the biggest set that is contained in X in U formed by certain equivalence classes $[x]_R$. Thus, for $\forall x \in U$: if $x \in R\_(X)$, then $x \in X$, that is, x is certainly a member of X; if $x \notin R^-(X)$, then $x \notin X$; that is, x is certainly not a member of X; if $x \in BN_R(X)$, then probably $x \in X$; that is, x may be a member of X.

Comparing the flexible set with the rough set, we can see that:

① The flexible set and the rough set are both determined by two subsets in corresponding universe.

② If an element in a flexible set of numerical values was viewed as an equivalence class, and the elements in a flexible set of entities were incorporated into one and another equivalence classes according to equivalence relation "measurement identical," then the core of flexible set is corresponding to the lower approximation $R\_(X)$ of rough set, the support set of flexible set is corresponding to the upper approximation $R^-(X)$ of rough set, and the boundary of flexible set is corresponding to $BN_R(X)$ of rough set.

For example, let A be a flexible set of numerical values (which represents a flexible attribute concept), O be a set of entities, and AO be a flexible set of entities, and let R be equivalence relation "measurement identical." Then, the support set and core of flexible set of entities, AO, can be rewritten as

$$\text{supp}(AO) = \{s_x | s_x \in O/R, 0 < m_{AO}(s_x) \leq 1\}$$
$$\text{core}(AO) = \{s_x | s_x \in O/R, m_{AO}(s_x) = 1\}$$
$$m_{AO}(s_x) = m_A(x)$$

here x is the measurement of an entity object, s_x is equivalence class $[o_x]_R$, and o_x is an object whose measurement is x. Thus, the cores of flexible sets A and AO, core(A) and core(AO), are corresponding to the lower approximation $R\_(X)$ of rough set, the support sets of flexible sets A and AO, supp(A) and supp(AO), are corresponding to the upper approximation $R^-(X)$ of rough set, and the boundaries of flexible sets A and AO, boun(A) and boun(AO), are corresponding to $BN_R(X)$ of rough set.

③ For the membership of objects inside the core and outside the support set, the flexible set and the rough set are completely the same.

For instance, let A be a flexible set in universe U. Then $\forall x \in U$; if $x \in$ core(A), then $m_A(x) = 1$; thus, x is certainly a member of A; if $x \notin$ supp(A), then $m_A(x) = 0$; thus, x is certainly not a member of A.

These are the similarities between flexible sets and rough sets. However, there are yet essential differences between the two.

First, flexible sets have the distinction of flexible sets of numerical values and flexible sets of entities, but rough sets have no such classification.

Second, the universe of discourse that a flexible set of numerical values belongs to must be a measurement space, while rough set has no such restriction.

Third, the rough set (method) uses some special subsets—equivalence classes—of a universe to portray and describe another subset (X) of the universe. But though the flexible set (method) can also be looked as using equivalence classes to describe the subset of a universe, more obviously it is using directly the elements of the universe and membership-degrees to portray and describe subsets of the universe. And the core and support sets of a flexible set are determined by human brain's flexible clustering for continuous quantities, while the lower approximation and upper approximation of a rough set are constructed by people using mathematical methods. The former has very strong subjectivity, while the latter is completely objective.

Finally, the key distinction is that in the treatment of the membership of objects in boundaries, flexible sets and rough sets are utterly different. As a matter of fact, for the flexible set, if $x \in$ boun(A), then $x \in A$ to a certain degree, but for the rough set, if $x \in BN_R(X)$, then $x \in X$ possibly. That is, an object in the boundary of a flexible set belongs to the flexible set with a membership-degree, while an object in the boundary of a rough set belongs to the rough set with a probability. In other words, the objects in the boundary of a flexible set certainly have the property possessed by members in the core and only the degree is large or small (greater than 0 and less than 1), while the objects in the boundary of a rough set do not certainly have the property possessed by members in the lower approximation, but once they have, then they have the property completely; that is, the degree is 1. In terms of logic, the former is somewhat true, the latter is possibly true. Further, the degree of an object in the boundary of a flexible set belonging to the flexible set is negatively related to the distance from the object to the core of the flexible set, but there is not any relationship between the probability of an object in the boundary of a rough set belonging to the rough set and the distance from the object to the lower approximation of the rough set.

By the above comparison, we see that flexible sets and rough sets have not only important similarities but also essential differences. Flexible sets are oriented at imprecise information while rough sets are oriented at uncertain information.

5.4   Basic Relations Between Flexible Sets

Like ordinary sets, there are also various relationships between flexible sets, such as intersection and inclusion. On the basis of containment, there is also the concept of subflexible sets. Since a flexible set of entities depends on the corresponding flexible set of numerical values, in the following we only discuss the relationships of flexible sets of numerical values.

Definition 5.5 Let U be an n-dimensional measurement space, and let A and B be two flexible subsets in U.

(1) If supp(A) $\cap$ supp(B) = $\varnothing$, then we say flexible sets A and B are disjoint.
(2) If supp(A) $\cap$ supp(B) $\neq \varnothing$, then we say the flexible sets A and B are intersectant.
(3) If supp(A) $\subseteq$ supp(B), then we say flexible set A is contained in flexible set B, or that flexible set B contains flexible set A; especially, if also core(A) $\subseteq$ core (B), then we say flexible set A is normally contained in flexible set B, or that flexible set B normally contains flexible set A, denote $A \subseteq B$.

Unless otherwise specified later, the inclusion relation between flexible sets would always refer to normal containment. The Venn diagram representations for

(a)   **(b)**   **(c)**

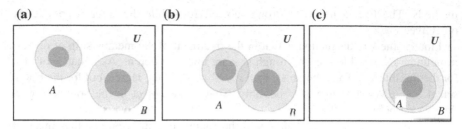

Fig. 5.1 Venn diagram representation for relationships between two-dimensional flexible sets. **a** Disjointing **b** intersection **c** containment

the disjointing, intersection, and containment of two-dimensional flexible sets are shown in Fig. 5.1 (here flexible sets are all flexible circles).

Theorem 5.1 *Let U be an n-dimensional measurement space, and let A and B be two flexible subsets of* **U**. *The sufficient and necessary condition for A⊆B is that for any* $x \in U$, $m_A(x) \leq m_B(x)$ *holds always.*

Proof Sufficiency: Assume that for any $x \in U$, always $m_A(x) \leq m_B(x)$. Suppose in such case, $A \not\subseteq B$. By Definition 5.5, it follows that supp(A) $\not\subseteq$ supp(B) or core (A) $\not\subseteq$ core(B). If supp(A) $\not\subseteq$ supp(B), then there at least exists a $x' \in$ supp(A) such that $m_A(x') > 0$ while $m_B(x') = 0$, but which is in contradiction with $m_A(x) \leq m_B(x)$. Likewise, if core(A) $\not\subseteq$ core(B), there at least exists a $x' \in$ core(A) such that $m_A(x') = 1$ while $m_B(x') < 1$, which is still in contradiction with $m_A(x) \leq m_B(x)$.

Necessity: Assume $A \subseteq B$, that is, supp(A) $\subseteq$ supp(B) and core(A) $\subseteq$ core(B). Then, for any $x \in U$:

If $x \in$ core(A), then $m_A(x) = 1$, additionally core(A) $\subseteq$ core(B), so certainly x core(B); thus, $m_B(x) = 1$, and it follows that $m_A(x) = m_B(x)$;

If $x \notin$ supp(A), then $m_A(x) = 0$, but from supp(A) $\subseteq$ supp(B), then $m_B(x) \geq 0$, and it follows that $m_A(x) \leq m_B(x)$;

If $x \in$ supp(A) − core(A), then $0 < m_A(x) < 1$, additionally supp(A) $\subseteq$ supp(B), so certainly $x \in$ supp(B); in this case, suppose that $m_A(x) > m_B(x)$, then since $m_A(x)$ and $m_B(x)$ are both linear functions, so there at least exists a $x' \in$ supp (A) such that $m_A(x') = 1$, while $m_B(x') < 1$, which is obviously contradictory to core (A) $\subseteq$ core(B).   ∎

Theorem 5.1 also means that $m_A(x) \leq m_B(x)$ is in fact equivalent to supp (A) $\subseteq$ supp(B) and core(A) $\subseteq$ core(B). Thus, we can also replace the latter by the former to define the inclusion relation of flexible sets.

Definition 5.6 Let U be an n-dimensional measurement space, and let A and B be two flexible subsets of U. If for any $x \in U$, always $m_A(x) \leq m_B(x)$, then we say flexible set A is contained in flexible set B or flexible set B contains flexible set A, denote $A \subseteq B$.

With inclusion relation $\subseteq$, we can further define the subflexible set.

Definition 5.7 Let U be an n-dimensional measurement space, and let A and B be two flexible subsets of U. If $A \subseteq B$, then A is called a subflexible set of B.

Definition 5.8 Let U be an n-dimensional measurement space, and let A and B be two flexible subsets of U. If for any $x \in U$, always $m_A(x) = m_B(x)$, then we say that flexible set A equals flexible set B, write $A = B$.

Theorem 5.2 *Let U be an n-dimensional measurement space, and let A and B be two flexible subsets of U. The sufficient and necessary condition for $A = B$ is supp $(A) = supp(B)$ and $core(A) = core(B)$.*

Theorem 5.3 *Let U be an n-dimensional measurement space, and let A and B be two flexible subsets of U. The sufficient and necessary condition for $A = B$ is $A \subseteq B$ and $B \subseteq A$.*

The proofs of these two theorems are similar to that of Theorem 5.1, so here are omitted.

5.5 Basic Operations on Flexible Sets

Just like ordinary sets, flexible sets also have intersection, union, complement, etc. We still use symbols $\cap$, $\cup$, and ^{c} to denote the intersection, union, and complement of flexible sets. Similarly, we only discuss the operations on flexible sets of numerical values.

5.5.1 Intersection, Union, Complement, and Difference of General Flexible Subsets

A flexible set is completely determined by its support set and core. Therefore, for flexible subsets in one and the same measurement space, the corresponding intersection, union, complement, and difference should also be determined by their support sets and cores.

1. Intersection of flexible sets

Definition 5.9 Let U be an n-dimensional measurement space, and let A and B be two flexible subsets of U. $A \cap B$ is called the intersection of A and B, whose support set and core are:

$$\text{supp}(A \cap B) = \text{supp}(A) \cap \text{supp}(B) \tag{5.2}$$

$$\text{core}(A \cap B) = \text{core}(A) \cap \text{core}(B) \tag{5.3}$$

From the definition, it can be explicitly seen that intersection $A \cap B \subset U$. Next we consider the membership function of $A \cap B$. Take arbitrary $x \in \text{supp}(A \cap B)$, then from Eq.(5.2), it follows that $x \in \text{supp}(A)$ and $x \in \text{supp}(B)$. Thus, the following two statements are equivalent.

① x belongs to $A \cap B$ with degree d.
② x belongs to A and to B both with degree d.

On the basis of this, we consider the relation between the membership function $m_{A \cap B}(x)$ of $A \cap B$ and membership functions $m_A(x)$ and $m_B(x)$ of A and B. Let $d = \min\{m_A(x), m_B(x)\}$, it can be seen that for any $d' \in (d, 1]$; the statement

③ x belongs to A and to B both with degree d'

is not correct; and for any $d' \in (0, d)$, though statement ③ cannot be considered wrong, it does not express sufficiently the degree of both x belonging to A and x belonging to B, while only when we take $d' = d$, statement ③ expresses just right and accurately the degree of both x belonging to A and x belonging to B. As a matter of fact, $d = \min\{m_A(x), m_B(x)\} = \inf\{d' \mid d' \in (d, 1]\} = \sup\{d' \mid d' \in (0, d)\}$. In addition, from the equivalence between statements ① and ②, x also belongs to $A \cap B$ with degree d.

The analysis above shows that $m_{A \cap B}(x)$ should be defined as $\min\{m_A(x), m_B(x)\}$; that is,

$$m_{A \cap B}(x) = \min\{m_A(x), m_B(x)\} \tag{5.4}$$

Note that the above analysis does not involve the relation between flexible sets A and B. However, in fact, there are relationships of disjointing, intersection, and containment between flexible sets. Then, has the relation between A and B any influence on membership function $m_{A \cap B}(x)$ or not? In other words, is Eq. (5.4) also true for flexible sets A and B with relations of disjointing, intersection, or containment? In the following, we analyze this problem specifically by using a two-dimensional measurement space as an example and Fig. 5.1 for reference.

(1) Suppose A and B intersect, or $\text{supp}(A) \cap \text{supp}(B) \neq \varnothing$. Then for $\forall (x, y) \in U$, there are the following cases:

(i) If $(x, y) \in \text{supp}(A) \cap \text{supp}(B)$, then $(x, y) \in \text{supp}(A)$ and $(x, y) \in \text{supp}(B)$. So in this case, $0 < m_A(x, y) \leq 1$ and $0 < m_B(x, y) \leq 1$. We may as well suppose $m_A(x, y) \leq m_B(x, y)$. Then just as the reason previously analyzed, now it would be most appropriate to only take $m_{A \cap B}(x, y) = \min\{m_A(x, y), m_B(x, y)\}$.

(ii) If $(x, y) \notin \text{supp}(A) \cap \text{supp}(B)$, but $(x, y) \in \text{supp}(A)$, then in this case, $m_{A \cap B}(x, y) = 0$, $0 < m_A(x, y) \leq 1$ and $m_B(x, y) = 0$. Thus, $\min\{m_A(x, y), m_B(x, y)\} = 0$. Therefore, $m_{A \cap B}(x, y) = \min\{m_A(x, y), m_B(x, y)\}$ too. Similarly, if $(x, y) \notin \text{supp}(A) \cap \text{supp}(B)$, but $(x, y) \in \text{supp}(B)$, then also $m_{A \cap B}(x, y) = \min\{m_A(x, y), m_B(x, y)\}$.

(iii) If $(x, y) \notin \text{supp}(A)$ and $(x, y) \notin \text{supp}(B)$, then $(x, y) \notin \text{supp}(A) \cap \text{supp}$
(B). In this case, $m_A(x, y) = 0$, $m_B(x, y) = 0$ and $m_{A \cap B}(x, y) = 0$. Thus, it
also follows that $m_{A \cap B}(x, y) = \min\{m_A(x, y), m_B(x, y)\}$.

In summary, when A and B intersect, $m_{A \cap B}(x, y) = \min\{m_A(x, y), m_B(x, y)\}$
holds always.

(2) Suppose A is contained in B, that is, $\text{supp}(A) \cap \text{supp}(B) = \text{supp}(A)$. Then for
$\forall (x, y) \in U$, there are the following cases:

(i) If $(x, y) \in \text{supp}(A)$, then $(x, y) \in \text{supp}(B)$. In this case, $0 < m_A(x, y) \le 1$
and $0 < m_B(x, y) \le 1$. We may as well suppose $m_A(x, y) \le m_B(x, y)$.
Then, just as the reason previously analyzed, now it would be most
appropriate to only take $m_{A \cap B}(x, y) = \min\{m_A(x, y), m_B(x, y)\}$.

(ii) If $(x, y) \notin \text{supp}(A)$, but $(x, y) \in \text{supp}(B)$, then $(x, y) \notin \text{supp}(A) \cap \text{supp}$
(B). In this case, $m_A(x, y) = 0$, $0 < m_B(x, y) \le 1$ and $m_{A \cap B}(x, y) = 0$.
Obviously also $m_{A \cap B}(x, y) = \min\{m_A(x, y), m_B(x, y)\}$.

(iii) If $(x, y) \notin \text{supp}(B)$, then certainly $(x, y) \notin \text{supp}(A)$. Thus
$(x, y) \notin \text{supp}(A) \cap \text{supp}(B)$. In this case, $m_B(x, y) = 0$, $m_A(x, y) = 0$
and $m_{A \cap B}(x, y) = 0$. Thus, we have also $m_{A \cap B}(x, y) = \min\{m_A(x, y),$
$m_B(x, y)\}$.

In summary, when A is contained in B, $m_{A \cap B}(x, y) = \min\{m_A(x, y), m_B(x, y)\}$
holds always.

(3) Suppose A and B are disjoint, or $\text{supp}(A) \cap \text{supp}(B) = \varnothing$. Then for
$\forall (x, y) \in U$, there are the following cases:

(i) If $(x, y) \in \text{supp}(A)$, then $(x, y) \notin \text{supp}(B)$. In this case, $0 < m_A(x, y) \le 1$
and $m_B(x, y) = 0$. Thus, $\min\{m_A(x, y), m_B(x, y)\} = 0$. And by supp
$(A) \cap \text{supp}(B) = \varnothing$, it follows that $m_{A \cap B}(x, y) = 0$. Therefore,
$m_{A \cap B}(x, y) = \min\{m_A(x, y), m_B(x, y)\}$.

(ii) If $(x, y) \in \text{supp}(B)$, then also $m_{A \cap B}(x, y) = \min\{m_A(x, y), m_B(x, y)\}$.

(iii) If $(x, y) \notin \text{supp}(A)$ and $(x, y) \notin \text{supp}(B)$, then $(x, y) \notin \text{supp}(A) \cap \text{supp}$
(B). In this case, $m_A(x, y) = 0$, $m_B(x, y) = 0$ and $m_{A \cap B}(x, y) = 0$. Thus,
we have also $m_{A \cap B}(x, y) = \min\{m_A(x, y), m_B(x, y)\}$.

In summary, when A and B are disjoint, $m_{A \cap B}(x, y) = \min\{m_A(x, y), m_B(x, y)\}$
holds always.

The above analysis shows that Eq. (5.4) is not related to the relations between
flexible sets A and B. That is to say, no matter whether flexible sets A and B are
disjoint, intersectant, or containment, it follows always that $m_{A \cap B}(x) = \min\{m_A(x),$
$m_B(x)\}$. Thus, from the relation between a flexible set and its membership function,
we then give the following definition.

Definition 5.9′ Let U be an n-dimensional measurement space, and let A and B be
two flexible subsets of U. $A \cap B$ is called the intersection of A and B, whose
membership function is

Fig. 5.2 An example of the
membership function of
intersection $A \cap B$

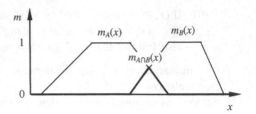

$$m_{A \cap B}(x) = \min\{m_A(x), m_B(x)\} \tag{5.5}$$

When U is one-dimensional measurement space $[a, b]$ and A and B intersect, the graph of membership function $m_{A \cap B}(x)$ is shown in Fig. 5.2.

2. Union of flexible sets

Definition 5.10 Let U be an n-dimensional measurement space, and let A and B be two flexible subsets of U. $A \cup B$ is called the union of A and B, whose support set and core are:

$$\text{supp}(A \cup B) = \text{supp}(A) \cup \text{supp}(B) \tag{5.6}$$

$$\text{core}(A \cup B) = \text{core}(A) \cup \text{core}(B) \tag{5.7}$$

By the definition, it can be explicitly seen that the union $A \cup B \subset U$. Making an analysis similar to that of intersection $A \cap B$ and its membership function, we can also have another definition of $A \cup B$.

Definition 5.10' Let U be an n-dimensional measurement space, and let A and B be two flexible subsets of U. $A \cup B$ is called the union of A and B, whose membership function is

$$m_{A \cup B}(x) = \max\{m_A(x), m_B(x)\} \tag{5.8}$$

When U is one-dimensional measurement space $[a, b]$ and A and B intersect, the graph of membership function $m_{A \cup B}(x)$ is shown in Fig. 5.3.

From Figs. 5.2 and 5.3, it can be observed that intersection $A \cap B$ and union $A \cup B$ therein do not fully meet the definition of a flexible set, so they are actually

Fig. 5.3 An example of the
membership function of union
$A \cup B$

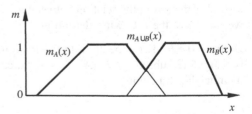

not real or standard flexible sets. That is to say, the set of flexible subsets of space U is not closed under intersection ($\cap$) and union ($\cup$) of flexible sets. Therefore, strictly speaking, intersection ($\cap$) and union ($\cup$) are not operations on flexible sets. But for habit's sake, we still call them the operations on flexible sets.

3. Complement of a flexible set

Similar to ordinary sets, when the whole universe U is partitioned into only two flexible subsets, one of them is just the complement of the other. According to the methods of flexible clustering and flexible partitioning in Sect. 2.1, we give the following definition.

Definition 5.11 Let U be an n-dimensional measurement space, and let A be a flexible subset of U. A^c is called the complement of A, whose support set and core are:

$$\text{supp}(A^c) = (\text{core}(A))^c \tag{5.9}$$

$$\text{core}(A^c) = (\text{supp}(A))^c \tag{5.10}$$

In the following, we take flexible class A of one-dimensional measurement space $U = [a, b]$ as an example to analyze the membership function of complement A^c. Let the core-boundary points and critical points of A be s_A^-, c_A^-, c_A^+ and s_A^+, and let the core-boundary points and critical points of A^c be $s_{A^c}^-, c_{A^c}^-, c_{A^c}^+$ and $s_{A^c}^+$. By Definition 5.11, it can be observed that these two groups of parameters should have the following relations:

$$c_{A^c}^+ = s_A^-, s_{A^c}^+ = c_A^-, s_{A^c}^- = c_A^+, c_{A^c}^- = s_A^+$$

Thus, for any $x \in (s_{A'}^-, c_{A'}^-)$, we have

$$m_{A^c}(x) = \frac{x - s_{A^c}^-}{c_{A^c}^- - s_{A^c}^-} = \frac{x - c_A^+}{s_A^+ - c_A^+}$$

$$m_A(x) = \frac{s_A^+ - x}{s_A^+ - c_A^+}$$

add the above two equations, we have

$$m_{A^c}(x) + m_A(x) = \frac{x - c_A^+}{s_A^+ - c_A^+} + \frac{s_A^+ - x}{s_A^+ - c_A^+} = 1$$

Likewise, for any $x \in (c_{A^c}^+, s_{A^c}^+)$, it also follows that

$$m_{A^c}(x) + m_A(x) = 1$$

And for any $x \in \text{core}(A^c)$, $m_{A^c}(x) = 1$ holds; on the other hand, since core $(A^c) = (\text{supp }(A))^{\ c} = U - \text{supp}(A)$, $m_A(x) = 0$. Therefore, also

$$m_{A^c}(x) + m_A(x) = 1$$

And for any $x \in \text{core}(A)$, $m_A(x) = 1$ holds; on the other hand, since core $(A) = U - \text{supp}(A^c)$, $m_{A^c}(x) = 0$. Thus, also

$$m_{A^c}(x) + m_A(x) = 1$$

The above analysis shows that for any $x \in U$, always

$$m_{A^c}(x) + m_A(x) = 1$$

Generalizing this equation, then it is as follows:
For any n-dimensional vector $\boldsymbol{x} \in U$, always

$$m_{A^c}(\boldsymbol{x}) + m_A(\boldsymbol{x}) = 1 \tag{5.11}$$

That is, the sum of the membership-degrees of one and the same object for a pair of relatively complemented flexible sets is always 1. Thus, we also have a definition of A^c.

Definition 5.11′ Let U be an n-dimensional measurement space, and let A be a flexible subset of U. A^c is called the complement of A, whose membership function is

$$m_{A^c}(\boldsymbol{x}) = 1 - m_A(\boldsymbol{x}) \tag{5.12}$$

When U is one-dimensional measurement space $[a, b]$, the graph of membership function $m_{A^c}(x)$ is shown in Fig. 5.4.

Actually, viewed from the angle of sets, that the sum of the membership-degrees of an object for relatively complemented flexible sets is 1 also means that there is a kind of complementation relation between membership-degrees of relatively complemented flexible sets. Therefore, we call Eqs. (5.11) and (5.12) to be the **complement law of membership-degrees**.

In the above, we gave, respectively, two definitions for each of the intersection, union, and complement operations of flexible sets. However, it can be proved that

Fig. 5.4 An example of the graph of the membership function of one-dimensional complement A^c

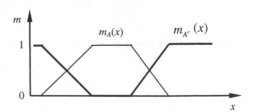

the two definitions are in fact equivalent. From this, we have immediately the following theorem.

Theorem 5.4 *Let U be an n-dimensional measurement space, and let A and B be two flexible subsets of U, then*

supp$(A \cap B)$ = supp$(A) \cap$ supp(B) and core$(A \cap B)$ = core$(A) \cap$ core$(B) \Leftrightarrow$

$$m_{A \cap B}(x) = \min\{m_A(x), m_B(x)\} \tag{5.13}$$

supp$(A \cup B)$ = supp$(A) \cup$ supp(B) and core$(A \cup B)$ = core$(A) \cup$ core$(B) \Leftrightarrow$

$$m_{A \cup B}(x) = \max\{m_A(x), m_B(x)\} \tag{5.14}$$

$$\text{supp}(A^c) = (\text{core}(A))^c \cap \text{core}(A^c) = (\text{supp}(A))^c \Leftrightarrow m_{A^c}(x) = 1 - m_A(x) \tag{5.15}$$

Since the intersection and union of flexible sets are defined by the intersection and union of their support sets and cores, the support sets and cores are ordinary sets whose operations satisfy associative laws, so the intersection and union of flexible sets in one-dimensional measurement space also satisfy associative laws. Therefore, these two operations can be generalized to the case of n flexible sets.

Definition 5.12 Let $A_1, A_2, \ldots, A_n$ be n flexible sets in n-dimensional measurement space U. $A_1 \cap A_2 \cap \ldots \cap A_n$ and $A_1 \cup A_2 \cup \ldots \cup A_n$ are separately the intersection and union of these n flexible sets, whose supports and cores are:

$$\text{supp}(A_1 \cap A_2 \cap \ldots \cap A_n) = \text{supp}(A_1) \cap \text{supp}(A_2) \cap \ldots \cap \text{supp}(A_n) \tag{5.16}$$

$$\text{core}(A_1 \cap A_2 \cap \ldots \cap A_n) = \text{core}(A_1) \cap \text{core}(A_2) \cap \ldots \cap \text{core}(A_n) \tag{5.17}$$

$$\text{supp}(A_1 \cup A_2 \cup \ldots \cup A_n) = \text{supp}(A_1) \cup \text{supp}(A_2) \cup \ldots \cup \text{supp}(A_n) \tag{5.18}$$

$$\text{core}(A_1 \cup A_2 \cup \ldots \cup A_n) = \text{core}(A_1) \cup \text{core}(A_2) \cup \ldots \cup \text{core}(A_n) \tag{5.19}$$

Definition 5.12′ Let $A_1, A_2, \ldots, A_n$ be n flexible sets in n-dimensional measurement space U. $A_1 \cap A_2 \cap \ldots \cap A_n$ and $A_1 \cup A_2 \cup \ldots \cup A_n$ are separately the intersection and union of these n flexible sets, whose membership functions are:

$$m_{A_1 \cap A_2 \cap \ldots \cap A_n}(x) = \min\{m_{A_1}(x), m_{A_2}(x), \ldots, m_{A_n}(x)\} \tag{5.20}$$

$$m_{A_1 \cup A_2 \cup \ldots \cup A_n}(x) = \max\{m_{A_1}(x), m_{A_2}(x), \ldots, m_{A_n}(x)\} \tag{5.21}$$

4. Difference of flexible sets

Definition 5.13 Let U be an n-dimensional measurement space, and let A and B be two flexible subsets of U. $A - B$ is called the difference of A minus B, whose support set and core are:

$$\text{supp}(A - B) = \text{supp}(A) - \text{supp}(B) \qquad (5.22)$$

$$\text{core}(A - B) = \text{core}(A) - \text{core}(B) \qquad (5.23)$$

5.5.2 Intersection and Union of Orthogonal Flexible Subsets

Let $U = [a, b]$ and $V = [c, d]$. Firstly, we consider the compound sets $A \cap B$ and $A \cup B$ of orthogonal flexible sets A and B in two-dimensional product measurement space $U \times V$.

Let flexible sets A and B be shown in Fig. 5.5. According to the above definition of intersection of flexible sets, only keeping the intersection of the support sets and the intersection of the cores of flexible sets A and B in the figure, we obtain a rectangular region as shown in Fig. 5.6, which is the intersection $A \cap B$ of orthogonal flexible sets A and B. Visually, the geometry of the intersection $A \cap B$ is a flexible block in space $U \times V$.

Similarly, according to the definition of union of flexible sets, keeping the support sets and cores of A and B in Fig. 5.5, we obtain the crisscross region as shown in Fig. 5.7, which is the union $A \cup B$ of orthogonal flexible sets A and B.

It can be observed that the compound sets $A \cap B$ and $A \cup B$ of orthogonal flexible sets A and B are still flexible sets in $U \times V$. Next, we analyze the membership functions of the two flexible sets.

(1) Membership function of flexible intersection A ∩ B

We enlarge separately the flexible intersection $A \cap B$ in Fig. 5.6 to that as shown in Fig. 5.8. It can be observed that the flexible boundary of $A \cap B$ (that is, the white part around the rectangular region) can be viewed as jointed together by the flexible boundary sections of A and those of B. And viewed from the direction of x-axis, the core of $A \cap B$ is also a part of the core of the original A; viewed from the direction of y-axis, the core of $A \cap B$ is also a part of the core of the original B. Thus, in the support

Fig. 5.5 Orthogonal flexible sets A and B

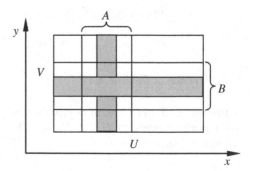

Fig. 5.6 Geometry of
intersection $A \cap B$

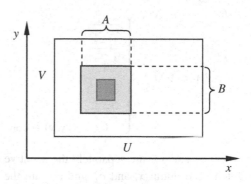

Fig. 5.7 Geometry of union
$A \cup B$

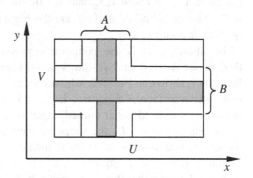

set of $A \cap B$, for points in the left and right boundaries (i.e., in x direction) of the core, their original membership-degrees for flexible set A is also their membership-degrees now for flexible set $A \cap B$; conversely, their membership-degrees for flexible set $A \cap B$ only be the membership-degrees to flexible set A. Similarly, in the support set, for points located at the upper and lower boundaries (i.e., in y direction) of the core, the membership-degrees to flexible set B are also the membership-degrees to flexible set $A \cap B$; conversely, the membership-degrees to flexible set $A \cap B$ only be those to flexible set B. That is equivalent to saying that for any point (x, y) in the support set, its membership-degree to flexible set $A \cap B$ can only be computed by the membership-degrees of its components x to A or y to B, but cannot be computed by (x, y) as a whole. Thus, the boundary of flexible set $A \cap B$ is divided into 4 parts (as shown in Fig. 5.8). Then, for $\forall (x, y) \in \text{core}(A \cap B)$, $m_{A \cap B}(x, y) = 1$; for $\forall (x, y) \in$ $\text{a}_1, m_{A \cap B}(x, y) = m_A(x) = \frac{x - s_A^-}{c_A^- - s_A^-}$; for $\forall (x, y) \in \text{a}_2, m_{A \cap B}(x, y) = m_A(x) = \frac{s_A^+ - x}{s_A^+ - c_A^+}$; for $\forall (x, y) \in \text{b}_1, m_{A \cap B}(x, y) = m_B(y) = \frac{y - s_B^-}{c_B^- - s_B^-}$; and for $\forall (x, y) \in \text{b}_2, m_{A \cap B}(x, y) = m_B(y) = \frac{s_B^+ - y}{s_B^+ - c_B^+}$.

To sum up the above analysis, we have

$$m_{A \cap B}(x,y) = \begin{cases} \dfrac{x-s_A^-}{c_A^- - s_A^-}, & (x,y) \in a_1 \\ \dfrac{s_A^+ - x}{s_A^+ - c_A^+}, & (x,y) \in a_2 \\ 1, & (x,y) \in \text{core}(A \cap B) \\ \dfrac{y - s_B^-}{c_B^- - s_B^-}, & (x,y) \in b_1 \\ \dfrac{s_B^+ - y}{s_B^+ - c_B^+}, & (x,y) \in b_2 \\ 0, & (x,y) \notin \text{core}(A \cap B) \cup a_1 \cup a_2 \cup b_1 \cup b_2 \end{cases} \tag{5.24}$$

where s_A^- and s_A^+ are separately the negative and positive critical points of flexible set $A \cap B$ about x, and c_A^- and c_A^+ are the negative and positive core-boundary points of $A \cap B$ about x; s_B^- and s_B^+ are the negative and positive critical points of $A \cap B$ about y, and c_B^- and c_B^+ are the negative and positive core-boundary points of $A \cap B$ about y.

Equation (5.24) is the common expression of the membership functions of flexible set $A \cap B$ on two-dimensional measurement space U. The graph of the function is shown in Fig. 5.9, whose shape is an edged (also called truncated square pyramidal) surface. Of course, Eq. (5.24) is the membership function of those flexible sets located at the non-edge part of space U. For those flexible sets located at the edge of space U, the shape of their membership functions is not standard terraces but semi-terraces, so these membership functional expressions should not be totally the same as Eq. (5.24), and here, we had rather not go to details.

It can be seen that the membership function of $A \cap B$ is also combined by the membership functions of flexible sets A and B. In fact, viewed from the graph, the membership function of $A \cap B$ is obtained from membership functions of A and B by cutting each other, while for the overlapping part of the two functions, in effect which just is tantamount to taking the smaller values of the functions from $m_A(x)$ and $m_B(y)$ as the value of function $m_{A \cap B}(x, y)$. And then, we see that such "taking the smaller" is actually also applicable to all points in the whole space. Thus, the membership function of $A \cap B$ can also be expressed by the following expression:

$$m_{A \cap B}(x,y) = \min\{m_A(x), m_B(y)\}, \quad x \in U, y \in V \tag{5.25}$$

Fig. 5.8 The flexible boundary of intersection $A \cap B$

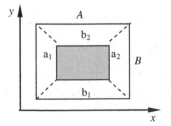

Fig. 5.9 Graph of
membership function
$m_{A \cap B}(x, y)$

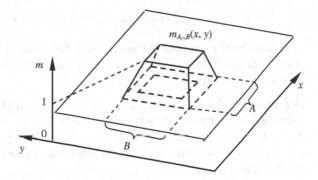

(2) Membership function of flexible union A ∪ B

From Fig. 5.7, it can be seen that for any point $(x, y) \in U \times V$, if $(x, y) \in$ supp (A) and $(x, y) \notin$ supp(B), then $m_{A \cup B}(x, y) = m_A(x, y)$; if $(x, y) \in$ supp(B) and $(x, y) \notin$ supp(A), then $m_{A \cup B}(x, y) = m_B(x, y)$; if $(x, y) \in$ supp$(A) \cap$ supp(B), then using the above analysis method of the membership function of flexible intersection $A \cap B$, we have $m_{A \cup B}(x, y) = \max\{m_A(x), m_B(y)\}$ (which is tantamount to only keeping the above curved surface of the overlapping part of the two function graphs); and for point $(x, y) \notin$ supp$(A) \cup$ supp(B), obviously, $m_{A \cup B}(x, y) = 0$. Thus, in summary, the membership function of flexible set $A \cup B$ is

$$m_{A \cup B}(x, y) = \max\{m_A(x), m_B(y)\}, \quad x \in U, y \in V \tag{5.26}$$

whose graph is shown in Fig. 5.10, with shape an orthogonal truncated ridged surface.

To sum up, we give the following definition.

Definition 5.14 Let A and B be two orthogonal flexible sets in two-dimensional measurement space $U \times V$. $A \cap B$ and $A \cup B$ are separately the intersection and union of A and B, and their membership functions are as follows:

$$m_{A \cap B}(x, y) = \min\{m_A(x), m_B(y)\}, \quad x \in U, y \in V \tag{5.27}$$

Fig. 5.10 Graph of
membership function
$m_{A \cup B}(x, y)$

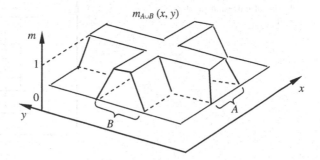

$$m_{A \cup B}(x, y) = \max\{m_A(x), m_B(y)\}, \quad x \in U, y \in V \qquad (5.28)$$

It is not hard to see that the membership functions of the intersection and union of the orthogonal flexible sets in the two-dimensional space can also be generalized to n-dimensional space.

Definition 5.15 Let A_1, A_2,..., A_n be pairwise orthogonal flexible subsets in n-dimensional product measurement space $U = U_1 \times U_2 \times ... \times U_n$, and let $A_1 \cap A_2 \cap ... \cap A_n$ and $A_1 \cup A_2 \cup ... \cup A_n$ be separately the intersection and union of the n flexible sets. The membership functions of the two compound flexible sets separately are:

$$m_{A_1 \cap A_2 \cap ... \cap A_n}(x_1, x_2, ..., x_n) = \min\{m_{A_1}(x_1), m_{A_2}(x_2), ..., m_{A_n}(x_n)\} \qquad (5.29)$$

$$m_{A_1 \cup A_2 \cup ... \cup A_n}(x_1, x_2, ..., x_n) = \max\{m_{A_1}(x_1), m_{A_2}(x_2), ..., m_{A_n}(x_n)\} \qquad (5.30)$$

5.6 Cartesian Product of Flexible Sets

In this section, we discuss the Cartesian product of flexible sets.

Definition 5.16 Let A and B be separately flexible sets in one-dimensional measurement spaces $U = [a, b]$ and $V = [c, d]$. $A \times B$ is called the Cartesian product of flexible sets A and B, whose core and support set are given as follows:

$$\text{core}(A \times B) = \text{core}(A) \times \text{core}(B) \qquad (5.31)$$

$$\text{supp}(A \times B) = \text{supp}(A) \times \text{supp}(B) \qquad (5.32)$$

From the definition, the support set and core of Cartesian product $A \times B$ are shown in Fig. 5.11. From the figure, it can be explicitly seen that the Cartesian product of flexible sets A and B, $A \times B \subset U \times V$, and which is still a flexible set. In the following, we consider the membership function of $A \times B$.

Fig. 5.11 The support set and core of cartesian product $A \times B$

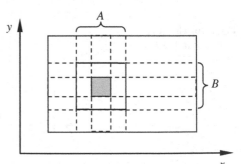

Now that $A \times B$ is the product of A and B, then for $\forall (x, y) \in U \times V, m_{A \times B}(x, y)$ should be a certain operation or function of $m_A(x)$ and $m_B(y)$. What kind of a function is it exactly?

Viewed from the shape, the core and support set of this product seem to be completely like those of the above intersection. But we notice that the two coordinate components x and y of point (x, y) in support set and core here are both related and contributive to the membership-degree of point (x, y) for the product $A \times B$, not like the point (x, y) in intersection, where only one coordinate component plays the role really. Therefore, the membership function of this product cannot be $\min\{m_A(x), m_B(y)\}$ like that of intersection. And in consideration of the constraint of the membership function itself, that is, for $\forall (x, y) \in U \times V$, always $0 \le m_{A \times B}(x, y) \le 1$, it seems that $m_{A \times B}(x, y)$ being taken as the weighted sum of $m_A(x)$ and $m_B(y)$ may be an appropriate choice. Thus, we tentatively suppose

$$m_{A \times B}(x, y) = \begin{cases} w_1 m_A(x) + w_2 m_B(y), & m_A(x) \ne 0 \wedge m_B(y) \ne 0 \\ 0, & m_A(x) = 0 \vee m_B(y) = 0 \end{cases} \qquad (*)$$

where $x \in U, y \in V, w_1, w_2 \in (0, 1)$ and $w_1 + w_2 = 1$.

Thus, for any $(x, y) \in U \times V$, there are then the following situations:

(i) If $(x, y) \in \text{core}(A \times B)$, then it follows by Definition 5.16 that $(x, y) \in \text{core}$ $(A) \times \text{core}(B)$. Thus, $x \in \text{core}(A)$ and $y \in \text{core}(B)$. Thus, by equation $(*)$ we have

$$m_{A \times B}(x, y) = w_1 m_A(x) + w_2 m_B(y) = w_1 \cdot 1 + w_2 \cdot 1 = w_1 + w_2 = 1$$

(ii) If $(x, y) \in \text{supp}(A \times B)$, then it follows by Definition 5.16 that $(x, y) \in \text{supp}$ $(A) \times \text{supp}(B)$. Thus, $x \in \text{supp}(A)$ and $y \in \text{supp}(B)$, and then $0 < m_A(x) \le 1$ and $0 < m_B(y) \le 1$. Therefore, by equation $(*)$ we have

$$0 < m_{A \times B}(x, y) = w_1 m_A(x) + w_2 m_B(y) \le 1$$

(iii) If $(x, y) \notin \text{supp}(A \times B)$, then it follows by Definition 5.16 that $(x, y) \notin \text{supp}$ $(A) \times \text{supp}(B)$. Thus, $x \notin \text{supp}(A)$ or $y \notin \text{supp}(B)$, that is, $m_A(x) = 0$ or $m_B(y) = 0$. Thus, by equation $(*)$ we have

$$m_{A \times B}(x, y) = 0$$

The analysis above shows that the function given above can serve as the membership function of product $A \times B$. Since in the argument above Eqs. (5.31) and (5.32) are used, this membership function can be said to have been derived from the two equations. Then, conversely, we consider whether Eqs. (5.31) and (5.32) of Definition 5.16 can be derived from this membership function.

Suppose there exists w_1, $w_2 \in (0, 1)$ and $w_1 + w_2 = 1$, such that $m_{A \times B}(x, y) = w_1 m_A(x) + w_2 m_B(y)$, $x \in U$, $y \in V$, then, for any $(x, y) \in U \times V$, there are then the following situations:

(i)   If $(x, y) \in \text{core}(A \times B)$, then $m_{A \times B}(x, y) = 1$; then, it would follow by $m_{A \times B}(x, y) = w_1 m_A(x) + w_2 m_B(y)$ that $w_1 m_A(x) + w_2 m_B(y) = 1$; from which it must follow that $m_A(x) = 1$ and $m_B(y) = 1$; thus, $x \in \text{core}(A)$ and $y \in \text{core}(B)$; thus, we have $(x, y) \in \text{core}(A) \times \text{core}(B)$. Conversely, let $(x, y) \subset \text{core}(A) \times \text{core}(B)$, this shows that $x \in \text{core}(A)$ and $y \in \text{core}(B)$, so it follows that $m_A(x) = 1$ and $m_B(y) = 1$, and then it follows that $m_{A \times B}(x, y) = w_1 m_A(x) + w_2 m_B(y) = w_1 \bullet 1 + w_2 \bullet 1 = w_1 + w_2 = 1$; thus, we have $(x, y) \in \text{core}(A \times B)$. That just proves that $\text{core}(A \times B) = \text{core}(A) \times \text{core}(B)$, namely Eq. (5.31).

(ii) When $(x, y) \in \text{supp}(A \times B)$, then $0 < m_{A \times B}(x, y) \leq 1$; also since $m_{A \times B}(x, y) = w_1 m_A(x) + w_2 m_B(y)$, while $m_A(x) \neq 0 \wedge m_B(y) \neq 0$, it follows that $0 < w_1 m_A(x) + w_2 m_B(y) \leq 1$, and $0 < m_A(x) \leq 1$ and $0 < m_B(y) \leq 1$; hence, $x \in \text{supp}(A)$ and $y \in \text{supp}(B)$; thus, we have $(x, y) \in \text{supp}(A) \times \text{supp}(B)$. Conversely, let $(x, y) \in \text{supp}(A) \times \text{supp}(B)$, then it shows that $x \in \text{supp}(A)$ and $y \in \text{supp}(B)$; thus, it follows that $0 < m_A(x) \leq 1$ and $0 < m_B(y) \leq 1$, and then it follows that $0 < w_1 m_A(x) + w_2 m_B(y) \leq 1$; thus, we have $(x, y) \in \text{supp}(A \times B)$. That just proves that $\text{supp}(A \times B) = \text{supp}(A) \times \text{supp}(B)$, namely Eq. (5.32).

To sum up the above analysis, we have also a definition below.

Definition 5.15′ Let A and B be separately flexible sets in one-dimensional measurement spaces $U = [a, b]$ and $V = [c, d]$. Flexible set $A \times B$ is the Cartesian product of A and B, whose membership function is:

$$m_{A \times B}(x, y) = \begin{cases} w_1 m_A(x) + w_2 m_B(y), & m_A(x) \neq 0 \wedge m_B(y) \neq 0 \\ 0, & m_A(x) = 0 \vee m_B(y) = 0 \end{cases} \quad (5.33)$$

where $x \in U$, $y \in V$, w_1, $w_2 \in (0,1)$ and $w_1 + w_2 = 1$.

By the above two definitions, we can also derive the following facts.

Corollary 5.1

(1) *When $m_A(x) > 0.5$ and $m_B(y) > 0.5$, for any w_1, $w_2 \in (0,1)$ ($w_1 + w_2 = 1$), always $m_{A \times B}(x, y) = w_1 m_A(x) + w_2 m_B(y) > 0.5$.*

(2) *Extended core $\text{core}(A \times B)^+$ of flexible product $A \times B$ cannot be expressed by an operational expression of $\text{core}(A)^+$ and $\text{core}(B)^+$ of extended cores of flexible sets A and B; we can only analyze it specifically according to specific weights. But there are always more points in $\text{core}(A \times B)^+$ than in $\text{core}(A)^+ \times \text{core}(B)^+$.*

(3) *$A \times B \neq A \times V \cap U \times B$. That is tantamount to saying that the intersection of two orthogonal flexible sets $A \times V$ and $U \times B$ in $U \times V$ is not equal to the product of the original sets A and B.*

In fact, for (x, y) that satisfies $m_A(x) > 0.5$ and $m_B(y) > 0.5$, let $m_A(x) = 0.5 + \varepsilon_1$ and $m_B(y) = 0.5 + \varepsilon_2$, then

$$
\begin{aligned}
w_1 m_A(x) + w_2 m_B(y) &= w_1 0.5 + \varepsilon_1 + w_2 0.5 + \varepsilon_2 \\
&= w_1 0.5 + w_2 0.5 + \varepsilon_1 + \varepsilon_2 \\
&= 0.5 + \varepsilon_1 + \varepsilon_2 > 0.5
\end{aligned}
$$

Thus, (1) has been proved.

Since $\forall (x, y) \in \text{core}(A)^+ \times \text{core}(B)^+$ satisfies $m_A(x) > 0.5$ and $m_B(y) > 0.5$, so $m_{A \times B}(x, y) > 0.5$. Consequently, $(x, y) \in \text{core}(A \times B)^+$. This shows core $(A)^+ \times \text{core}(B)^+ \subset \text{core}(A \times B)^+$ and also shows that points (x, y) within the median line of support set $\text{supp}(A \times B)$ all satisfy $m_{A \times B}(x, y) > 0.5$. However, it is not hard to see that between the median line and the boundary line of $\text{supp}(A \times B)$, there may still be points (x, y) which satisfy $m_{A \times B}(x, y) > 0.5$. For instance, take point $(x^*, y^*) \in \text{supp}(A \times B)$, suppose $m_A(x^*) = 1$ and $m_B(y^*) = 0.3$, so the point is located between the median line and the boundary line of $\text{supp}(A \times B)$. Then, when taking $w_1 = 0.8$ and $w_2 = 0.2$, we have

$$
m_{A \times B}(x, y) = w_1 m_A(x^*) + w_2 m_B(y^*) = 0.8 \times 1 + 0.2 \times 0.3 = 0.86 > 0.5
$$

That shows that point $(x^*, y^*) \in \text{core}(A \times B)^+$. Therefore, the points in core $(A \times B)^+$ are more than those in $\text{core}(A)^+ \times \text{core}(B)^+$. But, which points between the median line and the boundary line of $\text{supp}(A \times B)$ satisfy $m_{A \times B}(x, y) > 0.5$ need yet to be determined by specific weights w_1 and w_2. Clearly, for different weights w_1 and w_2, the points that satisfy $m_{A \times B}(x, y) > 0.5$ are different. That is to say, extended core $\text{core}(A \times B)^+$ cannot be expressed by a common expression but can only be analyzed specifically according to specific weights.

By (2), we can immediately have fact (3), which is an important distinction between flexible sets and rigid sets.

Like usual Cartesian products, the Cartesian product of flexible sets can also be generalized to the cases of multiple flexible sets.

Definition 5.17 Let $U_i = [a_i, b_i]$, and let A_i be a flexible set in U_i, $i = 1, 2, \ldots, n$, then flexible set $A = \overset{n}{\underset{i=1}{\times}} A_i$ is the Cartesian product of $A_1, A_2, \ldots, A_n$, whose core and support set are:

$$
\text{core}(A) = \overset{n}{\underset{i=1}{\times}} core(A_i) \tag{5.34}
$$

$$
\text{supp}(A) = \overset{n}{\underset{i=1}{\times}} \text{supp}(A_i) \tag{5.35}
$$

Definition 5.17' Let $U_i = [a_i, b_i]$, and let A_i be a flexible set in U_i, $i = 1, 2, \ldots, n$. Then, flexible set $A = \overset{n}{\underset{i=1}{\times}} A_i$ is the Cartesian product of $A_1, A_2, \ldots, A_n$, whose membership function is:

$$m_A(x) = \begin{cases} \sum_{i=1}^{n} w_i m_{A_i}(x_i), & m_{A_i}(x_i) \neq 0 \\ 0, & \text{for others} \end{cases} \tag{5.36}$$

where $x = (x_1, x_2,\ldots, x_n)$, $x_i \in U_i$, $w_i \in (0,1)$ and $\sum_{i=1}^{n} w_i = 1$.

By Definitions 5.17 and 5.17', we have the following theorem.

Theorem 5.6 *Let* $U_i = [a_i, b_i]$, *and let* A_i *be a flexible set in* U_i, $i = 1, 2,\ldots, n$, *then,*

$$\text{core}(A) = \underset{i=1}{\overset{n}{\times}} \text{core}(A_i) \text{ and } \text{supp}(A) = \underset{i=1}{\overset{n}{\times}} \text{supp}(A_i) \Leftrightarrow$$

$$m_A(x) = \begin{cases} \sum_{i=1}^{n} w_i m_{A_i}(x_i), & m_{A_i}(x_i) \neq 0 \\ 0, & \text{for others} \end{cases} \tag{5.37}$$

where $x = (x_1, x_2,\ldots, x_n)$, $x_i \in U_i$, $w_i \in (0,1)$ and $\sum_{i=1}^{n} w_i = 1$.

Now the problem remained is that the coefficient, i.e., weight w_i ($i = 1, 2,\ldots, n$), in this membership function is not determined. But the assignment of weights should be determined by flexible linguistic values that flexible sets $A_1, A_2,\ldots, A_n$ and $\underset{i=1}{\overset{n}{\times}} A_i$ correspond to. In fact, product $\underset{i=1}{\overset{n}{\times}} A_i$ is just corresponding to the synthetic linguistic value $A_1 \oplus A_2 \oplus \ldots \oplus A_n$ synthesized by the corresponding flexible linguistic values $A_1, A_2,\ldots, A_n$ (see Sect. 6.4.2 for details).

5.7 Flexible Relations

There are many kinds of relationships between things, of which some are rigid, that is, rigid relationship, but some are flexible, that is, flexible relationship. For instance, "equation," "parallel," and "father and son" are rigid relationship, while "similar," "analogous," "approximate," "approximately equal to," "far greater than," and "good friend" are all flexible relations. Like that a usual (rigid) relation can be stood for by a (rigid) set, a flexible relation can also be stood for by a flexible set.

We call the flexible relation stood for by a flexible set of numerical values to be the flexible relation of numerical values, and call the flexible relation stood for by a flexible set of entities to be the flexible relation of entities. For example, pure "similar" relation is a kind of flexible relation of numerical values, while the "similar" relation between persons is then of flexible relation of entities. Besides, the flexible relation between entity objects which cannot be represented by numerical values (e.g., the "friend" relation between persons) can also be classified as flexible relation of entities. Since a flexible set of entities depends on the corresponding flexible set of numerical values, a flexible relation of entities depends on the corresponding flexible relation of numerical values. Therefore, in this chapter we only discuss the flexible relations of numerical values.

Definition 5.18

(i) Let U and V be one-dimensional measurement spaces. Flexible subset R in product space $U \times V$ is called a flexible relation from U to V, which also called a binary flexible relation on $U \times V$.

(ii) Let U_i ($i = 1, 2,..., n$) be a k_i ($k_i \geq 1$)-dimensional measurement space. Flexible subset R in product space $U_1 \times U_2 \times ... \times U_n = U$ is called a flexible relation between $U_1, U_2,..., U_n$, which also called an n-ary flexible relation on U.

Example 5.2 The two flexible sets that Figs. 5.12 and 5.13 show can separately stand for two binary flexible relations of "approximately equal to" and "far greater than" between a range of positive numbers.

Since flexible relations are also a kind of flexible sets, they can also be represented by the method representing flexible sets. For example, binary flexible relation R can be represented as

$$\{((x,y), m_R(x,y)) | x \in U, y \in V\}$$

And a flexible relation between finite measurement spaces can also be represented as a matrix of the form

$$\begin{bmatrix} m_R(x_1, y_1) & m_R(x_1, y_2) & \cdots & m_R(x_1, y_n) \\ m_R(x_2, y_1) & m_R(x_2, y_2) & \cdots & m_R(x_2, y_n) \\ \cdots & \cdots & \cdots & \cdots \\ m_R(x_m, y_1) & m_R(x_m, y_2) & \cdots & m_R(x_m, y_n) \end{bmatrix}$$

or an arrow diagram or a directed graph. For example, Fig. 5.14 is an arrow diagram of binary flexible relation R from space $U = \{x_1, x_2, ..., x_m\}$ to space

Fig. 5.12 An example of the flexible set standing for "approximately equal to"

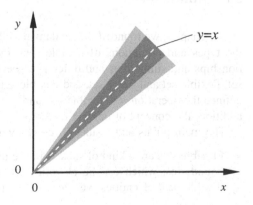

Fig. 5.13 An example of the
flexible set standing for "far
greater than"

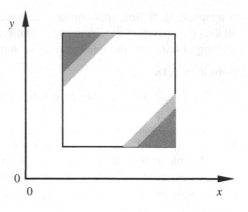

Fig. 5.14 An arrow diagram
of a flexible relation

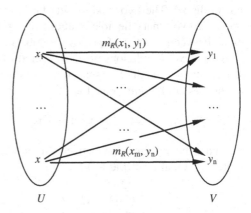

$V = \{y_1, y_2, \ldots, y_n\}$. Here, the numbers in the matrix and the numbers on the arrows
are the membership-degrees of corresponding ordered pair (x_i, y_j) for relation R.

5.8 Summary

In this chapter, we founded the fundamental theory of flexible sets. Firstly, we gave
the types and definitions of flexible sets and analyzed and expounded the relationships and similarities and differences between flexible set and ordinary (rigid)
set, flexible set and fuzzy set, and flexible set and rough set, respectively, and then
defined the operations on flexible sets and the relationships between flexible sets. In
addition, the concept of flexible relations is also presented.

The main points and results of the chapter are as follows:

- Flexible sets are a kind of subsets in the universe of discourse that have flexible
  boundaries, which can be classified as flexible sets of numerical values and
  flexible sets of entities, yet the latter depends on the former. We defined the

flexible set by using "core and support set" and membership function separately. Flexible sets are the extension of ordinary sets, which are a kind of special fuzzy sets, which have some important similarities with rough sets but also have essential differences.

- Flexible sets are a kind of mathematical models of flexible linguistic values (flexible concepts), the two are mutually correspondent, while some discrete fuzzy sets are only the subsets of instances of corresponding flexible linguistic values (flexible concepts).
- There are also relationships of disjointing, intersection, containment, and equality between flexible sets.
- Flexible sets also have the operations of intersection, union, complement, difference, and Cartesian product, which can be defined by the operations of cores and support sets of corresponding component sets, also can be defined by the operations of membership functions of component sets, and the two kinds of definitions can be deduced from each other.
- The sum of the membership-degrees of one and the same object to a pair of relatively complemented flexible sets is 1. We call this quantitative relation the complement law of membership-degrees.

References

1. Lian S (2009) Principles of imprecise-information processing. Science Press, Beijing
2. Zadeh LA (1965) Fuzzy sets. Inf Control 8:338–353
3. Pawlk Z (1982) Rough sets. Int J Inf Comput Sci 11(5):341–356

Chapter 6
Flexible Linguistic Values and Operations on Flexible Linguistic Values

Abstract This chapter founds the fundamental theory of flexible linguistic values. First, this chapter introduces the types of flexible linguistic values, then analyzes and defines the operations on flexible linguistic values, and in particular proposes the concepts and methods of algebraic composition and decomposition of flexible linguistic values. Meanwhile, this chapter also analyzes the properties and relations of relatively negative linguistic values and then proposes the complementary partition of a measurement space and the complementary relation of flexible linguistic values. Besides, this chapter also considers other relations between flexible linguistic values, especially analyzes and defines the approximation relation between flexible linguistic values, and presents the corresponding measuring method.

Keywords Flexible linguistic values · Consistency functions · Imprecise information

Flexible sets are the mathematical essence of flexible concepts, but which are only the denotations of flexible concepts, so using only the flexible sets, it is hard to reflect many semantic features and relations of flexible concepts. Flexible linguistic values are the semantic symbols of flexible concepts, and we also gave consistency functions the mathematical models for flexible linguistic values. Thus, in this chapter, we will focus on the flexible linguistic values, to discuss the operations on them and the types, relations, and properties of them, so as to set up relevant theoretical and technological bases.

6.1 Types of Flexible Linguistic Values

6.1.1 Atomic Linguistic Values, Basic Linguistic Values, and Composite Linguistic Values

The flexible linguistic value resulted from flexible clustering of a measurement space is an atomic linguistic value. The atomic flexible linguistic value corresponding to a group of flexible sets which form a flexible partition of a measurement space is a group of basic flexible linguistic values on the measurement space. A basic flexible linguistic value is certainly an atomic linguistic value, but the converse is not necessarily true. A composite flexible linguistic value is made of several flexible linguistic values (on the same space or distinct spaces) by logic operation of conjunction ($\wedge$) or disjunction ($\vee$), or algebraic operation of synthesis ($\oplus$).

6.1.2 One-Dimensional Linguistic Values and Multidimensional Linguistic Values

Viewed from the dimensions of space, flexible linguistic values can be separated into one-dimensional linguistic values and multidimensional linguistic values. That is, a linguistic value on the one-dimensional measurement space is called a one-dimensional linguistic value, and a linguistic value on the multidimensional measurement space is called a multidimensional linguistic value [1].

6.1.3 Full-Peak Linguistic Values and Semi-Peak Linguistic Values

According to the shapes of consistency functions, flexible linguistic values can be separated into full-peak linguistic values and semi-peak linguistic values. Specifically speaking, for a one-dimensional flexible linguistic value, if its consistency function is a full-triangle function, then it is called a full-peak linguistic value; if its consistency function is a semi-triangle function, then this flexible linguistic value is called a semi-peak linguistic value. In particular, a linguistic value A with a decreasing semi-triangle consistency function is called a positive semi-peak linguistic value, or a positive semi-peak value for short, denoted A^+; and a linguistic value A with an increasing semi-triangle consistency function is called a negative semi-peak linguistic value, or a negative semi-peak for short, denoted A^-. For a multidimensional flexible linguistic value, if the shape of its consistency function is a full wedge, full conical surface, or full conical hypersurface, then it is called a full-peak linguistic value; if the shape of its consistency function is a semi-wedge and semi-conical surface or semi-conical hypersurface, then this flexible linguistic value is called a semi-peak linguistic value.

6.1.4 Property-Type Linguistic Values and Relation-Type Linguistic Values

Property-type flexible linguistic values are a kind of flexible linguistic values that describes properties (also including states) of things, which is also the linguistic values characterizing flexible properties (concepts) of things. Relation-type flexible linguistic values are a kind of flexible linguistic values that describes relationships between things, which is also the linguistic values characterizing flexible relationships (concepts) between things.

6.2   Flexible Partition of a Space and Basic Flexible Linguistic Values

6.2.1   Flexible Partition of a One-Dimensional Space and Basic Flexible Linguistic Values

In Chap. 2, we have already done flexible partitioning of the one-dimensional measurement space. Now, we give its formal definition.

Definition 6.1 Let U be a one-dimensional measurement space, $\pi = \{C_1, C_2, ..., C_m\}$ is a non-empty group of flexible classes of U, if for $\forall x \in U$, there are the following facts:

(1)   There exists at least one $C_k \in \pi$ such that $m_{Ck}(x) \neq 0$;

(2) $\sum_{i=1}^{m} m_{C_i}(x) = 1$,

where π is called a flexible partition of U and flexible classes $C_1, C_2, ..., C_m$ are called the basic flexible classes of U, and they collectively form a group of basic flexible classes of U.

Next, we give the general method for flexible partitioning of one-dimensional measurement space $[a, b]$:

(1)   Determine the number of the flexible classes and the core of every flexible class (note that the negative core–boundary point $c_{C_1}^-$ of the core of the first flexible class should be the infimum a of the space, and the positive core–boundary point $c_{C_n}^+$ of the core of the last flexible class should be supremum b of the space);

(2)   From the left to the right, overlap one by one the positive critical point $s_{C_i}^+$ of the previous flexible class with the negative core–boundary point $c_{C_{i+1}}^-$ of the following flexible class and overlap the negative critical point $s_{C_{i+1}}^-$ of the following flexible class with the positive core–boundary point $c_{C_i}^+$ of the previous flexible class.

It can be proved that the group of flexible classes obtained by this way just forms a flexible partition of $[a, b]$.

In fact, let $\pi = \{C_1, C_2, ..., C_n\}$ be a group of flexible classes in $[a, b]$ obtained by the above-stated method. It is easy to see that π already forms a cover of $[a, b]$.

This is to say, condition (6.1) in the definition is already satisfied. From condition (6.2), we only need to prove:

For $\forall x \in C_i \cap C_{i+1} = (c_i^+, s_i^+) = (s_{i+1}^-, c_{i+1}^-)$ $(i = 1, 2, \ldots, n)$, then $m_{c_i}(x) + m_{c_{i+1}}(x) = 1$.

Let $x \in C_i \cap C_{i+1}$, then from Eq. (6.2–6.5), $m_{c_i}(x) = \frac{s_i^+ - x}{s_i^+ - c_i^+}$ and $m_{c_{i+1}}(x) = \frac{x - s_{i+1}^-}{c_{i+1}^- - s_{i+1}^-}$, since $c_i^+ = s_{i+1}^-$ and $s_i^+ = c_{i+1}^-$, thus

$$m_{c_i}(x) + m_{c_{i+1}}(x) = \frac{s_i^+ - x}{s_i^+ - c_i^+} + \frac{x - s_{i+1}^-}{c_{i+1}^- - s_{i+1}^-} = \frac{s_i^+ - x}{s_i^+ - s_{i+1}^-} + \frac{x - s_{i+1}^-}{s_i^+ - s_{i+1}^-} = 1$$

Therefore, π forms a flexible partition of $[a, b]$.

Definition 6.2 Let U be an n-dimensional measurement space, $A_1, A_2, \ldots, A_m$ be a group of basic flexible sets of U. Correspondingly, flexible linguistic values $A_1, A_2, \ldots, A_m$ is just a group of basic flexible linguistic values on U, which is also a group of basic flexible linguistic values of the corresponding feature of objects.

Example 6.1 Figure 6.1 is a flexible partition of range [0, 150] of human ages; the infancy, juvenile, young, middle-aged, and old ages are all basic flexible linguistic values on universe of discourse [0, 150], and they together form a group of basic flexible linguistic values on range [0, 150] of human ages.

6.2.2 Flexible Partition of a Multidimensional Space and Basic Flexible Linguistic Values

Previously, we gave the definition of the flexible partition of a one-dimensional measurement space, but which is hard to be generalized to multidimensional measurement space. In the following, we give a more general definition of the flexible partition.

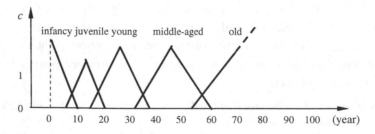

Fig. 6.1 A flexible partition of human age range

Definition 6.3 Let U be an n-dimensional measurement space and $\pi = \{A_1, A_2, \ldots, A_m\}$ is a non-empty group of flexible subsets of U. If

$$(1) \quad \bigcup_{i=1}^{m} \sup p(A_i) = U$$

$$(2) \quad \bigcap_{i=1}^{m} \operatorname{core}(A_i) = \varnothing$$

then π is called a flexible partition of U, and flexible sets $A_1, A_2, \ldots, A_m$ are called the basic flexible sets of U, and they together form a group of basic flexible sets of U. Correspondingly, flexible linguistic values $A_1, A_2, \ldots, A_m$ are the basic flexible linguistic values on U, and they together form a group of basic flexible linguistic values on U.

Obviously, this definition of flexible partition also applies to one-dimensional measurement space, and it also covers the previous Definition 6.1.

It is known from Chap. 3 that as far as shape is concerned, flexible classes on multidimensional measurement spaces are more plentiful than those on one-dimensional spaces. However, the flexible partitioning of a multidimensional space, generally speaking, can merely be bar flexible partitioning and square flexible partitioning. In the following, we take two-dimensional space as an example to make a brief description.

1. **Bar flexible partitioning**

   Suppose that space U is flexibly divided into m flexible classes $A_1, A_2, \ldots, A_m$ along the direction of x-axis (as shown in Fig. 6.2a, here take $m = 4$), and U flexibly divided as n flexible classes $B_1, B_2, \ldots, B_n$ along the direction of y-axis (as shown in Fig. 6.2b, here take $n = 3$). The grey rectangle areas in the figure are, respectively, the cores of the flexible classes, and the white rectangle areas by their sides are the boundaries of the corresponding flexible classes.

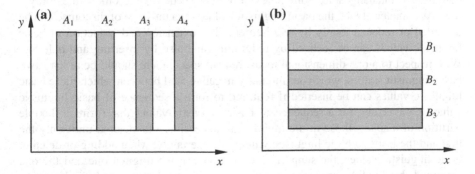

Fig. 6.2 Examples of bar flexible partitioning of two-dimensional measurement space

Fig. 6.3 An example of
square flexible partitioning of
two-dimensional
measurement space

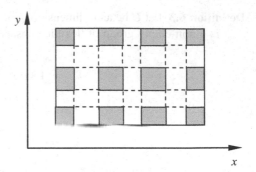

It can be easily seen that both $\pi_A = \{A_1, A_2, ..., A_m\}$ and $\pi_B = \{B_1, B_2, ..., B_n\}$ are flexible partitions of space U. Therefore, $A_1, A_2, ..., A_m$ and $B_1, B_2, ..., B_n$ can be two groups of orthogonal basic flexible linguistic values on space U.

2. **Square flexible partitioning**

   Let $U = U \times V$ be a two-dimensional measurement space, where $U = [a, b]$ and $V = [c, d]$. As shown in Fig. 6.3, we simultaneously divide flexibly U in two directions of x and y into $4 \times 3 = 12$ square flexible classes. The black squares in the figure are the cores of the flexible classes; the white areas between black squares are the public flexible boundaries of adjacent flexible classes; and the rectangles encircled by broken lines around every black square are the support set of the corresponding flexible classes. These 12 flexible classes stand for the 12 basic flexible linguistic values on U.

6.2.3 Extension and Reduction of Basic Flexible Linguistic Values

How many basic linguistic values should be defined on a measurement space is not definite and unchangeable. Sometimes, more are needed while other times only a few. As a matter of fact, the basic linguistic values on a universe of discourse can be extended or reduced totally by requirement. The extension and reduction of basic linguistic values can be realized by redefining, and also by inserting and deleting. With respect to a one-dimensional measurement space, there should be at least two basic linguistic values, which are mutually negative, and between which more basic linguistic values can be inserted if required to form a sequence of basic linguistic values. Conversely, for a sequence of basic linguistic values that forms a flexible partition of a space, those in the middle can unceasingly be deleted until only the first and the last two basic linguistic values left. Certainly, when adding or deleting basic linguistic values, the support sets and cores of the original one and the one remained should all be modified appropriately such that the current basic linguistic values can form a flexible partition of the universe of discourse.

6.3   Logical Operations on Flexible Linguistic Values on the Same Space

Logical operations on linguistic values on one and the same measurement space have conjunction, disjunction, and negation.

6.3.1   Conjunction and Disjunction

The mathematical essence of flexible linguistic values is flexible sets in corresponding measurement space. Therefore, the operations on flexible linguistic values are reduced to the operations on the corresponding flexible sets. The basic operations on flexible linguistic values are conjunction ($\wedge$) and disjunction ($\vee$), and according to the semantics, the operations on the corresponding flexible sets are intersection ($\cap$) and union ($\cup$). Then, from the membership functions of compound flexible sets $A \cap B$ and $A \cup B$ in Sect. 5.5, the consistency functions of the corresponding $A \wedge B$ and $A \vee B$ can be directly obtained.

Definition 6.4 Let A and B be two flexible linguistic values of feature $\mathscr{F}$ of objects, which is defined on n-dimensional measurement space U and whose consistency functions be $c_A(x)$ and $c_B(x)$. The $A \wedge B$ and $A \vee B$ connected by logical connectors $\wedge$ (conjunction) and $\vee$(disjunction) are separately called the conjunction and disjunction of A and B, whose consistency functions are as follows:

$$C_{A \wedge B}(x) = \min\{C_A(x), C_B(x)\} \quad x \in U \tag{6.1}$$

$$C_{A \vee B}(x) = \max\{C_A(x), C_B(x)\} \quad x \in U \tag{6.2}$$

On one-dimensional measurement space $U = [a, b]$, the graphs of the consistency functions of $A \wedge B$ and $A \vee B$ can be shown in Figs 6.4 and 6.5.

From the graphs, it can be seen that:

1. Conjunctive value $A \wedge B$ can have at most one region of support set, in which there may not be a region of core.
2. Disjunctive value $A \vee B$ may have two regions of support set and of core.

Fig. 6.4 An example of the graph of consistency function of $A \wedge B$

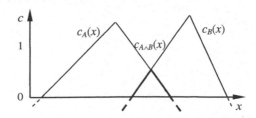

Fig. 6.5 An example of the
graph of consistency function
of $A \vee B$

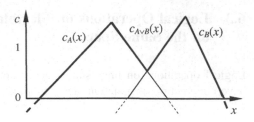

That shows that a set of flexible linguistic values on the same space is not close to the operations $\wedge$ and $\vee$ defined above. Therefore, strictly speaking, $\wedge$ and $\vee$ cannot be called the operations on flexible linguistic values on the same space. As a matter of fact, we seldom speak in this way in our daily communications.

The concepts of conjunctive values and disjunctive values on the same spaces can also be generalized to the case of n component linguistic values.

Definition 6.5 Let $A_1, A_2, \ldots, A_n$ be n flexible linguistic values of feature $\mathscr{F}$ of objects, which be defined on n-dimensional measurement space U, and whose consistency functions be $c_{A_1}(x), c_{A_2}(x), \ldots, c_{A_2}(x)(x \in U)$. $A_1 \wedge A_2 \wedge \cdots \wedge A_n$ and $A_1 \vee A_2 \vee \cdots \vee A_n$ are separately the conjunction and disjunction of these n flexible linguistic values, whose consistency functions are as follows:

$$c_{A_1 \wedge A_2 \wedge \ldots \wedge A_n}(x) = \min\{c_{A_1}(x), c_{A_2}(x), \ldots, c_{A_n}(x)\} \tag{6.3}$$

$$c_{A_1 \vee A_2 \vee \ldots \vee A_n}(x) = \max\{c_{A_1}(x), c_{A_2}(x), \ldots, c_{A_n}(x)\} \tag{6.4}$$

6.3.2 Negation

Likewise, according to the semantics of negative connective $\neg$, the flexible set to which $\neg A$ corresponds should be the complement A^c of flexible set A to which linguistic value A corresponds. Then, from the membership function of complement A^c in Sect. 5.5, the consistency function of $\neg A$ can be directly obtained.

Definition 6.6 Let A be a flexible linguistic value on n-dimensional measurement space U, whose consistency function be $c_A(x)$. $\neg A$ connected by logical connective $\neg$ (negation) is called the negation of A, whose consistency function is as follows:

$$c_{\neg A}(x) = 1 - c_A(x), \quad x \in U \tag{6.5}$$

Fig. 6.6 An example of the graph of consistency function of one-dimensional negative value ¬A

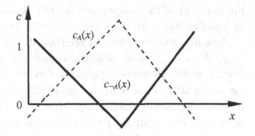

On one-dimensional measurement space $U = [a, b]$, the graph of the consistency function of the negation $\neg A$ of full-peak value A is shown in Fig. 6.6. The consistency function of negative value $\neg A$ on two-dimensional space $U \times V$ is as follows:

$$c_{\neg A}(x, y) = 1 - c_A(x, y) = 1 - c_A(x) \quad (x, y) \in U$$

namely

$$c_{\neg A}(x, y) = 1 - c_A(x) \quad = (x, y) \in U \tag{6.6}$$

The graph is shown in Fig. 6.7.
From Eq. (6.5), we have

$$c_A(x) + c_{\neg A}(x) = 1, \quad x \in U \tag{6.7}$$

That is, the sum of the consistency-degrees of an object having two relatively negative flexible linguistic values is 1. This shows that there is also complementation relation between consistency-degrees of relatively negative flexible linguistic values. We call Eqs. (6.6) and (6.7) to be the **complement law of consistency-degrees**.

Fig. 6.7 An example of the graph of consistency function of two-dimensional negative value ¬A

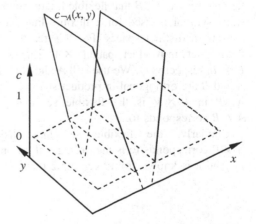

Further, we call the complement law of consistency-degrees and the complement law of membership-degrees in Sect. 5.5 to be the **complement law of degrees**.

Actually, the complement relation of degrees is also consistent with our intuition. We know that an object always has a flexible linguistic value A in a certain degree, which means that it also has flexible linguistic value $\neg A$ in another degree at the same time. That is to say, an object always has a pair of relatively negative flexible values simultaneously. Moreover, there is such a relation between the degrees of two relatively negative flexible linguistic values of one object: If the one is increasing, then the other is decreasing and vice versa.

6.4 Compositions and Decompositions of Flexible Linguistic Values on Distinct Spaces

6.4.1 Logical Composition and Decomposition, Combined Linguistic Values

The flexible linguistic values on distinct spaces can also perform operations to form a composite flexible linguistic value. For instance, "tall and big," "healthy and beautiful," and "knowledgeable or experienced" are just the flexible linguistic values composed by flexible linguistic values on distinct spaces.

6.4.1.1 Logical Composition of Flexible Linguistic Values

Let A and B be separately flexible linguistic values on measurement spaces U and V. Then, what shapes are the flexible sets that conjunctive flexible linguistic value $A \wedge B$ and disjunctive flexible linguistic value $A \vee B$ correspond to?

Obviously, $A \wedge B$ is not a flexible linguistic value on measurement spaces U or V, but should be a flexible linguistic value on product space $U \times V$. However, the flexible sets A and B that flexible linguistic values A and B correspond to (here the same symbol is used for a flexible linguistic value and its flexible set) are the subsets in distinct spaces. For this reason, A and B must be extended into the flexible sets in product space $U \times V$. That is, A and B be extended into $A \times V$ and $U \times B$, respectively. We use still A and B to denote $A \times V$ and $U \times B$. Obviously, A and B are orthogonal in product space $U \times V$. Thus, the orthogonal intersection $A \cap B$ in $U \times V$ is the flexible set that conjunctive flexible linguistic value $A \wedge B$ corresponds to.

Similarly, the flexible set that disjunctive flexible linguistic value $A \vee B$ corresponds to is the orthogonal union $A \cup B$ in product space $U \times V$.

And it is known from Sect. 5.5.2 that

$$m_{A \cap B}(x,y) = \min\{m_A(x), m_B(y)\}, \quad x \in U, y \in V$$
$$m_{A \cup B}(x,y) = \max\{m_A(x), m_B(y)\}, \quad x \in U, y \in V$$

Thus, the consistency functions of $A \wedge B$ and $A \vee B$ are as follows:

$$c_{A \wedge B}(x,y) = \min\{c_A(x), c_B(y)\}, \quad x \in U, y \in V \tag{6.8}$$

$$c_{A \vee B}(x,y) = \max\{c_A(x), c_B(y)\}, \quad x \in U, y \in V \tag{6.9}$$

Their graphs are shown in Fig. 6.8a, b, and the shape of the former is a square-tapered surface and the latter an orthogonal ridged surface.

Generally, we give the following definition.

Definition 6.7 Let $A_1, A_2, \ldots, A_n$ be separately flexible linguistic values of features $\mathscr{F}_1, \mathscr{F}_2, \ldots, \mathscr{F}_n$ of objects, which be separately defined on measurement spaces $U_1, U_2, \ldots, U_n$. Conjunctive value $A_1 \wedge A_2 \wedge \ldots \wedge A_n = C_a$ and disjunctive value $A_1 \vee A_2 \vee \ldots \vee A_n = C_o$ connected by logical connectives "and" ($\wedge$) and "or" ($\vee$), respectively, are the flexible linguistic values on product space $U = U_1 \times U_2 \times \cdots \times U_n$. We refer to C_a and C_o as the combined linguistic value of $A_1, A_2, \ldots, A_n$ or, simply, combined value, whose consistency functions are as follows:

$$c_{C_a}(x) = \min\{c_{A_1}(x_1), c_{A_2}(x_2), \ldots, c_{A_n}(x_n)\} \tag{6.10}$$

$$c_{C_o}(x) = \max\{c_{A_1}(x_1), c_{A_2}(x_2), \ldots, c_{A_n}(x_n)\} \tag{6.11}$$

where $x = (x_1, x_2, \ldots, x_n)$, $x_i \in U_i$ ($i = 1, 2, \ldots, n$), and $A_1, A_2, \ldots, A_n$ are called the component values of their combined value.

A combined value can be renamed, or not be renamed, but be said "A_1 and A_2 and ... and A_n" and "A_1 or A_2 or ...or A_n" separately. As a matter of fact, some combined linguistic values have already been renamed. For instance, people call the stature of "tall and big" to be "robust" and call "thin and small" to be "slim."

Now, we see that a combined linguistic value is formed by several flexible linguistic values on distinct spaces through logical operations conjunctive or disjunctive. We call this phenomenon the logical composition of flexible linguistic values.

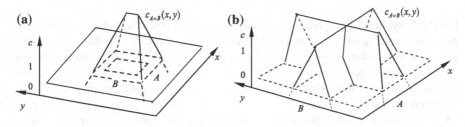

Fig. 6.8 Graphs of the consistency functions of $A \wedge B$ and $A \vee B$

6.4.1.2   Logical Decomposition of a Flexible Linguistic Value

Viewed backward the logical composition of flexible linguistic values, then, it is that a flexible linguistic value is decomposed into the conjunction or disjunction of several other flexible linguistic values. In fact, some flexible linguistic values themselves just can be decomposed or unfolded into the conjunction or disjunction of multiple flexible linguistic values. For instance, suppose that there are two sports events A and B. If it is thought that a sportsman is excellent only when the results of events A and B are both good, which is tantamount to "excellence" being logically decomposed into the conjunction of "A good" and "B good," namely excellence = A good $\wedge$ B good, the consistency function is $c_{\text{excellent}}(u) = \min \{c_A(x), c_B(y)\} (x \in U, y \in V)$. But if it is thought that a sportsman is excellent when the result of one at least of events A and B is good, which is tantamount to "excellence" being logically decomposed into the disjunction of "A good" and "B good," namely excellence = A good $\vee$ B good, the consistency function is $c_{\text{excellent}}(u) = \max\{c_A(x), c_B(y)\} (x \in U, y \in V)$.

Definition 6.8 Let C be a flexible linguistic value of feature $\mathscr{F}$ of objects, which can be defined on measurement space V. If there exists flexible linguistic value $A_i \subset U_i$ $(i = 1, 2, \ldots, n)$ such that $C = A_1 \wedge A_2 \wedge \cdots \wedge A_n$, or $C = A_1 \vee A_2 \vee \cdots \vee A_n$, namely

$$c_C(x) = \min\{c_{A_1}(x_1), c_{A_2}(x_2), \ldots, c_{A_n}(x_n)\}, \quad x = (x_1, x_2, \ldots, x_n), x_i \in U_i \quad (6.12)$$

or

$$c_C(x) = \max\{c_{A_1}(x_1), c_{A_2}(x_2), \ldots, c_{A_n}(x_n)\}, \quad x = (x_1, x_2, \ldots, x_n), x_i \in U_i \quad (6.13)$$

then we say that C can be logically decomposed into the conjunction or disjunction of flexible linguistic values $A_1, A_2, \ldots, A_n$ on measurement spaces $U_1, U_2, \ldots, U_n$.

6.4.2   Algebraical Composition and Decomposition, Synthetic Linguistic Value

6.4.2.1   Algebraical Composition of Flexible Linguistic Values

For the "excellent sportsman" in example above, suppose the judging criterion is not "one scope good" or "various scopes all good," but is "synthesizing scope good," which is the weighted sum of various scopes. Then, the relation between "A good" and "B good" and "excellent" is not logical "and" or "or" at this time, but should be a kind of numerical "plus." Considering that the synthesizing scope is the weighted sum of various scopes, "excellent" also should be the weighted sum of "A good" and "B good," which is represented by using consistency functions is as follows: $c_{excellent}(u) = w_1 c_A(x) + w_2 c_B(y)$ $(x \in U, \ y \in V, \ w_1, \ w_2 \in [0, 1], \ w_1 + w_2 = 1)$. Thus, "plus" is also a kind of operation of flexible linguistic values.

Definition 6.9 Let $A_1, A_2, ..., A_n$ be separately flexible linguistic values of features $\mathscr{F}_1, \mathscr{F}_2, ..., \mathscr{F}_n$ of objects, which is defined on measurement spaces $U_1, U_2, ..., U_n$, respectively. A_1 plus A_2 plus ... plus A_n, denote $A_1 \oplus A_2 \oplus \cdots \oplus A_n$, is a flexible linguistic value, denoted S, on product space $U_1 \times U_2 \times \cdots \times U_n = U$, which is called synthesis of $A_1, A_2, ..., A_n$, and its consistency function is as follows:

$$c_S(x) = \begin{cases} \sum_{i=1}^{n} w_i c_{A_i}(x_i), & \text{when } c_{A_i}(x_i) > 0 \\ 0, & \text{else} \end{cases} \tag{6.14}$$

where $x = (x_1, x_2, ..., x_n)$, $x_i \in U_i$, $w_i \in (0,1)$, $\sum_{i=1}^{n} w_i = 1$

At this time, $A_1, A_2, ..., A_n$ are called the ingredients of synthetic value S.

Now, in the judging criterion of "synthesizing scope good," A good $\oplus$ B good = excellence.

Thus, multiple flexible linguistic values from distinct spaces can form a synthetic linguistic value by operation of "plus." Considering that this "plus," i.e., weighted sum, is a kind of algebraic operation; therefore, we call this composition the algebraical composition of flexible linguistic values. The algebraical composition of linguistic values means objects can be described more roughly by more general languages.

It is necessary to note that, as we know, black color and white color can be mixed into gray color, and three primary colors of red, green, and blue can be mixed into various colors. Besides, sour taste and sweet taste can be mixed to a kind of flavor which is both sour and sweet. Viewed from the angle of linguistic value, the mixes of these colors and flavors can also be viewed as a kind of composition of linguistic values; however, this kind of composite linguistic values is really a kind of mixed linguistic values (but no synthetic linguistic values). Of course, we can yet regard logically (non-physically) them as synthetic linguistic values.

From Eq. (6.14) above, the corresponding expression of membership function is as follows:

$$m_S(x) = \sum_{i=1}^{n} w_i m_{A_i}(x_i), \quad m_{A_i}(x_i) \neq 0, \quad x_i \in U_i, W_i \in (0,1), \sum_{i=1}^{n} w_i = 1$$

It can be seen that the expression just is the membership function of flexible product set $A_1 \times A_2 \times \cdots \times A_n$. This shows that flexible product $A_1 \times A_2 \times \cdots \times A_n$ is the denotative mathematical model of synthetic linguistic value $A_1 \oplus A_2 \oplus ... \oplus A_n$.

6.4.2.2 Algebraical Decomposition of a Flexible Linguistic Value

Viewed backward the algebraical composition of flexible linguistic values, then it is that a flexible linguistic value is decomposed into the weighted sum of several other flexible linguistic values. For instance, viewing backward equation, A good $\oplus$ B good = excellence, then that is "excellence" be decomposed into the "sum" of "A good" and "B good." In fact, some flexible linguistic values themselves are just synthetic-type linguistic value, so they can be decomposed into the "sum" of several ingredient values.

In practical problems, in order to describe objects more carefully and accurately, a relatively abstract linguistic value sometimes needs to be specifically decomposed into multiple sublinguistic values. For example, the flexible linguistic value of "high" (of teaching level) can just be decomposed into three flexible linguistic values of "master" (to content of the course), "appropriate" (of teaching methods), and "good" (of effect of teaching). For another example, the flexible linguistic value of "beautiful" (of looks) can be just decomposed into three flexible linguistic values of "regular" (of facial features), "proper" (of facial structure), and "bright and clean" (of skin color).

It can be seen that this kind of decomposition of linguistic value is not the above-mentioned logical decomposition. Because the linguistic values obtained from decomposition are the ingredient values of the original linguistic value, and the relation between ingredient values is synthesis, i.e., weighted sum rather than simple logical conjunction or disjunction. For instance, although the three flexible linguistic values of "mastered," "appropriate," and "good" collectively form "high," their importance to "high" (of teaching level) is not the same, so there should be different weights. Likewise, though the three ingredient values of "regular," "proper," and "bright and clean" collectively form "beautiful," their contributions to "beautiful" are different, so each has its own weighted coefficient.

Definition 6.10 Let S be a flexible linguistic value of feature $\mathscr{F}$ of objects, which can be defined on measurement space V. If there exits flexible linguistic value $A_i \subset U_i$, such that $S = A_1 \oplus A_2 \oplus \ldots \oplus A_n$, namely

$$c_S(\pmb{x}) = \sum_{i=1}^{n} w_i c_{A_i}(x_i), c_{A_i}(x_i) > 0 \qquad (6.15)$$

where $\pmb{x} = (x_1, x_2, \ldots, x_n)$, $x_i \in U_i$, $w_i \in (0,1)$, and $\sum_{i=1}^{n} w_i = 1$, then S is a synthetic linguistic value, we say that which can be algebraically decomposed into the weighted sum of flexible linguistic values $A_1, A_2, \ldots, A_n$.

Example 6.2 Suppose that "beautiful" (of looks) can be algebraically decomposed into "regular" (of facial features), "proper" (of facial structure), and "bright and clean" (of skin color). Suppose the weights of the three ingredient values to their synthesis "beautiful" be separately: 0.4, 0.35, and 0.25, and the consistency functions of "regular," "proper," and "bright and clean" are separately $c_{regular}(x)$, $c_{proper}(y)$, and $c_{bright\ and\ clean}(z)$. Then, from Eq. (6.15), the consistency function of "beautiful" is as follows:

$$c_{beautiful}(x, y, z) = 0.4\, c_{regular}(x) + 0.35\, c_{proper}(y) + 0.25\, c_{bright\ and\ clean}(z)$$

Thus, a flexible linguistic value can also be algebraically decomposed into the weighted sum of multiple flexible linguistic values on distinct spaces. The algebraical decomposition of a flexible linguistic value means that objects can be described more detail by more accurate flexible linguistic values.

Note that in daily language, people sometimes also use connective "and" but strictly use "plus" to describe a synthetic linguistic value. For example, original "A plus B" is said as "A and B." In addition, the understanding and convention above about the synthesis of linguistic values are supposing there must exist all ingredient values for a synthetic value, that is, the consistency-degrees of all ingredient values that participate in the synthesis should all be greater than 0. But if we suppose or agree that it can be regarded as a synthetic value if there is only one ingredient value, that is, at least one of the consistency-degrees of all ingredient values that participate in a synthesis is greater than 0, then the consistency function of the synthetic value is as follows:

$$c_S(\pmb{x}) = \sum_{i=1}^{n} w_i c_{A_i}(x_i),\ \pmb{x} = (x_1, x_2, \ldots, x_n),\quad x_i \in U_i,\ w_i \in (0, 1),\ \sum_{i=1}^{n} w_i = 1$$

(6.16)

But note that the flexible set that this consistency function corresponds to, strictly speaking, is already not product $\overset{n}{\underset{i=1}{\times}} A_i$.

6.5 Relatively Negative Flexible Linguistic Values and Medium Point

From the definition of a negative value, it can be seen that the negation of the linguistic value is just the linguistic value itself. That is to say, the negation is in fact mutual. Therefore, mutual negation is also a kind of relation between linguistic values. As a matter of fact, any linguistic value has its negation.

Further examining relatively negative flexible linguistic values, we find that on one-dimensional space [a, b], there are the following three kinds of relatively negative flexible linguistic values.

1. Two relatively negative flexible linguistic values on a universe. As shown in Fig. 6.9, let A be a semi-peak flexible linguistic value on space [a, b], supp (A) = [a, b]. Then, from the above-mentioned consistency functional expression

Fig. 6.9 Relatively negative relation between two values on universe

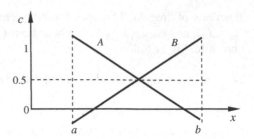

(6.5) of a negative value, negation $\neg A$ of A should be linguistic value B in Fig. 6.9, that is, $\neg A = B$. Conversely, the negation of B is just A, that is, $\neg B = A$. Thus, between A and B is just relatively negative relation. These two relatively negative flexible linguistic values are both semi-peak values, and they form a flexible partition of the universe.

2. Full-peak value and its negation. As shown in Fig. 6.10, let A be a full-peak flexible linguistic value on space $[a, b]$. Then, from the consistency functional expression (6.5) of a negative value, negation $\neg A$ of A should be as shown in Fig. 6.10. It is shown that the negation of a full-peak value is divided into two parts, and they are actually two values on space $[a, b]$. However, the flexible linguistic values having the relatively negative relation are not necessarily all basic flexible linguistic values.

3. Relatively negative relation between adjacent semi-peak values among basic flexible linguistic values.

Let $\pi = \{A_1, A_2, ..., A_n\}$ be a flexible partition of space $U = [a, b]$. Basic flexible linguistic values A_1, A_2, ..., A_n are adjacent one by one. Let the consistency functions of semi-peak values A_i^+ and A_{i+1}^- of A_i and A_{i+1} be separately as follows:

$$c_{A_i}(x) = \frac{s_{A_i}^+ - x}{s_{A_i}^+ - c_{A_i}^+}, \dots \xi_{Ai} \le x \le \xi_{Ai+1}$$

$$c_{A_{i+1}}(x) = \frac{x - s_{A_{i+1}}^-}{c_{A_{I+1}}^- - s_{A_{i+1}}^-}, \quad \xi_{Ai} \le x \le \xi_{Ai+1}$$

Fig. 6.10 Relatively negative relation between a full-peak value and its negation

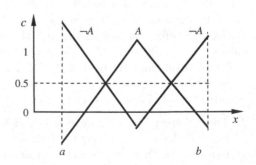

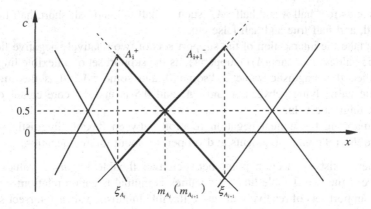

Fig. 6.11 Relatively negative relation between adjacent semi-peak values of basic flexible linguistic values

Their graphs are shown in Fig. 6.11.

From the figure, it can be seen that points $c_{A_i}^+$ and $s_{A_{i+1}}^-$ should be coincident, and $s_{A_i}^+$ and $c_{A_{i+1}}^-$ should be coincident, so $s_{A_i}^+ - c_{A_i}^+ = c_{A_{i+1}}^- - s_{A_{i+1}}^-$. Thus, for arbitrary $x \in [\xi_{Ai}, \xi_{Ai+1}]$,

$$c_{A_i}(x) + c_{A_{i+1}}(x) = \frac{s_{A_i}^+ - x}{s_{A_i}^+ - c_{A_i}^+} + \frac{x - s_{A_{i+1}}^-}{c_{A_{i+1}}^- - s_{A_{i+1}}^-} = \frac{s_{A_i}^+ - x}{s_{A_i}^+ - c_{A_i}^+} + \frac{x - c_{A_i}^+}{s_{A_i}^+ - c_{A_i}^+} = \frac{s_{A_i}^+ - c_{A_i}^+}{s_{A_i}^+ - c_{A_i}^+}$$
$$= 1$$

This shows that two adjacent basic flexible linguistic values A_i and A_{i+1} on space U, the positive semi-peak A_i^+ of the former A_i, and negative semi-peak A_{i+1}^- of the latter A_{i+1} are relatively negative on interval $[\xi_{Ai}, \xi_{Ai+1}]$.

The complement relation of consistency-degrees of relatively negative flexible linguistic values makes a pair of relatively negative flexible linguistic values be able to transform mutually. Thus, we can also unify or reduce a pair of relatively negative flexible linguistic values as one flexible linguistic value.

The complement relation of consistency-degrees has also another characteristic; that is, the consistency-degrees of an object with relatively negative flexible linguistic values are always symmetrical about 0.5. For two relatively negative linguistic values, this is just that the union of the range of values of the consistency functions of a pair of relatively negative flexible linguistic values is symmetrical about 0.5. We can visually see that from the above Figs. 6.9, 6.10, and 6.11. Whereas the numerical value that 0.5 corresponds to is just median points m_A, that also is, $m_{\neg A}$, of the common boundary of two relatively negative flexible linguistic values. That is to say, a median point has both two relatively negative flexible linguistic values with the same consistency-degree of 0.5. Thus, we call a median point the **medium point** between relatively negative flexible linguistic values, which stands for "half A and half $\neg A$," such as half tall and half short, half hot and half cold, and half true and half false etc.

If we take the intersection of the support sets of two relatively negative flexible linguistic values, e.g., supp(A) $\cap$ supp($\neg A$), as the support set of a flexible linguistic value, then this linguistic value is "some A and some $\neg A$." It is the **medium linguistic value** lying between A and $\neg A$, and 0.5 is just the core center of this medium linguistic value.

Summarizing the mutual negation relation between flexible linguistic values, there are the following judgments and properties about mutual negation:

- If there exists a medium point between two flexible linguistic values, then between the two flexible linguistic values is mutual negation relation;
- The support sets of relatively negative flexible linguistic values intersect surely; the intersection part is the common boundary of the two flexible linguistic values; and the center point of this common boundary region is the medium point between the two relatively negative flexible linguistic values.
- Two flexible classes are relatively negative if and only if there is a transition zone which is this and also is that between their cores.

6.6 Complementary Flexible Partition of a Space and Complementary Flexible Linguistic Values

6.6.1 Complementary Flexible Partition of a Space, Complementary Flexible Classes and Complementary Flexible Linguistic Values

From Definitions 6.1 and 6.3, it can be seen that flexible partition is actually a kind of complementary partition of a space, so between the flexible classes obtained is complementation relation, and between the corresponding flexible linguistic values also is complementation relation.

Definition 6.11 Let U be an n-dimensional measurement space, and let $A_1, A_2, ..., A_m$ be non-empty flexible subsets of U. If $\pi = \{A_1, A_2, ..., A_m\}$ is a flexible partition of U, then π is called a complementary flexible partition of space U, flexible classes $A_1, A_2, ..., A_m$ are called to be complementary, or say the corresponding flexible linguistic values $A_1, A_2, ..., A_m$ are complementary.

More simply, complementary flexible partition is that between the cores of two adjacent flexible sets there is a median point (line or plane) formed by medium points that are also this and also that. For instance, the two partitions in Fig. 6.12 are both complementary flexible partitions (the white broken line in the figures is the median line of adjacent flexible classes), and between the corresponding flexible

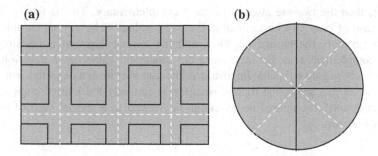

Fig. 6.12 Examples of complementary flexible partition, complementary flexible classes, and complementary flexible linguistic values. **a** Square space, **b** Circular space

classes and between the corresponding flexible linguistic values, all are complementation relation, respectively.

Comparing Definition 6.11 and Definition 6.3, it can be seen that the flexible linguistic values $A_1, A_2, \ldots, A_m$ on space U are of complementation relation if and only if $A_1, A_2, \ldots, A_m$ forms a group of basic flexible linguistic values on U.

Example 6.3 Suppose "small," "medium," and "large" be a group of basic flexible linguistic values on one-dimensional space [0, 100], then they are of complementation relation. If only to define two basic flexible linguistic values of "small" and "large" on space [0, 100], then "small" and "large" are also of complementation relation (of course, meantime, they are also of mutual negation relation).

6.6.2 Relationship Between Mutual Complementation and Mutual Negation

From the definition and examples of the complementation relation, it can be seen that the complementation relation and mutual negation relation between flexible linguistic values have a certain connection. As a mater of fact, the two have both something in common and difference.

What they are in common is there is a medium point (line or plane) between two relatively negative flexible linguistic values, while there is a medium point (line or plane) between complementary flexible linguistic values pairwise.

The difference between them is as follows: Mutual negation is only for two linguistic values and it is a relation between two linguistic values, while mutual complementation can be for more than two linguistic values, which is a relation among multiple linguistic values.

Thus, generally speaking, mutual complementation is not mutual negation. However, if there are only two basic flexible linguistic values which are mutually complementary on a universe, then the two are also of relatively negative. Conversely, if there are two relatively negative basic flexible linguistic values on a universe, then the two are also of mutually complementary. This is to say, in the special case of only two basic flexible linguistic values" The mutual complementation is certainly the mutual negation, and the mutual negation is certainly the mutual complementation. Besides, since any linguistic value has a negation, and two relatively negative flexible linguistic values can also form a flexible partition of a universe, in this sense, the mutual negation is certainly the mutual complementation. In a word, the mutual negation can be viewed as a special kind of mutual complementation.

6.7   Relations Between Flexible Linguistic Values

Mutual negation and mutual complementation are two relations among flexible linguistic values; besides, there are also some other relations among flexible linguistic values.

6.7.1   Order and Position

We use the order relation between peak value points of flexible linguistic values to make definite the order relation between corresponding flexible linguistic values.

Definition 6.12 Let A and B be two flexible linguistic values on one and the same numerical range $[a, b]$, and ξ_A and ξ_B be separately peak value points of A and B. Then, $A < B$ $(A > B)$ if and only if $\xi_A < \xi_B$ $(\xi_A > \xi_B)$. Where $A < B$ indicates that A is prior to B, or A is less than B, $A > B$ indicates that A is behind B, or A is greater than B.

Example 6.4 As shown in Fig. 6.13, "low," "medium," and "high" are three adjacent flexible linguistic values on range [0.5, 2.5] of human's heights, and from Definition 6.12, their order is "low" < "medium" < "high."

Similarly, we use the position of peak value point of a flexible linguistic value to make definite the position of corresponding flexible linguistic value.

Fig. 6.13 An example of order of relation between flexible linguistic values

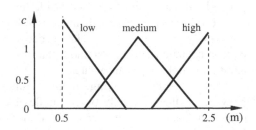

Definition 6.13 Let A be a flexible linguistic value on n-dimensional measurement space U, and ξ_A be peak value points of A, of which the coordinate is $(a_1, a_2, \ldots, a_n)$. Then, the coordinate $(a_1, a_2, \ldots, a_n)$ is also the position of the flexible linguistic value A in space U.

6.7.2 Composition–Decomposition Relation and Category–Subordination Relation

Speaking from the constitution of a linguistic value, we call relation between a combined value and its component values and that between a synthetic value and its ingredient values collectively to be the **composition–decomposition relation**.

Conceptually speaking, between flexible linguistic values there is a **category–subordination relation**. That is, relatively speaking, some flexible linguistic values are category concepts or higher level concepts, while some are subordinate concepts or lower level concepts. A subordinate concept is derived from corresponding category concept, and the category concept is a father concept or basic concept, while the subordinate concept is a son concept or a more special concept. For instance, flexible triangle is a category concept of flexible right-angled triangle; conversely, flexible right-angled triangle is a subordinate concept of flexible triangle. Similarly, flexible right-angled triangle and flexible right isosceles triangle are also of the category–subordination relation. Category–subordination relation is also called the derivative relation or the generalization relation.

6.7.3 Inclusion Relation and Same-Level Relation

Definition 6.14 Let A and B be flexible linguistic values on the same measurement space U. If the corresponding flexible set A is contained in flexible set B, then flexible linguistic value A is called to be contained in flexible linguistic value B, or B contains A.

Definition 6.15 Let A and B be flexible linguistic values on the same measurement space U. If A does not contain B and B does not contain A either, then we say flexible linguistic values A and B are same-level.

Example 6.5 What Fig. 6.14 shows is several flexible linguistic values having inclusion relation or same-level relation. From the figure, it can be visually seen that A_1 contains A_2, B_1 contains B_2, C_1 contains C_2 and C_3, and C_2 contains C_3, while A_1, B_1, and C_1 are of same-level.

From this example, it can be seen that the curves of consistency functions of the linguistic values having inclusion relation are not necessarily parallel.

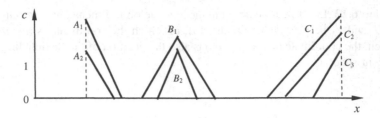

Fig. 6.14 Examples of same-level relation and inclusion relation between flexible linguistic values

6.8 Similarity and Approximation About Flexible Linguistic Values

6.8.1 Similarity and Approximation Relations Between Atom Flexible Linguistic Values and the Corresponding Measures

(1) Definition of the similarity and approximation relations between atom flexible linguistic values

Definition 6.16 Let A and B be atom flexible linguistic values on one and the same measurement space. A and B are similar, if and only if the width of each boundary of A and the width of corresponding boundary of B are equal.

It can be verified that the similarity relation between flexible linguistic values is an equivalence relation.

From the graph of consistency function, it is not hard to see that the similarity of one-dimensional flexible linguistic values A and B means the slopes of curves of consistency functions $c_A(x)$ and $c_B(x)$ are the same correspondingly.

Definition 6.17 Let A and B be atom flexible linguistic values on one-dimensional space $U = [a, b]$. A is approximate to B, if and only if

(i) A and B is similar;
(ii) $widt(\text{supp}(A)) = widt(\text{supp}(B))$, $widt(\text{core}(A)) = widt(\text{core}(B))$;
(iii) $\xi_A \in \text{core}(B)^+$, that is, $c_B(\xi_A) > 0.5$.

Example 6.6 Flexible linguistic value A shown in Fig. 6.15(a) is approximate to flexible linguistic value B, but flexible linguistic value A shown in Fig. 6.15(b) is not approximate to flexible linguistic value B.

It is not hard to see that the approximation relation between linguistic values does not satisfy symmetry.

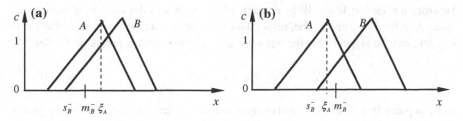

Fig. 6.15 Examples of the approximation relation between flexible linguistic values. **a** A is approximate to B, **b** A is not approximate to B

(2) Measure of the approximation of atom flexible linguistic values

For the convenience of narrating, in the following we use $d(*,*)$ to denote distance, where $*$ can be a point in a measurement space, or it can be a flexible linguistic value or its core, extended core or support set.

Definition 6.18 Let A and B be atom flexible linguistic values on one-dimensional space $U = [a, b]$, and set

$$d(A, B) = d(\xi_A, \xi_B)$$

to be called the distance between A and B, or the distance between their cores, between their extended cores or that between their support sets.

By this definition, for one-dimensional atom flexible linguistic values A and B, it follows that

$$d(A, B) = d(\text{core}(A), \text{core}(B)) = d(\text{core}(A)^+, \text{core}(B)^+) = d(\text{supp}(A), (\text{supp}(B))$$

By Definitions 6.18 and 6.17, we have the following theorem.

Theorem 6.1 For one-dimensional atom flexible linguistic values A and B, if A is approximate to B, then distance $d(\xi_A, \xi_B) = d(c_A^*, c_B^*) = d(s_A^*, s_B^*)$.

Generalizing Definition 6.18, we have the following definition.

Definition 6.18′ Let A and B be atom flexible linguistic values on n-dimensional measurement space U, and set

$$d(A, B) = d(\xi_A, \xi_B)$$

to be called the distance between flexible linguistic values A and B, or the distance between their cores, between their extended cores, or that between their support sets.

Since inclusion is not necessarily similar, inclusion is not necessarily approximate either.

Definition 6.19 Let W and W' be atom flexible linguistic values on one-dimensional space $U = [a, b]$, ξ_W and $\xi_{W'}$ be, respectively, the peak value points of W and W', and m_W^- and m_W^+ be respectively the negative and positive median points of W. Set

$$r_W^- = \xi_W - m_W^-, r_W^+ = m_W^+ - \xi_W \qquad (6.17)$$

to be separately called the negative approximate radius and the positive approximate radius of W;

$$D_{WW'} = \frac{d_{WW'}}{r_W^-} = \frac{\xi_W - \xi_{W'}}{\xi_W - m_W^-}, D_{WW'} = \frac{d_{WW'}}{r_W^+} = \frac{\xi_{W'} - \xi_W}{m_W^+ - \xi_W} \qquad (6.18)$$

to be called the difference-degree of linguistic values W' and W. If W' is approximate to W, then we say that

$$s_{WW'} = 1 - D_{WW'} \qquad (6.19)$$

is the approximation-degree of linguistic values W' and W.

It can be seen that what the difference-degree represents is the relative difference between peak value points $\xi_{W'}$ and ξ_W, while approximate radii r_W^- and r_W^+ of linguistic values W are just two unit distances of this relative difference. The reason why we take they as the unit distances is that, on the one hand, the two distances are, respectively, the maximum distance from all points at the two side of peak value point ξ_W in the extended core of flexible linguistic value W to the peak value point; on the other hand, from the previous definition of the approximation of flexible linguistic values, when linguistic value W' is approximate to linguistic value W, peak value point $\xi_{W'}$ of W' should fall within the extended core of W. Thus, the distance between two approximate flexible linguistic values will never exceed that maximum distance, so to take them as the unit distances is appropriate.

Relative difference reflects the degree of a difference, which can be independent of universes of discourse. Therefore, the relative differences between flexible linguistic values in different universes of discourse are completely comparable. From the meaning of the relative difference, we refer to relative difference between flexible linguistic values as the difference-degree of flexible linguistic values. While according to the definition of the distance between linguistic values, the distance between two approximate linguistic values will never exceed the approximate radius of each of the two linguistic values. Thus, the difference-degree of two approximate linguistic values should be between 0 and 1. Further, the approximation-degree should also be between 0 and 1.

Generalizing Definition 6.19 to n-dimensional atom flexible linguistic values, it is the following definition.

Definition 6.19′ Let W and W' be atom flexible linguistic values on n-dimensional measurement space U, ξ_W and $\xi_{W'}$ be, respectively, the peak value points of W and W'. Take

$$r_W = \min_{x \in c_W} d(\xi_W, x) \; (c_W \text{ is the median plane of } W)$$

as the approximate radius of W, and set

$$D_{WW'} = \frac{d_{WW'}}{r_w} \tag{6.20}$$

to be called the difference-degree of flexible linguistic values W' and W. If W' is approximate to W, then we say

$$s_{WW'} = 1 - D_{WW'} \tag{6.21}$$

is the approximation-degree of flexible linguistic values W' and W.

(3) Reduction and orientation of approximation

① **Reduction from full-peak-valued approximation to semi-peak-valued approximation**

Let A and A' be two full-peak flexible linguistic values on one-dimensional measurement space U, suppose A' is approximate to A from the negative side. Now, we divide A into two semi-peak values A^- and A^+. By the definition of approximation of flexible linguistic values, it should follow that $\xi_{A'} \in X_A = X_A^- \cup X_A^+$, so $\forall x \in X_{A'}^+$, always $x \in X_A^- \cup X_A^+$, and therefore, $X_{A'}^+ \subset X_A^- \cup X_A^+$. Thus, for the positive semi-peak value $A^{+\prime}$ of A', there would not exist the problem of two semi-peak A^- and A^+ being approximate to A. On the other hand, since A' is approximate to A from the negative side, so $\xi_{A'} < \xi_A$, this shows that $X_{A'}^- \not\subset X_A^- \cup X_A^+$. Therefore, the negative semi-peak value $A^{-\prime}$ of A' is just approximate to the negative semi-peak value A^- of A. The analysis above shows that full-peak value A' which is approximate to full-peak value A from the negative side actually is tantamount to its semi-peak value $A^{-\prime}$ to be approximate to the negative semi-peak value A^- of A from the negative side. Similarly, that full-peak value A' which is approximate to full-peak value A from the positive side is actually tantamount to its semi-peak value $A^{+\prime}$ to be approximate to the positive semi-peak value A^+ of A from the positive side. We say this fact to be the reduction from full-peak-valued approximation to semi-peak valued approximation, whose process is shown in Fig. 6.16.

② **The orientation of approximating to a semi-peak value**

From the definition of approximation relation of flexible linguistic values, it is not hard to see that to approximate to a positive semi-peak linguistic value can only be done from one side of the positive direction, while to approximate to a negative semi-peak linguistic value can only be done from one side of negative direction (as

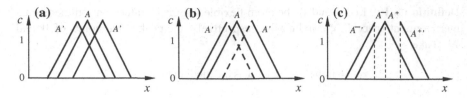

Fig. 6.16 Illustration of reduction from full-peak-valued approximation to semi-peak-valued approximation. (It is from **a** to **b** to **c**)

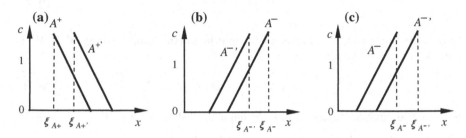

Fig. 6.17 Examples of the orientation of approximating to a semi-peak value where (a) and (b) are correct, while (c) is wrong

shown in Fig. 6.17(a) and (b)). In fact, for negative semi-peak value A^-, if there is a semi-peak value A^- being located at positive side of A^-, as shown in Fig. 6.17(c), then its peak value point is $\xi_{A^-} \notin \text{core}(A^-)^+$ surely, but this does not accord with Definition 6.5. Actually, if semi-peak value A^- is a boundary value, then it is obvious that at its one side of the positive direction, there would not exist a linguistic value which is same-level with A^-; if A^- is the negative semi-peak value of full-peak value A, then according to the reduction of approximation of a full-peak value in the above, at the side of the positive direction of A^-, the linguistic value close to A^- should be the approximate value $A^{+'}$ of positive semi-peak value A^+. The above situation is also analogues to positive semi-peak value A^+.

6.8.2 Similarity and Approximation Relations Between Composite Flexible Linguistic Values and the Corresponding Measures

(1) **Definition of the similarity and approximation relations between composite flexible linguistic values**

Definition 6.20 Let A and B be composite flexible linguistic values on one and the same space. A and B are similar, if and only if

(i)   The structures of A and B are same;
(ii) The corresponding component values or ingredient values of A and B are similar, respectively.

$$widt(\text{supp}(A_i')) = widt(\text{supp}(A_i)), widt(\text{core}(A_i')) = widt(\text{core}(A_i));$$

(iii) $\min_{i \in \{1,2,\ldots,n\}} c_{A_i}(\xi_{A_i'}) > 0.5$, that is, for $\forall i \in \{1, 2, \ldots, n\}$, always $\xi_{A_i'} \in \text{core}$
$(A_i)^+$.

Definition 6.21 Let A and A' be conjunctive flexible linguistic values on n-dimensional product space $U = U_1 \times U_2 \times \cdots \times U_n$. A' is approximate to A, if and only if

(i) A' and A are similar;
(ii) The corresponding component values A_i' and A_i $(i = 1, 2, \ldots, n)$ of A' and A satisfy

$$widt(\text{supp}(A_i')) = widt(\text{supp}(A_i)), widt(\text{core}(A_i')) = widt(\text{core}(A_i));$$

(iii) $\min_{i \in \{1,2,\ldots,n\}} c_{A_i}(\xi_{A_i'}) > 0.5$, that is, for $\forall i \in \{1, 2, \ldots, n\}$, always $\xi_{A_i'} \in \text{core}$
$(A_i)^+$.

Definition 6.22 Let A and A' be disjunctive flexible linguistic values on n-dimensional product space $U = U_1 \times U_2 \times \cdots \times U_n$. A' is approximate to A, if and only if

(i) A' and A are similar;
(ii) The corresponding component values A_i' and A_i $(i = 1, 2, \ldots, n)$ of A' and A satisfy

$$widt(\text{supp}(A_i')) = widt(\text{supp}(A_i)), widt(\text{core}(A_i')) = widt(\text{core}(A_i));$$

(iii) $\max_{i \in \{1,2,\ldots,n\}} c_{A_i}(\xi_{A_i'}) > 0.5$, that is, $\exists i \in \{1, 2, \ldots, n\}$, such that $\xi_{A_i'} \in \text{core}(A_i)^+$.

Definition 6.23 Let A and A' be synthetic flexible linguistic values on n-dimensional product space $U = U_1 \times U_2 \times \cdots \times U_n$. A' is approximate to A, if and only if

(i) A' and A are similar;
(ii) The corresponding ingredient values A_i' and A_i $(i = 1, 2, \ldots, n)$ of A' and A satisfy

$$widt(\operatorname{supp}(A_i')) = widt(\operatorname{supp}(A_i)), widt(\operatorname{core}(A_i')) = widt(\operatorname{core}(A_i));$$

(iii) $\displaystyle\sum_{i=1}^{n} w_i c_{A_i}(\xi_{A_i'}) > 0.5.$

(2) **Measure of approximation of composite flexible linguistic values**

Since a composite linguistic value is not like an atom linguistic value obtained directly by flexible clustering on measurement space, but is composed by flexible linguistic values from different measurement spaces through logical operation or algebraic operation, so we cannot directly define the distance between composite linguistic values on the corresponding product space, and then using the distance to define difference-degree and approximate-degree. However, just because a composite linguistic value is composed by flexible linguistic values from different measurement spaces through logical operation or algebraic operation, the approximation-degree of the composite flexible linguistic value should be the result of logical operation or algebraic operation of the approximation-degrees of their component values or ingredient values. According to this understanding, we give the definition of the approximation-degree of composite flexible linguistic values and then derive backward the computation formulas of their difference-degree and distance.

Definition 6.24 Let A and A' be conjunctive flexible linguistic values on n-dimensional product space $U = U_1 \times U_2 \times \cdots \times U_n$, and A' be approximate to A. Set

$$s_{A'A} = \min\{s_{A_1'A_1}, s_{A_2'A_2}, \ldots, s_{A_n'A_n}\} \tag{6.22}$$

to be called the approximation-degree of conjunctive flexible linguistic values A' and A.

Since

$$s_{A_i'A_i} = 1 - D_{A_i'A_i} \quad (i = 1, 2, \ldots, n),$$

therefore, $s_{A_i'A_i} = \min\{s_{A_1'A_1}, s_{A_2'A_2}, \ldots, s_{A_n'A_n}\} s_{A_i'A_i}$ is   equivalent   to $D_{A_k'A_k} = \max\{D_{A_1'A_1}, D_{A_2'A_2}, \ldots, D_{A_n'A_n}\}$. Thus, the difference-degree $D_{A'A}$ of A' and A is just

$$D_{A'A} = \max\{D_{A_1'A_1}, D_{A_2'A_2}, \ldots, D_{A_n'A_n}\} \tag{6.23}$$

Thus, when finding $s_{A'A}$, we can firstly find difference-degree $D_{A'A}$, then from $1 - D_{A'A}$, obtaining $s_{A'A}$, that is,

$$s_{A'A} = 1 - D_{A'A} = 1 - \max\{D_{A'_1A_1}, D_{A'_2A_2}, \ldots, D_{A'_nA_n}\} \tag{6.24}$$

and from

$$D_{A'_iA_i} = \frac{d_{A'_iA_i}}{r_{A_i}} \quad (i = 1, 2, \ldots, n)$$

so $d_{A'_kA_k}$ in max $\{\frac{d_{A'_1A_1}}{r_{A_1}}, \frac{d_{A'_2A_2}}{r_{A_2}}, \ldots, \frac{d_{A'_nA_n}}{r_{A_n}}\} = \frac{d_{A'_kA_k}}{r_{A_k}}$ is just distance $d_{A'A}$ between A' and A.

Definition 6.25 Let A and A' be disjunctive flexible linguistic values on n-dimensional product space $U = U_1 \times U_2 \times \cdots \times U_n$, and A' be approximate to A. Set

$$s_{A'A} = \max\{s_{A'_1A_1}, s_{A'_2A_2}, \ldots, s_{A'_nA_n}\} \tag{6.25}$$

to be called the approximation-degree of disjunctive flexible linguistic values A' and A.

Similarly, the difference-degree of disjunctive flexible linguistic values A' and A is as follows:

$$D_{A'A} = \min\{D_{A'_1A_1}, D_{A'_2A_2}, \ldots, D_{A'_nA_n}\} \tag{6.26}$$

Thus, approximation-degree is as follows

$$s_{A'A} = 1 - D_{A'A} = 1 - \min\{D_{A'_1A_1}, D_{A'_2A_2}, \ldots, D_{A'_2A_2}\} \tag{6.27}$$

While $d_{A'_kA_k}$ in min $\{D_{A'_1A_1}, D_{A'_2A_2}, \ldots, D_{A'_nA_n}\} = \min\{\frac{d_{A'_1A_1}}{r_{A_1}}, \frac{d_{A'_2A_2}}{r_{A_2}}, \ldots, \frac{d_{A'_nA_n}}{r_{A_n}}\} = \frac{d_{A'_kA_k}}{r_{A_k}}$ is just distance $d_{A'A}$.

Definition 6.26 Let A and A' be synthetic flexible linguistic values on n-dimensional product space $U = U_1 \times U_2 \times \cdots \times U_n$, and A' be approximate to A. Set

$$s_{A'A} = w_1 s_{A'_1A_1} + w_2 s_{A'_2A_2} + \ldots + w_n s_{A'_nA_n}, \quad \sum_{i=1}^{n} w_i = 1 \tag{6.28}$$

to be called the approximation-degree of synthetic flexible linguistic values A' and A.

From this definition, it is not hard to derive that the difference-degree of synthetic flexible linguistic values A' and A is as follows:

$$D_{A'A} = w_1 D_{A'_1 A_1} + w_{2D_{A'_2 A_2}} + \ldots + w_n D_{A'_n A_n} \tag{6.29}$$

Thus, approximation-degree is as follows:

$$s_{A'A} = 1 - D_{A'A} = 1 - \sum_{i-1}^{n} w_i D_{A'_I A_I} \tag{6.30}$$

However, distance $d_{A'A}$ cannot be obtained.

6.9 Summary

In this chapter, we found the fundamental theory of flexible linguistic values. First, we introduced the types of flexible linguistic values, then analyzed and defined the operations on flexible linguistic values, and, in particular, proposed the concepts and methods of algebraic composition and decomposition of flexible linguistic values. Meanwhile, we also analyzed the properties and relations of relatively negative linguistic values, and then proposed the complementary partition of a measurement space and the complementary relation of flexible linguistic values. Besides, we also considered other relations between flexible linguistic values, especially analyzed and defined the approximation relation between flexible linguistic values, and presented the corresponding measuring method.

The main points and the results of the chapter are as follows:

- There are many types of flexible linguistic values, among which the most frequently mentioned are atomic flexible linguistic values and composition flexible linguistic values.
- Operations on flexible linguistic values can be classified as the operation on flexible linguistic values on the same space (i.e., of the same feature) and the operation on flexible linguistic values on distinct spaces (i.e., of distinct features). The former has "conjunction," "disjunction," and "negation," and the corresponding set operations are intersection, union, and complement, while the latter has logical composition and algebraic composition. The logical composition includes "conjunction" and "disjunction"; the corresponding flexible set operations are orthogonal intersection and orthogonal union; and the composition values obtained are called the combined value. Algebraic composition is the weighted sum (which is called synthesis); the corresponding flexible set operation is Cartesian product; and the composition value obtained is called the synthetic value. This means that the synthetic value of flexible linguistic values corresponds to the Cartesian product of flexible sets, and the latter is the denotative mathematical model of the former.
- Any flexible linguistic value has its negation, and the sum of the consistency-degrees of an object having a pair of relatively negative flexible

linguistic values is 1. This relation is called the complement law of consistency-degrees. There is one and only one medium point (line or plane) of "half-this and half-that" between the flexible sets to which a pair of relatively negative flexible linguistic values corresponds. Such two flexible sets form a relatively negative partition of the corresponding measurement space. The generalization of relatively negative partition is complementary partition; that is, there is one and only one medium point (line or plane) of "half-this and half-that" between all two adjacent flexible sets in a space. A usual flexible partition is actually the complementary partition, and the relation between the corresponding flexible linguistic values is complementation relation, which form a group of complementary basic flexible linguistic values on the measurement space.

- Flexible linguistic values also have the relations of composition–decomposition, category–subordination, same-level, inclusion, similarity, approximation, and order.
- To approximate to a full-peak value can be reduced to approximate to a semi-peak value, and to approximate to a semi-peak value can only be done from one side of the direction of the peak value. The approximation-degree between two atomic flexible linguistic values is computed by directly using formula with distance, while the approximation-degree between two composite flexible linguistic values is a certain operation of the approximation-degrees between the corresponding component or ingredient linguistic values.

Reference

1. Lian S (2009) Principles of Imprecise-Information Processing. Science Press, Beijing

Chapter 7
Superposition, Quantification, Conversion, and Generalization of Flexible Linguistic Values

Abstract This chapter is the continuation of the basic theory of flexible linguistic values. First, the concepts of degree linguistic values and superposed linguistic values are proposed, and the types, levels, and mathematical models of superposed linguistic values are discussed. Next, the quantification of flexible linguistic values is considered, and the flexible linguistic value with degree and its notation is presented. And then, the relations between pure linguistic values, flexible linguistic values with degrees and numerical values are expounded, and the corresponding conversion principles and methods are presented. Lastly, one-dimensional flexible linguistic values are generalized to vector flexible linguistic values and flexible linguistic-valued vectors.

Keywords Flexible linguistic values · Consistency functions · Data conversion

7.1 Degree Linguistic Values and Superposed Linguistic Values

1. **Degree linguistic values**

   We call the usual adverbs portraying degrees such as comparatively, very, and extremely to be degree linguistic values. It can be seen that degree linguistic values should be the flexible linguistic values defined on the range $[1 - \beta, \beta]$ or $[-\beta, \beta]$ $(\beta \geq 1)$ of consistency-degrees. Figure 7.1 illustrates several common degree linguistic values. Of course, it still needs to be discussed how to most appropriately take the relevant parameters of (the consistency functions of) these degree linguistic values, and what here given only are reference models [1].

2. **Superposed flexible linguistic values**

   The so-called superposed flexible linguistic value is the flexible linguistic value that is formed by superposing a degree linguistic value and a flexible linguistic value, which is common occurrence in natural language. For instance, "very hot" is just a superposed flexible linguistic value formed by superposing degree linguistic value "very" and flexible linguistic value "hot." In a superposed

© Springer Science+Business Media Singapore 2016   161
S. Lian, *Principles of Imprecise-Information Processing*,
DOI 10.1007/978-981-10-1549-6_7

Fig. 7.1 Flexible linguistic values on range of consistency-degrees

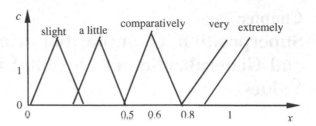

flexible linguistic value, the degree linguistic value in front has an effect of modifying and restricting upon flexible linguistic value behind, this can also be viewed as a kind of operation on the latter (can be said original language value), and the superposed flexible linguistic value obtained by the operation is also a flexible linguistic value, therefore, the degree linguistic values can be viewed as a kind of operator of flexible linguistic values. Then, viewed from the point of operation, a superposed flexible linguistic value can be considered to be obtained by the superposition operation from an original linguistic value.

Definition 7.1 Let A be a flexible linguistic value, and let c be a degree linguistic value, we call cA a superposed flexible linguistic value based on A, or a superposed value for short; and we call A the original linguistic value, or the original value for short; and call degree linguistic value c the degree operator.

For instance, "very tall" is a superposed value based on "tall," while "very" is a degree operator.

It can be seen that the function and effects of a degree operator are to weaken or strengthen the semantics of the original linguistic value, so degree operators can be separated into two classes: one is weakening operators, such as "slight," "a little," "comparatively," and "basically"; the other is strengthening operators, such as "very," "extremely," and "quite". We call superposed linguistic value cA formed by a weakening operator as a weakening superposed linguistic value, and denote it c^-A, and call superposed linguistic value cA formed by a strengthening operator as a strengthening superposed linguistic value and denote it c^+A.

If a superposed flexible linguistic value is again modified by a degree linguistic value, then a double superposed value is formed. For instance, "very very cold" is a double superposed value. A double superposed value can be denoted by c_1c_2A. Further, there can be n-fold superposed value $c_1...c_nA$.

3. **Relation between a superposed flexible linguistic value and the original value**

Here, we only discuss the relation between a one-dimensional onefold superposed linguistic value and the original value.

Examining and analyzing the relation between a superposed linguistic value and the original value, we find the following:

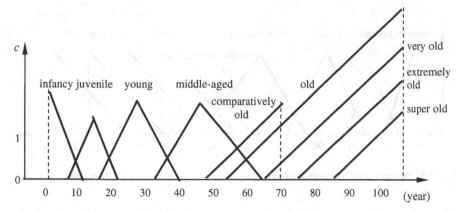

Fig. 7.2 Example 1 of the relation between superposed values and the original value

Firstly, a superposed linguistic value and the original value based on a full-peak value are the same-level relation, while a superposed linguistic value and the original value based on a semi-peak value can be the same-level relation as well as the inclusion relation.

Secondly, since a superposed value is a new linguistic value that modifies or restricts the original value somewhat, and one-dimensional measurement space is the linearly ordered set, so speaking from orientation, a weakening superposed value is certainly located at the negative side of the original value, while a strengthening superposed value is certainly located at the positive side of the original value.

Thirdly, a superposed value and its original value may be approximate relation or also may not be approximate relation.

Example 7.1 As shown in Fig. 7.2, flexible linguistic values on the range of human ages: infancy, juvenile, young, middle-aged, comparatively old, and old have the same-level relation, and old includes very old and extremely old. Here, "old" is an original value, while comparatively old, very old, extremely old, etc. are all superposed values based on "old." Among them, "comparatively old" is a weakening superposed value, which is located at the negative side of "old," "very old," and "extremely old", etc. are all strengthening superposed values, which are located at the positive side of "old."

Note that here the inclusion relation is theoretical or conceptual, while in engineering practice (such as approximate reasoning), sometimes the flexible linguistic values having inclusion relation needing to be designed as same-level relation. For instance, designing linguistic values having inclusion relation as shown in Fig. 7.2 into the same-level relation is shown in Fig. 7.3.

Generally speaking, there are only two basic linguistic values located at the boundary of a one-dimensional measurement space, for instance, small and large, short and tall, and cold and hot are all such linguistic values. However, in Fig. 7.3,

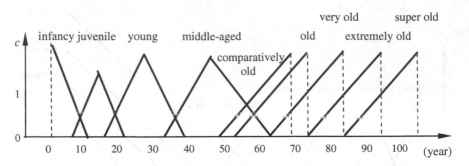

Fig. 7.3 Example 2 of the relation between superposed values and the original value

the original value "old" located at boundary is actually decomposed into multiple linguistic values, and also adding a "comparatively old" in front of "old," thus forming a sequence of approximate values.

4. **Indirect method to obtain the consistency function of a superposed flexible linguistic value**

A superposed value is also an independent linguistic value, so its consistency function can be written out according to its support set and core. However, since a superposed value is based on an original value, and there exist inclusion, parallel, or even similarity or approximation relations between the two, so the consistency function of a superposed value and that of its original value have a certain relationship. Through analysis of the relation between a superposed value and its original value, we find that for a one-dimensional flexible linguistic value, we can completely use the consistency function of original value to indirectly obtain the consistency function of the corresponding superposed value.

We have already known that the weakening superposed value of a one-dimensional full-peak linguistic value is sure located at its negative side, and the strengthening superposed value is sure located at its positive side. Then, if we suppose a superposed value is similar to its original value, then, translating appropriately the consistency function of original value, we can just obtain the consistency function of the superposed value. For brevity, we use A^- and A^+ to separately denote weakening superposed value c^-A and strengthening superposed value c^+A below.

Let A be an original linguistic value, and the consistency function of A be

$$
c_A(x) = \begin{cases} \dfrac{x - s_A^-}{c_A^- - s_A^-}, & a \le x \le \xi_A \\[2mm] \dfrac{s_A^+ - x}{s_A^+ - c_A^+}, & \xi_A \le x \le b \end{cases}
$$

Let the consistency function of weakening superposed value A^- be

$$c_{A^-}(x) \begin{cases} \dfrac{x - s_{A^-}^-}{c_{A^-}^- - s_{A^-}^-}, & a \leq x \xi_{A^-} \\[3mm] \dfrac{s_{A^-}^+ - x}{s_{A^-}^+ - c_{A^-}^+}, & \xi_{A^-} \leq x \leq b \end{cases}$$

We prescribe that A^- and A are similar, that is, the widths of the corresponding boundaries of A^- and A are equal, thus

$$\left| s_A^- - s_{A^-}^- \right| = \left| c_A^- - c_{A^-}^- \right| = \left| c_A^+ - c_{A^-}^+ \right| = \left| s_A^+ - s_{A^-}^+ \right| = \left| \xi_A - \xi_{A^-} \right|$$

while

$$\left| \xi_A - \xi_{A^-} \right| = d(A^-, A)$$

then, set

$$\delta = d(A^-, A)$$

Thus,

$$s_{A^-}^- = s_A^- - \delta, \quad c_{A^-}^- = c_A^- - \delta, \quad s_{A^-}^+ = s_A^+ - \delta, \quad c_{A^-}^+ = c_A^+ - \delta$$

Substitute these 4 expressions above into the consistency function of the above weakening superposed value A^-, we have

$$c_{A^-}(x) \begin{cases} \dfrac{x - (s_A^- - \delta)}{(c_A^- - \delta) - (s_A^- - \delta)}, & a \leq x \leq \xi_A - \delta \\[3mm] \dfrac{(s_A^+ - \delta) - x}{(s_A^+ - \delta) - (c_A^+ - \delta)}, & \xi_A - \delta \leq x \leq b \end{cases}$$

Modify the expression, getting

$$c_{A^-}(x) \begin{cases} \dfrac{(x + \delta) - s_A^-}{c_A^- - s_A^-}, & a \leq x \leq \xi_A - \delta \\[3mm] \dfrac{s_A^+ - (x + \delta)}{s_A^+ - c_A^+}, & \xi_A - \delta \leq x \leq b \end{cases} \tag{7.1}$$

From this, it is not hard to see that the relation between consistency functions $c_{A^-}(x)$ and $c_A(x)$ is

$$c_{A^-}(x) = c_A(x + \delta)$$

That is, $c_{A^-}(x)$ is the left shift of $c_A(x)$.

Making similar analysis, we also can obtain the relation between consistency functions of strengthening superposed value A^+ and its original value A as follows:

Fig. 7.4 An example of the consistency functions of superposed values of a full-peak value

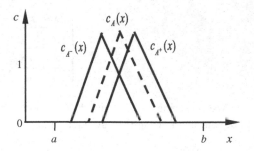

$$c_{A^+}(x) = c_A(x - \delta)$$

that is,

$$c_{A^+}(x) = \begin{cases} \frac{(x-\delta)-s_A^-}{c_A^--s_A^-}, & a \le x \le \xi_A + \delta \\ \frac{s_A^+-(x-\delta)}{s_A^+-c_A^+}, & \xi_A + \delta \le x \le b \end{cases} \quad (7.2)$$

The graphs of the above two consistency functions are shown in Fig. 7.4.

If original value A is a negative boundary value on a range of numerical values, then its superposed value A^- is located at the positive side of A, and A^+ is located at the negative side of A; thus, the consistency functions of A^- and A^+ are the following Eqs. (7.3, 7.4); if original value A is a positive boundary value, then superposed value A^- is located at the negative side of A, and A^+ is located at the positive side of A; thus, the consistency functions of A^- and A^+ are the following Eqs. (7.5, 7.6).

$$\begin{cases} c_{A^-}(x) = \frac{s_A^+-(x+\delta)}{s_A^+-c_A^+}, & x \in [a,b] \\ c_{A^+}(x) = \frac{s_A^+-(x-\delta)}{s_A^+-c_A^+}, & x \in [a,b] \end{cases} \quad (7.3, 7.4)$$

$$\begin{cases} c_{A^-}(x) = \frac{(x+\delta)-s_A^-}{c_A^--s_A^-}, & x \in [a,b] \\ c_{A^+}(x) = \frac{(x-\delta)-s_A^-}{c_A^--s_A^-}, & x \in [a,b] \end{cases} \quad (7.5, 7.6)$$

If the original value A is a non-boundary negative semi-peak value, then the consistency functions of superposed values A^- and A^+ are the following Eqs. (7.7, 7.8); if original value A is a non-boundary positive semi-peak value, then the consistency functions of superposed values A^- and A^+ are the following Eqs. (7.9, 7.10).

$$\begin{cases} c_{A^-}(x) = \frac{s_A^+-(x+\delta)}{s_A^+-c_A^+}, & x \in [a,b] \\ c_{A^+}(x) = \frac{s_A^+-(x-\delta)}{s_A^+-c_A^+}, & x \in [a,b] \end{cases} \quad (7.7, 7.8)$$

$$\begin{cases} c_{A^-}(x) = \frac{(x+\delta)-s_A^-}{c_A^- - s_A^-}, & x \in [a,b] \\ c_{A^+}(x) = \frac{(x-\delta)-s_A^-}{c_A^- - s_A^-}, & x \in [a,b] \end{cases} \qquad (7.9, 7.10)$$

Thus, as long as the distance between a superposed value and its original value is known, or as long as one of the critical points, core–boundary points, and peak value point of superposed value is determined, its consistency function can be obtained by doing translation transformation of the consistency function of original value. Therefore, we can call this kind of method to indirectly obtain a consistency function to be the **translation method**.

Example 7.2 For the linguistic value "tall" describing height, if we take 1.80 m as its negative core–boundary point and take 1.65 m as its negative critical point, then the consistency function of "tall" is

$$c_{\text{tall}}(x) = \frac{x - 1.65}{0.15}, \quad x \in [1.0, 2.50]$$

Additionally, we take "comparatively tall" as another linguistic value. Obviously, it is a weakening superposed value of "tall." We take 1.75 m as the negative core–boundary point of "comparatively tall," then the distance between "comparatively tall" and "tall" is 0.05. Thus, the consistency function of "comparatively tall" is

$$c_{\text{comparatively tall}}(x) = c_{\text{tall}}(x + 0.05) = \frac{x - 1.60}{0.15}, \quad x \in [1.0, 2.50]$$

Lastly, we point out, the relation between a onefold superposed value and its original value as well as the method to obtain indirectly the corresponding consistency function above can be generalized to n-fold superposed values by analogy.

7.2 Flexible Linguistic Value with Degree

We know that a linguistic value is the summarization and a collective name of a batch of continuous numerical values. However, the consistency-degrees of these numerical values with the corresponding flexible linguistic value are not all 1. Then, how to use corresponding flexible linguistic values to describe more accurately the features corresponding to the numerical values whose consistency-degrees are not 1? In natural language, this function is generally realized by using the flexible linguistic values modified and limited by degree adverbs "comparatively," "a little," "very," "extremely", etc. In this book's terminology, that is to use superposed linguistic values to describe the feature of an object more accurately. However, a superposed value is still a flexible linguistic value, so using a superposed value for description is still a kind of imprecise qualitative description. Therefore, we assume that if the flexible linguistic values can be quantified, then the features of objects can be described precisely by quantified flexible linguistic values. To this end, we introduce the concept and method of the flexible linguistic value with degree.

Definition 7.2 We call a flexible linguistic value portrayed by a number as degree to be the flexible linguistic value with degree.

A flexible linguistic value with degree can be represented by two-tuples

$$(A, d) \tag{7.11}$$

where A is a flexible linguistic value, $d \in [\alpha, \beta]$ ($\alpha \leq 0$, $1 \leq \beta$) is a degree, whose semantics is A with degree d.

For example, (fat, 0.6) is just a flexible linguistic value with degree, which represents "fat" with degree 0.6 and can be interpreted as "slightly fat." For another example, (hot, 1.2) is also a flexible linguistic value with degree, which represents "hot" with degree 1.2 and can be interpreted as "very hot."

Besides two-tuples, we can also use form

$$dA \tag{7.12}$$

or

$$A_d \tag{7.13}$$

to represent a flexible linguistic value with degree, here we use a coefficient or subscript to portray a linguistic value.

For instance, let A be "tall," then "0.8 tall" or "$\text{tall}_{0.8}$" just represents "tall" with degree 0.8, which can be interpreted as "comparatively tall." Similarly, "1.3 tall" or "$\text{tall}_{1.3}$" can represent "very tall," "1 tall" or "tall_1" represents a standard "tall." If we extend the scope of the degree to the range of equal to or less than 0, then "0 tall" or "tall_0" can be used to represent standard "not tall," and "−0.3 tall" or "$\text{tall}_{-0.3}$" can be used to represent "very not tall."

It can be seen that the flexible linguistic value with degree is actually the refining of the usual flexible linguistic value, in which the degree is the exact portrayal of the feature value an object has. Thus, for a feature of an object, there are two kinds of methods to describe it exactly: One is using a numerical value to describe directly, while the other is using a flexible linguistic value with degree to describe indirectly. For instance, for a person's height, we can use numerical value 1.68 m to represent it; and can also use flexible linguistic value with degree, (tall, 0.75) (degree 0.75 is a supposition), to represent it. So, **a flexible linguistic value with degree is in fact equivalent to a numerical value**.

In fact, the flexible linguistic value with degree is using both a linguistic value and a degree together to describe a feature of an object, which is a kind of description method that combines macro and micro and combines qualitative and quantitative. On the one hand, it uses a flexible linguistic value to orient roughly the feature of the object in macro; on the other hand, it also uses a degree to orient exactly the feature based on the macro-orientation. Therefore, the combination of the two just briefly and accurately characterizes the feature of the object. Such a representation is also consistent with the objective fact that an object always has certain flexible linguistic value in a certain degree.

From the relationship between flexible linguistic values and numerical values, it can be seen that **d in flexible linguistic value with degree, (A, d), is also consistency-degree $c_A(x_0)$ of a certain numerical object x_0 in the same universe of discourse with flexible linguistic value A, that is, $d = c_A(x_0)$.** Thus, through consistency function, the flexible linguistic values with degrees and numerical values on the same universe of discourse can be converted mutually.

Like pure flexible linguistic values, flexible linguistic values with degrees can also do composition operations. From the composition operations of flexible linguistic values in Sect. 6.4, the rules of composition operations of flexible linguistic values with degrees are as follows:

$$\bigwedge_{i=1}^{n} (A_i, d_i) = \left(\bigwedge_{i=1}^{n} A_i, \min_{i=1}^{n} d_i \right) \tag{7.14}$$

$$\bigvee_{i=1}^{n} (A_i, d_i) = \left(\bigvee_{i=1}^{n} A_i, \max_{i=1}^{n} d_i \right) \tag{7.15}$$

$$\bigoplus_{i=1}^{n} (A_i, d_i) = \left(\bigoplus_{i=1}^{n} A_i, \sum_{i=1}^{n} w_i d_i \right) \tag{7.16}$$

Of course, there may also be the situation that the consistency function is unknown. In that a case we can estimate the value of d based on relevant experience and knowledge.

7.3 Interconversion Between Flexible Linguistic Values and Numerical Values

Now, a feature that an object has can not only be characterized by a numerical value or a flexible linguistic value, but it can also be characterized by a flexible linguistic value with degree. Therefore, there will be problems: How are these three kinds of characterizations related? Can they be converted each other? And how to convert? This section will discuss these problems.

7.3.1 Interconversion Between Pure Flexible Linguistic Values and Numerical Values

We know that a flexible linguistic value is the summarization of a set of numerical values, while the numerical values are the instances of the corresponding flexible linguistic value. Then, for a numerical value, what is the flexible linguistic value it corresponds to? Or conversely, for a flexible linguistic value, what is the numerical value it corresponds to?

1. Converting from a numerical value to a pure flexible linguistic value

Firstly, we refer to the conversion from numerical values to pure flexible linguistic values as **N-L conversion**. N-L conversion can be viewed as a kind of flexible-ening (which is similar to softening) of numerical values.

Converting a numerical value x in the universe of discourse U to a flexible linguistic value on U has two case: One is that there are still not appropriate and ready flexible linguistic values on U to choose, or although there are ready flexible linguistic values, problem requires forming a flexible linguistic value with numerical value x as peak value point. The other is that on U, there are already ready flexible linguistic values for choose.

For the first case, a new flexible linguistic value needs to be constructed on U to generally represent numerical value x. The method to construct this flexible linguistic value is the method of flexible clustering stated in Chap. 2.

For the second case, we can choose one from the ready flexible linguistic values to replace the numerical value x, despite the x may have simultaneously multiple flexible linguistic values (e.g., a pair of relatively negative flexible linguistic values, A and $\neg A$) with a certain degree separately. Then, how is this flexible linguistic value to be chosen? It can be seen that this is actually to determine which flexible property object x more possesses or which flexible set x more should belong to, that is, to determine rigidly the possessive relation or membership relation of object x (actually, that human brain converts a numerical value to a flexible linguistic value just is so). Since the sum of degrees of a object having relatively negative linguistic values is 1, while the N-L transformation requires the corresponding flexible linguistic value to be unique, consistency-degree >0.5 is a basic condition for the N-L transformation; thus, the one with largest consistency-degree is the best choice. Further, the transformation in the situation can be still separated into the following two cases:

(1) Converting it into a basic flexible linguistic value

Let U be a one-dimensional measurement space, $x_0 \in U$ be a numerical value, and $A_1, A_2, \ldots, A_m$ be a group of basic flexible linguistic values on U. Since two adjacent basic flexible linguistic values are complementary, surely

$$\max\{c_{A_1}(x_0), c_{A_2}(x_0), \ldots, c_{A_m}(x_0)\} = c_{A_k}(x_0) \geq 0.5$$

Thus, if $c_{A_k}(x_0) > 0.5$, then flexible linguistic value A_k is the best basic flexible linguistic value to match numerical value x_0. Therefore, the general method to convert x_0 into a basic flexible linguistic value is:

Firstly, substitute x_0 into consistency functions $c_{A_1}(x), c_{A_2}(x), \ldots, c_{A_m}(x)$ separately, then take

$$c_{A_k}(x_0) = \max\{c_{A_1}(x_0), c_{A_2}(x_0), \ldots, c_{A_m}(x_0)\}$$

If $c_{A_k}(x_0) > 0.5$, then convert numerical value x_0 into flexible linguistic value A_k; if $c_{A_k}(x_0) = 0.5$, then or by the specific problem to decide whether to convert x_0 into A_k, or not to do the conversion. This conversion process is

$$x_0 \rightarrow c_{A_k}(x_0) \geq 0.5 \rightarrow A_k \tag{7.17}$$

(2) Converting it into an superposed linguistic value

Since our conversion principle is $c_{A_k}(x_0) > 0.5$, but not $c_{A_k}(x_0) = 1$, the above converted basic flexible linguistic value A_k may not be very accurate or very proper for numerical value x_0. Therefore, sometimes numerical value x_0 still needs to be converted into a more accurate and proper superposed linguistic value. This only needs to substitute separately consistency-degree $c_{A_k}(x_0)$ to consistency functions $c_{H_1}(y), c_{H_2}(y), \ldots, c_{H_n}(y)$ of relevant degree linguistic values $H_1, H_2, \ldots, H_n$, then take

$$c_{H_l}(c_{A_k}(x_0)) = \max\{c_{H_1}(c_{A_i}(x_0)), c_{H_2}(c_{A_i}(x_0)), \ldots, c_{H_n}(c_{A_i}(x_0))\}$$

then $H_l A_k$ is just the superposed linguistic value that numerical value x_0 corresponds to. This conversion process is

$$x_0 \rightarrow c_{A_k}(x_0) \rightarrow c_{H_k}(c_{A_k}(x_0)) \rightarrow H_l A_k \tag{7.18}$$

The above method converting a numerical value to a flexible linguistic value in one-dimensional space can also be generalized to multidimensional space.

Let U be an n-dimensional measurement space, $x_0 = (x_1, x_2, \ldots, x_n) \in U$ be an n-dimensional numerical vector.

(1) Forming directly a flexible linguistic value from x_0. The general method is the following: take x_0 as peak value point, determine corresponding core radius and support set radius according to requirement, forming a corresponding flexible linguistic value on space U by using flexible clustering. We denote the flexible linguistic value forming from vector x_0 by (x_0) later.

(2) Converting x_0 into a basic flexible linguistic value. The general method is as follows:

Let A_1, A_2, ..., A_m be a group of basic flexible linguistic values on U. Firstly, substitute separately x_0 into consistency functions $c_{A_1}(x), c_{A_2}(x)$, ..., $c_{A_m}(x)$, then take

$$c_{A_k}(x_0) = \max\{c_{A_1}(x_0), c_{A_2}(x_0), \ldots, c_{A_m}(x_0)\}$$

If $c_{A_k}(x_0) > 0.5$, then convert numerical vector r_0 into flexible linguistic value A_k; if $c_{A_k}(x_0) = 0.5$, then do not convert. This conversion process is

$$x_0 \rightarrow c_{A_k}(x_0) \rightarrow A_k \qquad (7.19)$$

In consideration that basic flexible linguistic values are all atom flexible linguistic values, and from the above-stated converting principle from a numerical value to a flexible linguistic value, it can be seen that $c_A(x_0) > 0.5$ is actually the sufficient and necessary condition for x_0 converting to flexible linguistic value A, while $c_A(x_0) > 0.5$ is also equivalent to $x_0 \in \text{core}(A)^+$, so we have the following theorem.

Theorem 7.1 *Numerical value x_0 can be converted to an atom flexible linguistic value A if and only if $x_0 \in \text{core}(A)^+$.*

It is not hard to see that this theorem is tantamount to giving a geometrical method for converting a numerical value into a flexible linguistic value.

It can be seen that converting from a numerical value to a basic flexible linguistic value is really tantamount to the classification of numerical values. Thus, the Theorem 7.1 is also equivalent to say, numerical value x_0 belongs fully to flexible set A if and only if $x_0 \in \text{core}(A)^+$. That is to say, from the application point of view, a flexible linguistic value (i.e., flexible set) is fully stood for by its extended core. Further, we have the following conclusion.

Proposition 7.1 *In concept, a flexible linguistic value (flexible set) is determined by its core and support set, but in practical, which is fully stood for by its extended core. In other words, core and support set are the conceptual model of a flexible linguistic value (flexible set), while extended core is its practical model.*

From the conversion from numerical values to flexible linguistic values, it can be seen that when numerical values x_0 is converted into pure flexible linguistic value A, the meaning of "x_0 is A" in daily language is already not s ambiguous "x_0 is A in a certain degree" but unambiguous "x_0 has A" or "x_0 belongs to A." Despite now the consistency-degree $c_A(x_0)$ and membership-degree $m_A(x_0)$ are not necessarily equal to or greater than 1 and they have still a certain elasticity in interval $(0.5, \beta]$ $(\beta \geq 1)$, but they certainly be greater than 0.5.

2. Converting from a pure linguistic values to a numerical value

Converting a flexible linguistic value into a numerical value is to select a number (or vector) from corresponding measurement space, such that it can replace or represent this flexible linguistic value. From the relation between flexible linguistic values and numerical values, this needs firstly to determine the extended core of the flexible linguistic value and then selecting a number (or vector) from which. It can be seen that speaking purely from consistency relation, in a extended core, the number (or vector) that can replace the corresponding flexible linguistic value is firstly the peak value point of the flexible linguistic value, and secondly is any number (or vector) in the core, then is any number (or vector) in the extended core. Thus, the method of converting a flexible linguistic value A on one-dimensional measurement space U into a numerical value x_0 is

$$A \rightarrow \text{core}(A)^+ \rightarrow x_0 \tag{7.20}$$

where $x_0 = \xi_A$ or $x_0 \in \text{core}(A)$ or $x_0 \in \text{core}(A)^+$.

Generally, the method of converting an atomic flexible linguistic value A on n-dimensional measurement space U into a vector $\boldsymbol{x_0}$ is

$$A \rightarrow \text{core}(A)^+ \rightarrow x_0 \tag{7.21}$$

where $\boldsymbol{x_0} = \xi_A$ or $\boldsymbol{x_0} \in \text{core}(A)$ or $\boldsymbol{x_0} \in \text{core}(A)^+$.

However, the conversion from a flexible linguistic value to a numerical value in practical problems is usually with respect to a certain specific object. For example, we know Zhang is tall but don't know exactly how many meters is he, in this case, it is needed that converting flexible linguistic value "tall" into a numerical value. It can be seen that this kind of conversion can only be done by guessing, so which is actually an uncertainty problem. Then, for this kind of conversion, if there is no guidance from relevant knowledge or information, then the accuracy of the conversion has no any assurance. And to increase the accuracy of the conversion, related heuristic information is required, such as the probability distribution or distribution density of numerical values or the relevant background information of the corresponding object. For example, if the distribution of human's heights is known, then the accuracy of Zhang's height would be effectively increased; while if we know Zhang is a player of the national basketball, then you would surely consider his height to be around 2 m.

Now we have seen that the conversions from flexible linguistic values to numerical values can be separated into the conversion based on the relation between linguistic values and numerical values and the conversion with respect to a certain object. We may as well call the former to be conceptual conversion and the latter specific conversion. From the above stated, conceptual conversion is purely related to the consistency function of a flexible linguistic value, which can be realized by using the previous expression (7.21); but specific conversion is an uncertainty problem, and to increase its accuracy, which should be guided by relevant heuristic information. If the density function of a numerical value is taken as the heuristic information, then the distinction between these two kinds of conversions can be visually seen from Fig. 7.5 (here, we now strictly put the density function and the consistency function of variable x on a certain interval in the same coordinate system).

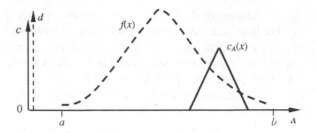

Fig. 7.5 Examples of the density function and consistency function about x, where $f(x)$ shown by the broken line is the density function and $c_A(x)$ shown by the real line is the consistency function

Besides, from Fig. 7.5, we can also clearly see that the consistency-degree of a numerical value x with a certain flexible linguistic value A and the probability or density that this numerical value x occurs are two different things, and they have no direct connection. Therefore, generally speaking, the probability or density that numerical value x occurs cannot be used to determine the consistency-degree of x with flexible linguistic value A and vice versa.

Lastly, we refer to conversion from flexible linguistic values to numerical values as **L-N conversion**. L-N conversion can be viewed as a kind of rigid-ening (which is similar to hardening) of flexible linguistic values. In the following, we use notation $[A]$ to represent the rigid-ening of flexible linguistic value A and the numerical value resulted from A being rigid-ened.

7.3.2 Interconversion Between Flexible Linguistic Values with Degrees and Numerical Values

1. Converting from a numerical value to a (or multiple) flexible linguistic value with degree

A numerical value can be converted as a flexible linguistic value with degree and can also be converted as multiple flexible linguistic values with degrees.

Let $x_0 \in U = [a, b]$, and $A_1, A_2, \ldots, A_m$ be flexible linguistic values on measurement space U. Set

$$d_j = \max\{c_{A_1}(x_0), c_{A_2}(x_0), \ldots, c_{A_m}(x_0)\} = c_{A_j}(x_0)$$

then, (A_j, d_j) is a flexible linguistic value with degree to which x_0 corresponds. This is a conversion of one to one. This converting process is

$$x_0 \rightarrow \max\{c_{A_1}(x_0), c_{A_2}(x_0), \ldots, c_{A_m}(x_0)\} = d_j \rightarrow (A_j, d_j) \qquad (7.22)$$

In general, the degree d_j in this flexible linguistic value with degree, (A_j, d_j), may be any number in corresponding range of degrees except for infimum α. But if A_1, $A_2, \ldots, A_m$ is a group of basic flexible linguistic values on space U, then certainly, $d_j \geq 0.5$; and if $d_j > 0.5$, then it is also unique; and if $d_j = 0.5$, then also $d_{j-1} = 0.5$

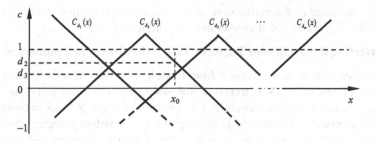

Fig. 7.6 Example of the degrees d_i and d_{i+1} satisfying >0 and <1

or $d_{j+1} = 0.5$. That is to say, x_0 actually can also be converted as $(A_{j-1}, 0.5)$ or $(A_{j+1}, 0.5)$. From this situation, converting x_0 into $(A_j, 0.5)$ or converting it into $(A_{j-1}, 0.5)$ or $(A_{j+1}, 0.5)$, or do not do converting, should be decided by specific problem.

Let $x_0 \in U = [a, b]$, and $A_1, A_2, \ldots, A_m$ be flexible linguistic values on measurement space U. Set $d_i = c_{A_i}(x_0)$ $(i = 1, 2, \ldots, m)$, then we have flexible linguistic values with degrees, $(A_1, d_1), (A_2, d_2), \ldots, (A_m, d_m)$, which are the flexible linguistic values with degrees corresponding to numerical value x_0. This is just a conversion of one to many. This converting process is

$$x_0 \rightarrow \begin{cases} c_{A_1}(x_0) = d_1 \\ c_{A_2}(x_0) = d_2 \\ \cdots \\ c_{A_m}(x_0) = d_m \end{cases} \rightarrow \begin{cases} (A_1, d_1) \\ (A_2, d_2) \\ \cdots \\ (A_m, d_m) \end{cases} \tag{7.23}$$

Note that since $(A_1, d_1), (A_2, d_2), \ldots, (A_m, d_m)$ are all from number x_0 by converting, that is, they stand for one and the same numerical value x_0, they are equivalent mutually.

Similarly, in general, the degrees $d_1, d_2, \ldots, d_m$ in these flexible linguistic values with degrees are also any m numbers in corresponding range of degrees, but if A_1, $A_2, \ldots, A_m$ is a group of basic flexible linguistic values on space U, and when numerical value x_0 is located in $\mathrm{supp}(A_i) \cap \mathrm{supp}(A_{i+1}) = (s_{A_{i+1}}^-, s_{A_i}^-)$ the boundary region of two adjacent flexible linguistic values A_i and A_{i+1} $(i \in \{1, 2, \ldots, m-1\})$, then d_i and d_{i+1} in the corresponding flexible linguistic values with degrees, (A_i, d_i) and (A_{i+1}, d_{i+1}), separately satisfy $0 < d_i < 1$ and $0 < d_{i+1} < 1$, and only d_i and d_{i+1} satisfy >0 and <1 (as shown in Fig. 7.6).

The conversion from numerical value x_0 to a one-dimensional flexible linguistic value with degree, obviously, can also be generalized to the conversion from n-dimensional vector x_0 to an n-dimensional flexible linguistic value with degree. The method is similar to the former, so it is unnecessary to go into detail here.

Actually, the conversion from a numerical value to a flexible linguistic value with degree in essence is to transform a numerical value into the consistency-degree of the numerical value with a certain flexible linguistic value by the mapping of consistency function.

Lastly, we refer to the conversion from numerical values to flexible linguistic values with degrees as **N-Ld conversion** for short.

2. Converting from a flexible linguistic value with degree to a numerical value

Firstly, we refer to the conversion from flexible linguistic values with degrees to numerical values as **Ld-N conversion** for short. Viewed from principle, Ld-N conversion is the inverse process of N-Ld conversion. Since N-Ld conversion is to transform a numerical value to a corresponding consistency-degree through a consistency function, Ld-N conversion is to transform the consistency-degree back to a numerical value (i.e., measurement) through the inverse function of the original consistency function. Thus, for a one-dimensional semi-peak linguistic value, Ld-N conversion is very easy. In fact, let $A \subset U$ be a one-dimensional semi-peak flexible linguistic value, and (A, d) be a flexible linguistic value with degree. Since its consistency function $c_A(x)$ is a 1-1 mapping, the inverse function $c_{A^{-1}}(y)$ of $c_A(x)$ is also a 1-1 mapping. Thus, substituting d into inverse function $c_{A^{-1}}(y)$ of $c_A(x)$, we have $c_{A^{-1}}(d) = x_0$. This converting process is

$$(A, d) \rightarrow c_{A^{-1}}(d) = x_0 \tag{7.24}$$

We then consider the conversion from a one-dimensional full-peak flexible linguistic value with degree to a numerical value.

Let $A \subset U$ be a one-dimensional full-peak flexible linguistic value and (A, d) be a flexible linguistic value with degree. Since consistency function $c_A(x)$ is a full triangular function, there are two of its inverse function $c_{A^{-1}}(y)$ and there would be two corresponding x_0, denote them by x_{0_1} and x_{0_2}. Then, which number should be chosen as x_0? Here, we present several schemes for the determination of x_0.

① Randomly take x_{0_1} or x_{0_2} as x_0.
② Take $x_0 = \xi_A$. That is, to treat peak value point ξ_A of flexible linguistic value A as x_0;
③ Take $x_0 = \overline{x_0} = (x_{0_1} + x_{0_2})/2$. That is, the average value of x_{0_1} and x_{0_2} is treated as x_0.

It can be seen that though taking peak value point ξ_A and average value $\overline{x_0}$ as the converted value x_0, there would occur some errors, since the two always are between x_{0_1} and x_{0_2}, it follows that

$$\max\{|x_{0_1} - \xi_A|, |x_{0_2} - \xi_A|\} < |x_{0_1} - x_{0_2}|$$
$$\max\{|x_{0_1} - \overline{x_0}|, |x_{0_2} - \overline{x_0}|\} < |x_{0_1} - x_{0_2}|$$

Thus, the error produced by taking ξ_A or $\overline{x_0}$ is less than those by randomly taking x_{0_1} or x_{0_2}.

④ Determine the selection of x_0 from the relevant background knowledge. For example, if it is known that in a certain condition, the probability of the value of x occurring at the half zone that x_{0_1} locates at is higher than occurring at the half zone that x_{0_2} locates at (it can be seen that in such a situation, x_{0_1} and x_{0_2}

certainly locate separately at the two half zones of support set supp(A)), then we can take $x_0 = x_{0_1}$, or we can take any number in the half zone as the value of x_0.

Since the consistency function of a multidimensional flexible linguistic value is irreversible, a multidimensional flexible linguistic value with degree is hard to be converted to a numerical vector.

7.3.3   Interconversion Between Pure Flexible Linguistic Values and Flexible Linguistic Values with Degrees

1. Converting from a flexible linguistic value with degree to a pure flexible linguistic value

The method of converting a flexible linguistic value with degree to a pure flexible linguistic value is as follows:

Let (A, d) be a flexible linguistic value with degree, and H_1, H_2, ..., H_n be degree linguistic values. Substitute separately degree d into consistency functions $c_{H_1}(x)$, $c_{H_2}(x)$, ..., $c_{H_n}(x)$, take

$$c_{H_k}(d) = \max\{c_{H_1}(d), c_{H_2}(d), \ldots, c_{H_n}(d)\}$$

then superposed linguistic value $H_k A$ is the pure flexible linguistic value that flexible linguistic value with degree, (A, d), corresponds to. The converting process is

$$(A, d) \rightarrow c_{H_k}(d) \rightarrow H_k A \tag{7.25}$$

We refer to the conversion from flexible linguistic values with degrees to pure flexible linguistic values as **Ld-L conversion** for short.

2. Converting from a pure flexible linguistic value to a flexible linguistic value with degree

Since the degree in a flexible linguistic value with degree is the image of corresponding object's measurement (i.e., numerical value) under the mapping of consistency function of the flexible linguistic value, the degree is closed connected with the numerical value that has this flexible linguistic value. Thus, to convert one-dimensional pure flexible linguistic value A into flexible linguistic value with degree, (A, d), corresponding measurement x_0 should be firstly known. When numerical value x_0 is known, then substitute it into consistency function $c_A(x)$; then, it follows that consistency-degree $c_A(x_0) = d$; and immediately, further, we have (A, d). The conversion process is

$$A \to c_A(x_0) \to (A, d) \tag{7.26}$$

However, if numerical value x_0 is not known, then degree d is hard to be determined. In such a situation, peak value point ξ_A of flexible linguistic value A can be taken as numerical value x_0 to compute corresponding degree d, that is, take $d = c_A(\xi_A)$. The conversion process is

$$A \to c_A(\xi_A) \to (A, d) \tag{7.27}$$

In particular, for superposed linguistic value cA, firstly find peak value point ξ_{cA} of this superposed linguistic value, then substitute ξ_{cA} into corresponding consistency function $c_A(x)$ of the original flexible linguistic value A to obtain consistency-degree $c_A(\xi_{cA})$, and take $d = c_A(\xi_{cA})$; Then, (A, d) is the flexible linguistic value with degree that flexible linguistic value cA corresponds to. The conversion process is

$$cA \to \xi_{cA} \to c_A(\xi_{cA}) \to (A, d) \tag{7.28}$$

For multidimensional flexible linguistic value A, the above-stated conversion method is also applicable, that is

$$A \to c_A(x_0) \to (A, d) \tag{7.29}$$

or

$$A \to c_A(\xi_A) \to (A, d) \tag{7.30}$$

Lastly, we refer to the conversion from pure flexible linguistic values to flexible linguistic values with degrees as **L-Ld conversion** for short.

7.4 Vector Flexible Linguistic Values and Flexible Linguistic-Valued Vectors

We know that an atom flexible linguistic value on one-dimensional space $[a, b]$ actually represents a flexible interval in space $[a, b]$, while one-dimensional atom flexible linguistic value "about x_0" or "near x_0" then represent a flexible interval with a center point, the consistency function of this linguistic value is a function for x. On multidimensional spaces, atom flexible linguistic value "about $P_0(x_{1_0}, x_{2_0}, ..., x_{n_0})$" or "near $P_0(x_{1_0}, x_{2_0}, ..., x_{n_0})$" then represents a flexible circle, flexible sphere, or flexible hyper sphere in the corresponding space, whose consistency function is a function for point P, that is, vector $(x_1, x_2, ..., x_n)$. Obviously, this type of multidimensional atom flexible linguistic values with a center point is the generalization of one-dimensional atom flexible linguistic values with a center point. In consideration of the characteristic of its consistency function being a function for vectors, we call this type of multidimensional atom flexible linguistic value to be **vector flexible linguistic value**.

Actually, the vector flexible linguistic value here is just the flexible linguistic value (x_0) forming from vector $x_0 = (x_1, x_2, ..., x_n)$ said in previous Sect. 7.3.1. Therefore, vector flexible linguistic value $((x_1, x_2, ..., x_n))$ and vector $(x_1, x_2, ..., x_n)$ can be converted mutually. The conversion method is also the method given in Sect. 7.3.1.

On the other hand, multiple one-dimensional atom flexible linguistic values can also form a flexible linguistic-valued vector. For instance, let $A_1, A_2, ..., A_n$ be separately atom flexible linguistic values on one-dimensional measurement spaces $U_1, U_2, ..., U_n$, then $(A_1, A_2, ..., A_n)$ is a flexible linguistic-valued vector. The flexible linguistic-valued vector is another kind of generalization of the one-dimensional atom flexible linguistic value. From the relation between vectors and points in a space, a flexible linguistic-valued vector $(A_1, A_2, ..., A_n)$ also denotes a point in corresponding flexible linguistic-valued vector space $L_1 \times L_2 \times \cdots \times L_n$, where $L_i = \{A_i| A_i \subset U_i\}$ is a set of atom flexible linguistic values on U_i, $i = 1, 2, ..., n$. Two-dimensional and three-dimensional flexible linguistic-valued vectors are shown in Fig. 7.7. Then, what does flexible linguistic-valued vector $(A_1, A_2, ..., A_n)$ denote in measurement space $U_1 \times U_2 \times \cdots \times U_n$?

It can be see that if the relation between components $A_1, A_2, ..., A_n$ in vector $(A_1, A_2, ..., A_n)$ is regarded as conjunction relation, *and*, then $(A_1, A_2, ..., A_n)$ is tantamount to conjunctive flexible linguistic value on distinct spaces, $A_1 \wedge A_2 \wedge \cdots \wedge A_n$. Thus, flexible linguistic-valued vector $(A_1, A_2, ..., A_n)$ denotes a flexible square region in measurement space $U_1 \times U_2 \times \cdots \times U_n$ to which conjunctive flexible linguistic value $A_1 \wedge A_2 \wedge \cdots \wedge A_n$ corresponds (the flexible square region denoted by two-dimensional flexible linguistic-valued vectors (A, B) is shown in Fig. 3.6). Conversely, a flexible square region in measurement space $U_1 \times U_2 \times \cdots \times U_n$ can also be denoted by a flexible linguistic-valued vector.

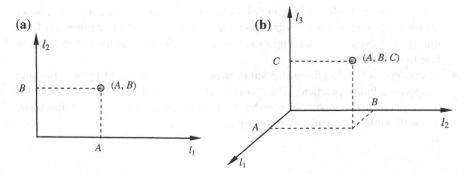

Fig. 7.7 Examples of flexible linguistic-valued vectors in flexible linguistic-valued vector spaces

7.5 Summary

The chapter is the continuation of the basic theory of flexible linguistic values. First, the concepts of degree linguistic values and superposed linguistic values are proposed, and the types, levels, and mathematical models of superposed linguistic values are discussed. Next, the quantification of flexible linguistic values is considered, and the flexible linguistic value with degree and its notation is presented. And then, the relations between pure linguistic values, flexible linguistic values with degrees, and numerical values are expounded, and the corresponding conversion principles and methods are presented. Lastly, one-dimensional flexible linguistic values are generalized to vector flexible linguistic values and flexible linguistic-valued vectors.

The main points and results of the chapter are the following:

- Degree linguistic values are what we usually call degree adverbs, and a superposed linguistic value is the linguistic value modified by a degree linguistic value. Degree linguistic values can be viewed as a kind of operator of linguistic values, which may be separated as two types of weakened ones and enhanced ones. A superposed value can be modified repeatedly to form multifold superposition. A superposed linguistic value is the same level with the original value; further, they can be of inclusion relation or approximate relation. The consistency function of a superposed value can be indirectly obtained by doing translation transformation of the consistency function of the original value.

- The linguistic value portrayed by a degree is called flexible linguistic value with degree, which describes the feature of an object by using both linguistic value and degree, which are equivalent to the corresponding numerical value in effect. For linguistic value A and object x_0, the corresponding flexible linguistic value with degree is $(A, c_A(x_0))$.

- Numerical values and flexible linguistic values as well as flexible linguistic values with degrees can be mutually converted. Of them, the conversion from numbers to flexible linguistic values has algebraic method and geometric method; the conversion from flexible linguistic values to numerical values can be separated as conceptual conversion and specific conversion; the former is related to the consistency function of the flexible linguistic value while the latter should be guided by the relevant heuristic information; and the conversion from numerical values to flexible linguistic values with degrees has the conversions of one to one and one to many.

- In concept, a flexible linguistic value (flexible set) is decided by its core and support set, but in practical, which is fully stood for by its extended core, that is to say, core and support set are the conceptual model of a flexible linguistic value (flexible set), while core is its practical model.

- A vector flexible linguistic value represents a flexible circle, flexible sphere or hyper flexible sphere in the multidimensional measurement space.
- A flexible linguistic-valued vector represents a point in corresponding flexible linguistic-valued vector space, but in measurement space, which represents a flexible square to which a corresponding conjunctive flexible linguistic value corresponds.

Reference

1. Lian S (2009) Principles of imprecise-information processing. Science Press, Beijing

Chapter 8
Relatively Opposite Flexible Linguistic Values and Relatively Opposite Flexible Sets

Abstract This chapter introduces the concepts of relatively opposite flexible linguistic values and relatively opposite flexible sets and founds the related theories.

Keywords Relatively opposite flexible linguistic values · Relatively opposite flexible sets

Flexible linguistic values discussed in the last two chapters are actually all flexible linguistic value with negation, that is, the flexible linguistic value having negation. In this chapter, we consider another kind of flexible linguistic values—flexible linguistic value with opposite, that is, the flexible linguistic value having a contradictory or opposite value. A flexible linguistic value with opposite and its opposite just form a pair of relatively opposite flexible linguistic values. Correspondingly, the two sets labeled by a pair of relatively opposite flexible linguistic values are just a pair of relatively opposite flexible sets. This chapter mainly discusses relatively opposite flexible linguistic values and relatively opposite flexible sets.

8.1 Relatively Opposite Flexible Linguistic Values and Their Types

1. Relatively opposite flexible linguistic values

Let A, B, and C be three basic flexible linguistic values that are adjacent in order on space $U = [a, b]$. As shown in Fig. 8.1, $B = \neg A \wedge \neg C$; that is, for A and C, B is just a neutral value, while A and C are opposite to each other relative to B. However, we see that this kind of contradiction relation between flexible linguistic values A and C is different from the relatively negative relation. There is a transition zone that is also this and also that between the cores of relatively negative flexible linguistic values, while the transition zone between the cores of this kind of contradictory flexible linguistic values is this but not that, and there is also a neutral

Fig. 8.1 Illustration of
relatively opposite flexible
linguistic values (1)

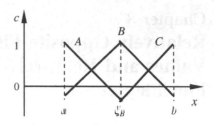

point—peak value point ξ_B of neutral flexible linguistic value B, that is neither this nor that. Therefore, this kind of contradiction relation between flexible linguistic values is actually a relatively opposite relation.

Definition 8.1 Let A, B, and C be three flexible linguistic values that are adjacent in order on space $U = [a, b]$. If $B = \neg A \wedge \neg C$, then we say A and C are relatively opposite about B, B is called the neutral value between A and C, the core core(B) of B is called the neutral zone between A and C, and peak value point ξ_B is called the neutral point between A and C.

Examining the relation among "small," "medium," and "big" as basic flexible linguistic values, it can be seen that "medium" is also "not big and not small," that is, a neutral value, so "small" and "big" are relatively opposite about "medium." Similarly, low–medium–high, young–middle-aged–old, cold–warm–hot, fast–intermediate–slow, small deficit–roughly balancing–small surplus, etc., are all the situation that the flexible linguistic value in the front and the one at the back are relatively opposite about the flexible linguistic value in the middle.

Actually, from the definition, for arbitrary three adjacent basic flexible linguistic values A_{i-1}, A_i, and A_{i+1} on space U, semi-peak values A_{i-1}^+ and A_{i+1}^- are relatively opposite about A_i, and A_i is the neutral value between A_{i-1}^+ and A_{i+1}^-.

As shown in Fig. 8.2, let A and C be two flexible linguistic values on space $U = [a, b]$, and $N_0 \in U$ be the boundary between A and C, and $N_0 \notin \text{supp}(A)$ and $N_0 \notin \text{supp}(C)$. It can be seen that $B = \{N_0\}$ is a transition zone (actually it is a transition point) between A and C that is neither this nor that. Thus, for A and C, B is a neutral value, while A and C are relatively opposite about B.

Definition 8.2 Let A and C be two flexible linguistic values on space U, and $N_0 \in U$ be the boundary between them. If $x_0 \notin \text{supp}(A)$ and $N_0 \notin \text{supp}(C)$, then we say

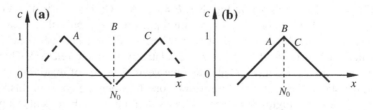

Fig. 8.2 Illustration of relatively opposite flexible linguistic values (2)

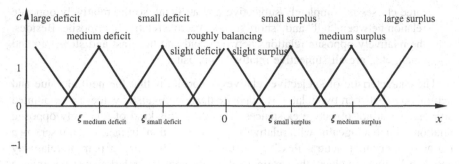

Fig. 8.3 Examples of various relatively opposite flexible linguistic values

A and C are relatively opposite about $B = \{N_0\}$, B is called the neutral value between A and C, and N_0 is called the neutral point between A and C.

Examining the relation between "small deficit," "balancing," and "small surplus" in marketing (as shown in Fig. 8.3, here positive is surplus, negative is deficit, and 0 is balancing), it can be seen that they are just tantamount to the linguistic values A, B, and C in Fig. 8.2a, while the linguistic values "slight deficit," "balancing," and "slight surplus" are separately then tantamount to the linguistic values A, B, and C in Fig. 8.2b. That is to say, "balancing" is the neutral value, "small deficit" and "small surplus" are relatively opposite, and "slight deficit" and "slight surplus" are relatively opposite.

Actually, the relatively opposite relation also existed in rigid linguistic values. For instance, "$\{0\}$" is just the neutral value between "positive" and "negative," so "positive" and "negative" are relatively opposite. Similarly, "deficit" and "surplus," "concave" and "convex," "rise" and "fall," "victory" and "defeat," "affirming" and "dissenting," etc., are all relatively opposite.

Later on, we denote the opposite of a flexible linguistic value A as $-A$, the neutral value as Neu, and a neutral point as N_0.

2. **Types of relatively opposite flexible linguistic values**
   After further examining, we find that the relatively opposite relation between flexible linguistic values is comparatively complex, and they can be separated into multiple types such as subjective relatively opposite, objective relatively opposite, face–face relatively opposite, back–back relatively opposite, symmetrical relatively opposite, standard relatively opposite, and normal relatively opposite.

(1) **Subjective relatively opposite and objective relatively opposite**
   Subjective relatively opposite is the relatively opposite relation that people think subjectively or that be artificially appointed. For instance, when we divide the heights of adults into two flexible classes of "tall" and "short," the "tall" and "short" are relatively negative relation; but when we divide it into three flexible classes of "tall," "medium" and "short," then "tall" and "short" are relatively opposite relation. Whether to divide heights into two classes or

three classes is completely subjective and artificial, so the relatively opposite relation between "tall" and "short" is a subjective relatively opposite. Besides, the relatively opposite relations between cold and hot, fast and slow, big and small, etc., are all subjective relatively opposite.

The characteristic of subjective relatively opposite is that the neutral value and neutral point between two relatively opposite flexible linguistic values are appointed artificially, but not objective existences. Therefore, this kind of relatively opposite relation is interchangeable with relatively negative relation. In fact, when inserting a neither-this-nor-that neutral flexible linguistic value between a pair of relatively negative linguistic values, the original relatively negative relation just becomes a relatively opposite relation. Conversely, removing neutral flexible linguistic value between a pair of relatively opposite flexible linguistic values, then the original relatively opposite relation just becomes a relatively negative relation. For example, if there is no "medium" between "big" and "small," then "big" and "small" are relatively negative; if there is "medium," then they are relatively opposite.

Objective relatively opposite is the relatively opposite relation that objectively existed and unchangeable. Its characteristic is that there exists objectively a neutral point that is neither this nor that between two relatively opposite linguistic values. This neutral point is just a turning point between two relatively opposite linguistic values. It is just this turning point that making there is no transition zone that is this and that between two relatively opposite linguistic values. Therefore, an objective relatively opposite cannot be changed into a relatively negative, and vice versa. For instance, "slight deficit" and "slight surplus" are just an objective relatively opposite, because there exists a neutral point "0" between the two. Similarly, "small deficit" and "small surplus," "large deficit" and "large surplus," "sharp rise" and "deep fall," "big victory" and "big failure," "firmly support" and "resolutely oppose," etc., are all objective-opposite. Besides, the relatively opposite relations between rigid linguistic values are all objective relatively opposite.

(2) **Face–face relatively opposite, back–back relatively opposite, and symmetrical relatively opposite**

Examining the characteristics of the relatively opposite flexible linguistic values shown in Figs. 8.1 and 8.2, it can be seen that A in Fig. 8.1 is a positive semi-peak value and C is a negative semi-peak value; thus, we visually refer to this kind of relatively opposite as face-to-face relatively opposite, or face–face relatively opposite for short; while A in Fig. 8.1b is a negative semi-peak value, C is a positive semi-peak value; therefore, we visually refer to this kind of relatively opposite as back-to-back relatively opposite, or back–back relatively opposite.

In a broad sense, we refer to all semi-peak flexible linguistic values that are symmetrical about a neutral value and have opposite peak types as face–face relatively opposite or back–back relatively opposite. Further, we call collectively the semi-peak and full-peak flexible linguistic values that are symmetrical about a neutral value as symmetrical relatively opposite.

Example 8.1 Let x = income payment and the range of x be $[a, b]$. As shown in Fig. 8.3, defining flexible linguistic values of "slight deficit," "slight surplus," "small deficit," "small surplus," etc., then "small deficit$^+$" and "small surplus$^-$," "medium deficit$^+$" and "medium surplus$^-$," and "large deficit" and "large surplus" are all face–face relatively opposite, and the corresponding neutral values are all "roughly balancing"; while "slight deficit" and "slight surplus," "small deficit$^-$" and "small surplus$^+$," and "medium deficit$^-$" and "medium surplus" are all back–back relatively opposite, the corresponding neutral values are all $\{0\}$. These relatively opposite relations are all symmetrical relatively opposite. In particular, full-peak values "small deficit" and "small surplus" and "medium deficit" and "medium surplus" are also symmetrical relatively opposite.

Note that as an independent flexible linguistic value,

$$\text{roughly balancing} = \text{slight deficit} \vee \{0\} \vee \text{slight surplus}$$

but as a neutral value,

$$\text{roughly balancing} = \neg \, \text{small deficit}^+ \wedge \neg \, \text{small surplus}^-$$

while "0" is the neutral point of all symmetrical relatively opposite linguistic values in the figure.

Besides, it also can be seen that if the peak value point of a full-peak value is treated as a neutral point, then from a full-peak flexible linguistic value, a pair of back–back relatively opposite flexible linguistic values can be constructed. For instance, with peak value points $\xi_{small\ deficit}$, $\xi_{medium\ deficit}$, $\xi_{small\ surplus}$, and $\xi_{medium\ surplus}$ in Fig. 8.3 as neutral points separately, 4 pairs of back–back relatively opposite flexible linguistic values can be constructed.

(3) **Global relatively opposite and local relatively opposite**

From Fig. 8.3, it also can be seen that besides the above-stated relatively opposite flexible linguistic values, "slight deficit" and "medium deficit" about "small deficit," "large deficit" and "small deficit" about "medium deficit," "slight surplus" and "medium surplus" about "small surplus," "small surplus" and "large surplus" about "medium surplus," etc., are all relatively opposite. But "small deficit," "medium deficit," "small surplus," and "medium surplus" as neutral values are merely local, while neutral values "roughly balancing" and $\{0\}$ are global. Similarly, neutral points $\xi_{small\ deficit}$, $\xi_{medium\ deficit}$, $\xi_{small\ surplus}$, and $\xi_{medium\ surplus}$ are also local, while neutral point 0 is global.

We refer to the relatively opposite relations relative to local neutral value as local relatively opposite, and the relatively opposite relations relative to global neutral value as global relatively opposite.

It can be seen that the global neutral value and the neutral point in a space are unique, and also objective and absolute, while the local neutral value and the neutral point are subjective and relative. In fact, a global neutral point is tantamount to dividing a whole space into two "half space" that are relatively opposite, while defining relatively opposite flexible linguistic values on either "half space" (of course, we can even more define relatively negative flexible linguistic values; for instance, in Fig. 8.3, "large deficit," "medium deficit," "small deficit," and "slight deficit" are just relatively negative pairwise, and "slight surplus," "small surplus," "medium surplus," and "large surplus" are also relatively negative pairwise). Viewed conversely, the universe space that has a global neutral point is a "full space" connected by two "half spaces" with a neutral point.

8.2 Relation Between Consistency-Degrees of Relatively Opposite Flexible Linguistic Values

Two flexible linguistic values that are face–face relatively opposite are separated by a neutral point, but because of the continuity of space, the difference between objects in the support sets of these two linguistic values is still related to the distance. Therefore, for those objects located at neutral zone and located at the support set of an opposite value, we can use negative numbers to represent the consistency-degrees of them having the corresponding another opposite value. In this sense, the consistency functions of two face–face relatively opposite flexible linguistic values can be extended to the whole universe space. Thus, we can discuss the relation between degrees of one and the same object having a pair of face–face relatively opposite flexible linguistic values.

1. **Relation between the consistency-degrees of normal face–face relatively opposite flexible linguistic values**

   We call a pair of face–face relatively opposite flexible linguistic values that the corresponding neutral point is located at the center of the neutral zone and that their core widths and boundary widths are separately equal, to be the **normal face–face relatively opposite flexible linguistic values**. It can be seen that the curves of consistency functions of a pair of normal face–face relatively opposite flexible linguistic values are axis-symmetrical about straight line $x = N_0$. As shown in Fig. 8.4, suppose that flexible linguistic values A and B on universe U are normally face–face relatively opposite, and C and D are also normally face–face relatively opposite, that is, $B = -A$, $D = -C$. Then, straight lines $y = c_A(x)$ and $y = c_B(x)$ are axis-symmetrical about vertical line $x = N_0$. Also let the intersection point of straight lines $y = c_A(x)$ and $y = c_B(x)$ be (x^*, y^*). Thus, from geometry knowledge, it is known that $y = c_A(x)$ and $y = c_B(x)$ are symmetrical about horizontal line $y = y^*$. Now, arbitrarily take $x_0 \in U$ to construct line $x = x_0$, which intersects $y = c_A(x)$ at point (x_0, y_1), and intersects $y = c_B(x)$ at point (x_0, y_2). Then, from $y = c_A(x)$ and $y = c_B(x)$ being symmetrical about $y = y^*$, we can obtain that y_1 and y_2 are symmetrical about y^*. Therefore, we have

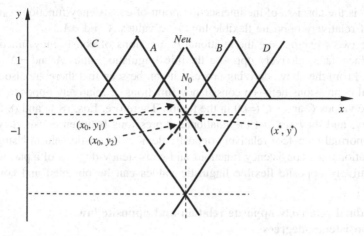

Fig. 8.4 Illustration of normal face–face relatively opposite

$$|y_1 - y_2| = 2|y^* - y_2|$$

It can be seen that when $y_1 > y_2$, $y^* > y_2$; thus

$$y_1 - y_2 = 2(y^* - y_2)$$

when $y_1 < y_2$, $y^* < y_2$; thus, still

$$y_1 - y_2 = 2(y^* - y_2)$$

It follows by this equation that

$$y_1 + y_2 = 2y^*$$

while

$$y_1 = c_A(x_0), y_2 = c_{-A}(x_0), y^* = c_{-A}(x^*) = c_A(x^*)$$

Thus

$$c_A(x_0) + c_{-A}(x_0) = 2c_{-A}(x^*)$$

Since x_0 is arbitrary, we have

$$c_A(x) + c_{-A}(x) = 2c_A(x^*) \tag{8.1}$$

$$c_A(x) + c_{-A}(x) = 2c_{-A}(x^*) \tag{8.2}$$

where x^* is the abscissa of the intersection point of consistency functions $c_A(x)$ and $c_{-A}(x)$ of relatively opposite flexible linguistic values A and $-A$.

These two equations are the relational expressions of consistency functions of normal face–face relatively opposite flexible linguistic values A and $-A$ $(=B)$ in Fig. 8.4. From the above deriving process, it can be seen that there are also similar relational expressions between consistency functions of relatively opposite flexible linguistic values C and $-C$ $(=D)$ in the figure. Therefore, Equs. (8.1) and (8.2) have generality, and they are just the relational expressions between consistency functions of normal face–face relatively opposite flexible linguistic values. Using these two equations, the consistency functions and consistency-degrees of a pair of face–face relatively opposite flexible linguistic values can be obtained and converted mutually.

2. Standard relatively opposite relation and opposite law of consistency-degrees

Definition 8.3 We call such a pair of normal face–face relatively opposite flexible linguistic values that there is only one point—neutral point in corresponding neutral zone to be the standard face–face relatively opposite flexible linguistic values, or **standard relatively opposite flexible linguistic values** for short.

For instance, the flexible linguistic values A and $-A$ in Fig. 8.5 are just a pair of standard relatively opposite flexible linguistic values.

It can be seen that neutral point N_0 between standard relatively opposite flexible linguistic values is also the common boundary point of the supports sets of these two linguistic values, and the union of these two support sets is symmetrical about neutral point N_0; the flexible set to which neutral value Neu corresponds is a flexible set with core containing single point.

We now consider the relation between the consistency functions of standard relatively opposite flexible linguistic values.

It can be seen that the ordinate of intersection point (x^*, y^*) of the graphs of the consistency functions of standard relatively opposite flexible linguistic values is 0, that is, $y^* = 0$. Therefore, standard relatively opposite flexible linguistic values are also normal face–face relatively opposite flexible linguistic values, the intersection

Fig. 8.5 Illustration of standard relatively opposite flexible linguistic values

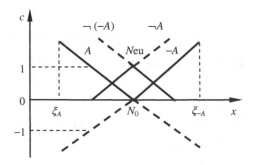

point of their consistency functions being $(x^*, 0)$, while $y^* = c_A(x^*) = c_{-A}(x^*)$. Thus, from Eqs. (8.1) and (8.2), we have

$$c_A(x) + c_{-A}(x) = 0 \qquad (8.3)$$

$$c_{-A}(x) = -c_A(x) \qquad (8.4)$$

This is the relation between consistency functions of standard relatively opposite flexible linguistic values.

Equation (8.4) means that the consistency-degrees of one and the same object with two standard relatively opposite flexible linguistic values are opposite to each other. We call Eq. (8.4) to be the opposite law of consistency-degrees of standard relatively opposite flexible linguistic values, or simply, **opposite law of consistency-degrees of relatively opposite flexible linguistic values** [1].

The opposite law is concerned with the consistency-degree of standard relatively opposite flexible linguistic values. Then, for the membership degrees of the corresponding flexible sets, how is the two related? It can be seen that if viewed from the range [0, 1] of membership degrees, the membership degrees of the corresponding flexible sets of standard relatively opposite flexible linguistic values have no relation. However, if the range of membership degrees is extended to [−1, 1], then the membership degrees of one and the same object belonging to the flexible sets corresponding to two standard relatively opposite flexible linguistic values are also mutually opposite number. The graphs of corresponding two membership functions are shown in Fig. 8.6.

Since the standard relatively opposite relation has such an important property of degrees being relatively opposite, we will mainly consider standard relatively opposite later. If there is no special note later, "relatively opposite" always refers to the standard relatively opposite. In particular, we call two flexible linguistic values that are standard relatively opposite to be the **flexible linguistic value with opposite**.

Fig. 8.6 Membership functions of the corresponding flexible sets of standard relatively opposite flexible linguistic values

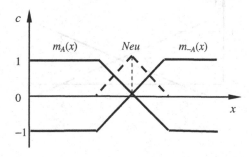

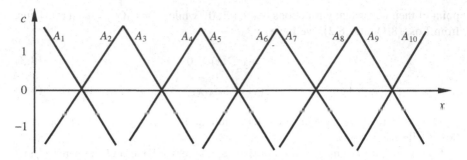

Fig. 8.7 Examples of flexible linguistic values with opposites

Thus, A and $-A$ in Fig. 8.5 are just two flexible linguistic values with opposites. Besides, the flexible linguistic values like $A_1, A_2, \ldots, A_{10}$ in Fig. 8.7 are also flexible linguistic values with opposites.

8.3   Interchange Between Relatively Negative Relation and Relatively Opposite Relation

In the above from the standpoint of relatively opposite relation, we referred to the relation between flexible linguistic values that can be relatively opposite as well as relatively negative as the subjective relatively opposite relation. Then, if viewed from the standpoint of relatively negative relation, this kind of relation is also the **subjective relatively negative relation**. In fact, many flexible concepts that are usually thought to be relatively negative are all the subjective relatively negative. That means the relatively negative relation between these flexible linguistic values can also be changed into the relatively opposite relation.

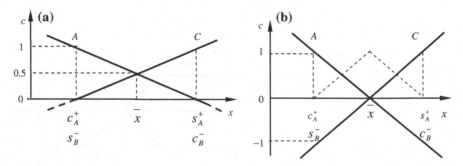

Fig. 8.8 Illustration of interchanging between subjective relatively negative relation and subjective relatively opposite relation

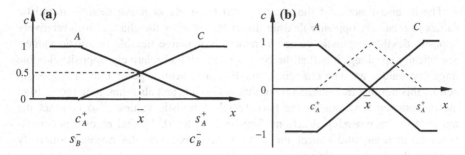

Fig. 8.9 Illustration of interchanging between subjective relatively negative relation and subjective relatively opposite relation (membership functions)

Let flexible linguistic values A and C be relatively negative, the graphs of whose consistency functions are shown in Fig. 8.8a. It can be seen that point $\bar{x} = \frac{s_A^+ + c_A^+}{2} = \frac{s_C^- + c_C^-}{2}$ is the median point of the common boundary of A and C. Then, now we take $\bar{x}$ as the neutral point between A and C, while core-boundary points c_A^+ and c_C^- of A and C do not change; then, the consistency functions of A and C become

$$c_{A'}(x) = \frac{\bar{x} - x}{\bar{x} - c_A^+}, \quad c_A^- \le x$$

$$c_{C'}(x) = \frac{x - \bar{x}}{c_C^- - \bar{x}}, \quad s_C^- \le x$$

Since $\bar{x}$ is the midpoint between core(A) and core(C), $\bar{x} - c_A^+ = c_C^- - \bar{x}$; therefore,

$$c_{A'}(x) + c_{C'}(x) = \frac{\bar{x} - x}{\bar{x} - c_A^+} + \frac{x - \bar{x}}{c_C^- - \bar{x}} = \frac{\bar{x} - x}{\bar{x} - c_A^+} + \frac{x - \bar{x}}{\bar{x} - c_A^+} = 0$$

That is to say, A and C already become relatively opposite relation, whose graphs are shown in Fig. 8.8b.

The membership functions that flexible linguistic values A and C are changed from relatively negative to relatively opposite are shown in Fig. 8.9.

It can be visually seen from Figs. 8.8 and 8.9 that the general method of changing relatively negative flexible linguistic values into relatively opposite flexible linguistic values is as follows: Do not change the cores of the original relatively negative flexible linguistic values, but to treat the original median point that is this and also that as the neutral point that is neither this nor that, reconstruct the consistency functions of two flexible linguistic values; also extend the range of the membership function from [0, 1] to [−1, 1] to reconstruct the membership function, and insert a set with core containing single point with the neutral point as center in the original boundary region to treat as the neutral value between two relatively opposite flexible linguistic values that are newly constructed.

The inverse process of the change from relatively negative flexible linguistic values to relatively opposite flexible linguistic values is the change from relatively opposite flexible linguistic values to relatively negative flexible linguistic values. Specifically speaking, it is that the cores of the original relatively opposite flexible linguistic values are not changing, but treat the neutral point that is originally neither this nor that as a median point that is also this and also that, then reconstruct the consistency functions of the two flexible linguistic values; also contract the range of the membership functions from [−1, 1] to [0, 1], and reconstruct membership functions, and cancel the neutral value between the original relatively opposite flexible linguistic values.

Thus, when relatively negative flexible linguistic values are changed into relatively opposite flexible linguistic values, the original medium point becomes a neutral point, and thus, the original extended core becomes a support set; conversely, when relatively opposite flexible linguistic values are changed into relatively negative flexible linguistic values, the original neutral point becomes a medium point, and thus, the original support set becomes an extended core. That is to say, with respect to the subjective relatively opposite and subjective relatively negative, the support set of a flexible linguistic value with opposite is tantamount to the extended core of a flexible linguistic value with negation, and vice versa.

8.4 Relevant Theories About Flexible Linguistic Value with Opposite

Viewed from a linguistic value itself in isolation, there is no difference between a flexible linguistic value with opposite and a flexible linguistic value with negation. Therefore, some basic theories about flexible linguistic values with negations in Chaps. 6 and 7 such as composition and decomposition, inclusion relation, similarity relation, flexible linguistic value with degree, superposed linguistic values, and interconversion between flexible linguistic values and numerical values are also applicable and tenable for the flexible linguistic values with opposites. So here repetition is omitted.

But, it should be specified and noted that the condition of the conversion from a numerical value to a flexible linguistic value with opposite should be $c_{A_k}(x_0) > 0$ but not $c_{A_k}(x_0) > 0.5$. Thus, in dual, from Theorem 7.1 we have the theorem below.

Theorem 8.1 Numerical value (vector) x_0 can be converted into flexible linguistic value with opposite, A, if and only if $x_0 \in \text{supp}(A)$.

Certainly, the operations of flexible linguistic values with opposites also have their uniqueness. For instance, because a flexible linguistic value with opposite also has its negative value, it is also a flexible linguistic value with negation at the same time. Thus, there can be the compound flexible linguistic value that has both opposite operation and negative operation.

Let A be a flexible linguistic value with opposite on universe $U = [a, b]$, then $-A$ is also a flexible linguistic value with opposite. Thus, it follows that compound linguistic values $\neg A$, $\neg(-A)$, and $\neg A \wedge \neg(-A)$. From the definitions of related operations of flexible linguistic values with opposites and flexible linguistic values with negations, the consistency functions of these compound linguistic values are:

$$c_{\neg A}(x) = 1 - c_A(x) \tag{8.5}$$

$$c_{\neg(-A)}(x) = 1 - c_{-A}(x) = 1 - (-c_A(x)) = 1 + c_A(x) \tag{8.6}$$

$$c_{\neg A \wedge \neg(-A)}(x) = \min\{c_{\neg A}(x), c_{\neg(-A)}(x)\} = \min\{1 - c_A(x), 1 + c_A(x)\} \tag{8.7}$$

While $\neg A \wedge \neg(-A) = Neu$, then

$$c_{Neu}(x) = \min\{1 - c_A(x), 1 + c_A(x)\} \tag{8.8}$$

Of course, we can have more such compound linguistic values that have opposite-negation double operations

It can be seen that these consistency functions are all functions of the consistency functions about flexible linguistic value with opposite, A. Thus, using these consistency functions we can calculate the consistency-degree of x with corresponding compound flexible linguistic values in the situation when the consistency-degree $c_A(x)$ of object x with a certain flexible linguistic value with opposite, A, is known.

Example 8.2 As shown in Fig. 8.10, we define three basic flexible linguistic values of "affirming," "dissenting" and "abstention" on score range $[-100, 100]$. Then, "affirming" and "dissenting" are relatively opposite, and "abstention" is the neutral value. Now it is already known that the degree of someone P affirming something is 0.8. Find the corresponding degrees of the attitude of this person to other 4 flexible

Fig. 8.10 Example of relation between degrees of relatively negative and relatively opposite flexible linguistic values

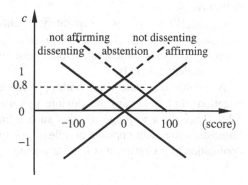

linguistic values. What is degree when the degrees of "affirming" are separately 1, 0, and −1?

Solution: From the figure, it can be seen that here "affirming" and "dissenting" are a pair of standard relatively opposite flexible linguistic values, while "affirming" and "not affirming" and "dissenting" and "not dissenting" are two pairs of relatively negative flexible linguistic values; neutral value "abstention" is neither affirming nor dissenting, that is, "abstention" = "not affirming" ∧ "not dissenting." From this, when $c_{\text{affirming}}(P) = 0.8$, we have

$c_{\text{not affirming}}(P) = 1 - 0.8 = 0.2$ (by complement law of consistency − degrees)

$c_{\text{dissenting}}(P) = -0.8$ (by opposite law of consistency − degrees)

$c_{\text{not dissenting}}(P) = 1 - (-0.8) = 1.8$ (by complement law of consistency − degrees)

$c_{\text{abstention}}(P) = \min(c_{\text{not affirming}}(P), c_{\text{not dissenting}}(P))$

$\qquad\qquad = \min(0.2, 1.8) = 0.2$ (by conjunction operation of flexible linguistic values)

Thus, $c_{\text{affirming}}(P) = 0.8$ is separately tantamount to $c_{\text{not affirming}}(P) = 0.2$, $c_{\text{dissenting}}(P) = -0.8$, $c_{\text{not dissenting}}(P) = 1.8$, and $c_{\text{abstention}}(P) = 0.2$. Then, if we interpret (affirming, 0.8), (not affirming, 0.2), (dissenting, −0.8), (not dissenting, 1.8), and (abstention, 0.2) separately as "affirming on the whole," "a bit not affirming," "totally opposite with dissenting on the whole," "quite not dissenting," and "slightly neutral," then the 5 versions are equivalent.

With the same reason, we have the following results:

$c_{\text{affirming}}(P) = 1$ is separately tantamount to $c_{\text{not affirming}}(P) = 0$, $c_{\text{dissenting}}(P) = -1$, $c_{\text{not dissenting}}(P) = 2$, and $c_{\text{abstention}}(P) = 0$; thus, "totally affirming," "totally not disaffirming," "quite not dissenting," and "totally not neutral" are equivalent mutually.

$c_{\text{affirming}}(P) = 0$ is separately tantamount to $c_{\text{not affirming}}(P) = 1$, $c_{\text{dissenting}}(P) = 0$, $c_{\text{not dissenting}}(P) = 1$, and $c_{\text{abstention}}(P) = 1$; thus, "totally not affirming," "totally not un-dissenting," "totally not dissenting," and "remain strictly neutral" are equivalent mutually.

$c_{\text{affirming}}(P) = -1$ is separately tantamount to $c_{\text{not affirming}}(P) = 2$, $c_{\text{dissenting}}(P) = 1$, $c_{\text{not dissenting}}(P) = 0$, and $c_{\text{abstention}}(P) = 0$; thus, "absolutely not affirming," "totally dissenting," "totally not un-dissenting," and "totally not neutral" are equivalent mutually.

A flexible linguistic value with opposite is also a flexible linguistic value with negation; that is to say, the flexible linguistic value with opposite plays double roles. Then, how to judge whether such a linguistic values is treated as a flexible linguistic value with opposite or a flexible linguistic value with negation in practical application? Obviously, it is hard to distinguish it from the flexible linguistic value itself, but if we consider from the background of problem, then it is easy to determine whether it should be with opposite or negation.

Here, it also should be noted that a flexible linguistic value with negation is not necessarily a flexible linguistic value with opposite, so generally cannot be written as $-(\neg A)$.

8.5   Exclusive Flexible Partition of a Space and Exclusive Flexible Linguistic Values

1. **Exclusive flexible partition of a space, exclusive flexible classes, and exclusive flexible linguistic values**

Examining the characteristics of the support sets of standard relatively opposite flexible linguistic values, it can be seen that two flexible linguistic values are of standard relatively opposite if and only if the intersection of the complement of their support sets is a single-point set. That is,

$$A = -B \wedge B = -A \Leftrightarrow \operatorname{supp}(A)^c \cap \operatorname{supp}(B)^c = \{n_0\} \tag{8.9}$$

Definition 8.4 Let A_1, A_2, ..., A_m be non-empty flexible sets in one-dimensional measurement space U, if

$$\operatorname{supp}(A_i) \cap \operatorname{supp}(A_{i+1}) = \emptyset \quad (i = 1, 2, \ldots, m - 1)$$
$$\operatorname{supp}(\neg A_i) \cap \operatorname{supp}(\neg A_{i+1}) = \{n_{0_{i+1}}\} \quad (i = 1, 2, \ldots, m - 1)$$
$$\bigcup_{i=1}^{m} \operatorname{supp}(A_i) \cup \bigcup_{i=1}^{m-1} \{n_{0_{i+1}}\} = U$$

then we say $\pi = \{A_1, A_2, \ldots, A_m\}$ is an exclusive flexible partition of space U and flexible sets A_1, A_2,...,A_m are mutually exclusive, or that the corresponding flexible linguistic values A_1, A_2,...,A_m are mutually exclusive and A_1, A_2,...,A_m form a group of mutually exclusive basic flexible linguistic values on space U.

Example 8.3 As shown in Fig. 8.11a, $\pi_1 = \{A_1, A_2, \ldots, A_6\}$ is an exclusive flexible partition of one-dimensional space $U = [a, b]$, flexible sets A_1, A_2, ..., A_6 are mutually exclusive, and flexible linguistic values A_1, A_2, ..., A_6 are mutually exclusive. As shown in Fig. 8.11b, $\pi_2 = \{A_1, A_2, \ldots, A_{10}\}$ is also a exclusive flexible partition of space $U = [a, b]$, flexible sets A_1, A_2, ..., A_{10} are mutually exclusive, and flexible linguistic values A_1, A_2, ..., A_{10} are mutually exclusive.

Definition 8.5 Let A_1, A_2, ..., A_m be non-empty flexible sets in n-dimensional measurement space U, if for any adjacent A_i and A_j $(i < j, 1 \leq i < m, 1 < j \leq m)$,

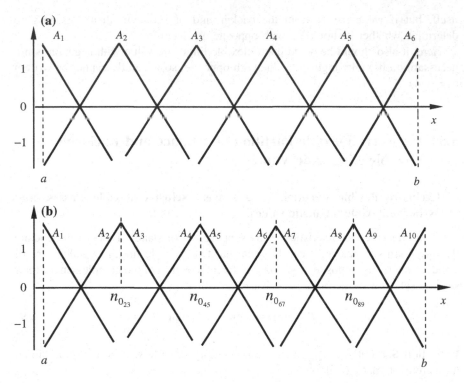

Fig. 8.11 a Example 1 of exclusive flexible partition of one-dimensional space and the corresponding exclusive flexible linguistic values. **b** Example 2 of exclusive flexible partition of one-dimensional space and the corresponding exclusive flexible linguistic values

$$\mathrm{supp}(A_i) \cap \mathrm{supp}(A_j) = \emptyset$$
$$\mathrm{supp}(\neg A_i) \cap \mathrm{supp}(\neg A_j) = \{e_{0_{ij}}\}$$

and

$$\bigcup_{i=1}^{m} \sup p(A_i) \cup \bigcup_{i,j \in \{1,2,\dots,m\}} \{e_{0_{ij}}\} = U$$

then we say $\pi = \{A_1, A_2, \dots, A_m\}$ is an exclusive flexible partition of space U and flexible sets $A_1, A_2, \dots, A_m$ are mutually exclusive and the corresponding flexible linguistic values $A_1, A_2, \dots, A_m$ are mutually exclusive, and $A_1, A_2, \dots, A_m$ form a group of mutually exclusive basic flexible linguistic values on space U. Here, $\{e_{0_{ij}}\}$ denotes a single-element set, which can be a single-point set $\{n_0\}$, single-line set $\{l_0\}$, or single-plane set $\{p_0\}$.

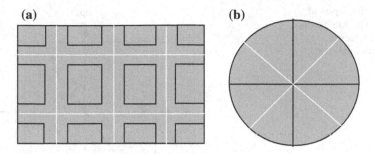

Fig. 8.12 Examples of exclusive flexible partition of multidimensional space and the corresponding exclusive flexible linguistic values

More simply speaking, a mutually exclusive flexible partition is that there is one and only one demarcation point (line or plane) formed by neutral point(s) that is neither this nor that between the support sets of two adjacent flexible sets.

Example 8.4 Figure 8.12 shows the examples of the exclusive flexible partition of two-dimensional space and the corresponding mutually exclusive flexible sets and mutually exclusive flexible linguistic values, where the white lines are the demarcation lines of adjacent flexible classes.

Note that the exclusive flexible partition is much like usual rigid partition, yet actually it is not rigid partition but is still flexible partition. Because the exclusive flexible partition is still a partition based on flexible clustering, the classes obtained are still flexible classes; only that there is only a boundary point (line or place) between the support sets of two adjacent flexible classes, not like complementary flexible partition, that the support sets of two flexible classes are intersected and that there is a region as boundary between the cores.

2. **Relation between mutual exclusion and relative opposite**

From the definition and examples of mutual exclusion relation of flexible linguistic values, it can be seen that the mutual exclusion has a certain connection with relative opposite of flexible linguistic values, and the two have both something in common and differences.

Resemblance: There is a neutral point (line or place) between two relatively opposite flexible linguistic values, while there is a neutral point (line or plane) between adjacent two of mutually exclusive flexible linguistic values.

Difference: Relative opposite is only concerned with two linguistic values, which is a relation between two linguistic values, while mutual exclusion can be concerned with two or more than two linguistic values, which is a relation among multiple linguistic values.

Thus, generally speaking, mutual exclusion is not relative opposite. However, if there are only two basic flexible linguistic values that are mutually exclusive on a universe space, then the two are also relatively opposite. Conversely, if there are only two basic flexible linguistic values that are relatively opposite on a universe

Fig. 8.13 Examples of
relatively opposite flexible
linguistic values in
two-dimensional space

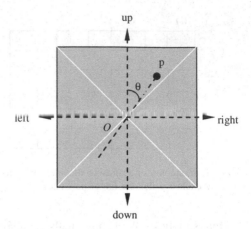

space, then the two are also mutually exclusive. That is to say, relative opposite is a
special mutual exclusion; in the situation that there are only two basic flexible
linguistic values, relative opposite and mutual exclusion are just identical.

Further, we find that though generally speaking, mutual exclusion is not relative
opposite, and some flexible linguistic values that are mutually exclusive contain
flexible linguistic valued pairs that are relatively opposite.

As a matter of fact, as shown in Fig. 8.13, starting from point O of square region
$U \times V$, draw forth 4 rays to up, down, left, and right (as shown by the broken lines
in the figure) such that the upper and lower two rays are on the same vertical line
and that left and right two rays are on the same horizontal line; then, take point O as
the origin, the vertical line as the vertical axis, and the horizontal line as the lateral
axis; next, construct separately the angle bisectors of 4 quadrants through origin
O (as shown by the white line in the figure, here they are just the diagonals of 4
squares), and then, the original square region is divided into 4 triangular sub
regions. Take the white line as the neutral line of the adjacent regions; then, the
above-stated partition is a mutually exclusive partition. Thus, we get 4 basic flexible
linguistic values on the corresponding space, and name them one by one as "up,"
"down," "left," and "right" according to the direction and denote them one by one
as U, D, L, and R.

It can be seen that the original 4 rays that pass origin O are also separately the
core centers of these 4 flexible linguistic values. As shown in Fig. 8.13, for point P
$(x, y) \in U \times V$, construct straight line l through $P(x, y)$ and origin O, and let the
included angle of l and the core center line of flexible linguistic value U (i.e., the
included angle of l and the vertical axis) be θ. Obviously, θ can be determined by
the polar coordinates of point $P(x, y)$, so θ should be the function of x and y. Thus,
from the geometric characteristic of flexible linguistic value U, we obtain the
consistency function of it about included angle θ as follows:

$$c_U(\theta) = \begin{cases} \frac{45-\theta}{45}, & \text{when} \quad y > 0 \\ -\frac{45-\theta}{45}, & \text{when} \quad y < 0 \end{cases}$$

Let $\theta = \varphi(x, y)$, thus, we have

$$c_U(x, y) = \begin{cases} \frac{45-\varphi(x,y)}{45}, & \text{when} \quad y > 0 \\ -\frac{45-\varphi(x,y)}{45}, & \text{when} \quad y < 0 \end{cases}$$

Similarly, we have

$$c_D(x, y) = \begin{cases} \frac{45-\varphi(x,y)}{45}, & \text{when} \quad y < 0 \\ -\frac{45-\varphi(x,y)}{45}, & \text{when} \quad y > 0 \end{cases}$$

Now, take $\forall (x_0, y_0) \in U \times V$, let $y_0 > 0$, and suppose that the included angle of the straight line l through point $P(x_0, y_0)$ and the core center line of U is θ_0, that is, $\varphi(x_0, y_0) = \theta_0$, then

$$c_U(x_0, y_0) = \frac{45 - \varphi(x_0, y_0)}{45} = \frac{45 - \theta_0}{45}$$

$$c_D(x_0, y_0) = -\frac{45 - \varphi(x_0, y_0)}{45} = -\frac{45 - \theta_0}{45}$$

Thus,

$$c_U(x_0, y_0) + c_D(x_0, y_0) = \frac{45 - \theta_0}{45} - \frac{45 - \theta_0}{45} = 0$$

This shows that U and D are relatively opposite; that is, the two linguistic values of "up" and "down" are relatively opposite flexible linguistic values.

Similarly, we can derive that L and R are relatively opposite; that is, the two linguistic values of "left" and "right" are relatively opposite flexible linguistic values.

Thus, we obtain a group of mutually exclusive flexible linguistic values that contains relatively opposite flexible linguistic values by conducting appropriate mutually exclusive flexible partition of plane region $U \times V$.

Note here the appropriate mutually exclusive flexible partition, the characteristic of which is to exclusively divide the plane region $U \times V$ into 4 regions that are strictly symmetrical pairwise. We may as well refer to this kind of partition that is strictly symmetrical as **normal partition**. That is to say, the normal partition can produce a group of mutually exclusive flexible linguistic values that contains relatively opposite flexible linguistic values. However, if the partition of a universe space is not normal, then it cannot be guaranteed that the exclusive flexible linguistic values obtained contain relatively opposite flexible linguistic values. For instance, though the partition as shown in Fig. 8.14 may be an exclusive flexible

Fig. 8.14 An example of
non-normal partition of a
space

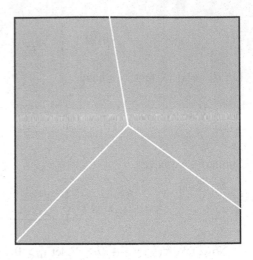

partition, it is not a normal partition, so relatively opposite flexible linguistic values
cannot be produced.

In the above, plane region $U \times V$ is normally divided into 4 regions, obtaining
two pairs of relatively opposite flexible linguistic values. Similarly, plane region
$U \times V$ can also be exclusively divided into $2 \times 2^2 = 8$, $2 \times 2^3 = 16$, …, 2×2^n
regions that are strictly symmetrical pairwise, obtaining 2^n pairs of relatively
opposite flexible linguistic values.

On the other hand, using the same method, three-dimensional space region
$U \times V \times W$ can also be normally divided into 6 regions, obtaining 3 pairs of
relatively opposite flexible linguistic values, and further to exclusively divide space
$U \times V \times W$ into $3 \times 2^2 = 12$, $3 \times 2^3 = 24$, …, 3×2^n regions that are strictly
symmetrical pairwise, obtaining $3 \times 2^{n-1}$ pairs of relatively opposite flexible linguistic values.

From the above examples, we see that there seems to be such a law: A group of
mutually exclusive basic flexible linguistic values whose total number is an even
number may contain relatively opposite flexible linguistic values, and a group of
mutually exclusive basic flexible linguistic values whose total number is an odd
number cannot contain relatively opposite flexible linguistic values.

8.6 Relatively Opposite Flexible Sets and Flexible Set
with Opposite

From the relation between a flexible linguistic value and its corresponding flexible
set, two flexible sets labeled by a pair of relatively opposite flexible linguistic values
are also relatively opposite flexible sets. Next, we give the definition of relatively
opposite flexible sets starting from sets directly.

Definition 8.6 Let A and C be two flexible sets of one-dimensional space $U = [a, b]$, and $\text{supp}(A)^c \cap \text{supp}(C)^c = B$. If sets $\text{supp}(A)$, B, and $\text{supp}(C)$ are adjacent one by one, and $\text{supp}(A) \cap B \cap \text{supp}(C) = \varnothing$, then we say flexible sets A and C are relatively opposite about B, and B is called the neutral set between A and C. If $B = \{N_0\}$ ($N_0 \in U$), then we say flexible sets A and C are the standard relatively opposite.

Definition 8.7 Let A and C be two flexible sets of one-dimensional space $U = [a, b]$. If set $\text{supp}(A)^c \cap \text{supp}(C)^c = \{N_0\}$ ($N_0 \in U$), then we say flexible sets A and C are relatively opposite about $\{N_0\}$, $\{N_0\}$ is called the neutral set between flexible sets A and C, and N_0 is called the neutral point between A and C.

We call two flexible sets that are standard relatively opposite to be the **flexible set with opposite** (dually, we call two complementary flexible sets to be the **flexible set with complement**).

Viewed from a flexible set itself in isolation, there is no distinction between a flexible set with opposite and a flexible set with complement, and a flexible set with opposite can also be a flexible set with complement at the same time. Therefore, the discussion and conclusions about flexible set with complement in Chap. 5 are also applicable and tenable for flexible set with opposite. In the following, we directly give the membership functions of flexible sets A^c, $(-A)^c$, $A^c \cap (-A)^c$, and Neu:

$$m_{A^c}(x) = 1 - m_A(x) \tag{8.10}$$

$$m_{(-A)^c}(x) = 1 + m_A(x) \tag{8.11}$$

$$m_{A^c \cap (-A)^c}(x) = \min\{1 - m_A(x), 1 + m_A(x)\} \tag{8.12}$$

$$m_{Neu}(x) = \min\{1 - m_A(x), 1 + m_A(x)\} \tag{8.13}$$

8.7 Summary

In this chapter, we introduced the concepts of relatively opposite flexible linguistic values and relatively opposite flexible sets and founded the related theories. The main points and results are as follows:

- If there is a neither-this-nor-that neutral point (line or plane) between the corresponding flexible sets of two flexible linguistic values, then these two flexible linguistic values are of relatively opposite relation, and they are called relatively opposite values; the corresponding flexible sets are also relatively opposite relation, and the two sets are called relatively opposite flexible sets.
- Relatively opposite flexible linguistic values have the types of subjective relatively opposite and objective relatively opposite, face–face relatively opposite and back–back relatively opposite, global relatively opposite and local relatively

opposite, normal face–face relatively opposite, standard relatively opposite, etc. The consistency-degrees of one and the same numerical value with a pair of standard relatively opposite flexible linguistic values are mutually opposite number, and the relation is called the relatively opposite principle of the consistency-degrees of relatively opposite flexible linguistic values.

- The subjective relative opposite and the subjective relative negation can be mutually changed When a relative opposite relation is changed into a relative negation relation, the corresponding neutral point becomes the medium point, and the support set becomes the extended core; conversely, when a relative negation relation is changed into a relative opposite relation, the corresponding medium point becomes the neutral point, and the extended core becomes the support set.
- Flexible linguistic values with opposites have the operations and relations as well as superposing and conversions, etc., similar to flexible linguistic values with negations.
- The flexible sets to which a pair of relatively opposite flexible linguistic values correspond form a relatively opposite partition of the corresponding measurement space. The generalization of relatively opposite partition is mutually exclusive partition; that is, there is one and only one neither-this-nor-that neutral point (line or plane) between adjacent flexible sets in a space. Between flexible linguistic values obtained from a mutually exclusive partition is mutually exclusive relation, which forms a group of mutually exclusive basic flexible linguistic values on the corresponding space.

Reference

1. Lian S (2009) Principles of imprecise-information processing. Science Press, Beijing

Chapter 9
Correspondence Between Flexible Sets, and Flexible Linguistic Functions

Abstract This chapter analyzes firstly the mathematical backgrounds and relational representations of a set correspondence and a flexible-set correspondence, thus revealing the mathematical essence, mathematical background, and relational representation of a flexible-linguistic-valued correspondence, then proposes and discusses flexible linguistic functions and flexible linguistic correlations, presents their types and representation, analyzes their characteristics, properties, and evaluations, and in particular, discovers and proposes a quantitative description and numerical model of flexible linguistic functions and flexible linguistic correlations.

Keywords Set correspondence · Flexible-set correspondence · Flexible-linguistic-valued correspondence · Flexible linguistic functions · Flexible linguistic correlations · Numerical-model representative

Now that the numerical values in a measurement space can be summarized into one and another flexible linguistic value; then, a function or correlation between numerical values in measurement spaces would be summarized into a function or correlation described by flexible linguistic values. In fact, flexible linguistic functions and correlations are just a kind of model or supplementary or necessary means of modeling of some complex systems. In this chapter, we first talk about the correspondence between sets, then discuss the correspondences between flexible sets and between flexible linguistic values, and then will introduce and discuss flexible linguistic functions and flexible linguistic correlations.

9.1 Correspondence Between Flexible Sets

In this section, we will examine another relation between flexible sets—correspondence. We consider firstly the correspondence between sets.

© Springer Science+Business Media Singapore 2016
S. Lian, *Principles of Imprecise-Information Processing*,
DOI 10.1007/978-981-10-1549-6_9

205

9.1.1 Correspondence Between Two Sets

Let A and B be two subsets. Conceive that if to each $x \in A$ there corresponds to a $y \in B$, then viewed from the level of set, there occurs a correspondence (relation) between sets A and B.

1. Definition and type of correspondence between sets

Definition 9.1 Let U and V be two sets, and $A \subset U$ and $B \subset V$. If to each $x \in A$ there corresponds to at least one $y \in B$, then we say that set A is corresponded to set B, or set B corresponds to set A, write as $A \longmapsto B$.

From this definition, if there is a function f from a set A to a set B, then set A is corresponded to set B. The large amounts of function and correlation in mathematics and practical problems show that the correspondence (relation) between sets is existent.

Definition 9.2 Let A and B be two sets and $A \longmapsto B$. If each element in B is the image of a certain element or some elements in A under correspondence $A \longmapsto B$, then we say the correspondence $A \longmapsto B$ to be an onto (or surjective) correspondence.

Example 9.1 There are some examples of correspondence between sets in the following Fig. 9.1 (here A and B are sets of real numbers), of them (b) and (d) are onto correspondence.

Actually, set $B^s = \{y | y = f(x) \in B, x \in A\}$, then generally, $B^s \subseteq B$, while when $B^s = B$, $A \longmapsto B$ is an onto correspondence.

2. Relationship between the correspondence between sets and the correspondence between elements

From the definitions and examples above, it can be seen that ① a correspondence between sets is really the covering of a function or correlation between elements of sets; conversely, it is just the function or correlation between elements in microscopic that forms the correspondence between sets in macroscopic; ② one and the same set correspondence covers simultaneously many or even an infinite of functions or correlations. For example, when the set correspondences $A \longmapsto B$ in (a), (b), (c), and (d) in Fig. 9.1 are one and the same set correspondence, those two functions in (a) and (b) and those two correlations in (c) and (d) are covered by the same set correspondence $A \longmapsto \in B$.

3. Relational representation and graph of a set correspondence

By definition, that set correspondence $A \longmapsto B$ is represented by a set is

$$\{(x, y) | x \in A, y = f(x) \in B\}$$

here f denotes a function or correlation covered by set correspondence $A \longmapsto B$.

Fig. 9.1 Examples of set correspondences and the binary relations summarized by which and the relational representations and graphs of correspondences $A \longmapsto B$

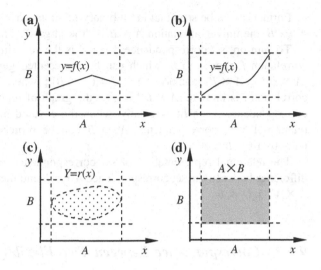

It can be seen that this set is a subset of product $A \times B$; we denote it as $(A \times B)^s$. Thus, $(A \times B)^s \subseteq A \times B$. This is to say, $(A \times B)^s$ is a binary relation from set A to set B. However, this binary relation is not general binary relation, which is a function or correlation from A to B. Consequently, $(A \times B)^s = f$.

Now we have seen that

① When the function or correlation covered is known, set correspondence $A \longmapsto B$ can be represented as that function or correlation f as binary relation covered by which, it is just the graph of f in visual. For example, if the four correspondences, $A \longmapsto B$, in Fig. 9.1 are not the same each other, then the correspondence $A \longmapsto B$ in (a) can be represented as the corresponding binary relation—function $y = f(x)$; in visual, it is also that function curve in the figure; and the correspondence $A \longmapsto B$ in (b) is similar; but the correspondence $A \longmapsto B$ in (c) can be represented as the corresponding binary relation—correlation $Y = f(x)$ (Y is a set of images of x); in visual, it is also that set of points or region whose shape is irregular in the figure; and the correspondence $A \longmapsto B$ in (d) can be represented as the corresponding binary relation—correlation—universal relation $A \times B$; in visual, it is also that whole rectangular region $A \times B$ in the figure.

② When the function or correlation covered is not known, correspondence $A \longmapsto B$ cannot be definitely gave in the form of a specific binary relation. Because one and the same set correspondence $A \longmapsto B$ covers many or even an infinite of functions or correlations, which may be various (merely the correspondence between x and y have many cases of one-to-one, one-to-many, many-to-one, etc.), therefore, the corresponding subsets $(A \times B)^s$ of correspondence $A \longmapsto B$ cannot have a unified expressing form and visual graph.

Further, it can be seen that in all binary relations covered by set correspondence $A \mapsto B$, the universal relation $A \times B$ is "the largest". This is an extreme case.

To sum up, set correspondence $A \mapsto B$ is the covering of a certain function or correlation f from A to B, which can be represented generally as binary relation $(A \times B)^s \subseteq A \times B$, namely $A \mapsto B = (A \times B)^s$; when the covered function or correlation f is known, $(A \times B)^s = f$, so the graph of relation f is also the graph of correspondence $A \mapsto B$; especially, when the covered binary relation is universal relation $A \times B$, correspondence $A \mapsto B$ can be represented as product $A \times B$, namely $A \mapsto B = A \times B$.

The relational representation of set correspondence $A \mapsto B$ shows clearly the difference between the set correspondence $A \mapsto B$ and the set operations $A \times B$ and $A \times V \cap U \times B$.

9.1.2 Correspondence Between Two Flexible Sets

Let U and V be two measurement spaces, and A and B be separately the flexible subsets of U and V. Conceive that if to each $x \in U$ belonging to A with a degree there corresponds a $y \in V$ belonging to B with a degree, then viewed from the level of set, there occurs a correspondence (relation) between flexible sets A and B.

1. **Definition and type of correspondence between flexible sets**
   The members of a flexible set can only belong to the flexible set with a degree, and only the degrees greater than 0 are meaningful. Thus, we give the definition below.

Definition 9.3 Let U and V be two measurement spaces, and A and B be separately the flexible subsets of U and V. If to each $x \in U$ with $m_A(x) > 0$ there corresponds to at least one $y \in V$ with $m_B(y) > 0$, then we say that flexible set A is corresponded to flexible set B, or flexible set B corresponds to flexible set A, write as $A \mapsto B$.

It can be seen that membership degrees $m_A(x)$ and $m_B(y)$ only show or bound separately the scopes of x and y, but they bear no relation to whether y corresponds to x. That is to say, in the sense of correspondence, the membership degree lose effect, the status of all members in support sets supp(A) and supp(B) are all equal. Then, since $m_A(x) > 0$ is just $x \in$ supp(A), and $m_B(y) > 0$ is just $y \in$ supp(B), thus, the correspondence from flexible set A to flexible set B can be translated into or reduced as the correspondence from support set supp(A) to support set supp(B); conversely, the correspondence from support set supp(A) to support set supp(B) is also the correspondence from flexible set A to flexible set B. While the correspondence between support sets is the correspondence between rigid sets, from last section, which is existent, therefore, the correspondence between flexible sets is also existent.

Fig. 9.2 Examples of
flexible-set correspondences
and the binary relations
summarized by which and the
relational representations and
graphs of correspondences
$A \longmapsto B$

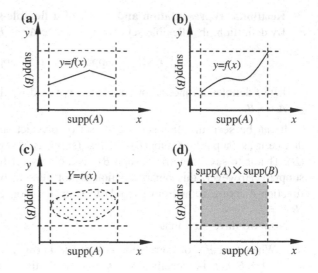

Definition 9.3′ Let U and V be two measurement spaces, and A and B be separately the flexible subsets of U and V. If to each $x \in \text{supp}(A)$ there corresponds to at least one $y \in \text{supp}(B)$, then we say that flexible set A is corresponded to flexible set B, or flexible set B corresponds to flexible set A, write as $A \longmapsto B$.

Obviously, the membership degrees $m_A(x)$ and $m_B(y)$ bear more no relation to whether y corresponds to x.

Definition 9.4 Let A and B be two flexible sets and $A \longmapsto B$. If each element in supp (B) is the image of a certain element or some elements in supp(A) under correspondence $A \longmapsto B$, we say the correspondence $A \longmapsto B$ to be onto (or surjective) correspondence.

Example 9.2 There are some examples of flexible-set correspondence in the following Fig. 9.2, of them (b) and (d) are onto correspondence.

2. **Relationship of the correspondence between flexible sets and the correspondence between elements**

From the definitions and examples above, it can be seen that ① a flexible-set correspondence is really the covering of a function or correlation between elements of flexible sets; conversely, it is just the function or correlation between elements in microscopic that forms the correspondence between flexible sets in macroscopic; ② one and the same flexible-set correspondence covers simultaneously many or even an infinite of functions or correlations. For example, when flexible-set correspondences $A \longmapsto B$ in (a), (b), (c), and (d) in Fig. 9.2 are one and the same flexible-set correspondence, those two functions in (a) and (b) and those two correlations in (c) and (d) are covered by the one and the same flexible-set correspondence $A \longmapsto B$.

3. **Relational representation and graph of a flexible-set correspondence**

By definition, that flexible-set correspondence $A \mapsto B$ is represented by a set is

$$\{(x,y)|x \in \text{supp}(A), y = f(x)\text{supp(B)}\}$$

here f denotes a function or correlation covered by flexible-set correspondence $A \mapsto B$.

It can be seen that this set is a subset of product $\text{supp}(A) \times \text{supp}(B)$, and we denote it as $(\text{supp}(A) \times \text{supp}(B))^s$. Thus, $(\text{supp}(A) \times \text{supp}(B))^s \subseteq \text{supp}(A) \times \text{supp}(B)$. That is to say, $(\text{supp}(A) \times \text{supp}(B))^s$ is a binary relation from set $\text{supp}(A)$ to set $\text{supp}(B)$. However, this binary relation is not general binary relation, which is a function or correlation from $\text{supp}(A)$ to $\text{supp}(B)$. Consequently, $(\text{supp}(A) \times \text{supp}(B))^s = f$.

Now we have seen that

① When function or correlation covered is known, flexible-set correspondence $A \mapsto B$ can be specifically represented as that function or correlation f as binary relation covered by which, it is just the graph of f in visual. For example, if the four correspondences, $A \mapsto B$, in Fig. 9.2 are not the same each other, then the correspondence $A \mapsto B$ in (a) can be represented as the corresponding binary relation—function $y = f(x)$; in visual, it is also that function curve in the figure; and the correspondence $A \mapsto B$ in (b) is similar; but the correspondence $A \mapsto B$ in (c) can be represented as the corresponding binary relation—correlation $Y = f(x)$ (Y is a set of images of x); in visual, it is also that set of points or region whose shape is irregular in the figure; and the correspondence $A \mapsto B$ in (d) can be represented as the corresponding binary relation—correlation—universal relation $\text{supp}(A) \times \text{supp}(B)$; in visual, it is also that whole rectangular region $\text{supp}(A) \times \text{supp}(B)$ in the figure.

② When function or correlation covered is not known, correspondence $A \mapsto B$ cannot be definitely given by the form of a specific binary relation. Because one and the same flexible-set correspondence $A \mapsto B$ covers many or even an infinite of functions or correlations, which may be various (merely the correspondence between x and y have many cases of one-to-one, one-to-many, many-to-one, etc.), therefore, the corresponding subsets $(\text{supp}(A) \times \text{supp}(B))^s$ of correspondence $A \mapsto B$ cannot have a unified expressing form and visual graph.

Further, it can be seen that in all binary relations covered by flexible-set correspondence $A \mapsto B$, the universal relation $\text{supp}(A) \times \text{supp}(B)$ is "the largest," which is an extreme case.

To sum up, flexible-set correspondence $A \mapsto B$ is the covering of a certain function or correlation f from A to B, which can be represented generally as binary relation $(\text{supp}(A) \times \text{supp}(B))^s \subseteq \text{supp}(A) \times \text{supp}(B)$, namely $A \mapsto B = (\text{supp}(A) \times \text{supp}(B))^s$; when the covered function or correlation f is known, $(\text{supp}(A) \times \text{supp}(B))^s = f$, so the graph of relation f is also the graph of correspondence

$A \mapsto B$; especially, when the covered binary relation is universal relation supp $(A) \times$ supp(B), correspondence $A \mapsto B$ can be represented as product supp $(A) \times$ supp(B), namely $A \mapsto B =$ supp$(A) \times$ supp(B).

The relational representation of flexible-set correspondence $A \mapsto B$ shows clearly the difference between the flexible-set correspondence $A \mapsto B$ and the flexible set operations $A \times B$ and $A \times V \cap U \times B$.

Note that although the correspondence between numbers is always rigid, the correspondence between sets (including flexible sets) can be rigid as well as flexible. In fact, the correspondence $\mapsto$ between (flexible) sets that we define above is just rigid, while the "partial correspondence" between flexible sets, such as "basically corresponding", "some corresponding" and so on, is then flexible, For "partial correspondence" between two (flexible) sets, we will discuss in Sect. 20.6.

9.1.3 The Correspondence with a Compound Flexible Set

What the set correspondence and flexible-set correspondence above cover are the functions and correlations of a variable, but there are also the functions and correlations of multiple variables in mathematics and practical problems. A function or correlation of multiple variables forms a correspondence from one compound set to another set.

Definition 9.5 Let $A_1, A_2, \ldots, A_n$ be separately flexible subsets of measurement spaces $U_1, U_2, \ldots, U_n$, and B be a flexible subset of V. If to each $(x_1, x_2, \ldots, x_n) \in$ supp$(A_1 \cap A_2 \cap \cdots \cap A_n)$ there corresponds to at least one $y \in$ supp(B), then we say that compound flexible set $A_1 \cap A_2 \cap \cdots \cap A_n$ is corresponded to flexible set B, or flexible set B corresponds to flexible set $A_1 \cap A_2 \cap \cdots \cap A_n$, write $A_1 \cap A_2 \cap \cdots \cap A_n \mapsto B$.

Definition 9.5' Let $A_1, A_2, \ldots, A_n$ be separately flexible subsets of measurement spaces $U_1, U_2, \ldots, U_n$, and B be flexible subsets of V. If there exists a function or correlation from supp$(A_1 \cap A_2 \cap \cdots \cap A_n)$ to supp(B), then we say that compound flexible set $A_1 \cap A_2 \cap \cdots \cap A_n$ is corresponded to flexible set B, or flexible set B corresponds to flexible set $A_1 \cap A_2 \cap \cdots \cap A_n$, write $A_1 \cap A_2 \cap \cdots \cap A_n \mapsto B$.

Definition 9.6 Let $A_1, A_2, \ldots, A_n$ be separately flexible subsets of measurement spaces $U_1, U_2, \ldots, U_n$, and B be flexible subsets of V. If to each $x_1 \in$ supp(A_1) or $x_2 \in$ supp(A_2) or ... or $x_n \in$ supp(A_n) there corresponds to at least one $y \in$ supp(B), then we say that compound flexible set $A_1 \cup A_2 \cap \cdots \cap A_n$ is corresponded to flexible set B, or flexible set B corresponds to flexible set $A_1 \cap A_2 \cap \cdots \cap A_n$, write $A_1 \cap A_2 \cap \cdots \cap A_n \mapsto B$.

This definition is to say that correspondence $A_1 \cap A_2 \cap \cdots \cap A_n \mapsto B$ is a union of $A_1 \mapsto B, A_2 \mapsto B, \ldots$ and $A_n \mapsto B$, that is, $A_1 \cap A_2 \cap \cdots \cap A_n \mapsto B$ is equivalent to $A_1 \mapsto B$ or $A_2 \mapsto B$ or ... or $A_n \mapsto B$.

Definition 9.7 Let A_1, A_2, ..., A_n be separately flexible subsets of measurement spaces U_1, U_2, ..., U_n, and B be flexible subsets of V. If to each $(x_1, x_2, ..., x_n) \in$ supp $(A_1 \times A_2 \times \cdots \times A_n)$ there corresponds to at least one $y \in$ supp(B), then we say that compound flexible set $A_1 \times A_2 \times \cdots \times A_n$ is corresponded to flexible set B, or flexible set B corresponds to flexible set $A_1 \times A_2 \times \cdots \times A_n$, write $A_1 \times A_2 \times \cdots \times A_n \mapsto B$.

From the definitions above, it can be see that

- Correspondence $A_1 \cap A_2 \cap \cdots \cap A_n \mapsto B$ is formed by a function or correlation from $A_1 \cap A_2 \cap \cdots \cap A_n$ to B and in turn which covers the function or correlation, and also covers simultaneously all functions and correlations from $A_1 \cap A_2 \cap \cdots \cap A_n$ to B, of them the largest is (local) universal relation supp $(A_1 \cap A_2 \cap \cdots \cap A_n) \times$ supp(B);
- Correspondence $A_1 \cap A_2 \cap \cdots \cap A_n \mapsto B$ is formed by n functions or correlations from A_1 to B, from A_2 to B, ..., from A_n to B and in turn which covers these functions or correlations, and also covers simultaneously all function and correlations from A_1 to B, from A_2 to B, ..., from A_n to B, of them the largest is (local) universal relation

$$\text{supp}(A_1) \times \text{supp}(B) \cup \text{supp}(A_2) \times \text{supp}(B) \cup \cdots \cup \text{supp}(A_n) \times \text{supp}(B)$$
$$= \text{supp}(A_1 \cup A_2 \cup \cdots \cup A_n) \times \text{supp}(B);$$

- Correspondence $A_1 \times A_2 \times \cdots \times A_n \mapsto B$ is formed by a function or correlation from $A_1 \times A_2 \times \cdots \times A_n$ to B and in turn which covers the function or correlation, and also covers simultaneously all functions and correlations from $A_1 \times A_2 \times \cdots \times A_n$ to B, of them the largest is (local) universal relation supp $(A_1 \times A_2 \times \cdots \times A_n) \times$ supp(B).

By the definitions of intersections and Cartesian products of flexible sets and rigid sets, we have

$$\text{supp}(A_1 \cap A_2 \cap \cdots \cap A_n) = \text{supp}(A_1) \cap \text{supp}(A_2) \cap \cdots \cap \text{supp}A_n)$$
$$= \text{supp}(A_1) \times \text{supp}(A_2) \times \cdots \times \text{supp}A_n)$$
$$\text{supp}(A_1 \times A_2 \times \cdots \times A_n) = \text{supp}(A_1) \times \text{supp}(A_2) \times \cdots \times \text{supp}A_n)$$

Thus, the largest of numerical relations covered by correspondence $A_1 \cap A_2 \cap \cdots \cap A_n \mapsto B$ and that covered by correspondence $A_1 \times A_2 \times \cdots \times A_n \mapsto B$ are the same, that is, those are all universal relation supp$(A_1) \times \cdots \times$ supp$(A_n) \times$ supp(B).

9.1.4 Flexible Relations and Flexible-Set Correspondences

First, it is not hard to see that from a binary flexible relation, we can obtain a flexible-set correspondence (or a rigid-set correspondence). For example, it is such

Fig. 9.3 Illustration
of a flexible-set
(set) correspondence obtained
from a binary flexible

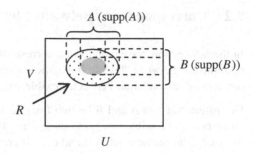

an example that shown in Fig. 9.3, here we obtain flexible-set correspondence
$A \longmapsto B$ [and rigid-set correspondence supp(A) $\longmapsto$ supp(B)] from binary flexible
relation R.

Then, conversely, from two sections above we see that since the functions and
correlations between numbers are all rigid relations, so binary relations covered by
flexible-set correspondences are all binary rigid relation. That is to say, a
flexible-set correspondence can and merely can be represented as a binary rigid
relation, but a binary flexible relation (thus, there exist no "flexible mappings,"
"flexible functions," or "flexible correlations" between two flexible sets in the sense
of (complete) correspondence $\longmapsto$). For example, see Fig. 9.3 reversely, the binary
relation covered by obtained flexible-set correspondence $A \longmapsto B$ are then various
binary rigid relations that include that oval region in the figure, but not original
binary flexible relation R. This is to say, even if such a flexible-set correspondence
$A \longmapsto B$ obtained by binary flexible relation R cannot be represented as original
binary flexible relation R either.

From the analyses above, we find that a flexible subset of a Cartesian product has
dual role and dual interpretation.

In fact, let U and V be two measurement spaces. Then, a subset R of product
space $U \times V$ stands for a certain binary flexible relation, such as approximately
equal, far greater than, approximate, similar, and so forth. However, on the other
hand, since a ordered pair (x, y) in flexible subset R itself represents the correspondence between two elements, so viewed from correspondence, the flexible
subset R also stands for a correspondence relation (may be function or correlation)
between elements, and no matter the R standing for what practical flexible relation,
viewing abstractly, which stands always for the correspondence relation between
elements. That is to say, one and the same flexible subset R of product set
$U \times V$ stands really for two kinds of binary relations—a practical binary relation in
semantics and a correspondence relation between elements at the same time. Since
the correspondence between numbers is always rigid, thus, one and the same
flexible subset R of product space $U \times V$ can stand for a certain practical binary
flexible relation as well as "correspondence relation" the binary rigid relation.

Actually, a rigid subset of a Cartesian product also has dual role and dual
interpretation, only practical relation, and correspondence relation stood for by
which are all rigid relation.

9.2   Correspondence Between Flexible Linguistic Values

In the last section, we examined the correspondence between flexible sets. From the relationship between flexible linguistic values and flexible sets, there also occurs correspondence relation between flexible linguistic values.

Definition 9.8 Let A and B be two flexible linguistic values, which also label the corresponding flexible sets at the same time. Flexible linguistic value B corresponds to flexible linguistic value A if and only if corresponding flexible set B corresponds to flexible set A.

Thus, from the conclusions about flexible-set correspondence in Sect. 9.1, we can have immediately the following conclusion about flexible-linguistic-valued correspondence:

- Flexible-linguistic-valued correspondence $A \mapsto B$ is the summarization of a certain function or correlation from flexible set A to flexible set B, which also summarizes simultaneously all functions and correlations from A to B, of them the largest is universal relation supp(A) $\times$ supp(B).
- Correspondence with composite flexible linguistic value, $A_1 \wedge A_2 \wedge \cdots \wedge A_n \mapsto B$, is the summarization of a certain function or correlation from compound flexible set $A_1 \cup A_2 \cap \cdots \cap A_n$ to flexible set B, which also summarizes simultaneously all functions and correlations from $A_1 \cap A_2 \cap \cdots \cap A_n$ to B, of them the largest is (local) universal relation supp $(A_1 \cap A_2 \cap \cdots \cap A_n) \times$ supp(B).
- Correspondence with composite flexible linguistic value, $A_1 \vee A_2 \vee \cdots \vee A_n \mapsto B$, is the summarization of a certain function or correlation from compound flexible set $A_1 \cup A_2 \cup \cdots \cup A_n$ to flexible set B, which also summarizes simultaneously all functions and correlations from $A_1 \cup A_2 \cup \cdots \cup A_n$ to B, of them the largest is (local) universal relation supp $(A_1 \cup A_2 \cup \cdots \cup A_n) \times$ supp(B).
  Of course, correspondence $A_1 \vee A_2 \vee \cdots \vee A_n \mapsto B$ can also be viewed as a union of $A_1 \mapsto B$, $A_2 \mapsto B$, ..., $A_n \mapsto B$, that is, $A_1 \vee A_2 \vee \cdots \vee A_n \mapsto B$ is equivalent to $A_1 \mapsto B$ or $A_2 \mapsto B$ or ... or $A_n \mapsto B$.
- Correspondence with composite flexible linguistic value, $A_1 \oplus A_2 \oplus \cdots \oplus A_n \mapsto B$, is the summarization of a certain function or correlation from compound flexible set $A_1 \times A_2 \times \cdots \times A_n$ to flexible set B, which also summarizes simultaneously all functions and correlations from $A_1 \times A_2 \times \cdots \times A_n$ to B, of them the largest is (local) universal relation supp $(A_1 \times A_2 \times \cdots \times A_n) \times$ supp(B).
- A flexible-linguistic-valued correspondence can be represented as that function or correlation between corresponding measurement subspaces summarized by which, that is also its background function or background correlation.

In practical problems, the function or correlation summarized by a flexible-linguistic-valued correspondence is the mathematical background of the flexible-linguistic-valued correspondence. We call it the **background function** and **background correlation** of the flexible-linguistic-valued correspondence.

In the following, we also consider incidentally the relationship between flexible-linguistic-valued vectors and flexible-linguistic-valued correspondences.

As numerical vector (x, y) can also be regarded as denoting the correspondence $x \longmapsto y$ from x to y, flexible-linguistic-valued vector (A, B) can also be regarded as denoting the correspondence $A \longmapsto B$ from A to B. In fact, since (A, B) is an ordered pair, so (A, B) is tantamount to correspondence $A \longmapsto B$ between flexible linguistic values in the corresponding measurement space $U \times V$. From last section, we have known that correspondence $A \longmapsto B$ is the summarization of a certain function or correlation from corresponding flexible set A to flexible set B, of them the biggest is (local) universal relation $\text{supp}(A) \times \text{supp}(B)$. Thus, in space $U \times V$, flexible-linguistic-valued vector (A, B) represents a region whose upper bound is $\text{supp}(A) \times \text{supp}(B)$.

Similarly, the flexible linguistic values in n-dimensional flexible-linguistic-valued vector $(A_1, A_2, \ldots, A_n)$ are ordered. Therefore, in n-dimensional measurement space $U_1 \times U_2 \times \cdots \times U_n$, $(A_1, A_2, \ldots, A_n)$ is tantamount to $A_1 \longmapsto (A_2, \ldots, A_n)$, $(A_2, \ldots, A_n)$ is tantamount to $A_2 \longmapsto (A_3, \ldots, A_n)$, $\ldots$, the rest may be inferred, and consequently $(A_1, A_2, \ldots, A_n)$ is tantamount to $A_1 \longmapsto (A_2 \longmapsto (\ldots(A_{n-2} \longmapsto (A_{n-1} \longmapsto A_n) \ldots)$. Also, by the logical equivalence $A \longmapsto (B \longmapsto C) \Longleftrightarrow A \wedge B \longmapsto C$. Therefore, expression $A_1 \longmapsto (A_2 \longmapsto (\ldots(A_{n-2} \longmapsto (A_{n-1} \longmapsto A_n) \ldots)$ is tantamount to flexible-linguistic-valued correspondence $A_1 \wedge A_2 \wedge \cdots \wedge A_{n-1} \longmapsto A_n$. Actually, from $(A_1, A_2, \ldots, A_n) = ((A_1, A_2, \ldots, A_{n-1}), A_n)$, we can also obtain immediately correspondence $A_1 \wedge A_2 \wedge \cdots \wedge A_{n-1} \longmapsto A_n$. Thus, flexible-linguistic-valued vector $(A_1, A_2, \ldots, A_n)$ is tantamount to flexible-linguistic-valued correspondence $A_1 \wedge A_2 \wedge \ldots \wedge A_{n-1} \longmapsto A_n$, while from last section we have known that which is the summarization of a certain function or correlation from corresponding flexible set $A_1 \cap A_2 \cap \cdots \cap A_{n-1}$ to flexible set A_n, of them the biggest is (local) universal relation $\text{supp}(A_1) \times \text{supp}(A_2) \times \cdots \times \text{supp}(A_n)$. Thus, in space $U_1 \times U_2 \times \cdots \times U_n$, flexible-linguistic-valued vector $(A_1, A_2, \ldots, A_n)$ represents a region whose super bound is $\text{supp}(A_1) \times \text{supp}(A_2) \times \cdots \times \text{supp}(A_n)$.

From Sect. 7.4, we have known that a flexible-linguistic-valued vector $(A_1, A_2, \ldots, A_n)$, in measurement space $U_1 \times U_2 \times \cdots \times U_n$, represents a flexible square region corresponded to conjunctive flexible linguistic value $A_1 \wedge A_2 \wedge \cdots \wedge A_n$, now which also represents a region whose super bound is $\text{supp}(A_1) \times \text{supp}(A_2) \times \cdots \times \text{supp}(A_n)$, namely the product of the support sets of all component linguistic values.

9.3 Flexible Linguistic Functions

9.3.1 Definitions and Types of Flexible Linguistic Functions

Definition 9.9

(1) We call the variable taking on linguistic values to be a linguistic variable and denote it by capital letters X, Y, Z, $\ldots$, etc.

(2) We call the function in which the independent variable(s) or dependent variable take on flexible linguistic value(s) to be a flexible linguistic function.

From the definition, flexible linguistic functions can be classified as:

① The flexible-linguistic-valued function of flexible linguistic variable(s), short for Language-Language-type function, written L-L function;
② The flexible-linguistic-valued function of numerical variable(s), short for Number-Language-type function, written N-L function;
③ The numerical-valued function of flexible linguistic variable(s), short for Language-Number-type function, written L-N function.

Besides, there may be the flexible-linguistic-valued function of hybrid variable(s).

In the above, we introduce flexible linguistic functions on the basis of existing function concept. In order to facilitate the further study, in the following we define again the flexible linguistic function from the perspective of relation.

Definition 9.10 Let U and V be two measurement spaces, $\mathscr{L}_U$ be a set of flexible linguistic values on U, and $\mathscr{L}_V$ be a set of flexible linguistic values on V, and let $f \subseteq \mathscr{L}_U \times \mathscr{L}_V$ be a relation from $\mathscr{L}_U$ to $\mathscr{L}_V$. If for each flexible linguistic value X in $\mathscr{L}_U$, there always exists one and only one Y in $\mathscr{L}_V$ such that $(X, Y) \in f$, then we say f is a flexible linguistic function from $\mathscr{L}_U$ to $\mathscr{L}_V$, write $Y = f(X)$.

Example 9.3 Let $U = [a, b]$ and $V = [c, d]$ be two measurement spaces, and let $L_U = \{A_1, A_2, A_3, A_4, A_5, A_6 \mid A_1, A_2, A_3, A_4, A_5, A_6$ are flexible linguistic values on $U\}$, and $\mathscr{L}_V = \{B_1, B_2, B_3 \ B_4 \mid B_1, B_2, B_3 \ B_4$ are flexible linguistic values on $V\}$. Then, $f = \{(A_1, B_1), (A_2, B_4), (A_3, B_2), (A_4, B_1), (A_5, B_3), (A_6, B_2)\}$ is a flexible linguistic function from $\mathscr{L}_U$ to $\mathscr{L}_V$.

The Definition 9.10 is only a basic definition of flexible linguistic function. Taking into account the type and number of independent variables and the type of values of functions, we give more specific definitions of flexible linguistic functions below.

Definition 9.11 Let U_i be a one-dimensional measurement space, and $\mathscr{L}_i = \{X_i \mid X_i$ is an atomic flexible linguistic value on $U_i\}$ $(i = 1, 2, \ldots, n)$, let V be a one-dimensional measurement space, and $\mathscr{L}_V = \{Y \mid Y$ is an atomic flexible linguistic value on $V\}$, and let $f \subseteq (\mathscr{L}_1 \times \mathscr{L}_2 \times \cdots \times \mathscr{L}_n) \times \mathscr{L}_V$ be a relation from $\mathscr{L}_1 \times \mathscr{L}_2 \times \cdots \times \mathscr{L}_n$ to $\mathscr{L}_V$. If for each $X = (X_1, X_2, \ldots, X_n) \in \mathscr{L}_1 \times \mathscr{L}_2 \times \cdots \times \mathscr{L}_n$, there always exists one and only one $Y \in \mathscr{L}_V$ such that $(X, Y) \in f$, then we say f is an L-L function from $\mathscr{L}_1 \times \mathscr{L}_2 \times \cdots \times \mathscr{L}_n$ to $\mathscr{L}_V$, write $Y = f(X)$.

Definition 9.12 Let U_i be a one-dimensional measurement space $(i = 1, 2, \ldots, n)$, V be a one-dimensional measurement space, $\mathscr{L}_V = \{Y \mid Y$ is an atomic flexible linguistic value on $V\}$, and let $f \subseteq (U_1 \times U_2 \times \cdots \times U_n) \times \mathscr{L}_V$ be a relation from $U_1 \times U_2 \times \cdots \times U_n$ to $\mathscr{L}_V$. If for each $x = (x_1, x_2, \ldots, x_n) \in U_1 \times U_2 \times \cdots \times U_n$, there always exists one and only one $Y \in L_V$ such that $(x, Y) \in f$, then we say f is an N-L function from $U_1 \times U_2 \times \cdots \times U_n$ to $\mathscr{L}_V$, write $Y = f(x)$.

Definition 9.13 Let U_i be a one-dimensional measurement space, $L_i = \{X_i \mid X_i$ is an atomic flexible linguistic value on $U_i\}$ $(i = 1, 2, \ldots, n)$, let V be a one-dimensional measurement space, and let $f \subseteq (\mathscr{L}_1 \times \mathscr{L}_2 \times \cdots \times \mathscr{L}_n) \times V$ be a relation from $\mathscr{L}_1 \times \mathscr{L}_2 \times \cdots \times \mathscr{L}_n$ to V. If for each $X = (X_1, X_2, \ldots, X_n) \in \mathscr{L}_1 \times \mathscr{L}_2 \times \cdots \times \mathscr{L}_n$, there always exists one and only one $y \in V$ such that $(X, y) \in f$, then we say f is an L-N function from $\mathscr{L}_1 \times \mathscr{L}_2 \times \cdots \times \mathscr{L}_n$ to V, write $y = f(X)$.

In the above definitions, the spaces U_i $(i = 1, 2, \ldots, n)$ and V are both one-dimensional, and the values that linguistic variables X and Y take are both one-dimensional atomic flexible linguistic values, and the values that numerical variables $x_1, x_2, \ldots, x_n$ and y take are all scalars. Therefore, these flexible linguistic functions are a kind of most simple flexible linguistic function. This kind of flexible linguistic functions is directly based on the measurement space and is the most common flexible linguistic function, so we call which as typical flexible linguistic function.

Generalizing the typical flexible linguistic function, we give the more general flexible linguistic function.

Definition 9.11′ Let U_i be a k_i $(k_i \geq 1)$-dimensional measurement space, $\mathscr{L}_i = \{X_i \mid X_i$ is an atomic flexible linguistic value on $U_i\}$ $(i = 1, 2, \ldots, n)$, let V be an m $(m \geq 1)$-dimensional measurement space, $\mathscr{L}_V = \{Y \mid Y$ is an atomic flexible linguistic value on $V\}$, and let $f \subseteq (\mathscr{L}_1 \times \mathscr{L}_2 \times \cdots \times \mathscr{L}_n) \times \mathscr{L}_V$ be a relation from $\mathscr{L}_1 \times_2 \times \cdots \times \mathscr{L}_n$ to $\mathscr{L}_V$. If for each $X = (X_1, X_2, \ldots, X_n) \in \mathscr{L}_1 \times \mathscr{L}_2 \times \cdots \times \mathscr{L}_n$, there always exists one and only one $Y \in \mathscr{L}_V$ such that $(X, Y) \in f$, then we say f is an L-L function from $\mathscr{L}_1 \times \mathscr{L}_2 \times \cdots \times \mathscr{L}_n$ to $\mathscr{L}_V$, write $Y = f(X)$.

Definition 9.12′ Let U_i be a k_i $(k_i \geq 1)$-dimensional measurement space, $(i = 1, 2, \ldots, n)$, let V be an m $(m \geq 1)$-dimensional measurement space, $\mathscr{L}_V = \{Y \mid Y$ is an atomic flexible linguistic value on $V\}$, and let $f \subseteq (U_1 \times U_2 \times \cdots \times U_n) \times \mathscr{L}_V$ be a relation from $U_1 \times U_2 \times \cdots \times U_n$ to $\mathscr{L}_V$. If for each $x = (x_1, x_2, \ldots, x_n) \in U_1 \times U_2 \times \cdots \times U_n$, there always exits one and only one $Y \in \mathscr{L}_V$ such that $(x, Y) \in f$, then we say f is an N-L function from $U_1 \times U_2 \times \cdots \times U_n$ to $\mathscr{L}_V$, write $Y = f(x)$.

Definition 9.13′ Let U_i be a k_i $(k_i \geq 1)$-dimensional measurement space, $\mathscr{L}_i = \{X_i \mid X_i$ is an atomic flexible linguistic value on $U_i\}$ $(i = 1, 2, \ldots, n)$, let V be an m $(m \geq 1)$-dimensional measurement space, and let $f \subseteq (\mathscr{L}_1 \times \mathscr{L}_2 \times \cdots \times \mathscr{L}_n) \times V$ be a relation from $\mathscr{L}_1 \times \mathscr{L}_2 \times \cdots \times \mathscr{L}_n$ to V, If for each $X = (X_1, X_2, \ldots, X_n) \in \mathscr{L}_1 \times \mathscr{L}_2 \times \cdots \times \mathscr{L}_n$, there always exits one and only one $y = (y_1, y_2, \ldots, y_m) \in V$ such that $(X, y) \in f$, then we say f is an L-N function from $\mathscr{L}_1 \times \mathscr{L}_2 \times \cdots \times \mathscr{L}_n$ to V, write $y = f(X)$.

It can be seen that Definition 9.11′ is the generalization of Definition 9.11, but actually it is already a function from a set of linguistic-valued vectors to a set of linguistic values; Definition 9.12′ is the generalization of Definition 9.12, but when there is at least one space U_k, whose dimension $k_k > 1$, the function actually is already a function from a set of vectors formed by the numerical vectors to a set of linguistic values; Definition 9.13′ is the generalization of Definition 9.13, but it actually is already a function from a set of linguistic-valued vectors to a set of numerical vectors.

Besides, flexible linguistic functions, like numerical functions, can also be separated as one-to-one and non-one-to-one as well as simple and compound.

9.3.2 Representations of Flexible Linguistic Functions

Like usual numerical functions, there are also two representation methods of enumeration and formulation for flexible linguistic functions.

The enumeration representation of flexible linguistic functions is using a set of the pairs of the values of independent variable and function to represent a flexible linguistic function. For example, the flexible linguistic function in Example 9.3 above is just using enumeration method to represent.

The formulation representation of flexible linguistic functions is using an operational expression of flexible linguistic variable(s) to represent a flexible linguistic function. For example,

$$Y = (X_1 \wedge X_2) \oplus (X_3 \vee X_4) \tag{9.1}$$

is just a flexible linguistic function represented by formulation method, where X_1, X_2, X_3, X_4, and Y are all linguistic variable.

The equation above is an L-L function. Generally, the formulation representation of an L-L function is

$$Y = E(X) \tag{9.2}$$

and the formulation representations of an N-L function and an L-N function are given separately

$$Y = (f(x)) \tag{9.3}$$

and

$$y = [E(X)] \tag{9.4}$$

Here parentheses () and brackets [] denotes separately N-L conversion and L-N conversion. For example, there are an N-L function and an L-N function represented by formulation method in the following:

$$Z = (3x^2 + 4y^3 + 1) \tag{9.5}$$

$$y = [X \wedge Y] \tag{9.6}$$

Note that since the operations of flexible linguistic values only have ¬, ∧, ∨, and ⊕, so ¬X, $X \wedge Y$, $X \vee Y$, and $X \oplus Y$ the four operational expressions are most basic and most simple flexible linguistic functions, and other flexible linguistic functions represented by formulations are all their certain combination or compound. For example, the above L-L function shown as Eq. (9.1) is actually being compounded by $Y_1 = X_1 \wedge X_2$, $Y_2 = X_3 \vee X_4$, and $Y = Y_1 \oplus Y_2$. As for what are the outcomes of these most basic operations needs to determine according to the actual problems. That is to say, the flexible linguistic functions represented by formulations, besides the functional expressions, also need to have a group of operational definitions such as $A_1 = B_2 \wedge C_3$, $A_2 = B_1 \vee C_3$ and $D_1 = A_2 \oplus E_3$, and so on.

Like numerical functions, although the representation of flexible linguistic functions can be separated as enumeration method and formulation method, speaking from theory, any flexible linguistic function can be represented by using enumeration method.

9.3.3 Quantitative Description and Numerical Model of a Flexible Linguistic Function

In the above flexible linguistic functions represented by using enumeration and formulation are only a kind of descriptive definitions, which can only be treated as a kind of qualitative models, of which the "geometric graphs" can only be "sets of points" in the spaces of flexible linguistic values (speaking generally, it is difficult to draw these point sets). Obviously, for many practical problems the qualitative models are not able to meet the requirements. Then, can the flexible linguistic functions be quantitatively described? And then, can the flexible linguistic functions be translated into a kind of pure numerical object?

Considering that the flexible linguistic value with degree is the quantitative representation of flexible linguistic values, we can use flexible linguistic value with degree to quantitatively describe a flexible linguistic function. That is, $(Y, d_y) = f(X, d_x)$. It can be seen that the key of the quantitative description is to know the correspondence relation (function or correlation) between the degrees d_x and d_y. The correspondence relation between the degrees is also the quantitative model. But it is not easy to obtain a precise quantitative model, Chap. 14 will discuss the approaches to obtain approximate quantitative model.

In the following, we consider the numerical model of flexible linguistic functions.

It can be seen that ordered pair (A, B) of flexible linguistic values is tantamount to flexible-linguistic-valued correspondence $A \mapsto B$. Thus, viewed from enumeration representation, a flexible linguistic function is a set of flexible-linguistic-valued correspondences. And from the relation between flexible-linguistic-valued correspondence and flexible-set correspondence, we know that a flexible-linguistic-valued correspondence can be represented as that function or correlation summarized by which, that is, its background function or background correlation. The background function or background correlation is the numerical model of a flexible-linguistic-valued correspondence. Then, does putting together the background functions or background correlations of all flexible-linguistic-valued correspondences (i.e., ordered pairs of flexible linguistic values) of a flexible linguistic function not just form a numerical model, i.e., the background function or background correlation, of the flexible linguistic function?

However, the problem is that in practical problem, the background function or background correlation of a flexible-linguistic-valued correspondence is often unknown. That is to say, it is also very difficult to obtain the numerical model of a flexible linguistic function.

From Sect. 9.2, we have known that a flexible-linguistic-valued correspondence $A \longmapsto B$ summarizes simultaneously all functions and correlations from A to B, of them the largest is (local) universal relation $supp(A) \times supp(B)$. That is to say, universal relation $supp(A) \times supp(B)$ not only is a correlation summarized by flexible-linguistic-valued correspondence $A \longmapsto B$, but also the least upper bound, namely supremum, of all functions and correlations from A to B. Thus, we can take universal relation $supp(A) \times supp(B)$ as a representative of numerical model of flexible-linguistic-valued correspondence $A \longmapsto B$.

From the Proposition 7.1 in Sect. 7.3.1, we have known that in concept, a flexible linguistic value (flexible set) is decided by its core and support set, but in practical, which is fully stood for by its extended core. Then, conceptual flexible-linguistic-valued correspondence $A \longmapsto B$, i.e., set correspondence $supp(A) \longmapsto supp(B)$, is practically set correspondence $core(A)^{+} \longmapsto core(B)^{+}$. Thus, we further use smaller universal relation $core(A)^{+} \times core(B)^{+}$ instead of $supp(A) \times supp(B)$ to represent the numerical model of flexible-linguistic-valued correspondence $A \longmapsto B$.

Generally, we call a local universal relation summarized by a flexible-linguistic-valued correspondence to be the **numerical model representative** of the flexible-linguistic-valued correspondence.

Thus, a set of all local universal relations summarized by a flexible linguistic function is just a representative of numerical model of the flexible linguistic function, that is, the **numerical model representative** of the flexible linguistic function.

Further, we call the numerical model representative formed by support sets to be the **conceptual representative**, the numerical model representative formed by extended cores to be the **practical representative**. In usual, the numerical representatives we say refer to practical representative.

A local universal relation is also a region of corresponding product space. Regarding that this region is a "square," we call might as well which a "block point" in corresponding measurement space. Thus, the geometric graph of the numerical model representative of a flexible linguistic function is a block-point curve in corresponding product measurement space. For example, the graph of the

Fig. 9.4 Example 1 of numerical representative of a flexible linguistic function

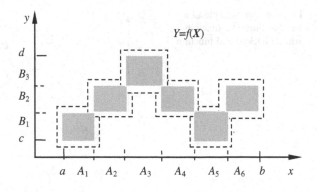

Fig. 9.5 Example 2 of numerical model representative of a flexible linguistic function

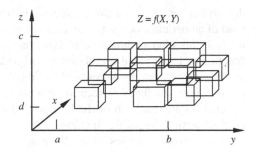

numerical model representative of the flexible linguistic function in Example 9.3 is shown in Fig. 9.4.

More generally, the numerical model representative of a multivariate flexible linguistic function is a set of local universal relations such as core $(A_1 \cap A_2 \cap \cdots \cap A_n)^+ \times \text{core}(B)^+$, its graph is a block-point surface or block-point hypersurface in corresponding measurement space (see the example in Fig. 9.5).

Now, we can translate a flexible linguistic function qualitatively described by a group of flexible-linguistic-valued correspondences (i.e., ordered pairs of flexible linguistic values) into its numerical model representative, that is, a group of local universal relations. Though, the latter is only a special case of the functions and correlations summarized by former, speaking generally, and which is not the background function or background correlation of the former, since each local universal relation covers the background function or background correlation of corresponding ordered pair of flexible linguistic values, so the flexible linguistic function represented by local universal relations covers the background function or background correlation of original flexible linguistic function (as shown in Fig. 9.6). It can be seen that this kind of big-granule function of block points is just convenient for characterization of macro-characteristics of corresponding systems (this coincided just with the original intention of linguistic-valued function). Therefore, using it in place of original flexible linguistic function is more suitable for and more convenient for macro-analysis of corresponding systems. In fact, for

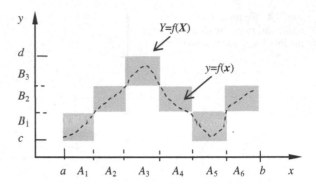

Fig. 9.6 An example of a flexible linguistic function and its background function

the systems on two- and three-dimensional measurement spaces, according to this kind of numerical models we can conduct visual examination and analysis. Further, through this kind of block-point function, we also can estimate the background function or background correlation of original flexible linguistic function. In fact, since it covers the background function or background correlation of original flexible linguistic function section by section, so it provides a basis and framework for analysis and research of the background function or background correlation of original flexible linguistic function.

In the above examples are all the examples of numerical model representatives of L-L functions, in the following we give again an example of numerical model representatives of N-L functions (see Fig. 9.7). But note that L-N functions do not require numerical-model representatives because they themselves are also their background functions (an example as shown in Fig. 9.8).

9.3.4 Characteristics, Properties, and Evaluations of Flexible Linguistic Functions

The flexible linguistic functions shown in Figs. 9.4 and 9.5 are all ideal cases—a basic flexible linguistic value on domain just corresponds to a basic flexible

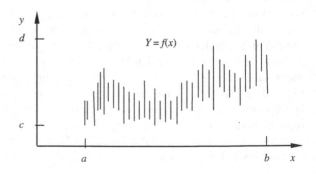

Fig. 9.7 An example of numerical model representatives of N-L functions

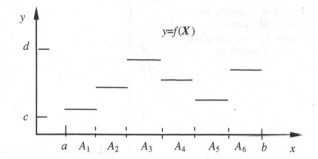

Fig. 9.8 An example of L-N functions

linguistic value on range and the block-point curves are also unbroken; however, the flexible linguistic functions in practical problems are not necessarily so, that is, a basic flexible linguistic value on domain does not necessarily correspond to the basic flexible linguistic value on range but may correspond to the non-basic flexible linguistic value (as shown in Fig. 9.9), and there also may occur discrete block points (as shown in Fig. 9.10). Certainly, that a flexible linguistic function should consist of what kind of block points is completely determined by its characteristics. Besides, since one and the same range of numerical values can have multiple different partitions thus can result multiple different sets of linguistic values, therefore one and the same numerical function can also be summarized by multiple linguistic functions. Obviously, among the multiple linguistic functions summarizing one and the same numerical function, the block-point curve or surface whose block points are smaller would be more close to the curve or surface of the numerical function. Especially, if the numerical function is a usual single-valued function, then when (the size of) block-point approach usual points, the numerical function is just the limit of such a sequence of linguistic functions. Thus, in theory, a numerical function (include multivalued function and vectored function) can be approached by a sequence of linguistic functions. Or in other words, we can use a sequence of linguistic functions in which the sizes of linguistic values become gradually smaller and smaller to approach to a numerical function (here the size,

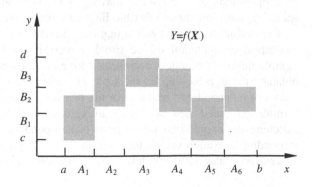

Fig. 9.9 Example 3 of numerical model representatives of flexible linguistic functions

Fig. 9.10 Example 4 of
numerical model
representatives of flexible
linguistic functions

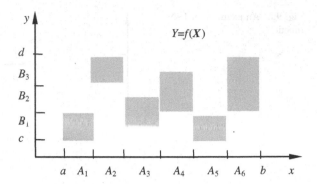

that is, the granule size, of a linguistic value refers to the size of the set or flexible set (see Chap. 20) labeled by the linguistic value).

Like the usual numerical functions, the flexible linguistic functions would also have some relevant properties, such as monotony, continuity, periodicity, etc. It is not hard to see that the monotony of flexible linguistic functions is similar to that of usual numerical functions, which can be defined by orders of independent variable and function value, that is, when the value of independent variable increases, the value of function also increases and the function is monotonously increased; when the value of independent variable increases, the value of function decreases and the function is monotonously decreased. The continuity of flexible linguistic functions, for the flexible linguistic function that defined on a set of basic flexible linguistic values on a measurement space, the continuity can be defined by the succession of independent variable and function value, that is, when independent variable takes next flexible linguistic value of present flexible linguistic value, the function also takes next flexible linguistic value of present flexible linguistic value; for the flexible linguistic function that defined on set of all flexible linguistic values on a measurement space, the continuity can be defined by the continuity of peak value points of independent variable(s) and function value. We can compute even the rate of change of a flexible linguistic function from the peak value points of related linguistic values. On the properties of flexible linguistic functions, we need to do further study. Of course, because the granule size of linguistic values is larger and their operations are very finite (having only ∧, ∨, and ⊕), therefore, generally speaking, the properties of flexible linguistic functions are relatively little.

The evaluation of the flexible linguistic functions, in principle, is similar to the evaluation computation of the usual numerical functions. Specifically, for the flexible linguistic functions represented by enumeration, the function value can be obtained through looking up the table of valued pairs of a function according to values of independent variables; for the flexible linguistic functions represented by formulation, need first according to values of independent variables through matching the expressions of the most basic operations defined to obtain the corresponding operation values, then according to expression of the function to match successively and level by level the definitional expressions of corresponding operations to obtain the corresponding operation values, until the final function value is obtained. Consequently, once evaluation process of a flexible linguistic function forms a tree of linguistic values, and many times evaluation process form a net of linguistic values. For example, suppose there is a flexible linguistic function $Y = (X_1 \wedge X_2) \oplus (X_3 \vee X_4)$, and the corresponding definitional expressions of basic operations are $A_1 \wedge B_1 = E_1$, $C_1 \vee D_2 = F_2$, $E_1 \oplus F_2 = G_1$.... Then, when $X_1 = A_1$, $X_2 = B_1$, $X_3 = C_1$ and $X_4 = D_2$, from the definitional expressions of first and second operations we can have E_1 and F_2, also from E_1, F_2 and the definitional expression of third operation we have the value G_1 of the function. It is not hard to see that the evaluation of a flexible linguistic function represented by formulation actually does not involve real computations but only is many times looking up the table of definitions of operations. That is to say, the evaluation of a flexible linguistic function is actually all looking up tables, but the flexible linguistic function represented by enumeration requires a time looking up table while a flexible linguistic functions represented by formulation requires many times looking up table.

9.4 Flexible Linguistic Correlations

Similar to correlations between numbers, there are correlations between flexible linguistic values.

Definition 9.14 Let U and V be two measurement spaces, $\mathcal{L}_U$ and $\mathcal{L}_V$ be separately the sets of flexible linguistic values on U and V, and C be a relation from $\mathcal{L}_U$ to $\mathcal{L}_V$. If for each flexible linguistic value $X \in \mathcal{L}_U$, there always exists at least one $Y \in \mathcal{L}_V$ such that $(X, Y) \in C$, then we say C is a flexible linguistic correlation from L_U to L_V, write $Y = C(X)$.

Example 9.4 Let $U = [a, b]$ and $V = [c, d]$ be two measurement spaces, and let $\mathcal{L}_U = \{A_1, A_2, A_3, A_4, A_5, A_6 | A_1, A_2, A_3, A_4, A_5, A_6$ are flexible linguistic values on $U\}$, and $\mathcal{L}_V = \{B_1, B_2, B_3 B_4 | B_1, B_2, B_3 B_4$ are flexible linguistic values on $V\}$. Then, $C = \{(A_1, B_1), (A_2, B_2), (A_2, B_4), (A_3, B_1) (A_3, B_2), (A_4, B_1), (A_5, B_3), (A_6, B_2)\}$ is a flexible linguistic correlation from $\mathcal{L}_U$ to $\mathcal{L}_V$.

Obviously, flexible linguistic correlations only can be represented by enumeration method, that is, by a set of pairs of flexible linguistic values. Of course, a flexible linguistic correlation represented by a set of pairs of flexible linguistic values can only treated as a kind of qualitative model of corresponding system.

It can be seen that a flexible linguistic correlation is also formed by correlation between numerical values and in turn which summarizes corresponding numerical correlation. Similarly, in theory, a flexible linguistic correlation can be represented as a numerical correlation summarized by which between corresponding measurement space, that is, its background correlation, but in practical problems the background correlation of a flexible linguistic correlation is not known in general. This case is similar to the problem meted by flexible linguistic function. Therefore,

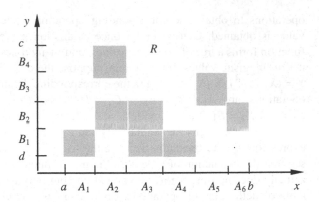

Fig. 9.11 An example of numerical representatives of flexible linguistic correlations

we can take corresponding set of local universal relations as a numerical model representative to represent the flexible linguistic correlation. For example, the graph of the numerical model representative of the flexible linguistic correlation in Example 9.4 is shown in Fig. 9.11.

Similarly, although, speaking generally, this kind of numerical model representative is not the correlation summarized by original flexible linguistic correlation, that is, background correlation, it covers original flexible linguistic correlation, thus through it we can estimate the background correlation of original flexible linguistic correlation. Similarly, this kind of big-granule correlation of block points just easily characterizes macro-characteristics of corresponding systems. Therefore, using it in place of original flexible linguistic correlation is more suitable for and more convenient for macro-analysis of corresponding systems. In fact, for the systems on two- and three-dimensional measurement spaces, based on this kind of numerical model representative, we can conduct visual examination and analysis.

From Definition 9.14 and Fig. 9.11, we can see that a flexible linguistic correlation can also be viewed as a multivalued flexible linguistic function, while flexible linguistic functions are then a kind of special flexible linguistic correlations. Besides, we can also see that a flexible linguistic correlation is generally formed by pairs of basic flexible linguistic values. Then, if we unite properly the multiple basic flexible linguistic values that correspond to the value of one and the same independent variable, then some flexible linguistic correlations would become flexible linguistic functions.

9.5 Summary

In this chapter, we analyzed firstly the mathematical backgrounds and relational representations of a set correspondence and a flexible-set correspondence, thus revealing the mathematical essence, mathematical background, and relational representation of a flexible-linguistic-valued correspondence; then, we proposed and discussed flexible linguistic functions and flexible linguistic correlations, presented their types and representation, analyzed their characteristics, properties, and evaluations, and in particular, discovered and proposed a quantitative description and numerical model of flexible linguistic functions and flexible linguistic correlations.

The main points and results of the chapter are:

- Flexible-set correspondence $A \longmapsto B$ is formed by a certain function or correlation from flexible set A to flexible set B and in turn covers the function or correlation, and which also covers simultaneously all functions and correlations from A to B, of them the largest is universal relation $\mathrm{supp}(A) \times \mathrm{supp}(B)$.
- A flexible-set correspondence can be specifically represented as that function or correlation covered by which.
- For one and the same pair of flexible sets A and B, the orthogonal intersection $A \cap B$, the Cartesian product $A \times B$ and the correspondence $A \longmapsto B$ are not the same each other.
- Flexible-linguistic-valued correspondence $A \longmapsto B$ is a summarization of a certain function or correlation from flexible set A to flexible set B, which also summarizes simultaneously all functions and correlations from A to B, of them the largest is universal relation $\mathrm{supp}(A) \times \mathrm{supp}(B)$.
- In theory, a flexible-linguistic-valued correspondence can be represented as that function or correlation summarized by which, that is also its background function or background correlation.
- A local universal relation summarized by a flexible-linguistic-valued correspondence can be treated as the numerical model representative of the flexible-linguistic-valued correspondence, and its geometrical graph is a block point in corresponding measurement space.
- A flexible linguistic function is generally a summarization of a certain (global) numerical function or correlation without explicit expression, which can be classified as the types of L-L, N-L, and L-N ones.
- The basic representation method of flexible linguistic functions is the enumeration method, but some of them can also be represented by formulation method.
- Flexible linguistic functions have some characteristics like numerical functions, but because the granule sizes of the independent variable and function value of which are larger, so their properties are less.
- A flexible linguistic function can be quantitatively described by flexible linguistic values with degrees, but one needs to know the corresponding quantitative model, that is, the correspondence relation between degrees.
- A set of local universal relations summarized by a flexible linguistic function can be treated as the numerical model representative of the flexible linguistic function, and its graph is a block-point curve, surface, or hypersurface.
- A flexible linguistic correlation can only be represented by enumeration method, and we can also take a set of corresponding local universal relations as it is numerical model representative.
- The flexible linguistic function or correlation summarizing a numerical function or correlation is not unique; in general, the flexible linguistic function or

correlation whose linguistic value's sizes are smaller, then which more approaches to the corresponding numerical function or correlation.

- Flexible-set correspondence $A \longmapsto B$ can and only can be represented as a binary rigid relation, but cannot be represented as a binary flexible relation.
- There exist no "flexible mappings," "flexible functions," or "flexible correlations" between two sets or flexible sets in the sense of (complete) correspondence $\longmapsto$.
- One and the same flexible subset of product measurement space stands for a certain practical binary flexible relation as well as "correspondence relation" the binary rigid relation.

Chapter 10
Flexible Numbers and Flexible Functions

Abstract This chapter proposes the concepts of flexible numbers, flexible-numbered vectors, and flexible vectors based on flexible linguistic values and flexible sets; gives their definitions and representations; and analyzes their geometric characteristics. Further, this chapter defines the arithmetic operations, scalar multiplication, and exponentiation of flexible numbers; defines the addition and scalar multiplication of flexible vectors; and points out the properties of these operations. Then, on the bases of flexible numbers and flexible linguistic functions, this chapter proposes the concept of flexible functions, gives their definitions and types, analyzes their analytic expressions, and further discusses the flexible-vector functions.

Keywords Flexible numbers · Flexible functions

In this chapter, we will discuss a kind of special flexible linguistic value—flexible numbers—and a kind of special flexible linguistic function—flexible functions.

10.1 Definition and Notation of Flexible Numbers

Examining the numbers described by flexible linguistic values "about 5," "near 100," and so on in usual, it can be seen that this type of numbers represents really a flexible interval in real number field **R**, which can be called the flexible numbers.

Definition 10.1 A flexible interval with real number $r \in \mathbf{R}$ as the center point standing for "about r" is called a flexible real number on real number field **R**, or a flexible number for short, denoted (r).

From the definition, a flexible number can be represented as a 5-tuple:

$$\left(s_r^-, c_r^-, r, c_r^+, s_r^+\right) \tag{10.1}$$

© Springer Science+Business Media Singapore 2016
S. Lian, *Principles of Imprecise-Information Processing*,
DOI 10.1007/978-981-10-1549-6_10

where r is the center number of flexible number (r); c_r^-, c_r^+, s_r^-, and s_r^+ are separately the negative and positive core–boundary points and negative and positive critical points of (r).

Obviously, the core–boundary points and critical points of (r) are related to the core radius and support set radius of (r). So flexible number (r) can also be represented as a 3-tuple:

$$(r, r_c, r_s) \tag{10.2}$$

where r_c and r_s are separately the core radius and support set radius of flexible number (r).

Since the core radius and support set radius of "about r" actually change with the size of the absolute value $|r|$ of r, all the flexible numbers cannot be designated with the same pair of core radius and support set radius. For this reason, we introduce two parameters of core radius ratio c_r and support set radius ratio s_r. "Core radius ratio" and "support set radius ratio" are separately meant the ratios of core radius r_c and support set radius r_s to the absolute value $|r|$ of center number r, namely

$$c_r = \frac{r_c}{|r|}, \; s_r = \frac{r_s}{|r|}$$

From this, we have

$$r_c = c_r |r|, \; r_s = s_r |r|$$

Thus, with the two radius ratios, we can represent flexible numbers according to positive and negative separately as follows:

$$(r(1 - s_r), r(1 - c_r), r, r(1 + c_r), r(1 + s_r)), \quad r > 0 \tag{10.3}$$

$$(r(1 + s_r), r(1 + c_r), r, r(1 - c_r), r(1 - s_r)), \quad r < 0 \tag{10.4}$$

where r is the center number of flexible number (r); $r(1 - c_r)$ and $r(1 + c_r)$ are core–boundary points of (r); and $r(1 - s_r)$ and $r(1 + s_r)$ are the critical points of (r).

From the determination of radius ratios c_r and s_r, the approaches of "personal preference," "statistics from a group," or "derivation with instances" in Sect. 4.1 can be employed. Certainly, once radius ratios c_r and s_r are determined, then they are common for all flexible numbers on the universe.

Since radius ratios c_r and s_r are meaningless to flexible 0, (0), we represent (0) singly as follows:

$$(-r_c, -r_s, 0, r_c, r_s) \tag{10.5}$$

where r_c and r_s are separately the core radius and support set radius of (0), the determination of which is the same as that of radius ratios c_r and s_r.

Example 10.1 Taking core radius ratio $c_r = 0.02$ and support set radius ratio $s_r = 0.04$, then the representations of flexible numbers (1) and (5) are as follows:

$$(1) = (0.96, 0.98, 1, 1.02, 1.04)$$
$$(5) = (4.80, 4.90, 5, 5.10, 5.20)$$
$$(10) = (10.6, 10.8, 10, 10.2, 10.4)$$
$$(100) = (96, 98, 100, 102, 104)$$
$$(1000) = (960, 980, 1000, 1020, 1.040)$$
$$(-100) = (-104, -102, -100, -98, -96)$$

The meanings of them are "about 1" and "about 5."

For the given radius ratios c_r and s_r, the four numbers of $1 - s_r$, $1 - c_r$, $1 + c_r$, and $1 + s_r$ are fixed. So setting $s_1 = 1 - s_r$, $c_1 = 1 - c_r$, $c_2 = 1 + c_r$, and $s_2 = 1 + s_r$, then the expression (10.3) becomes

$$(rs_1, rc_1, r, rc_2, rs_2) \tag{10.6}$$

Thus, the consistency function and membership function of the corresponding positive flexible number (r) are as follows:

$$c_{(r)}(x) = \begin{cases} \dfrac{x - rs_1}{rc_1 - rs_1}, & x \le r \\ \dfrac{rs_2 - x}{rs_2 - rc_2}, & r \le x \end{cases} \tag{10.7}$$

$$m_{(r)}(x) = \begin{cases} 0, & x \le rs_1 \\ \dfrac{x - rs_1}{rc_1 - rs_1}, & rs_1 < x < rc_1 \\ 1, & rc_1 \le x \le rc_2 \\ \dfrac{rs_2 - x}{rs_2 - rc_2}, & rc_2 < x < rs_2 \\ 0, & rs_2 \le x \end{cases} \tag{10.8}$$

The corresponding function graph is shown in Fig. 10.1.

Similarly, the consistency function and membership function of negative flexible number (r) can also be obtained. It would be more direct as for the consistency function and membership function of (0).

In addition to "about r," "near r" and "slightly exceeding r" can also be treated as flexible numbers. But the flexible intervals that the two flexible numbers correspond to are a kind of "semiflexible interval." Therefore, the two flexible numbers can be called **semiflexible number** (in comparison, "about r" is a **full-flexible number**).

Fig. 10.1 Graphs of the consistency function and membership function of flexible number (r)

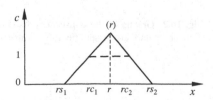

Definition 10.2 A semiflexible interval with real number $r \in \mathbf{R}$ as the supremum (but does not contain r) standing for "near r" is called a weak semiflexible number on real number field $\mathbf{R}$, denoted as $(r]$; a semiflexible interval with real number $r \in \mathbf{R}$ as the infinum (but does not contain r) standing for "slightly exceeding r" is called a strong semiflexible number on real number field $\mathbf{R}$, denoted $[r)$.

From the definition, a weak semiflexible number and a strong semi-number can be separately represented as

$$(rs_1, rc_1, r] \tag{10.9}$$

$$[r, rc_2, rs_2) \tag{10.10}$$

It is easy to see that the consistency functions and membership functions of a weak semiflexible number and a strong semi-number are, respectively, as follows:

$$c_{(r]}(x) = \frac{x - rs_1}{rc_1 - rs_1}, \quad x < r \tag{10.11}$$

$$m_{(r]}(x) = \begin{cases} 0, & x \le rs_1 \\ \dfrac{x - rs_1}{rc_1 - rs_1}, & rs_1 < x < rc_1 \\ 1, & rc_1 \le x < r \end{cases} \tag{10.12}$$

$$c_{[r)}(x) = \frac{rs_2 - x}{rs_2 - rc_2}, \quad r < x \tag{10.13}$$

$$m_{[r)}(x) = \begin{cases} 1, & r < x \le rc_2 \\ \dfrac{rs_2 - x}{rs_2 - rc_2}, & rc_2 < x < rs_2 \\ 0, & rs_2 \le x \end{cases} \tag{10.14}$$

The corresponding function graphs are separately shown in Fig. 10.2a, b.

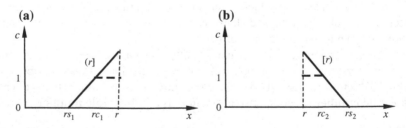

Fig. 10.2 Graphs of the consistency functions and membership functions of weak semiflexible number $(r]$ and strong semiflexible number $[r)$

10.2   Operations on Flexible Numbers

Since flexible numbers are also a kind of flexible linguistic values, the flexible numbers can also have logical operations and algebraic operations such as general flexible linguistic values. However, flexible numbers are also a kind of numbers at the same time, so flexible numbers should also have numerical operations of addition, subtraction, multiplication, and division like usual numbers.

1. Arithmetic operations on flexible numbers

Definition 10.3 Let $\mathscr{N}_R$ be a set of flexible numbers on real number field $\mathbf{R}$, (x), $(y) \in \mathscr{N}_R$. $(x) + (y)$, $(x) - (y)$, $(x) \times (y)$, and $(x)/(y)$ are in order called the operations of addition, subtraction, multiplication, and division on flexible numbers, and the operation rules are as follows:

$$(x) + (y) = (x+y) \tag{10.15}$$

$$(x) - (y) = (x - y) \tag{10.16}$$

$$(x) \times (y) = (x \times y) \tag{10.17}$$

$$(x)/(y) = (x/y) \tag{10.18}$$

This definition means that the results of arithmetic operations of two flexible numbers are still flexible numbers and the center numbers of the flexible numbers are the results of the corresponding real number operations with the center numbers of the two flexible numbers that take part in the operations.

Example 10.2 Let core radius ratio $c_r = 0.02$ and support set radius ratio $s_r = 0.04$, and taking flexible numbers (7), $(8) \in \mathscr{N}_R$, then

$(7) = (7 \times 0.96, 7 \times 0.98, 7, 7 \times 1.02, 7 \times 1.04) = (6.72, 6.86, 7, 7.14, 7.28)$
$(8) = (8 \times 0.96, 8 \times 0.98, 8, 8 \times 1.02, 8 \times 1.04) = (7.68, 7.84, 8, 8.16, 8.32)$

Thus,

$(7) + (8) = (7+8) = (15)$
$\qquad\qquad = (15 \times 0.96, 15 \times 0.98, 15, 15 \times 1.02, 15 \times 1.04)$
$\qquad\qquad = (14.40, 14.7, 15, 15.3, 15.6)$
$(7) - (8) = (7 - 8) = (-1)$
$\qquad\qquad = (-1 \times 1.04, -1 \times 1.02, -1, -1 \times 0.98, -1 \times 0.96)$
$\qquad\qquad = (-1.04, -1.02, -1, -0.98, -0.96)$
$(7) \times (8) = (7 \times 8) = (56) = (53.76, 54.88, 56, 57.12, 58.24)$
$\qquad (7)/(8) = (7/8) = (0.875) = (0.840, 0.856, 0.875, 0.893, 0.910)$

Now that the operations of two flexible numbers are reduced to the operations of their center numbers, while the latter are completely the arithmetic operations of real numbers, the operations of flexible numbers also satisfy all properties such as commutative laws, associative laws, and distributive laws of real operations, which means that we can found various kinds of **algebraic systems of flexible numbers** on the set $\mathcal{N}_\mathbf{R}$ of flexible numbers or its subsets.

2. **Scalar multiplication and exponentiation on flexible numbers**

Definition 10.4 Let $\mathcal{N}_\mathbf{R}$ be a set of flexible numbers on real number field $\mathbf{R}$, $\alpha \in \mathbf{R}$, $(x) \in \mathcal{N}_\mathbf{R}$. $\alpha(x)$ and $(x)^\alpha$ are separately called the scalar multiplication and exponentiation on flexible numbers, whose operation rules are as follows:

$$\alpha(x) = (\alpha x) \tag{10.19}$$

$$(x)^\alpha = (x^\alpha) \tag{10.20}$$

This definition means that the product of a number and a flexible number is a flexible number whose center number is the product of the number and the center number of the flexible number, and the power of a flexible number is a flexible number whose center number is the power of the center number of the flexible number. Thus, it follows by the definition that

$$2(x) = (2x), \; n(x) = (nx)$$
$$(x)^2 = (x^2), \; (x)^n = (x^n)$$

On the other hand, from the addition of flexible numbers, it follows that $(x) + (x) = (x + x) = (2x)$. And by the above scalar multiplication, thus, $(x) + (x) = 2(x)$. Generally,

$$\underbrace{(x) + (x) + \cdots + (x)}_{n(x)_s} = n(x)$$

Similarly, generally,

$$\underbrace{(x) \times (x) \times \cdots \times (x)}_{n(x)_s} = (x)^n$$

3. **Operations on semiflexible numbers**

Since semiflexible numbers have the separation of strong and weak, the operations of semiflexible numbers are relatively complex. Further, we find that there still exits one problem in the operations on semiflexible numbers; that is, except the addition of semiflexible numbers with same sign, other operations cannot be defined by the operations of the corresponding supremum or infinum.

For instance, let core radius ratio $c_r = 0.02$ and support set radius ratio $s_r = 0.04$, and taking semiflexible numbers (100] and (200] and expanding them, we have

$$(100] = (96, 98, 100] \quad \text{(whose median point is 97)}$$
$$(200] = (192, 196, 200] \quad \text{(whose median point is 194)}$$

Suppose we can define $(x] - (y] = (x - y]$, then $(100] - (200] = (100 - 200] = (-100]$. Thus, for $\forall \ x \in (100]$ and $\forall \ y \in (200]$, $x - y \in (100] - (200] = (-100]$. While $(-100] = (-104, -102, -100]$, median point is -103. Now, we take 99 and 194.1, and then from the core–boundary points and median points of (100] and (200], it can be seen that it should follow that $99 \in (100]$ and $194.1 \in (200]$; however, $910.194.1 = -95.1 > -100$, so $-95.1 \notin (-100]$, that is, $910.194.1 \notin (100 - 200]$. This counterexample shows that, generally speaking, $(x] - (y] \neq (x - y]$.

For another example, for the above radius ratio, taking semiflexible numbers (0.1] and (0.2] and expanding them, we have

$$(0.1] = (0.096, 0.098, 0.1] \quad \text{(whose median point is 0.097)}$$
$$(0.2] = (0.192, 0.196, 0.2] \quad \text{(whose median point is 0.194)}$$

Suppose we can define $(x] \times (y] = (x \times y]$, then $(0.1] \times (0.2] = (0.1 \times 0.2] = (0.02]$. Thus, for $\forall x \in (0.1]$ and $\forall y \in (0.2]$, it should follow that $x \times y \in (0.1] \times (0.2] = (0.02]$. While $(0.02] = (0.0192, 0.0196, 0.02]$, median point is 0.0194. Now, taking $0.0971 \in (0.1]$ and $0.1941 \in (0.2]$, however, $0.0971 \times 0.1941 = 0.01884711 < 0.0194$. That shows that $0.01884711 \notin (0.02]$, that is, $0.0971 \times 0.1941 \notin (0.1] \times (0.2]$. This counterexample shows that, generally speaking, $(x] \times (y] \neq (x \times y]$.

In view of the analysis above, we only present the addition operation on the semiflexible numbers with the same sign.

Definition 10.5 Let $M_{\mathbf{R}}$ be a set of semiflexible numbers on real number field **R**, $(x]$, $(y]$, $[x)$, and $[y) \in M_{\mathbf{R}}$, and their signs are the same. The addition operation rules are as follows:

$$(x] + (y] = (x + y] \tag{10.21}$$

$$[x) + [y) = [x + y) \tag{10.22}$$

$$(x] + [y) = (x + y) \tag{10.23}$$

$$[x) + (y] = (x + y) \tag{10.24}$$

10.3   Flexible-Numbered Vectors

Definition 10.6 A vector $((x_1), (x_2), \ldots, (x_n))$ with flexible numbers $(x_1), (x_2), \ldots, (x_n)$ as components is called a flexible-numbered vector.

The flexible-numbered vectors are an extension of the flexible numbers (of course, which is also an extension of the usual vectors), which is also a kind of special flexible linguistic-valued vector. Therefore, a flexible-numbered vector $((x_1), (x_2), \ldots, (x_n))$ represents a point in flexible-numbered vector space (two-dimensional and three-dimensional flexible-numbered vectors are similar to the flexible linguistic-valued vectors as shown in Fig. (7.7). However, if the relation between components (x_1), $(x_2), \ldots, (x_n)$ is regarded as conjunction relation, *and*, then $((x_1), (x_2), \ldots, (x_n))$ is tantamount to $(x_1) \wedge (x_2) \wedge \cdots \wedge (x_n)$. Thus, $((x_1), (x_2), \ldots, (x_n))$ represents a flexible square in n-dimensional space $\mathbf{R}^n$ of which the label is $(x_1) \wedge (x_2) \wedge \ldots \wedge (x_n)$, that is, the intersection of orthogonal "bar-shaped" flexible squares $(x_1) \times \mathbf{R} \times \cdots \times \mathbf{R}$, $\mathbf{R} \times (x_2) \times \cdots \mathbf{R}, \cdots, \mathbf{R} \times \mathbf{R} \times \cdots \times (x_n)$ in $\mathbf{R}^n$ (the flexible square represented by two-dimensional flexible-numbered vector $((x_1), (x_2))$ is shown in Fig. 10.3).

On the other hand, since the components in flexible-numbered vector $((x_1), (x_2), \ldots, (x_n))$ are ordered, so, like flexible linguistic-valued vector, flexible-numbered vector $((x_1), (x_2), \ldots, (x_n))$ is tantamount to flexible-numbered correspondence $(x_1) \wedge (x_2) \wedge \cdots \wedge (x_{n-1}) \mapsto (x_n)$, which is the summarization of a certain function or correlation from corresponding flexible set $(x_1) \cap (x_2) \cap \cdots \cap (x_{n-1})$ to flexible set (x_n), of them the biggest is (local) universal relation $\text{supp}((x_1)) \times \text{supp}((x_2)) \times \cdots \times \text{supp}((x_n))$. Thus, in space $\mathbf{R}^n$, flexible-numbered vector $((x_1), (x_2), \ldots, (x_n))$ represents a region whose super bound is $\text{supp}((x_1)) \times \text{supp}((x_2)) \times \cdots \times \text{supp}((x_n))$.

Example 10.3 $((1), (2), (3))$ is just a flexible-numbered vector, which is a point in flexible-numbered space $(\mathbf{R})$; but in three-dimensional space $\mathbf{R}^3$ represents a flexible square whose label is $(1) \wedge (2) \wedge (3)$ as well as a region whose super bound is $\text{supp}((1)) \times \text{supp}((2)) \times \text{supp}((3))$.

Fig. 10.3 An example of the flexible squares represented by two-dimensional flexible-numbered vectors

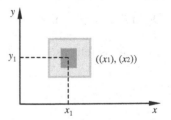

10.4 Flexible Vectors and the Operations on Flexible Vectors

Definition 10.7 A flexible point with vector $r = (r_1, r_2, ..., r_n) \in \mathbf{R}^n$ as center point standing for "about r" is called a flexible vector on n-dimensional space $\mathbf{R}^n$, denoted (r).

By the definition, flexible vector (r) can be represented as follows:

$$(r, r_c, r_s) \tag{10.25}$$

where r_c and r_s are separately the core radius and support set radius of flexible vector (r).

The geometry of flexible vector (r) is a flexible circle, flexible sphere, or flexible hypersphere with point r as the center point in n-dimensional space $\mathbf{R}^n$, whose membership function and consistency function can directly adopt expressions (3.25) and (3.26) in Sect. 3.3.2. But note that the parameters r_c and r_s here are the same for all flexible vectors in one and the same universe. Actually, the flexible vectors here are also the vector linguistic values called in Sect. 7.4.

Example 10.4 $((1, 2, 3))$ is a flexible vector, which represents "about $(1, 2, 3)$" and whose geometry is a flexible sphere with center point $(1, 2, 3)$ in three-dimensional space $\mathbf{R}^3$.

Note that viewed from the form, the flexible vectors can be viewed as a kind of extension of flexible numbers, but in essence, there is great difference between the flexible vectors and the flexible numbers. In fact, the core radius and support set radius of flexible numbers change with the center numbers, but in one and the same space, the core radius and support set radius of all flexible vectors are fixed and unchanged. That is to say, a flexible number is related to its center number, while a flexible vector is related to the position of its center point. So speaking strictly, the flexible vectors are not the real extension of the flexible numbers. Thus, for one and the same (x), the meaning of (x) as a flexible number is actually not same to that as a one-dimensional flexible point.

Just the same as usual vectors, we can also define the operations of addition and scalar multiplication on flexible vectors.

Definition 10.8 Let $(\mathbf{R}^n)$ be a flexible-vector space formed by all flexible vectors on space $\mathbf{R}^n$, and let $(x), (y) \in (\mathbf{R}^n)$. $(x) + (y)$ is called the addition of flexible vectors, whose operation rule is as follows:

$$(x) + (y) = (x + y)$$

This definition means that the sum of two flexible vectors is also a flexible vector, whose center vector is the sum of the original two center vectors. For instance,

$$((1, 2, 3)) + ((4, 5, 6)) = ((1, 2, 3) + (4, 5, 6)) = ((5, 7, 9))$$

Definition 10.9 Let $(\mathbf{R}^n)$ be a flexible-vector space formed by all flexible vectors on space $\mathbf{R}^n$, let $\alpha \in \mathbf{R}$, and $(x) \in (\mathbf{R}^n)$. $\alpha(x)$ is called the scalar multiplication of flexible vectors, whose operation rule is as follows:

$$\alpha(x) = (\alpha x)$$

This definition means that the product of a number and a flexible vector is also a flexible vector, whose center vector is the product of this number and the center vector of the factor flexible vector. For instance,

$$5((1, 2, 3)) = (5(1, 2, 3)) = ((5, 10, 15))$$

It can be seen that the addition and scalar multiplication of flexible vectors are actually reduced the addition and scalar multiplication of usual (rigid) vectors. By the operation laws that the addition and scalar multiplication of usual n-dimensional vectors satisfy, it is not hard to verify the following:

① Addition of flexible vectors satisfies commutative laws and associative laws;
② The sum of any flexible vector $(x) \in (\mathbf{R}^n)$ and flexible zero vector $(0) \in (\mathbf{R}^n)$ is still (x), that is, $(x) + (0) = (0) + (x) = (x)$;
③ For any flexible vector $(x) \in (\mathbf{R}^n)$, there exits negative vector $-(x) \in (\mathbf{R}^n)$ such that $(x) + ((x)) = (0)$.

Therefore, the flexible-vector space $(\mathbf{R}^n)$ with addition and scalar multiplication of flexible vectors can form a linear space on real number field $\mathbf{R}$.

Definition 10.10 A flexible void point with center point $r = (r_1, r_2, \ldots, r_n) \in \mathbf{R}^n$ which stands for "near r" is called a flexible void vector on n-dimensional space $\mathbf{R}^n$, denoted (r).

It is not hard to see that the flexible void vectors have the operations and operation rules like flexible vectors, so it is unnecessary to go into details here.

10.5   Interconversion Between Flexible Numbers (Flexible Vectors) and Rigid Numbers (Rigid Vectors)

1. Interconversion between flexible numbers and rigid numbers

Compared with the flexible numbers, usual numbers are just "rigid numbers." So the flexible numbers can be viewed as the extension of rigid numbers, and the rigid numbers can be viewed as the contraction of flexible numbers. Since flexible numbers are also a kind of flexible linguistic values, the flexible numbers and rigid numbers can be converted mutually. The conversion method can employ that between flexible linguistic values and numerical values as given in Sect. 7.3.1.

Actually, a flexible number (x) converted into a rigid number is also a kind of rigid-ening of the flexible number, which can be denoted by $[(x)]$; and a rigid number x converted into a flexible number is also a kind of flexible-ening of the rigid number, which can be denoted by (x).

2. Interconversion between flexible vectors and rigid vectors

Compared with the flexible vectors, usual vector is just "rigid vectors." The flexible vectors can be viewed as the extension of rigid vectors, while rigid vectors can be viewed as the contraction of flexible vectors. Since flexible vectors are also a kind of flexible linguistic values, the flexible vectors and rigid vectors can be converted mutually. The conversion method can employ that between flexible linguistic values and vectors given in Sect. 7.3.1.

Similarly, a flexible vector $((x_1, x_2, \ldots, x_n))$ converted into a rigid vector is also a kind of rigid-ening (hardening) of the flexible vector, which can be denoted by $[((x_1, x_2, \ldots, x_n))]$; and a rigid vector $(x_1, x_2, \ldots, x_n)$ converted into a flexible vector is also a kind of flexible-ening (softening) of the rigid vector, which can be denoted by $((x_1, x_2, \ldots, x_n))$.

10.6   Flexible Functions

10.6.1   Definitions and Types of Flexible Functions

Definition 10.11

(1) We call the variable taking on flexible numbers to be a flexible-numbered variable or flexible variable in short.
(2) We call the function in which the independent variable(s) or dependent variable take on flexible number(s) to be a flexible-numbered function or flexible function in short.

From the definition, flexible functions can be classified as follows:

- Flexible-number-valued function of flexible variable(s), short for flexible number–flexible number-type function, written as FN–FN function;
- Numerical-valued function of flexible variable(s), short for flexible number–number-type function, written as FN–N function;
- Flexible-number-valued function of numerical variable(s), short for number–flexible number-type function, written as N–FN function.

In the above, we introduce flexible functions on the basis of existing function concept. In the following, we define again the flexible function from the perspective of relation.

Definition 10.12 Let **R** be real number field, $N_{\mathbf{R}}$ be a set of flexible numbers on **R**, and $f \subseteq N_{\mathbf{R}} \times N_{\mathbf{R}}$ be a relation on $N_{\mathbf{R}}$. If for each flexible number $(x) \in N_{\mathbf{R}}$, there always exists one and only one $(y) \in N_{\mathbf{R}}$ such that $((x), (y)) \in f$, then we say f is a flexible function on $N_{\mathbf{R}}$, write $(y) = f((x))$.

The above Definition 10.12 is a basic definition of flexible functions. Taking into account the dimension of domain (i.e., the number of independent variables), we give more types of flexible functions below.

Definition 10.13 Let **R** be real number field, $N_{\mathbf{R}}$ be a set of flexible numbers on **R**, and $f \subseteq (N_{\mathbf{R}})^n \times N_{\mathbf{R}}$ be a relation from $(N_{\mathbf{R}})^n$ to $N_{\mathbf{R}}$. If for each flexible-numbered vector $((x_1), (x_2),\ldots, (x_n)) \in (N_{\mathbf{R}})^n$, there always exists one and only one flexible number $(y) \in N_{\mathbf{R}}$ such that $(((x_1), (x_2),\ldots, (x_n)), (y)) \in f$, then we say f is a FN-FN function from $(N_{\mathbf{R}})^n$ to $N_{\mathbf{R}}$, write $(y) = f((x_1), (x_2),\ldots, (x_n))$.

Definition 10.14 Let **R** be real number field, $N_{\mathbf{R}}$ be a set of flexible numbers on **R**, and $f \subseteq (\mathcal{N}_{\mathbf{R}})^n \times \mathbf{R}$ be a relation from $(\mathcal{N}_{\mathbf{R}})^n$ to **R**. If for each flexible-numbered vector $((x_1), (x_2),\ldots, (x_n)) \in (\mathcal{N}_{\mathbf{R}})^n$, there always exists one and only one real number $y \in \mathbf{R}$ such that $(((x_1), (x_2),\ldots, (x_n)), y) \in f$, then we say f is a FN–N function from $(\mathcal{N}_{\mathbf{R}})^n$ to **R**, write $y = f((x_1), (x_2),\ldots, (x_n))$.

Definition 10.15 Let **R** be real number field, $\mathcal{N}_{\mathbf{R}}$ be a set of flexible numbers on **R**, and $f \subseteq \mathbf{R}^n \times N_{\mathbf{R}}$ be a relation from $\mathbf{R}^n$ to $\mathcal{N}_{\mathbf{R}}$. If for each vector $(x_1, x_2,\ldots, x_n) \in \mathbf{R}^n$, there always exists one and only one flexible number $(y) \in \mathcal{N}_{\mathbf{R}}$ such that $((x_1, x_2,\ldots, x_n), (y)) \in f$, then we say f is a N–FN function from $\mathbf{R}^n$ to $\mathcal{N}_{\mathbf{R}}$, write $(y) = f(x_1, x_2,\ldots, x_n)$.

It can be seen that the flexible function is a kind of special flexible linguistic function; on the other hand, the flexible function is also an extension of the usual "rigid" function; conversely, the rigid function is the contraction of the flexible function.

10.6.2 Analytic Expressions and Properties of Flexible Functions

Since flexible functions are a kind of special flexible linguistic function, in theory, a flexible function can be represented by a set of pairs of values of independent variable(s) and function, that is, by using enumeration method, just as general flexible linguistic functions. However, since the operations of flexible numbers are really numerical operations, therefore, flexible function is more suitable for the formula representation.

(1) **Analytic expression of FN–FN function**

Firstly, since the arithmetic operations, scalar multiplication, and exponentiation operation of flexible numbers all reduce to the corresponding operations of the corresponding center numbers, the expression of a flexible function may be an arbitrary expression for the arithmetic operations, scalar multiplication, and exponentiation operation of flexible numbers. For example,

$$3(x)^2 + (x) = (3x^2 + x)$$

Secondly, we know that the transcendental function (such as exponential function, log function, trigonometric function, and hyperbolic function) on the real number field can be expanded into a power series. Thus, a transcendental function of flexible numbers can also be represented by the form of a power series of flexible numbers. For example, the expansion of power series of exponential function e^x is as follows

$$e^x = 1 + \frac{x}{1!} + \frac{x^2}{2!} + \frac{x^3}{3!} + \cdots + \frac{x^n}{n!} + \cdots$$

Thus, the flexible exponential function $e^{(x)}$ can be represented as follows:

$$e^{(x)} = (1) + \frac{(x)}{1!} + \frac{(x)^2}{2!} + \frac{(x)^3}{3!} + \cdots + \frac{(x)^n}{n!} + \cdots$$

The power series on the right-hand side of the equation is an expression of arithmetic operations of flexible numbers, so

$$(1) + \frac{(x)}{1!} + \frac{(x)^2}{2!} + \frac{(x)^3}{3!} + \cdots + \frac{(x)^n}{n!} + \cdots$$
$$= (1 + \frac{x}{1!} + \frac{x^2}{2!} + \frac{x^3}{3!} + \cdots + \frac{x^n}{n!} + \cdots) = (e^x)$$

Therefore,

$$e^{(x)} = (e^x)$$

Similarly, we can also have

$$\ln(x) = (\ln x)$$
$$\sin(x) + \cos(y) = (\sin x) + (\cos y) = (\sin x + \cos y)$$

Consequently, generally,

$$f((x_1), (x_2), \ldots, (x_n)) = (f(x_1, x_2, \ldots, x_n)) \tag{10.26}$$

This is to say, the analytic expression of a FN–FN function can be a expression of operations of flexible variables (x_1), (x_2), ..., (x_n), that is,

$$(y) = f((x_1), (x_2), \ldots, (x_n)) \tag{10.27}$$

(2) Analytic expression of N–FN function

In consideration of that, N–FN function is a function for usual numerical variables x_1, x_2, ..., x_n, while the value of the function is a flexible number, so its expression should be

$$(y) = (f(x_1, x_2, \ldots, x_n)) \tag{10.28}$$

where $f(x_1, x_2, \ldots, x_n)$ can be a certain functional expression for x_1, x_2, ..., x_n.

For instance, (y) = about $f(x)$ is a typical and very useful N–FN function. Here, $f(x)$ can be any function for x, such as $x^2 + 1$, $\sin x$, and e^x; thus, the corresponding N–FN functions are as follows:

$$(y) = \text{about } \text{``}^2 + 1\text{''}$$
$$(y) = \text{near ``}\sin x\text{''}$$
$$(y) = \text{slightly exceeding ``}e^x\text{''}$$

Note that the graphs of the N–FN functions in space of real numbers are also a kind of flexible line or flexible plane, and their geometries are also a kind of flexible band, flexible rope, or flexible plate, but the width, diameter, or thickness of these flexible bands, flexible ropes, and flexible plates are not uniform. The reason is that the core radii and support set radii of flexible numbers vary with the sizes of the center numbers. This is to say, the graph of an N–FN function in space of real numbers is differential with flexible band, flexible rope, and flexible plate obtained by flexible clustering in Chap. 3 (the width, diameter, or thickness of the latter are uniform). For instance, the following Fig. 10.4 is just an example of the graph of an N–FN function.

(3) Analytic expression of FN–N function

From the characteristics of FN–N functions, their analytic expression can be

$$y = [f((x_1), (x_2), \ldots, (x_n))] \tag{10.29}$$

Fig. 10.4 An example of the graphs of N–FN functions

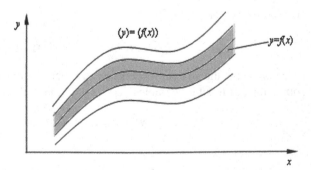

That is, firstly, it represents the function as a FN–FN function and then rigid-en (harden) as the value of corresponding flexible function $f((x))$.

For instance, $y = [(x)^3 + (y)^3]$, $y = [\ln(x)]$.

From the analytic expressions of flexible functions, it can be seen that the properties and evaluation of a flexible function can be reduced to the properties and evaluation of the corresponding numerical function. Therefore, flexible functions have almost all properties of the corresponding numerical functions, and the evaluation process of them includes the evaluation of corresponding numerical functions and the conversion between rigid numbers and flexible numbers. Thus, we can examine the properties (such as monotonicity, periodicity, continuity, and differentiability) of a flexible function by the corresponding numerical function.

10.6.3 Flexible-Vector Functions and Flexible-Vector-Valued Functions

Extending flexible-numbered function, we obtain the flexible-vector function and flexible-vector-valued functions.

Definition 10.16 Let **R** be real number field, $\mathcal{N}_{\mathbf{R}}$ be a set of flexible numbers on **R**, $V_{\mathbf{R}}^n$ be a set of flexible vectors on $\mathbf{R}^n$, and $f \subseteq V_{\mathbf{R}}^n \times N_{\mathbf{R}}$ be a relation from $V_{\mathbf{R}}^n$ to $\mathcal{N}_{\mathbf{R}}$. If for each flexible vector $(x) \in V_{\mathbf{R}}^n$, there always exists one and only one flexible number $(y) \in N_{\mathbf{R}}$ such that $((x), (y)) \in f$, then we say f is a flexible-vector function from $V_{\mathbf{R}}^n$ to $\mathcal{N}_{\mathbf{R}}$, write $(y) = f((x))$.

Definition 10.17 Let **R** be real number field, $N_{\mathbf{R}}$ be a set of flexible numbers on **R**, $V_{\mathbf{R}}^n$ be a set of flexible vectors on $\mathbf{R}^n$, and $f \subseteq \mathcal{N}_{\mathbf{R}} \times V_{\mathbf{R}}^n$ be a relation from $\mathcal{N}_{\mathbf{R}}$ to $V_{\mathbf{R}}^n$. If for each flexible number $(x) \in \mathcal{N}_{\mathbf{R}}$, there always exists one and only one flexible vector $(y) \in V_{\mathbf{R}}^n$ such that $((x), (y)) \in f$, then we say f is a flexible-vector-valued function from $\mathcal{N}_{\mathbf{R}}$ to $V_{\mathbf{R}}^n$, write $(y) = f((x))$.

Definition 10.18 Let **R** be real number field, $V_{\mathbf{R}}^n$ be a set of flexible vectors on $\mathbf{R}^n$, $V_{\mathbf{R}}^m$ be a set of flexible vectors on $\mathbf{R}^m$, and $f \subseteq V_{\mathbf{R}}^n \times V_{\mathbf{R}}^m$ be a relation from $V_{\mathbf{R}}^n$ to $V_{\mathbf{R}}^m$. If for each flexible vector $(x) \in V_{\mathbf{R}}^n$, there always exists one and only one flexible vector $(y) \in V_{\mathbf{R}}^m$ such that $((x), (y)) \in f$, then we say f is a flexible-vector-flexible-vector-valued function from $V_{\mathbf{R}}^n$ to $V_{\mathbf{R}}^m$, write $(y) = f((x))$.

It can be seen that the flexible-vector functions and flexible-vector-valued functions are also suitable for the formula representation, which also have some properties of usual vector functions and vector-valued functions.

10.7 Summary

In this chapter, we proposed the concepts of flexible numbers, flexible-numbered vectors, and flexible vectors based on flexible linguistic values and flexible sets, gave their definitions and representations, and analyzed their geometric character-istics. Further, we defined the arithmetic operations, scalar multiplication, and exponentiation of flexible numbers, defined the addition and scalar multiplication of flexible vectors, and pointed out the properties of these operations. Then, on the bases of flexible numbers and flexible linguistic functions, we proposed the concept of flexible functions, gave their definitions and types, analyzed their analytic expressions, and further discussed flexible-vector functions.

The main points and the results of this chapter are as follows:

- The flexible number (r) is a flexible interval standing for "about r" in real number field $\mathbf{R}$, and the weak semiflexible number $(r]$ and strong semiflexible number $[r)$ are separately a semiflexible interval standing for "near r" and "slightly exceeding r." The center number, core radius ratio, and support set radius ratio are three key parameters of a flexible number and a semiflexible number, which completely determine a flexible number or semiflexible number.
- Since flexible numbers are both a kind of flexible linguistic value and a kind of number, flexible numbers can have logical operations and algebraic operations as common flexible linguistic values, and they can also have numerical opera-tions such as addition, subtraction, multiplication, division, and scalar multi-plication just as the usual numbers. But the operations of flexible numbers are reduced to the operations of their center numbers. Therefore, the operations of flexible numbers also satisfy all properties of real number operations; thus, we can found various algebraic systems on sets of flexible numbers.
- Flexible-numbered vector is an extension of flexible numbers and also a special case of usual linguistic-valued vectors. A flexible-numbered vector represents a point in flexible-numbered vector space but in real number space, which rep-resents a flexible square to which a corresponding conjunctive flexible-number corresponds as well as a region whose upper bound is the product of the support sets of all component flexible numbers.
- Flexible vector (r) is a flexible point representing "about r" in n-dimensional space $\mathbf{R}^n$, which is another kind of extension of flexible numbers, and also a special case of common vector linguistic values. We can also define addition and scalar multiplication operations of flexible vectors. The flexible-vector space $(\mathbf{R}^n)$ with addition and scalar multiplication of flexible vectors can form a linear space on real number field $\mathbf{R}$. Besides, flexible void vector (r) is the flexible vector that does not contain center point r, which represents "near r."
- Flexible functions are a kind of flexible linguistic functions whose independent variable(s) or function value take on flexible numbers, which can be classified as FN–FN function, N–FN function, and FN–N function, whose analytical expressions are written as $(y) = f((x_1), (x_2),\ldots, (x_n))$, $(y) = (f(x_1, x_2,\ldots, x_n))$, and $y = [f((x_1), (x_2),\ldots, (x_n))]$, respectively.

- Flexible functions can almost satisfy all properties of corresponding numerical functions, and we can examine the properties of a flexible function by the corresponding numerical function.
- Flexible-vector functions and flexible-vector-valued functions are the extension of flexible-numbered functions.

Part IV
Truth-Degreed Logic and Flexible-Linguistic-Truth-Valued Logic: Logic Theories on Imprecise Information

Chapter 11
Truth-Degreed Logic and Corresponding Inference

Abstract This chapter proposes the concept of flexible propositions and founds the corresponding truth-degreed logic and inference theories. First, it considers the representations and truth values of flexible propositions; next, it defines the numerical truth value; that is, the truth-degree, of a flexible proposition, introduces the formal representations of flexible propositions and the flexible propositions with composite linguistic values, particularly, introduces the algebraic compound flexible proposition and the flexible proposition with a synthetic linguistic value, and then analyzes and expounds the computation principles and methods of truth-degrees of the various flexible propositions; then, it founds the corresponding truth-degreed logic algebras, and then, it introduces flexible-propositional formulas and flexible-predicate formulas, extends true and false and proposes the concepts of degree-true, degree-false, near-true, and near-false, extends the validity of argument forms and proposes the concepts of degree-valid argument form and near-valid argument form, and extends the concepts of tautology and logical implication and proposes the concepts of degree-true tautology, degree-true logical implication, near-true tautology, and near-true logical implication, thus obtaining rules of degree-true inference and rules of near-true inference and establishing the principles and methods called degree-true inference and near-true inference in truth-degreed logic.

Keywords Flexible propositions · Truth-degreed logic · Degree-true inference · Near-true inference

A proposition is a statement that describes object(s) having a certain property or relationship. Then, when the property or relation described by a proposition is flexible property or flexible relation, this proposition would be a "flexible proposition". In our information world, flexible propositions can be found everywhere. In this chapter, we will examine flexible propositions and their numerical truth values as well as the logic and inference at the level of numerical truth value.

© Springer Science+Business Media Singapore 2016
S. Lian, *Principles of Imprecise-Information Processing*,
DOI 10.1007/978-981-10-1549-6_11

11.1 Representations and Truth Values of Flexible Propositions

11.1.1 Flexible Propositions and Their Formal Representations

We call a proposition that contain flexible concepts (that is, flexible linguistic values) to be a flexible proposition. For example:

① Zhang is an excellent student.
② Li is tall.
③ 20000 is far greater than 2.

are all flexible propositions. Of them, ① and ② can also be said the property-type flexible propositions by the linguistic values contained being property-type flexible linguistic values, and ③ can be said a relation-type flexible proposition. The representation of these three flexible propositions in first-order predicates are usually

① excellent student (Zhang)
② tall (Li)
③ far greater than (20000, 2)

The predicate names of these three predicates are all flexible linguistic values, but of them, the "tall" and "far greater than" are flexible attribute concepts, while "excellent student" is an flexible entity concept made of flexible attribute concept "excellent" and entity concept "student". Besides, the "Zhang" and "Li" are entity objects.

We know that flexible linguistic values (flexible concepts) are formed on the numerical feature values (measurements) of entity objects, whose mathematical models are just established on the corresponding measurement spaces. On the other hand, imprecise-information processing based on flexible propositions only involves flexible linguistic values and corresponding numerical values in general. Therefore, we use the corresponding measurements in place of those entity objects and use the corresponding flexible attribute linguistic values in place of those flexible linguistic values, which represent flexible entity concepts. Thus, flexible propositions are uniformly represented as a kind of mathematical propositions. Thus, the flexible propositions writing in predicates are uniformly expressed as mathematical predicates; for example, those three flexible propositions above can be mathematically written as follows:

① excellent (x_0)
② tall (y_0)
③ far greater than (20000, 2)

where x_0 is the grade of Zhang, y_0 is the height of Li, while the flexible entity linguistic value "excellent student" is simplified as flexible attribute linguistic value "excellent".

It can be seen that the semantics of this kind of propositions represented by predicates is that the related object has a certain property or there is a certain relationship between the related objects, while the latter is also that the set of the objects has a certain property, so this representation is actually a form of possession.

Definition 11.1 Let U be an n-dimensional measurement space, $x_0 = (x_{1_0}, x_{2_0}, \ldots, x_{n_0}) \in U$, and A be a flexible linguistic value on U. Expression

$$A(x_0) \tag{11.1}$$

is called the representation of a flexible proposition in the form of possession, which means that x_0 has flexible linguistic value A [1].

It can be seen that using the form of possession to represent a proposition, a flexible linguistic value A just determines a cluster of flexible propositions, $\{A(x_0) \mid x_0 \in U\}$, while $A(x)$ $(x \in U)$ is also a propositional form. In order to distinguish it from a proposition, we denote simply the propositional form $A(x)$ $(x \in U)$ as p_x.

The form of possession is actually a representation of a proposition based on the connotation of a concept. In the following, we give again a kind of representation of a flexible proposition based on the denotation of a flexible concept.

Definition 11.2 Let U be an n-dimensional measurement space, $x_0 = (x_{1_0}, x_{2_0}, \ldots, x_{n_0}) \in U$, and A be a flexible linguistic value on U. Expression

$$x_0 \in A \tag{11.2}$$

is called the representation of a flexible proposition in the form of membership, which means that x_0 belongs to flexible set A.

Thus, the above three propositions can also be represented in the form of membership as follows:

① $x_0 \in A$
② $y_0 \in B$
③ $(20000, 2) \in R$

here, x_0 and y_0 are separately the measured values representing the original entity objects, A, B, and R are flexible sets that the flexible linguistic values "excellent," "tall," and "far greater than" correspond separately to.

Similarly, using the form of membership to write a proposition, a flexible linguistic value A also determines a flexible cluster of propositions, $\{x_0 \in A \mid x_0 \in U\}$, while $x \in A$ $(x \in U)$ is also a kind of propositional form. In order to distinguish it from a proposition, we also denote simply propositional form $x \in A$ $(x \in U)$ as p_x.

In the above, we discuss flexible propositions and their formal representations. It is not hard to see that any flexible proposition can be represented as or be translated into the two mathematical forms of flexible propositions stated above. Therefore, we will consider mainly the two forms of flexible propositions and, in general situation, also only use the two forms of flexible propositions.

11.1.2   Numerical Truth Value of a Flexible Proposition— Truth-Degree

We see that the characteristic of the flexible propositions is that the linguistic values therein are flexible linguistic values. Therefore, the authenticity of a flexible proposition should not use the usual rigid linguistic truth values "true" and "false" to describe. Then, how to describe the truth of a flexible proposition? In the following, we consider and solve this problem.

1. Denotation-truth-degree

Let p: $x_0 \in A$ be a flexible proposition.

Although viewed from the representation, x_0 in the proposition belongs to flexible set A, actually x_0 belongs to flexible set A only with degree $m_A(x_0)$. Membership degree $m_A(x_0)$ reflects the degree of x_0 being a member of flexible set A, so number $m_A(x_0)$ also represents the degree of truth of flexible proposition p. Therefore, we take number $m_A(x_0)$ as a kind of numerical truth value of proposition p and call it the degree of truth of proposition p, and simply write **truth-degree**.

Definition 11.3 Let U be an n-dimensional measurement space, A be a flexible subset of U, $x_0 = (x_{1_0}, x_{2_0}, \ldots, x_{n_0}) \in U$, and let p: $x_0 \in A$, be a flexible proposition. Set

$$t(p) = m_A(x_0) \tag{11.3}$$

to be called the denotation-truth-degree of flexible proposition p.

Since denotation-truth-degree $t(p)$ of flexible proposition p is defined by membership function $m_A(x)$ of flexible linguistic value A in the proposition, range $[0, 1]$ of membership function $m_A(x)$ is also the range of denotation-truth-degrees of all flexible propositions in flexible proposition cluster $\{x_0 \in A \mid x_0 \in U\}$. And since the range of membership function of any flexible linguistic value is all interval $[0, 1]$, the ranges of the denotation-truth-degrees of all flexible propositions are also real interval $[0, 1]$.

2. Connotation-truth-degree

Let p: $A(x_0)$, be a flexible proposition.

Just the same, viewed from the representation, x_0 in the proposition has flexible linguistic value A, but actually x_0 has A only with degree $c_A(x_0)$. Consistency-degree $c_A(x_0)$ reflects the degree of x_0 having flexible linguistic value A, so number $c_A(x_0)$ also represents the truth-degree of the flexible proposition p. Thus, we take number $c_A(x_0)$ as another kind of truth-degree of flexible proposition p.

Definition 11.4 Let U be an n-dimensional measurement space, A be a flexible linguistic value on U, $x_0 = (x_{1_0}, x_{2_0}, \ldots, x_{n_0}) \in U$, and let p: $A(x_0)$ be a flexible proposition. Set

$$t(p) = c_A(x_0) \tag{11.4}$$

to be called the connotation-truth-degree of flexible proposition p.

Example 11.1 Let $U = [1, 2.5]$ be the height range of adults, and we define flexible linguistic value "tall" on U as

$$c_{\text{tall}}(x) = \frac{x - 1.5}{0.3}$$

and let

$$p_1: 1.70\,\text{m be tall}, \quad p_2: 1.80\,\text{m be tall}, \quad p_3: 1.90\,\text{m be tall},$$

$$p_4: 1.50\,\text{m be tall}, \quad p_5: 1.40\,\text{m be tall}, \quad p_6: 1.30\,\text{m be tall}.$$

Then

$$t(p_1) = c_{\text{tall}}(1.70) = 0.67,$$
$$t(p_2) = c_{\text{tall}}(1.80) = 1,$$
$$t(p_3) = c_{\text{tall}}(1.90) = 1.3,$$

$$t(p_4) = c_{\text{tall}}(1.50) = 0,$$
$$t(p_5) = c_{\text{tall}}(1.40) = -0.33,$$
$$t(p_6) = c_{\text{tall}}(1.30) = -1.$$

Since connotation-truth-degree $t(p)$ of flexible proposition p is defined by consistency function $c_A(x)$ of flexible linguistic value A in the proposition, range $[\alpha_A, \beta_A]$ $(\alpha_A \leq 0, \ 1 \leq \beta_A)$ of consistency function $c_A(x)$ is also the range of connotation-truth-degrees of propositions in proposition cluster $\{A(x_0) \mid x_0 \in U\}$. Since the ranges of the consistency function of different flexible linguistic values are not all the same, generally speaking, flexible propositions in different flexible proposition clusters have different range of connotation-truth-degrees. But all ranges $[\alpha_X, \beta_X]$ $(X \subset U)$ of connotation-truth-degrees satisfy $\alpha_X \leq 0$ and $1 \leq \beta_X$.

In this section, we define the denotation-truth-degree and connotation-truth-degree of a flexible proposition, and the two sometimes are generally called the truth-degree of a proposition. Truth-degree is the numerical truth value of a flexible proposition.

11.2   Formal Representations and Computation Models of Truth-Degrees of Compound Flexible Propositions

Let p: $A(x)$ and q: $B(y)$ be atom flexible propositions (we temporarily view x and y as constants), and let $p \wedge q$, $p \vee q$, $\neg p$, $p \rightarrow q$, and $p \longleftrightarrow q$ be compound flexible propositions made of p and q, which are referred to as the conjunctive proposition, disjunctive proposition, negative proposition, implicational proposition and equivalent proposition one by one. In this section, we examine the formal representations and computations of truth-degrees of these basic compound flexible propositions.

11.2.1   Representation in the Form of Possession, and Computation Models of Connotation-Truth-Degrees

Let proposition p: $A(x)$, whose literal meaning is that x has flexible linguistic value A, and let proposition q: $B(y)$, whose literal meaning is that y has flexible linguistic value B. Then,

(1) $p \wedge q$ literally means that x has A and y has B. From this literal meaning, (x, y) should have compound or combined flexible linguistic value $A \wedge B$. Thus, the representation of $p \wedge q$ in the form of possession should be

$$A \wedge B(x, y) \tag{11.5}$$

By this expression and the definition of the connotation-truth-degree of a flexible proposition, it follows that $t(p \wedge q) = c_{A \wedge B}(x, y)$. And by Eq. (5.10), we have

$$c_{A \wedge B}(x, y) = \min\{c_A(x), c_B(y)\}$$

(Actually, since the degree of x having A is $c_A(x)$, and the degree of y having B is $c_B(y)$, the degree of (x, y) having $A \wedge B$ should be $\min\{c_A(x), c_B(y)\}$). Thus, the connotation-truth-degree of compound proposition $p \wedge q$ is

$$t(p \wedge q) = \min\{c_A(x), c_B(y)\} \tag{11.6}$$

(2) $p \vee q$ literally means that x has A or y has B. From this literal meaning, (x, y) should has compound or combined flexible linguistic value $A \vee B$. Thus, the representation of $p \vee q$ in the form of possession should be

$$A \vee B(x, y) \qquad (11.7)$$

By this expression and the definition of the connotation-truth-degree of a flexible proposition, it follows that $t(p \vee q) = c_{A \vee B}(x, y)$. And by Eq. (5.11), we have

$$c_{A \vee B}(x, y) = \max\{c_A(x), c_B(y)\}$$

(Actually, since the degree of x has A is $c_A(x)$, and the degree of y has B is $c_B(y)$, the degree of (x, y) having $A \vee B$ should be $\max\{c_A(x), c_B(y)\}$) Thus, the connotation-truth-degree of compound proposition $p \vee q$ is

$$t(p \vee q) = \max\{c_A(x), c_B(y)\} \qquad (11.8)$$

(3) $\neg p$ literally means that x does not have A, which is tantamount to x having $\neg A$. From that, the representation of $\neg p$ in the form of possession is

$$\neg A(x) \qquad (11.9)$$

By this expression and the definition of the connotation-truth-degree of a flexible proposition, it follows that $t(\neg p) = c_{\neg A}(x)$. And by Eq. (5.5), we have

$$c_{\neg A}(x) = 1 - c_A(x)$$

(Actually, this equation can also be obtained by the degree of x having A being $c_A(x)$ and the complement law of degrees.) Thus, the connotation-truth-degree of compound proposition $\neg p$ is

$$t(\neg p) = 1 - c_A(x) \qquad (11.10)$$

(4) $p \rightarrow q$ is that p implies q (namely, $p \Rightarrow q$), or in other words, p is the sufficient condition for q. That means that if there is p, then there is q; but if there is no p, there might be q or might not be q.
Obviously, p and $\neg p$ cannot occur at the same time. Thus, $p \rightarrow q$ is just equivalent to $(p \wedge q) \vee (\neg p \wedge (q \vee \neg q))$; that is,

$$p \rightarrow q \Leftrightarrow (p \wedge q) \vee (\neg p \wedge (q \vee \neg q)) \qquad (11.11)$$

We use T to denote the true proposition of truth-degree >0, that is, the proposition being true with a certain degree. Thus,

$$q \vee \neg q = T \qquad (11.12)$$

and for any proposition r,

$$r \wedge T = T \wedge r = r \tag{11.13}$$

In fact, if $t(r) > 0$, then $r = T$; thus,

$$r \wedge T = T \wedge T$$
$$= T \text{ (by the above Eq. (11.6),}$$
$$t(T \wedge T) = \min\{t(T), t(T)\} = t(T) > 0)$$
$$= r \text{ (by } r = T)$$

and if $t(r) \leq 0$, then $r \wedge T = r$ (because $t(r \wedge T) = \min\{t(r), t(T)\} = t(r)$). Consequently,

$$(p \wedge q) \vee (\neg p \wedge (q \vee \neg q)) = (p \wedge q) \vee (\neg p \wedge T)$$
$$= (p \wedge q) \vee \neg p$$
$$= (p \vee \neg q) \wedge (q \vee \neg p)$$
$$= T \wedge (q \vee \neg p)$$
$$= \neg p \vee q$$

Thus, we have

$$p \rightarrow q \Leftrightarrow \neg p \vee q \tag{11.14}$$

This equivalence relation translates the representation and truth-degree computation of relational compound proposition $p \rightarrow q$ to that of operational compound proposition $\neg p \vee q$, while the representation and the truth values of the latter can be indirectly obtained from the representations and truth values of basic compound propositions $p \vee q$ and $\neg p$. In fact, by expressions (11.7) and (11.9), we can obtain that the representation of $\neg p \vee q$ in the form of possession is $\neg A \vee B\ (x, y)$, while

$$c_{\neg A \vee B}(x, y) = \max\{c_{\neg A}(x), c_B(y)\} = \max\{1 - c_A(x), c_B(y)\}$$

Consequently, the representation of $p \rightarrow q$ in the form of possession is

$$\neg A \vee B(x, y) \tag{11.15}$$

whose connotation-truth-degree is

$$t(p \rightarrow q) = \max\{1 - c_A(x), c_B(y)\} \tag{11.16}$$

(5) $p \longleftrightarrow q$ means that p and q are equivalent; that is, p and q imply each other, or in other words, they are mutually sufficient conditions. Thus, we have

$$p \longleftrightarrow q \Leftrightarrow (p \to q) \wedge (q \to p)$$

And then from expression (11.14), we have

$$p \longleftrightarrow q \Leftrightarrow (\neg p \vee q) \wedge (\neg q \vee p) \tag{11.17}$$

Thus, from this equivalence relation and the above expressions (11.5) and (11.15), the representation of $p \longleftrightarrow q$ in the form of possession is

$$(\neg A \vee B) \wedge (A \vee \neg B)(x, y) \tag{11.18}$$

while

$$c_{(\neg A \vee B) \wedge (A \vee \neg B)}(x, y) = \min\{\max\{1 - c_A(x), c_B(y)\},$$
$$\max\{c_A(x), 1 - c_B(y)\}\}$$

so the connotation-truth-degree of $p \longleftrightarrow q$ is

$$t(p \longleftrightarrow q) = \min\{\max\{1 - c_A(x), c_B(y)\},$$
$$\max\{c_A(x), 1 - c_B(y)\}\} \tag{11.19}$$

Of course, $p \longleftrightarrow q$ can also be explained as: p is the sufficient and necessary condition for q (or q is the sufficient and necessary condition for p). That means: If there is p, then there is q, while if there is no p, then there is no q, that is, q if and only if p. Thus, $p \longleftrightarrow q$ is also equivalent to

$$(p \wedge q) \vee (\neg p \wedge \neg q)$$

that is,

$$p \longleftrightarrow q \Leftrightarrow (p \wedge q) \vee (\neg p \wedge \neg q) \tag{11.20}$$

From this equivalence relation and the above expressions (11.5) and (11.7), the representation of $p \longleftrightarrow q$ in the form of possession is

$$(A \wedge B) \vee (\neg A \wedge \neg B)(x, y) \tag{11.21}$$

Consequently,

$$c_{(A \wedge B) \vee (\neg A \wedge \neg B)}(x, y) = \max\{\min\{c_A(x), c_B(y)\},$$
$$\min\{1 - c_A(x), 1 - c_B(y)\}\}$$

and then, the connotation-truth-degree of $p \longleftrightarrow q$ is

$$t(p \longleftrightarrow q) = \max\{\min\{c_A(x), c_B(y)\}, \\ \min\{1 - c_A(x), 1 - c_B(y)\}\} \qquad (11.22)$$

Thus, $p \longleftrightarrow q$ has two expressions and two computation formulas of connotation-truth degree. But it can be proved that

$$(p \wedge q) \vee (\neg p \wedge \neg q) \Leftrightarrow (\neg p \vee q) \wedge (\neg q \vee p)$$

Thus,

$$\max\{\min\{c_A(x), c_B(y)\}, \min\{1 - c_A(x), 1 - c_B(y)\}\} \\ = \min\{\max\{1 - c_A(x), c_B(y)\}, \max\{c_A(x), 1 - c_B(y)\}\}$$

Now, we view the x and y in the above propositions as variables. Thus, we then obtain a group of connotation-truth-degree computation formulas for basic compound flexible propositions:

$$t(p \wedge q) = \min\{c_A(x), c_B(y)\} \qquad (11.23)$$

$$t(p \vee q) = \max\{c_A(x), c_B(y)\} \qquad (11.24)$$

$$t(\neg p) = 1 - c_A(x) \qquad (11.25)$$

$$t(p \rightarrow q) = \max\{1 - c_A(x), c_B(y)\} \qquad (11.26)$$

$$t(p \longleftrightarrow q) = \min\{\max\{1 - c_A(x), c_B(y)\}, \\ \max\{c_A(x), 1 - c_B(y)\}\} \qquad (11.27)$$

where $x \in U$ and $y \in V$ (U and V can be the same).

It can be seen that this group of formulae are a group of functions defined on the corresponding measurement ranges. We may as well call them the functions of connotation-truth-degrees of basic compound flexible propositions, and the examples of graphs of first three functions are shown in Fig. 11.1.

11.2.2 Representation in the Form of Membership, and Computation Models of Denotation-Truth-Degrees

From the representation in the forms of possession and the connotation-truth-degree computation functions of the basic compound flexible propositions above, we can

(a)   **(b)**

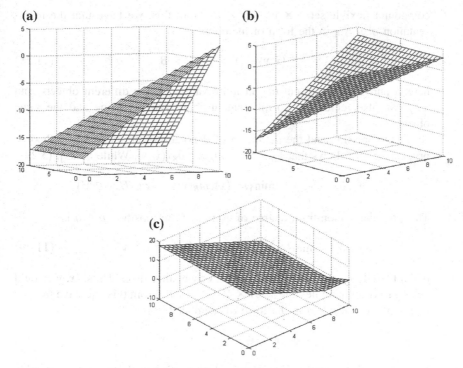

(c)

Fig. 11.1 Examples of graphs of the functions of connotation-truth-degrees of compound flexible propositions. **a** $t(p \wedge q) = \min\{c_A(x), c_B(y)\}$, **b** $t(p \vee q) = \max\{c_A(x), c_B(y)\}$, **c** $t(p \rightarrow q) = \max\{1 - c_A(x), c_B(y)\}$

obtain their representations in the form of membership and denotation-truth-degree computation functions. But in the following we still start directly from the membership relation to consider the representations in the form of membership and the computations of denotation-truth-degree of the compound propositions. Similarly, temporarily we view x and y in the following expressions as constants. In the following, we do discussion in three situations.

Situation 1: Let proposition p: $x \in A$, whose literal meaning is that x belongs to flexible class A; and let proposition q: $y \in B$, whose literal meaning is that y belongs to flexible class B. Then, there are two cases:

1. $A \subset U$ and $B \subset V$

(1) $p \wedge q$ literally means that x belongs to A and y belongs to B. Since A and B are orthogonal, $(x, y) \in A \times B$ or $(x, y) \in A \times V \cap U \times B$, but since p is related to q by conjunction, $A \times V \cap U \times B$ corresponds to the logical conjunction of linguistic values A and B, while $A \times B$ corresponds to the algebraic synthesis of linguistic values A and B; on the other hand, for the flexible sets, $A \times V \cap U \times B \neq A \times B$, and $A \times B$ does not have such extended core and support set like usual flexible sets. Therefore, (x, y) should belong to

compound flexible set $A \times V \cap U \times B$. From this, we have that the representation of $p \wedge q$ in the form of membership is

$$(x, y) \in A \times V \cap U \times B \tag{11.28}$$

here, x and y can represent the attribute values of two different objects, and they can also represent two different attribute values of one and the same object.

By this expression and the definition of denotation-truth-degree of a flexible proposition, it follows that $t(p \wedge q) = m_{A \times V \cap U \times B}(x, y)$. While by Eq. (5.29),

$$m_{A \times V \cap U \times B(x,y)} = \min\{m_A(x), m_B(y)\} \quad (x \in U, y \in V)$$

Thus, the denotation-truth-degree of compound proposition $p \wedge q$ is

$$t(p \wedge q) = \min\{m_A(x), m_B(y)\} \quad (x \in U, y \in V) \tag{11.29}$$

(2) $p \vee q$ literally means that x belongs to A or y belongs to B. Thus, (x, y) should belong to compound flexible set $A \times V \cup U \times B$. From this, we have that the representation of $p \vee q$ in the form of membership is

$$(x, y) \in A \times V \cup U \times B \tag{11.30}$$

By this expression and the definition of denotation-truth-degree of a flexible proposition, it follows that $t(p \vee q) = m_{A \times V \cup U \times B}(x, y)$, while by Eq. (5.30),

$$m_{A \times V \cup U \times B}(x, y) = \max\{m_A(x), m_B(y)\}$$

Thus, the denotation-truth-degree of compound proposition $p \vee q$ is

$$t(p \vee q) = \max\{m_A(x), m_B(y)\} \tag{11.31}$$

(3) $\neg p$ literally means that x does not belong to A, which is equivalent to x belonging to A^c. From this, we have that the representation of $\neg p$ in the form of membership is

$$x \in A^c \tag{11.32}$$

By this expression and the denotation-truth-degree definition of a flexible proposition, it follows that $t(\neg p) = m_{A^c}(x)$, while by Eq. (5.31),

$$m_{A^c}(x) = 1 - m_A(x)$$

(Actually, from the degree of x belonging to A, $m_A(x)$, and the complement law of degrees, this equation can also be obtained.) Thus, the denotation-truth-degree of compound proposition $\neg p$ is

$$t(\neg p) = 1 - m_A(x) \tag{11.33}$$

(4) For $p \rightarrow q$, from

$$p \rightarrow q \Leftrightarrow \neg p \vee q$$

and expressions (11.30) and (11.32), we have the representation of $p \rightarrow q$ in the form of membership is

$$(x,y) \in A^c \times V \cup U \times B \tag{11.34}$$

while the membership function is

$$m_{A^c \times V \cup U \times B}(x,y) = \max\{1 - m_A(x), m_B(y)\}$$

thus, the denotation-truth-degree of compound proposition $p \rightarrow q$ is

$$t(p \rightarrow q) = \max\{1 - m_A(x), m_B(y)\} \tag{11.35}$$

(5) For $p \longleftrightarrow q$, from

$$p \longleftrightarrow q \Leftrightarrow (\neg p \vee q) \wedge (\neg q \vee p)$$

we have immediately that the representation of $p \longleftrightarrow q$ in the form of membership is

$$(x,y) \in (A^c \times V \cup U \times B) \cap (A \times V \cup U \times B^c) \tag{11.36}$$

while the membership function is

$$m_{(A^c \times V \cup U \times B) \cap (A \times V \cup U \times B^c)}(x,y) = \min\{\max\{1 - m_A(x), m_B(y)\},$$
$$\max\{m_A(x), 1 - m_B(y)\}\}$$

Thus, the denotation-truth-degree of compound proposition $p \longleftrightarrow q$ is

$$t(p \longleftrightarrow q) = \min\{\max\{1 - m_A(x), m_B(y)\}, \atop \max\{m_A(x), 1 - m_B(y)\}\} \tag{11.37}$$

2. $A \subset U$ and $B \subset U$

In this case, A and B stand for different flexible linguistic values of one and the same feature, and $x, y \in U$ are then different numerical values of one and the same feature. Obviously, here x and y cannot be different numerical values of one and the same object (for instance, x and y cannot be separately one and the same person's

height and weight), but they can only be numerical values of one and the same
feature of different objects (for instance, x is the height of P_1, and y is the height of
P_2). Therefore, a compound proposition in such a case can be viewed as a special
case of $A \subset U$ and $B \subset V$ in the above, that is, the compound proposition when
$V = U$. So viewed from the form, the representations in the forms of membership
and the computation formulas of denotation-truth-degrees of $p \wedge q$, $p \vee q$, $\neg p$,
$p \rightarrow q$, and $p \longleftrightarrow q$ are completely the same as those the above.

But in this case, there exists question in the meanings of the representations in
the forms of membership of compound propositions; for example, let p: 1.50 m is
short, and q: 1.80 m is tall, then how is $p \wedge q$: $(x, y) \in A \cap B$ to be explained?

Situation 2: Let proposition p: $x \in A$, and proposition q: $y \in A$. Then, there is
only one case of $A \subseteq U$. Hence, here x and y also can only are numerical values of
one and the same feature of different objects (for instance, p: 1.75 m is tall, and q:
1.76 m is tall).

Making an analysis similar to (1.1), we can have:

$$p \wedge q: (x, y) \in A \cap A \tag{11.38}$$

$$p \vee q: (x, y) \in A \cup A \tag{11.39}$$

$$\neg p: x \in A^c \tag{11.40}$$

$$p \rightarrow q:(x, y) \in A^c \cup A \tag{11.41}$$

$$p \longleftrightarrow q: (x, y) \in (A^c \cup A) \cap (A \cup A^c) \tag{11.42}$$

But in consideration that between x and y in propositions p and q, there is
originally the relation of "greater than" or "less than", that is, $x < y$ or $x > y$, so
propositions p and q have implication relation. For instance, suppose $x < y$, then
$p \rightarrow q$. Therefore, it is not fit for this type of propositions to be represented in the
form of membership.

Situation 3: Let proposition p: $x \in A$, and proposition q: $x \in B$. Then,

(1) When $A, B \subset U$, making an analysis similar to (1.1), we can have:

$$p \wedge q: x \in A \cap B \tag{11.43}$$

$$p \vee q: x \in A \cup B \tag{11.44}$$

$$\neg p: x \in A^c \tag{11.45}$$

$$p \rightarrow q: x \in A^c \cup B \tag{11.46}$$

$$p \longleftrightarrow q: x \in (A^c \cup B) \cap (A \cup B^c) \tag{11.47}$$

(2) When $A \subset U$ and $B \subset V$, we can rewrite the original proposition q as: y is B, and thus, the problem is reduced to be the previous (1.1).

Now we view the x and y in all the propositions above as variables. Thus, we have a group of computation formulas of denotation-truth-degrees of basic compound propositions:

$$t(p \wedge q) = \min\{m_A(x), m_B(y)\} \tag{11.48}$$

$$t(p \vee q) = \max\{m_A(x), m_B(y)\} \tag{11.49}$$

$$t(\neg p) = 1 - m_A(x) \tag{11.50}$$

$$t(p \rightarrow q) = \max\{1 - m_A(x), m_B(y)\} \tag{11.51}$$

$$t(p \longleftrightarrow q) = \min\{\max\{1 - m_A(x), m_B(y)\},$$
$$\max\{m_A(x), 1 - m_B(y)\}\} \tag{11.52}$$

where $x \in U$ and $y \in V$ (U and V can be the same).

It can be seen that this group of formulae are a group of functions defined on the corresponding measurement ranges, which are the functions of denotation-truth-degrees of 5 basic compound propositions. In the following, we give the examples of graphs of first three functions (see Fig. 11.2).

11.2.3 Indirect Computation Models Based on the Truth-Degrees of Component Propositions

1. Indirect computation models of denotation-truth-degrees

In the above, we use membership degrees to define directly the denotation-truth-degree of compound propositions. But it can be seen that the membership function of a compound proposition is also the function of the membership functions $m_A(x)$ and $m_B(y)$ of its component propositions, while $m_A(x) = t(p)$ and $m_B(y) = t(q)$. Therefore, we have

$$t(p \wedge q) = \min\{m_A(x), m_B(y)\}$$
$$= \min\{t(p), t(q)\}$$

namely

$$t(p \wedge q) = \min\{t(p), t(q)\} \tag{11.53}$$

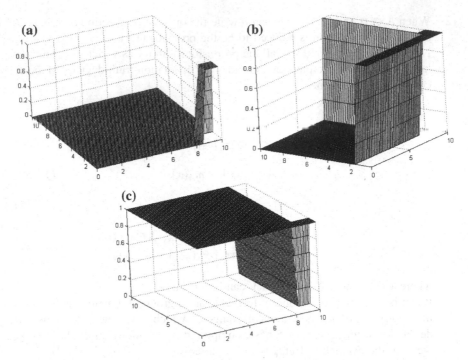

Fig. 11.2 Examples of graphs of the functions of denotation-truth-degrees of compound flexible propositions.　　**a**　　$t(p \wedge q) = \min\{m_A(x), m_B(y)\}$,　　**b**　　$t(p \vee q) = \max\{m_A(x), m_B(y)\}$, **c** $t(p \rightarrow q) = \max\{1 - m_A(x), m_B(y)\}$

Similarly, we have

$$t(p \vee q) = \max\{t(p), t(q)\} \tag{11.54}$$

$$t(\neg p) = 1 - t(p) \tag{11.55}$$

$$t(p \rightarrow q) = \max\{1 - t(p), t(q)\} \tag{11.56}$$

$$t(p \longleftrightarrow q) = \min\{\max\{1 - t(p), t(q)\}, \\ \max\{t(p), 1 - t(q)\}\} \tag{11.57}$$

This group of equations also separately describe the relationship between the denotation-truth-degrees of compound propositions $p \wedge q$, $p \vee q$, $\neg p$, $p \rightarrow q$, $p \longleftrightarrow q$, and the denotation-truth-degrees of their component propositions p and q. Thus, when the denotation-truth-degrees of component propositions p and q are known, we can use these equations to indirectly obtain the denotation-truth-degrees of the corresponding compound propositions. Thus, this group of formulas is also another group of functions of the denotation-truth-degrees of basic compound propositions. They are a group of functions defined on the corresponding ranges of denotation-truth-degrees, whose graphs are shown in Fig. 11.3.

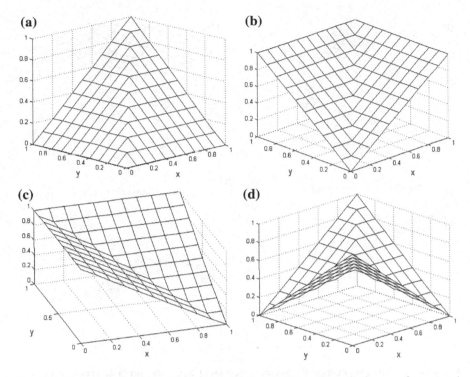

Fig. 11.3 Examples of graphs of the indirect functions of denotation-truth-degrees of compound propositions. **a** $t(p \wedge q) = \min\{t(p), t(q)\}$, **b** $t(p \vee q) = \min\{t(p), t(q)\}$, **c** $t(p \rightarrow q) = \min\{t(p), t(q)\}$, **d** $t(p \longleftrightarrow q) = \min\{\max\{1 - t(p), t(q)\}, \max\{t(p), 1 - t(q)\}\}$

2. Indirect computation models of connotation-truth-degrees

Similarly, from $c_A(x) = t(p)$ and $c_B(y) = t(q)$, the following equations can be deduced from the definition expressions of the connotation-truth-degrees of compound propositions:

$$t(p \wedge q) = \min\{t(p), t(q)\} \tag{11.58}$$

$$t(p \vee q) = \max\{t(p), t(q)\} \tag{11.59}$$

$$t(\neg p) = 1 - t(p) \tag{11.60}$$

$$t(p \rightarrow q) = \max\{1 - t(p), t(q)\} \tag{11.61}$$

$$t(p \longleftrightarrow q) = \min\{\max\{1 - t(p), t(q)\}, \\ \max\{t(p), 1 - t(q)\}\} \tag{11.62}$$

These 5 equations are the relational expressions between the connotation-truth-degrees of compound propositions $p \wedge q$, $p \vee q$, $\neg p$, $p \rightarrow q$, and $p \longleftrightarrow q$ and

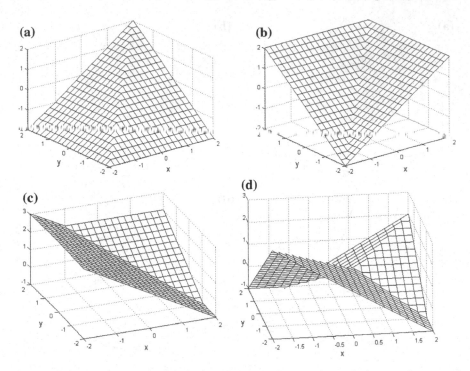

Fig. 11.4 Examples of graphs of the indirect functions of connotation-truth-degrees of compound propositions.   **a** $t(p \wedge q) = \min\{t(p), t(q)\}$,   **b** $t(p \vee q) = \min\{t(p), t(q)\}$, **c** $t(p \rightarrow q) = \min\{t(p), t(q)\}$, **d** $t(p \longleftrightarrow q) = \min\{\max\{1 - t(p), t(q)\}, \ \max\{t(p), \ 1 - t(q)\}$

those of their component propositions p and q. Thus, when the connotation-truth-degrees of p and q are known, these formulas can be used to indirectly obtain the connotation-truth-degrees of corresponding compound propositions $p \wedge q$, $p \vee q$, $\neg p$, $p \rightarrow q$, and $p \longleftrightarrow q$. Therefore, this group of formulas is another group of functions of the connotation-truth-degrees of basic compound propositions. They are a group of functions defined on the corresponding ranges of connotation-truth-degrees, whose graphs are shown in Fig. 11.4.

In the above, we derived the indirect computation formulas or functions of the denotation-truth-degrees and the connotation-truth-degrees of basic compound propositions. Although viewed from the expressions, the two groups of formulas are completely the same, and the domain and range of the two are different. In fact, the domain and range of the denotation-truth-degree functions are separately $[0, 1] \times [0, 1]$ and $[0, 1]$, while those of the connotation-truth-degree functions are separately $[\alpha_A, \beta_A] \times [\alpha_B, \beta_B]$ ($\alpha_A \leq 0, 1 \leq \beta_A$; $\alpha_B \leq 0, 1 \leq \beta_B$) and $[\alpha, \beta]$ ($\alpha \leq 0, 1 \leq \beta$).

Note that originally, the definition domain of the operation min of denotation-truth-degrees is $[0, 1] \cap [0, 1]$, and that of max is $[0, 1] \cup [0, 1]$, but because $[0, 1]$ is a universe for the denotation-truth-degrees and the two $[0, 1]$ are of orthogonal relation, $[0, 1] \cap [0, 1] = [0, 1] \cup [0, 1] = [0, 1] \times [0, 1]$. Similarly, the definition domain of the operation min of connotation-truth-degrees is

$[\alpha_A, \beta_A] \cap [\alpha_B, \beta_B]$, and that of max is $[\alpha_A, \beta_A] \cup [\alpha_B, \beta_B]$, but because $[\alpha_A, \beta_A]$ and $[\alpha_B, \beta_B]$ are both universe for the connotation-truth-degrees and the two are of orthogonal relation, $[\alpha_A, \beta_A] \cap [\alpha_B, \beta_B] = [\alpha_A, \beta_A] \cup [\alpha_B, \beta_B] = [\alpha_A, \beta_A] \times [\alpha_B, \beta_B]$.

Besides, it is not hard to verify that those truth-degree computation formulas of conjunctive and disjunctive compound propositions above can also be generalized to more than 2 propositions.

From the above truth-degree computation formula $t(\neg p) = 1 - t(p)$ of the negative proposition, we have

$$t(p) + t(\neg p) = 1 \qquad (11.63)$$

This is the relation between the degrees of a pair of relatively negative propositions. We call this equality the **complement law of truth-degrees**.

11.2.4 Range of Truth-Degrees of Flexible Propositions and Its Symmetry

Since the range of denotation-truth-degrees of all flexible propositions is real interval [0, 1], generally, we call interval [0, 1] the denotation-truth-degree range of flexible propositions. The complement relation of truth-degrees of relatively negative propositions shows that the truth-degrees of a pair of relatively negative propositions are just symmetrical about 0.5, while interval [0, 1] contains the denotation-truth-degrees of all relatively negative propositions, which are just symmetrical with center 0.5.

Next, we examine what a kind of real interval is the connotation-truth-degree range of flexible propositions and whether it also has symmetry.

We know that one flexible linguistic value A determines a flexible proposition cluster $\{A(x_0) \mid x_0 \in U\}$ and a connotation-truth-degree range $[\alpha_A, \beta_A]$ $(\alpha_A \leq 0, 1 \leq \beta_A)$. Then, the negative value $\neg A$ of A also determines a flexible proposition cluster $\{\neg A(x_0) \mid x_0 \in U\}$ and a connotation-truth-degree range $[\alpha_{\neg A}, \beta_{\neg A}]$ $(\alpha_{\neg A} \leq 0, 1 \leq \beta_{\neg A})$. Thus,

$$[\alpha, \beta] = [\alpha_A, \beta_A] \cup [\alpha_{\neg A}, \beta_{\neg A}]$$
$$= [\min\{\alpha_A, \alpha_{\neg A}\}, \max\{\beta_A, \beta_{\neg A}\}]$$

is just the connotation-truth-degree range of propositions in flexible proposition cluster $\{A(x_0) \mid x_0 \in U\} \cup \{\neg A(x_0) \mid x_0 \in U\}$.

Now, take $\forall p \in \{A(x_0) \mid x_0 \in U\} \cup \{\neg A(x_0) \mid x_0 \in U\}$, then correspondingly, $\neg p \in \{A(x_0) \mid x_0 \in U\} \cup \{\neg A(x_0) \mid x_0 \in U\}$. By the complement law of truth-degrees, it follows that $t(p) + t(\neg p) = 1$, which shows truth-degrees $t(p)$ and $t(\neg p)$ are symmetrical about 0.5. Also, since the proposition p is taken arbitrarily, the truth-degree range $[\alpha, \beta]$ ($\alpha \leq 0$, $1 \leq \beta$) is a real interval with symmetrical center 0.5. In fact, the symmetry of truth-degrees $t(A(x_0))$ and $t(\neg A(x_0))$ about 0.5 originates from the symmetry of consistency-degrees $c_A(x_0)$ and $c_{\neg A}(x_0)$ about 0.5, which we have already known in Sect. 6.5.

Since the range $[\alpha, \beta]$ of truth-degrees is symmetrical about 0.5, $\alpha + \beta = 1$. Thus, $\alpha = 1 - \beta$, while $\beta = 1 - \alpha$. Therefore,

$$[\alpha, \beta] = [1 - \beta, \beta] = [\alpha, 1 - \alpha]$$

In view of this characteristic, from now on, we denote connotation-truth-degree range as $[1 - \beta, \beta]$ ($1 \leq \beta$).

The characteristic of a truth-degree range being symmetrical about 0.5 guarantees that truth-degree $t(\neg p) = 1 - t(p)$ is always computable; that is, truth-degree range is the closed under negation operation of truth-degree (which will be defined later).

11.3 Algebraic Compound Flexible Proposition/Flexible Proposition with a Synthetic Linguistic Value and Computation Model of Its Truth-Degree

The compound flexible propositions stated in the last section are usual compound flexible propositions formed by using connectives "and" and "or". However, we find that besides these compound flexible propositions, there are compound flexible propositions that formed by using "plus"; for example, "he is gifted plus he is studious" is such a compound flexible proposition. In consideration that the relation between the component propositions of this kind of compound flexible proposition is not logic conjunction or disjunction but algebraic synthesis, we call this kind of compound flexible proposition to be the **algebraical compound flexible proposition**. In order to distinguish, we call usual compound flexible propositions to be the **logical compound flexible proposition**.

Of course, in daily language, people sometimes also use connective "and" but not use strictly "plus" to describe an algebraic compound flexible proposition. For instance, original "$A(x_0)$ plus $B(y_0)$" is said "$A(x_0)$ and $B(y_0)$".

Still like Sect. 6.4.2, we denote "plus" by $\oplus$. Then, in general, "$A(x_0)$ plus $B(y_0)$" can be symbolized as "$A(x_0) \oplus B(y_0)$".

Let $A_1(x_{1_0}) \oplus A_2(x_{2_0}) \oplus \cdots \oplus A_n(x_{n_0})$ be an algebraic compound flexible proposition. Then, its representation in the form of membership is

$$A_1 \oplus A_2 \oplus \cdots \oplus A_n(x_{1_0}, x_{2_0}, \ldots, x_{n_0}) \tag{11.64}$$

It can be seen that the linguistic value in expression (11.64) is a synthetic linguistic value. So we call the proposition of such form to be the **flexible proposition with a synthetic linguistic value**.

Similarly, we call the proposition of forms

$$A_1 \wedge A_2 \wedge \cdots \wedge A_n(x_{1_0}, x_{2_0}, \ldots, x_{n_0}) \tag{11.65}$$

and

$$A_1 \vee A_2 \vee \cdots \vee A_n(x_{1_0}, x_{2_0}, \ldots, x_{n_0}) \tag{11.66}$$

to be the **flexible proposition with a combined linguistic value**.

And then, we call together combined linguistic value and synthetic linguistic value to be the **flexible proposition with a composite linguistic value**.

The computation formulas of truth-degrees of flexible propositions with a combined linguistic value have been really given in the above section. In the following, we consider the computing formulas of truth-degrees of flexible propositions with a synthetic linguistic value.

Let A_i be a flexible linguistic value on the measurement space U_i, $x_{i_0} \in U_i$ ($i = 1, 2, \ldots, n$), and $A_1 \oplus A_2 \oplus \cdots \oplus A_n(x_{1_0}, x_{2_0}, \ldots, x_{n_0})$ be a flexible propositions with a synthetic linguistic value. By the definition of the connotation-truth-degree of a flexible proposition, it follows that

$$t(A_1 \oplus A_2 \oplus \cdots \oplus A_n(x_{1_0}, x_{2_0}, \ldots, x_{n_0})) = c_{A_1 \oplus A_2 \oplus \cdots \oplus A_n}(x_{1_0}, x_{2_0}, \ldots, x_{n_0})$$

Also from Eq. (6.17), we have

$$c_{A_1 \oplus A_2 \oplus \cdots \oplus A_n}(x_{1_0}, x_{2_0}, \ldots, x_{n_0}) = \sum_{i=1}^{n} w_i c_{A_i}(x_{i_0}), \quad c_{A_i}(x_{i_0}) > 0$$

where $x_{i_0} \in U_i$, $w_i \in [0, 1]$, $\sum_{i=1}^{n} w_i = 1$.

Consequently, the connotation-truth-degree of flexible proposition with a synthetic linguistic value, $A_1 \oplus A_2 \oplus \cdots \oplus A_n(x_{1_0}, x_{2_0}, \ldots, x_{n_0})$, is

$$t(A_1 \oplus A_2 \oplus \cdots \oplus A_n(x_{1_0}, x_{2_0}, \ldots, x_{n_0})) = \sum_{i=1}^{n} w_i c_{A_i}(x_{i_0}), c_{A_i}(x_{i_0}) > 0 \tag{11.67}$$

where $x_{i_0} \in U_i$, $w_i \in [0, 1]$, $\sum_{i=1}^{n} w_i = 1$.

From expression (11.64) and Eq. (11.67), we give directly the representation of the proposition in the form of membership

$$(x_{1_0}, x_{2_0}, \ldots, x_{n_0}) \in A_1 \times A_2 \times \cdots \times A_n \tag{11.68}$$

Fig. 11.5 An example of graph of the function of truth-degree of flexible proposition with a synthetic linguistic value, $A_1 \oplus A_2(x_{1_0}, x_{2_0})$

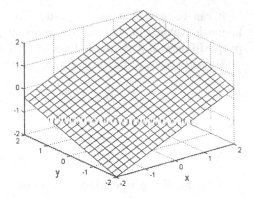

and the corresponding computation formula of denotation-truth-degree

$$t((x_{1_0}, x_{2_0}, \ldots, x_{n_0}) \in A_1 \times A_2 \times \cdots \times A_n)$$
$$= \sum_{i=1}^{n} m_{A_i}(x_{i_0}), \quad m_{A_i}(x_{i_0}) > 0 \tag{11.69}$$

where $x_{i_0} \in U_i$, $w_i \in [0, 1]$, $\sum_{i=1}^{n} w_i = 1$.

Further, we have

$$t(A_1 \oplus A_2 \oplus \cdots \oplus A_n(x_{1_0}, x_{2_0}, \ldots, x_{n_0})) = \sum_{i=1}^{n} w_i t(p_{x_{i_0}}), \quad t(p_{x_{i_0}}) > 0 \tag{11.70}$$

where $p_{x_{i_0}} = c_{A_i}(x_{i_0})$. This truth-degree computation formula is also a function on the corresponding region $[\alpha_{A_1}, \beta_{A_1}] \times [\alpha_{A_2}, \beta_{A_2}] \times \cdots \times [\alpha_{A_n}, \beta_{A_n}]$. When $n = 2$, the graph of truth-degree computation function of flexible proposition with a synthetic linguistic value, $A_1 \oplus A_2(x_{1_0}, x_{2_0})$, is shown in Fig. 11.5.

11.4   Truth-Degree-Level Logic Algebras

Let p: $A(x_1)$, and q: $A(x_2)$, $x_1, x_2 \in U$, and let $[1 - \beta, \beta]$ ($\beta \geq 1$) be the connotation-truth-degree range of flexible proposition cluster $\{A(x_0) \mid x_0 \in U\} \cup \{\neg A(x_0) \mid x_0 \in U\}$. Denote $t(p)$ and $t(q)$ as t_1 and t_2 separately. Examining the indirect computation formula $t(p \wedge q) = \min\{t(p), t(q)\}$ of connotation-truth-degree of compound proposition $p \wedge q$, it can be seen that since the interval $[1 - \beta, \beta]$ is closed under operation $\wedge$, this formula determines a operation or function •:

$$t_1 \bullet t_2 = \min\{t_1, t_2\}, \quad t_1, t_2 \in [1 - \beta, \beta]$$

on the truth-degree range $[1 - \beta, \beta]$ really.

Similarly, formulas $t(p \vee q) = \max\{t(p), t(q)\}$ and $t(\neg p) = 1 - t(p)$ also determine operations:

$$t_1 + t_2 = \max\{t_1, t_2\}$$
$$\bar{t} = 1 - t$$

on the $[1 - \beta, \beta]$.

Definition 11.5 Let $\mathscr{T}_c = [1 - \beta, \beta]$ ($\beta \geq 1$) be a connotation-truth-degree range to which a certain flexible proposition cluster corresponds, and t_1 and t_2 be two truth-degreed variables taking values from T_c. we define three operations on T_c:

$$t_1 \bullet t_2 = \min\{t_1, t_2\} \tag{11.71}$$

$$t_1 + t_2 = \max\{t_1, t_2\} \tag{11.72}$$

$$\bar{t} = 1 - t \tag{11.73}$$

They are one by one called connotation-truth-degreed multiplication, connotation-truth-degreed addition, and connotation-truth-degreed complement, and they are collectively called connotation-truth-degreed operations.

It can be verified that operations $+$ and $\bullet$ satisfy commutative laws, associative laws, idempotent laws, and absorption laws, so $\langle \mathscr{T}_c, +, \bullet \rangle$ forms a lattice. At the same time, they also satisfy distributive laws, so $\langle \mathscr{T}_c, +, \bullet \rangle$ is also a distributive lattice. Besides, operations $\bullet$ and $+$ also satisfy identity law and zero-one laws (here the zero element is α, and the one element is β), complement operation $-$ also satisfies double complement law and De Morgan's laws, so $\langle [1 - \beta, \beta], +, \bullet, - \rangle$ also forms a Morgan algebra.

Similarly, we can also define truth-degreed operations $+$, $\bullet$, and $-$ on range $T_d = [0, 1]$ of denotation-truth-degrees, they are one by one called denotation-truth-degreed multiplication, denotation-truth-degreed addition, and denotation-truth-degreed complement, and they are collectively called denotation-truth-degreed operations. And $\langle \mathscr{T}_d, +, \bullet \rangle$ forms also distributive lattice, and $\langle \mathscr{T}_d, \bullet, +, - \rangle$ forms also Morgan algebra.

From stated above, based on truth-degrees and their operations we can set up the logic algebras at the level of numerical truth value, that is, truth-degree. But it can be seen that the connotation-truth-degree-level logic systems above actually include denotation-truth-degree-level logic systems. Therefore, we only discuss connotation-truth-degree-level logic and inference in what follows.

11.5   Degree-True Inference

11.5.1   *Flexible-Propositional Formulas and Flexible-Predicate Formulas*

Definition 11.6 We call the propositional formula (or propositional form) that denotes a flexible proposition to be a flexible-propositional formula (or flexible-propositional form); and call the predicate formula (or predicate form) in which the predicate names are flexible linguistic values to be a flexible-predicate formula (or flexible-predicate form).

Definition 11.7 Let $P(p_1, p_2, \ldots, p_n)$ and $Q(p_1, p_2, \ldots, p_n)$ be two flexible-propositional formulas defined on truth-degree space $[1 - \beta, \beta]^n$ ($\beta \geq 1$). If for arbitrary $(\varepsilon_1, \varepsilon_2, \ldots, \varepsilon_n) \in [1 - \beta, \beta]^n$ ($\beta \geq 1$), always

$$P(\varepsilon_1, \varepsilon_2, \ldots, \varepsilon_n) = Q(\varepsilon_1, \varepsilon_2, \ldots, \varepsilon_n)$$

and then $P(\varepsilon_1, \varepsilon_2, \ldots, \varepsilon_n)$ and $Q(\varepsilon_1, \varepsilon_2, \ldots, \varepsilon_n)$ are called to be logically equivalent at the level of truth-degree, or truth-degree-level logically equivalent for short, denoted by writing

$$P(p_1, p_2, \ldots, p_n) \Leftrightarrow {}_t Q(p_1, p_2, \ldots, p_n) \tag{11.74}$$

or

$$P(p_1, p_2, \ldots, p_n) = {}_t Q(p_1, p_2, \ldots, p_n) \tag{11.74$'$}$$

The two expressions are called truth-degree logical equivalence or truth-degree identity.

Those equalities in Table 11.1 are some important truth-degree identities.

11.5.2   *Degree-Valid Argument Form*

We know that logical inference must follow a valid argument form. A valid argument form is the argument form that can guarantee that a true conclusion follows from true premise(s). A valid argument form in logic is called a rule of inference. There are many rules of inference in the traditional two-valued logic, such as modus ponens, modus tollens, hypothetical syllogism.

In the traditional two-valued logic, rules of inference and the implicational tautologies are one-to-one correspondence. That is to say, in the traditional two-valued logic, an argument form is valid when and only when the corresponding implication is always true. However, we find that there exists no tautology in the traditional sense in truth-degreed logic; that is, there exists no such flexible-propositional formula P

Table 11.1 Some important truth-degree identities

E_1	$\neg\neg P = P$	(Double negative law)
E_2	$P \wedge P = P, P \vee P = P$	(Idempotent laws)
E_3	$P \wedge Q = Q \wedge P, P \vee Q = P \vee Q$	(Commutative laws)
E_4	$(P \wedge Q) \wedge R = P \wedge (Q \wedge R),$ $(P \vee Q) \vee R = P \vee (Q \vee R)$	(Associative laws)
E_5	$P \wedge (Q \vee R) = P \wedge Q \vee P \wedge R,$ $P \vee (Q \wedge R) = (P \vee Q) \wedge (P \vee R)$	(Distributive laws)
E_6	$P \wedge (P \vee Q) = P,$ $P \vee (P \wedge Q) = P$	(Absorption laws)
E_7	$\neg(P \wedge Q) = \neg P \vee \neg Q,$ $\neg(P \vee Q) = \neg P \wedge \neg Q$	(De Morgan's laws)
E_8	$P \rightarrow Q = \neg P \vee Q$	(Implicational expression)
E_9	$P \longleftrightarrow Q = (P \rightarrow Q) \wedge (Q \rightarrow P)$	(Equivalent expression)

$(p_1, p_2, \ldots, p_n)$, for any $(\varepsilon_1, \varepsilon_2, \ldots, \varepsilon_n) \in [1 - \beta, \beta]^n$ ($\beta \geq 1$), $P(\varepsilon_1, \varepsilon_2, \ldots, \varepsilon_n) = 1$ always.

In fact, a flexible-propositional formula on $[1 - \beta, \beta]^n$ ($\beta \geq 1$) is only a expression of truth-degreed operations, which is made of some propositional variables joined by connectives $\neg, \wedge, \vee, \rightarrow$ or $\longleftrightarrow$, and thus, though each ε_i in $(\varepsilon_1, \varepsilon_2, \ldots, \varepsilon_n)$ is all from respective subrange $(0, 1)$, they are tantamount to be all from one and the same subrange $(0, 1)$ when doing operations, while from the operator definitions of connectives $\neg, \wedge, \vee, \rightarrow$ and $\longleftrightarrow$, the open interval $(0, 1)$ is all closed under these operations; that is, for any $p, q \in (0, 1)$, it follows that $\neg p, p \wedge q$, $p \vee q, p \rightarrow q$ and $p \longleftrightarrow q \in (0, 1)$. Thus, when $(\varepsilon_1, \varepsilon_2, \ldots, \varepsilon_n) \in (0, 1)^n \subset [1 - \beta, \beta]^n$, $P(\varepsilon_1, \varepsilon_2, \ldots, \varepsilon_n) \in (0, 1)$ but $\notin \{0, 1\}$ surely; that is to say, when $(\varepsilon_1, \varepsilon_2, \ldots, \varepsilon_n) \in (0, 1)^n$, $P(\varepsilon_1, \varepsilon_2, \ldots, \varepsilon_n) \neq 1$.

There exists no tautology in truth-degreed logic, which means that there exist no rules of inference in the sense of "true" (that is, truth-degree = 1) in truth-degreed logic, and then means that the concept of validity of argument forms in traditional two-valued logic is not applicable to truth-degreed logic. Then, for truth-degreed logic, what argument form is valid?

Considering that a truth-degree is the degree of a proposition being true, while a flexible proposition is always true with a certain degree, and the range of truth-degrees is $[0, 1]$ or more general $[1 - \beta, \beta]$ ($\beta \geq 1$), the "true" in truth-degreed logic should not refer only to absolute truth, i.e., truth-degree 1, but should be "truth with a certain degree with infimum 0", i.e., truth-degree >0. With this understanding, we can now define the validity of an argument form in truth-degreed logic. In order to facilitate description, we call collectively the "true" with a certain degree with infimum 0, i.e., the truth-degree >0, to be **degree-true**. In dual, we call the "false" with a certain degree with supremum 1, i.e., the truth-degree <1, to be **degree-false**.

Definition 11.8 In truth-degreed logic, an argument form is valid, which means that which can always guarantee that a conclusion of truth-degree >0 (degree-true) follows from premise(s) of truth-degree >0 (degree-true).

Of course, this definition is only an intuitive definition. Comparing this definition with the definition of a valid argument form in traditional two-valued logic, we can also say that the valid argument form is valid with a certain degree, or in other words, which is a valid argument form in the sense of degree-true, we might as well call it the **degree-valid argument form**.

11.5.3 Degree-True Tautologies and Degree-True Logical Implication

1. Degree-true tautologies

Definition 11.9 Let $P(p_1, p_2, \ldots, p_n)$ be a flexible-propositional formula on truth-degree space $[1 - \beta, \beta]^n$ ($\beta \geq 1$).

(1) If for each $(\varepsilon_1, \varepsilon_2, \ldots, \varepsilon_n) \in [1 - \beta, \beta]^n$, always $P(\varepsilon_1, \varepsilon_2, \ldots, \varepsilon_n) > 0$, then P $(p_1, p_2, \ldots, p_n)$ is called a tautology in the sense of degree-true on $[1 - \beta, \beta]^n$, or degree-true tautology for short.

(2) If for each $(\varepsilon_1, \varepsilon_2, \ldots, \varepsilon_n) \in [1 - \beta, \beta]^n$, always $P(\varepsilon_1, \varepsilon_2, \ldots, \varepsilon_n) < 1$, then P $(p_1, p_2, \ldots, p_n)$ is called a contradiction in the sense of degree-false on $[1 - \beta, \beta]^n$, or degree-false-contradiction for short.

When a flexible-propositional formula is a degree-true tautology, the propositional formula also is said to be the **logically degree-valid**.

With Definition 11.9, we can look for degree-true tautologies on truth-degree space $[1 - \beta, \beta]^n$. But we find, viewed from the form, these degree-true tautologies are all tautologies in traditional two-valued logic, and vice versa.

In fact, let $P(p_1, p_2, \ldots, p_n)$ be a degree-true tautology on truth-degree space $[1 - \beta, \beta]^n$, then on the truth-degree range $\{0, 1\}$, necessarily $t(P(p_1, p_2, \ldots, p_n)) > 0$, but only $1 > 0$ in $\{0, 1\}$, necessarily $t(P(p_1, p_2, \ldots, p_n)) = 1$, which shows that $P(p_1, p_2, \ldots, p_n)$ is an degree-true tautology on $\{0, 1\}$.

On the other hand, we find that the truth-degree of a degree-true tautology on $\{0, 1\}$ is always >0 (precisely, which should be ≥ 0.5) on truth-degree range $[0, 1]$ or on more general $[1 - \beta, \beta]$ ($\beta \geq 1$); for example, for any $p, q \in [1 - \beta, \beta]$, always $\neg p \vee p \geq 0.5$. Then, on the truth-degree range $[1 - \beta, \beta]$, a degree-true tautology on original $\{0, 1\}$ is surely a degree-true tautology.

The above facts show that the degree-true tautologies in truth-degreed logic and the tautologies in traditional two-valued logic have the same logical expressions. Thus, we can obtain indirectly degree-true tautologies from known tautologies.

Degree-true tautologies are closely related with degree-valid argument forms.

Theorem 11.1 *In truth-degreed logic, an argument form is degree-valid if and only if its corresponding implication formula is always degree-true, that is, which is a implicational degree-true tautology.*

Proof Let $P(p_1, p_2, ..., p_n)$ and $Q(p_1, p_2, ..., p_n)$ be two flexible-propositional formulas on truth-degree space $[1 - \beta, \beta]^n$ ($\beta \geq 1$), $P \Rightarrow Q$ be an argument form in truth-degreed logic, and $P \rightarrow Q$ be corresponding implication.

Suppose $P \Rightarrow Q$ is degree-valid. Then, when $P > 0$, necessarily $Q > 0$. This is to say, in the condition of $P \Rightarrow Q$ being degree-valid, the case of $P > 0$ and $Q \leq 0$ would not occur, while when $P \leq 0$, $1 - P > 0$, and so no matter what value of Q, always $P \rightarrow Q = \max\{1 - P, Q\} > 0$. Thus, for any values of propositional variables $p_1, p_2, ..., p_n$, always $P \rightarrow Q = \max\{1 - P, Q\} > 0$; that is, $P \rightarrow Q$ is always degree-true, or logically degree-valid.

Conversely, suppose that $P \rightarrow Q$ is a degree-true tautology; that is, for any values of propositional variables $p_1, p_2, ..., p_n$, always $P \rightarrow Q > 0$. Since $P \rightarrow Q = \max\{1 - P, Q\}$, always $\max\{1 - P, Q\} > 0$, this inequality shows that of $1 - P$ and Q, at least one is >0; that is, there are three cases: ① $1 - P > 0$, $Q \leq 0$; ② $1 - P \leq 0$, $Q > 0$; ③ $1 - P > 0$, $Q > 0$. It can be seen that cases ② and ③ do not affect $\max\{1 - P, Q\} > 0$; but for case ①, if $P > 0$, then $1 - P > 0$ has not guarantee, and then $\max\{1 - P, Q\} > 0$ would have no guarantee. Therefore, in the condition of $P \rightarrow Q$ being a degree-true tautology, the case of $P > 0$ but $Q \leq 0$ would not occur. That is to say, when $P > 0$, necessarily $Q > 0$. Thus, $P \Rightarrow Q$ is a degree-valid argument form. ∎

From Theorem 11.1, we can obtain a degree-valid argument form through an always degree-true implication, i.e., an implicational degree-true tautology; or conversely, by judging whether the corresponding implication is always degree-true, i.e., whether which is an implicational degree-true tautology, we can judge the validity of an argument form.

2. Degree-true logical implication

Definition 11.10 Let $P(p_1, p_2, ..., p_n)$ and $Q(p_1, p_2, ..., p_n)$ be two flexible-propositional formulas on truth-degree space $[1 - \beta, \beta]^n$ ($\beta \geq 1$). For arbitrary truth-degree vectors $(\varepsilon_1, \varepsilon_2, ..., \varepsilon_n) \in [1 - \beta, \beta]^n$, if $P(\varepsilon_1, \varepsilon_2, ..., \varepsilon_n) > 0$ (degree-true), then also $Q(\varepsilon_1, \varepsilon_2, ..., \varepsilon_n) > 0$ (degree-true), and then we call that $P(p_1, p_2, ..., p_n)$ logically implies $Q(p_1, p_2, ..., p_n)$ in the sense of degree-true, symbolically,

$$P(p_1, p_2, ..., p_n) \Rightarrow_{>0} Q(p_1, p_2, ..., p_n) \tag{11.75}$$

This expression is called the degree-true logical implication.

Those expressions in Table 11.2 are some important degree-true logical implications.

Table 11.2 Some important degree-true logical implications

I_1	$P \Rightarrow _{>0}P \vee Q$	(Law of addition)
I_2	$P \wedge Q \Rightarrow _{>0}P, P \wedge Q \Rightarrow _{>0}Q$	(Law of reduce)
I_3	$(P \rightarrow Q) \wedge P \Rightarrow _{>0}Q$	(Modus ponens)
I_4	$(P \rightarrow Q) \wedge \neg Q \Rightarrow _{>0}\neg P$	(Modus tollens)
I_5	$(P \vee Q) \wedge \neg P \Rightarrow _{>0}Q$	(Disjunctive syllogism)
I_0	$(P \rightarrow Q) \wedge (Q \rightarrow R) \Rightarrow _{>0}P \rightarrow R$	(Hypothetical syllogism)
I_7	$(P \longleftrightarrow Q) \wedge (Q \leftarrow \rightarrow R) \Rightarrow _{>0}P \longleftrightarrow R$	

We only prove I_3, the rest are left for the readers.

Proof Left side of

$$I_3 = (P \rightarrow Q) \wedge P$$
$$= (\max(1 - P, Q)) \wedge P \quad \text{(by definition of } \rightarrow)$$
$$= \min(\max(1 - P, Q), P)) \quad \text{(by definition of } \wedge)$$

Suppose $\min(\max(1 - P, Q), P) > 0$, then necessarily $\max(1 - P, Q) > 0$ and $P > 0$; whereas when $\max(1 - P, Q) > 0$, then necessarily $1 - P > 0$ or $Q > 0$. Now we know $P > 0$, so it can not be guaranteed that $1 - P > 0$, therefore it can only be $Q > 0$. Consequently, when $P > 0$ and $Q > 0$, left-hand side of $I_3 > 0$. Thus, also right-hand side of $I_3 > 0$. ∎

It can be seen that a degree-true logical implication also represents actually a degree-valid argument form. Thus, from Theorem 11.1, we have immediately the following theorem.

Theorem 11.2 Let P and Q be two flexible-propositional formulas. $P \Rightarrow _{>0}Q$, if and only if always $P \rightarrow Q > 0$ (i.e., $P \rightarrow Q$ is an implicational degree-true tautology).

Definition 11.11 Let $P(p_1, p_2, ..., p_n)$ and $Q(p_1, p_2, ..., p_n)$ be two flexible-propositional formulas on truth-degree space $[1 - \beta, \beta]^n$ ($\beta \geq 1$). For arbitrary $(\varepsilon_1, \varepsilon_2, ..., \varepsilon_n) \in [1 - \beta, \beta]^n$, if $P(\varepsilon_1, \varepsilon_2, ..., \varepsilon_n)$ and $Q(\varepsilon_1, \varepsilon_2, ..., \varepsilon_n)$ are always >0 (degree-true) or <1 (degree-false) at the same time, then we call that P $(p_1, p_2, ..., p_n)$ is logically equivalent to $Q(p_1, p_2, ..., p_n)$ in the sense of degree-true, symbolically,

$$P(p_1, p_2, ..., p_n) \Leftrightarrow _{>0}Q(p_1, p_2, ..., p_n) \tag{11.76}$$

or

$$P(p_1, p_2, ..., p_n) = _{>0}Q(p_1, p_2, ..., p_n) \tag{11.77}$$

The two expressions are called the degree-true logical equivalence.

Table 11.3 Some important degree-true logical equivalence

E_1	$\neg\neg P \Leftrightarrow_{>0} P$	(Double negative law)
E_2	$P \wedge P \Leftrightarrow_{>0} P,\ P \vee P \Leftrightarrow_{>0} P$	(Idempotent laws)
E_3	$P \wedge Q \Leftrightarrow_{>0} Q \wedge P,\ P \vee Q \Leftrightarrow_{>0} P \vee Q$	(Commutative laws)
E_4	$(P \wedge Q) \wedge R \Leftrightarrow_{>0} P \wedge (Q \wedge R),$ $(P \vee Q) \vee R \Leftrightarrow_{>0} P \vee (Q \vee R)$	(Associative laws)
E_5	$P \wedge (Q \vee R) \Leftrightarrow_{>0} P \wedge Q \vee P \wedge R,$ $P \vee (Q \wedge R) \Leftrightarrow_{>0} (P \vee Q) \wedge (P \vee R)$	(Distributive laws)
E_6	$P \wedge (P \vee Q) \Leftrightarrow_{>0} P,\ P \vee (P \wedge Q) \Leftrightarrow_{>0} P$	(Absorption laws)
E_7	$\neg(P \wedge Q) \Leftrightarrow_{>0} \neg P \vee \neg Q,$ $\neg(P \vee Q) \Leftrightarrow_{>0} \neg P \wedge \neg Q$	(De Morgan's laws)
E_8	$P \rightarrow Q \Leftrightarrow_{>0} \neg P \vee \neg Q$	(Implicational expression)
E_9	$P \longleftrightarrow Q \Leftrightarrow_{>0} (P \rightarrow Q) \wedge (Q \rightarrow P)$	(Equivalent expression)

Those expressions in Table 11.3 are some important degree-true logical equivalence.

The proofs for these equivalence in Table 11.3 are waited still readers to do later.

Theorem 11.3 *Let P and Q be two flexible-propositional formulas. $P \Leftrightarrow_{>0} Q$ if and only if $P \Rightarrow_{>0} Q$ and $Q \Rightarrow_{>0} P$.*

Proof Suppose $P \Leftrightarrow_{>0} Q$. From Definition 11.11, when $P > 0$, $Q > 0$, and when $Q > 0$, $P > 0$, and thus by Definition 11.10, it follows that $P \Rightarrow_{>0} Q$ and $Q \Rightarrow_{>0} P$.

Suppose $P \Rightarrow_{>0} Q$ and $Q \Rightarrow_{>0} P$. From Definition 11.10, when $P > 0$, $Q > 0$, and when $Q > 0$, $P > 0$, and this shows that both P and Q are >0 at the same time. Thus, by Definition 11.11, it follows that $P \Leftrightarrow_{>0} Q$. ■

Theorem 11.4 *Let P and Q be two flexible-propositional formulas. If $P \Leftrightarrow_{t} Q$, then $P \Rightarrow_{>0} Q$ and $Q \Rightarrow_{>0} P$.*

Proof Suppose $P \Leftrightarrow_{t} Q$. From Definition 11.7, the truth-degrees of P and Q are always equal, so any $(\varepsilon_1, \varepsilon_2, ..., \varepsilon_n) \in [1 - \beta, \beta]^n$ ($\beta \geq 1$) which can make $P(\varepsilon_1, \varepsilon_2, ..., \varepsilon_n) > 0$ also can make $Q(\varepsilon_1, \varepsilon_2, ..., \varepsilon_n) > 0$ necessarily, and thus it follows that $P \Rightarrow_{>0} Q$; and, any $(\varepsilon_1, \varepsilon_2, ..., \varepsilon_n) \in [1 - \beta, \beta]^n$ which can make $Q(\varepsilon_1, \varepsilon_2, ..., \varepsilon_n) > 0$ also can make $P(\varepsilon_1, \varepsilon_2, ..., \varepsilon_n) > 0$ necessarily, and thus it follows that $Q \Rightarrow_{>0} P$. ■

Theorem 11.5 *Let P and Q be two flexible-propositional formulas. If $P \Leftrightarrow_{t} Q$, then $P \Leftrightarrow_{>0} Q$.*

Proof Suppose $P \Leftrightarrow_{t} Q$. From Definition 11.7, the truth-degrees of P and Q are always equal, and then, in truth-degree range $[1 - \beta, \beta]^n$ ($\beta \geq 1$), the both P and Q are >0 at the same time at the same time. Thus, by Definition 11.11, it follows that $P \Leftrightarrow_{>0} Q$. ■

11.5.4   Rules of Degree-True Inference and Degree-True Inference

1. Rules of degree-true inference

We know that a valid argument form just is a rule of inference. From the stated above, the degree-true logical implication to which an always degree-true implication, i.e., an implicational degree-true tautology, corresponds is actually a rule of inference in truth-degreed logic; conversely, such a rule of inference in truth-degreed logic can also be written as a degree-true logical implication. On the other hand, a degree-true tautology in truth-degreed logic and a tautology in traditional two-valued logic have the same logical expression. Thus, in the sense of degree-true, logical implications in traditional two-valued logic are just the degree-true logical implications in truth-degreed logic, and rules of inference in traditional two-valued logic are also, or can be treated as, the rules of inference in truth-degreed logic. But as a rule of inference in truth-degreed logic, its premises and conclusion are both degree-true, namely, their truth-degrees are all >0. So, for definiteness, the corresponding argument forms should give clear indication of this characteristic; for example, now, the universal modus ponens in truth-degreed logic is

$$\frac{\begin{array}{c} A(x) \rightarrow B(y), t(A(x) \rightarrow B(y)) > 0 \\ A(x_0), t(A(x_0)) > 0 \end{array}}{\therefore B(y_0), t(B(y_0)) > 0} \tag{11.78}$$

We call the rules of inference in the sense of degree-true to be the **rules of degree-true inference**. Thus, the rule of inference shown by expression (11.78) above is **degree-true-universal modus ponens (degree-true-UMP** for short). With rules of degree-true inference, the reasoning in truth-degreed logic would be conducted. We call this kind of reasoning following rules of degree-true inference to be the **degree-true inference**.

The degree-true-UMP demands that major premise $A(x) \rightarrow B(y)$ satisfies $t(A(x) \rightarrow B(y)) > 0$. Then, how can we logically judge whether $t(A(x) \rightarrow B(y)) > 0$?

We know that $t(A(x) \rightarrow B(y)) = \max\{1 - t(A(x)), t(B(y))\}$, so $t(A(x) \rightarrow B(y)) > 0$ is also $\max\{1 - t(A(x)), t(B(y))\} > 0$, the latter is tantamount to that $1 - t(A(x)) > 0$ or $t(B(y)) > 0$, and then it also is tantamount to that $t(A(x)) \leq 0$ or $t(B(y)) > 0$. Thus, the following three cases can all make major premise $A(x) \rightarrow B(y)$ satisfying demand "$t(A(x) \rightarrow B(y)) > 0$":

① $t(A(x)) \leq 0$, $t(B(y)) > 0$
② $t(A(x)) \leq 0$, $t(B(y)) \leq 0$
③ $t(A(x)) > 0$, $t(B(y)) > 0$

However, on the other hand, degree-true-UMP demands also minor premise $A(x_0)$ satisfying $t(A(x_0)) > 0$. Thus, the $A(x)$ in major premise $A(x) \rightarrow B(y)$ can only take $t(A(x)) > 0$.

The analysis above shows that so long as $t(A(x)) > 0$ and $t(B(y)) > 0$, then $t(A(x) \to B(y)) > 0$; that is, the former implies the latter. Thus, $(A(x), t(A(x)) > 0) \to (B(y), t(B(y)) > 0)$ also implies $(A(x) \to B(y), t(A(x) \to B(y)) > 0)$. Therefore, we use the former in place of the latter to make degree-true-UMP becoming

$$(A(x), t(A(x)) > 0) \to (B(y), t(B(y) > 0)$$
$$\frac{A(x_0), t(A(x_0)) > 0}{\therefore B(y_0), t(B(y_0)) > 0} \tag{11.79}$$

It can be seen that we do not need to judge whether $t(A(x) \to B(y)) > 0$ actually, while only we need to know $A(x) \to B(y)$ satisfying correspondence relation $(A(x), t(A(x)) > 0) \to (B(y), t(B(y)) > 0)$.

From the relation between the truth-degree of a flexible proposition and the consistency-degree of object in the proposition with the corresponding flexible linguistic value (as $t(A(x)) = c_A(x)$), the above degree-true-UMP is also tantamount to the following universal modus ponens with consistency-degrees shown by expression (11.80), and vice versa.

$$(A(x), c_A(x) > 0) \to (B(y), c_B(y) > 0)$$
$$\frac{A(x_0), c_A(x_0) > 0}{\therefore B(y_0), c_B(y_0) > 0} \tag{11.80}$$

Example 11.2 Suppose there is a proposition: If the degree of fat of a person is >0, then the degree of heavy of the person is >0 and also known that the degree of fat of Zhang is 0.35. Since $0.35 > 0$, we have the conclusion: The degree of heavy of Zhang is >0. This argument is also a degree-true inference, and its formal version is

$$(\text{fat}(x), c_{\text{Fat}}(x) > 0) \to (\text{heavy}(y), c_{\text{Heavy}}(y) > 0)$$
$$\frac{\text{fat}(x_{\text{Zhang}}), c_{\text{Fat}}(x_{\text{Zhang}}) = 0.35 > 0}{\therefore \text{heavy}(y_{\text{Zhang}}), c_{\text{Heavy}}(y_{\text{Zhang}}) > 0}$$

here, x and y denote separately the degree of fat and the degree of heavy of one and the same person, fat(x) and heavy(y) denote separately "x is fat" and "y is heavy", and x_{Zhang} and y_{Zhang} are separately the degree of fat and the degree of heavy.

The natural language version of this argument is:

If a person is fat to a certain degree, then the person is heavy to a certain degree

Zhang is fat with a certain degree

∴ Zhang is heay with a certain degree

Conversely, the argument in natural language similar to this expression way is just a degree-true inference based on degree-true-UMP, or in other words, the arguments of this kind of expression way are the deductive reasoning suitable for degree-true inference.

From expression (11.80) and the example, we see that when conducting a degree-true inference, the computing and judging of the truth-degrees (which are really consistency-degrees) of minor premise would be involved. For instance, if the $A(x)$ in expression (11.80) is a compound proposition form (as $A_1(x_1) \wedge A_2(x_2)$), then need to compute the overall truth-degree of minor premise in argument, then judge whether it >0. And in outcome of the argument, only a scope of $t(B(y_0))$ (i.e., $c_B(y_0)$) is pointed out, but the specific number is not given, as to that the numerical object y_0 is specifically which number also cannot be known. In fact, the meaning of the conclusions is only: $\exists y_0 \in \text{supp}(B)$ such that $c_B(y_0) > 0$, i.e., $t(B(y_0)) > 0$ (degree-true). Therefore, degree-true inference is only an inference about scope of truth-degrees, which is not the exact reasoning at the level of truth-degree.

Then, if requiring, whether the $c_B(y_0)$ or y_0 can be found out? It can be seen to find $c_B(y_0)$ we need to know the correspondence relation between $c_A(x)$ and $c_B(y)$. However, major premise merely gives a macro-correspondence relation between the two, and obviously, from this relation, it is hard to obtain the $c_B(y_0)$ by $c_A(x_0)$. Consistency-degree $c_B(y_0)$ cannot be obtained, and then number y_0 also cannot be obtained, because to find y_0 needs $c_B(y_0)$ and inverse function $c_B^{-1}(u)$ of $c_B(y)$. Then, in the condition of $c_B(y_0)$ being unknown, if requiring y_0, we can only select a number from the support set of linguistic value B as an approximate value of y_0.

11.6   Near-True Inference

11.6.1   Near-Valid Argument Form

Although degree-true inference is a kind of reasoning method, it is only suitable to such unconventional sentences as "if x is A to a certain degree, then y is B to a certain degree" whose meanings are ambiguous, and since the scope of >0 is too large, the conclusions obtained are also not precision. And, those flexible propositions involved in usual arguments are all the flexible propositions whose meanings are unambiguous such as "if x is A, then y is B". From the logical semantics of the propositions (see Sect. 12.4), the truth-degrees of this kind of flexible propositions are all >0.5. This is to say, in usual arguments, whether the truth-degrees of premise >0.5 (but not merely >0) is should to be considered. Thus, the condition "truth-degree >0" in the definition of degree-valid argument form appears too broad and not practical.

Actually, although a truth-degree of >0 is true to a certain degree, a truth-degree of >0.5 is more tending toward true. And from the complement law of truth-degrees, the truth-degrees of <0.5 are more tending toward false, as to truth-degree 0.5 is just "half true and half false". In order to facilitate description, we call collectively truth-degrees of >0.5 to be **near-true** and call collectively truth-degrees of <0.5 to be **near-false**. Thus, 0.5 is just located in the intermediate between near-true and near-false, which is a watershed between near-true and near-false. On the other hand, we see that the truth-degree of a degree-true tautology actually is always ≥0.5 and the truth-degree of a degree-false tautology is then always ≤0.5. These facts seem to tip us to take truth-degree 0.5 as another datum to define the validity of argument forms in truth-degreed logic.

Definition 11.8′ In truth-degreed logic, an argument form is valid means that which can always guarantee that a conclusion of truth-degree >0.5 (near-true) follows from premise(s) of truth-degree >0.5 (near-true).

Similarly, this definition is only an intuitive definition. We might as well call it the **near-valid argument form**.

Note that actually, we can also use "truth-degree ≥0.5" to define the validity of an argument form and "true" tautology in truth-degreed logic to obtain a **quasi-near-valid argument form**. But since 0.5 is a neutral truth-degree not near-true as well as not near-false, while usual reasoning is the reasoning in the sense of near-true (truth-degree >0.5), and therefore, we use "truth-degree >0.5" rather than "truth-degree ≥0.5" in Definition 11.9′.

Besides, we can also use "≥0.5" in place of ">0.5" in Definitions 11.10 and 11.11 above to obtain quasi-near-true tautologies and quasi-near-true logical implication, and then to obtain the rules of quasi-near-true inference. However, similarly, since usual reasoning is the reasoning in the sense of near-true (truth-degree >0.5), in the following we only discuss the near-true formulas, near-true logical implication, and near-true inference in the sense of truth-degree >0.5.

11.6.2   Near-True Tautologies and Near-True Logical Implication

1. Near-true tautologies

Since near-true denotes subrange $(0.5, \beta]$ $(\beta \geq 1)$ of truth-degrees, and near-false denotes subrange $[1 - \beta, 0.5)$ of truth-degrees, to make a propositional formula is always near-true or near-false, medium truth-degree 0.5 must be removed in $[1 - \beta, \beta]$ firstly. From three kinds of basic operations of truth-degrees, it can be known that the truth-degree of a propositional formula is 0.5 only when the truth values its variables take contain 0.5. Thus, the truth-degree range $[1 - \beta, \beta]$ in which 0.5 is removed becomes $[1 - \beta, \beta] - \{0.5\} = [1 - \beta, 0.5) \cup (0.5, \beta]$, and the corresponding truth-degree space becomes $([1 - \beta, \beta] - \{0.5\})^n$. Now, we can define corresponding near-true tautologies and near-false-contradictions on the truth-degree space $([1 - \beta, \beta] - \{0.5\})^n$ without medium truth-degree 0.5.

Definition 11.9′ Let $P(p_1, p_2, ..., p_n)$ be a flexible-propositional formula on truth-degree space $[1 - \beta, \beta]^n$ $(\beta \geq 1)$.

(1) If for each $(\varepsilon_1, \varepsilon_2, ..., \varepsilon_n) \in ([1 - \beta, \beta] - \{0.5\})^n$, always $P(\varepsilon_1, \varepsilon_2, ..., \varepsilon_n) > 0.5$, then $P(p_1, p_2, ..., p_n)$ is called a tautology in the sense of near-true on the truth-degree space without middle (0.5), $([1 - \beta, \beta] - \{0.5\})^n$, or a near-true tautology for short.

(2) If for each $(\varepsilon_1, \varepsilon_2, ..., \varepsilon_n) \in ([1 - \beta, \beta] - \{0.5\})^n$, always $P(\varepsilon_1, \varepsilon_2, ..., \varepsilon_n) < 0.5$, then $P(p_1, p_2, ..., p_n)$ is called a contradiction in the sense of near-false on the truth-degree space without middle, $([1 - \beta, \beta] - \{0.5\})^n$, or a near-false-contradiction for short.

When a flexible-propositional formula is a near-true tautology, the propositional formula also is said to be the **logically near-valid**.

With Definition 11.9′, we can look for near-true tautologies on truth-degree space $([1 - \beta, \beta] - \{0.5\})^n$. But we find, in forms, these near-true tautologies are all tautologies in traditional two-valued logic, and vice versa.

In fact, let $P(p_1, p_2, ..., p_n)$ be an tautology on truth-degree space $([1 - \beta, \beta] - \{0.5\})^n$, then, on the truth-degree range $\{0, 1\}$, necessarily $t(P(p_1, p_2, ..., p_n)) > 0.5$. While only $1 > 0.5$ in $\{0, 1\}$, necessarily $t(P(p_1, p_2, ..., p_n)) = 1$, which shows that $P(p_1, p_2, ..., p_n)$ is a tautology on $\{0, 1\}$.

On the other hand, we find that the truth-degree of a tautology on $\{0, 1\}$ is always ≥ 0.5 on truth-degree range $[0, 1]$ or on more general $[1 - \beta, \beta]$ $(\beta \geq 1)$; for example, for any $p, q \in [1 - \beta, \beta]$, always $\neg p \vee p \geq 0.5$. Then, on the truth-degree range $[1 - \beta, \beta] - \{0.5\}$ that does not contain medium truth-degree 0.5, a tautology on original $\{0, 1\}$ is surely a near-true tautology.

The above facts show that the near-true tautologies in truth-degreed logic and the tautologies in traditional two-valued logic have the same logical expressions. Thus, we can obtain indirectly near-true tautologies from known tautologies.

Near-true tautologies are closely related with near-valid argument forms.

Theorem 11.1′ *In truth-degreed logic, an argument form is near-valid if and only if its corresponding implication is always near-true, that is, which is a near-true tautology.*

Proof Let $P(p_1, p_2, ..., p_n)$ and $Q(p_1, p_2, ..., p_n)$ be two flexible-propositional formulas on truth-degree space $([1 - \beta, \beta] - \{0.5\})^n$ $(\beta \geq 1)$, $P \Rightarrow Q$ be an argument form in truth-degreed logic, and $P \rightarrow Q$ be corresponding implication.

Suppose $P \Rightarrow Q$ is near-valid. Then, when $P > 0.5$, necessarily $Q > 0.5$. This is to say, in the condition of $P \Rightarrow Q$ being near-valid, the case $P > 0.5$ and $Q < 0.5$ would not occur. While when $P < 0.5$, $1 - P > 0.5$, no matter what value of Q, always $P \rightarrow Q = \max\{1 - P, Q\} > 0.5$. Thus, for any values of propositional variables $p_1, p_2, ..., p_n$, always $P \rightarrow Q = \max\{1 - P, Q\} > 0.5$; that is, $P \rightarrow Q$ is always near-true, or logically near-valid.

Conversely, suppose that $P \to Q$ is a near-true tautology; that is, for any values of propositional variables p_1, p_2, ..., p_n, always $P \to Q > 0.5$. Since $P \to Q = \max\{1 - P, Q\}$, always $\max\{1 - P, Q\} > 0.5$, this inequality shows that of $1 - P$ and Q, at least one is >0.5; that is, there are three cases: ① $1 - P > 0.5$, $Q < 0.5$; ② $1 - P < 0.5$, $Q > 0.5$; ③ $1 - P > 0.5$, $Q > 0.5$. It can be seen that cases ② and ③ do not affect $\max\{1 - P, Q\} > 0.5$; but for case ①, $P > 0.5$ does not guarantee $1 - P > 0.5$, and then does not guarantee $\max\{1 - P, Q\} > 0.5$. Therefore, in the condition of $P \to Q$ a near-true tautology, the case of $P > 0.5$ but $Q < 0.5$ would not occur. That is to say, when $P > 0.5$, necessarily $Q > 0.5$. Thus, $P \Rightarrow Q$ is a near-valid argument form. ∎

From Theorem 11.1′, we can obtain a near-valid argument form through an always near-true implication, i.e., an implicational near-true tautology; or conversely, by judging whether the corresponding implication is always near-true, i.e., whether which is an implicational near-true tautology, we can judge the validity of an argument form.

2. Near-true logical implication

Definition 11.10′ Let $P(p_1, p_2, ..., p_n)$ and $Q(p_1, p_2, ..., p_n)$ be two flexible-propositional formulas on truth-degree space $([1 - \beta, \beta] - \{0.5\})^n$ ($\beta \geq 1$). For arbitrary truth-degree vectors $(\varepsilon_1, \varepsilon_2, ..., \varepsilon_n) \in ([1 - \beta, \beta] - \{0.5\})^n$, if $P(\varepsilon_1, \varepsilon_2, ..., \varepsilon_n) > 0.5$ (near-true), then also $Q(\varepsilon_1, \varepsilon_2, ..., \varepsilon_n) > 0.5$ (near-true), and then we call that $P(p_1, p_2, ..., p_n)$ logically implies $Q(p_1, p_2, ..., p_n)$ in the sense of near-true, symbolically,

$$P(p_1, p_2, ..., p_n) \Rightarrow {}_{>0.5}Q(p_1, p_2, ..., p_n) \tag{11.75′}$$

This expression is called the near-true logical implication.

Similarly, there are near-true logical implications similar to those in Table 11.2 in truth-degree logic.

It can be seen that a near-true logical implication also represents actually a near-valid argument form. Thus, from Theorem 11.1′, we have immediately the following theorem.

Theorem 11.2′ *Let P and Q be two flexible-propositional formulas. $P \Rightarrow {}_{>0.5}Q$, if and only if always $P \to Q > 0.5$ (i.e., $P \to Q$ is an implicational near-true tautology).*

Proof Suppose $P \Rightarrow {}_{>0.5}Q$. Then, when $P > 0.5$, necessarily $Q > 0.5$, thus, $P \to Q = \max\{1 - P, Q\} = Q > 0.5$, namely, $P \to Q$ near-true; when $P < 0.5$, $P \to Q = \max\{1 - P, Q\} = 1 - P > 0.5$, namely, $P \to Q$ also near-true.

Conversely, suppose $P \to Q$ is a implicational near-true tautology; that is, always $P \to Q > 0.5$. Since $P \to Q = \max\{1 - P, Q\}$, always $\max\{1 - P, Q\} > 0.5$, and this inequality shows that the case of $P > 0.5$ and $Q < 0.5$, that is, P near-true and Q near-false, would not occur. Therefore, P logically implies Q in the sense of near-true, namely, $P \Rightarrow {}_{>0.5}Q$. ∎

Definition 11.11′ Let $P(p_1, p_2, \ldots, p_n)$ and $Q(p_1, p_2, \ldots, p_n)$ be two flexible-propositional formulas on truth-degree space $[1 - \beta, \beta]^n$ ($\beta \geq 1$). For arbitrary $(\varepsilon_1, \varepsilon_2, \ldots, \varepsilon_n) \in ([1 - \beta, \beta] - \{0.5\})^n$, if $P(\varepsilon_1, \varepsilon_2, \ldots, \varepsilon_n)$ and $Q(\varepsilon_1, \varepsilon_2, \ldots, \varepsilon_n)$ are always >0.5 (near-true) or <0.5 (near-false) at the same time, then we call that $P(p_1, p_2, \ldots, p_n)$ is logically equivalent to $Q(p_1, p_2, \ldots, p_n)$ in the sense of near-true, symbolically,

$$P(p_1, p_2, \ldots, p_n) \Leftrightarrow {}_{>0.5}Q(p_1, p_2, \ldots, p_n) \tag{11.76′}$$

or

$$P(p_1, p_2, \ldots, p_n) = {}_{>0.5}Q(p_1, p_2, \ldots, p_n) \tag{11.77′}$$

The two expressions are called the near-true logical equivalence.

Similarly, there are near-true logical equivalences similar to those in Table 11.3 in truth-degree logic.

Theorem 11.3′ *Let P and Q be two flexible-propositional formulas. $P \Leftrightarrow {}_{>0.5}Q$ if and only if $P \Rightarrow {}_{>0.5}Q$ and $Q \Rightarrow {}_{>0.5}P$.*

Proof Suppose $P \Leftrightarrow {}_{>0.5}Q$. From Definition 11.11′, when $P > 0.5$, $Q > 0.5$, and when $Q > 0.5$, $P > 0.5$, thus by Definition 11.10′, it follows that $P \Rightarrow {}_{>0.5}Q$ and $Q \Rightarrow {}_{>0.5}P$.

Suppose $P \Rightarrow {}_{>0.5}Q$ and $Q \Rightarrow {}_{>0.5}P$. From Definition 11.10′, when $P > 0.5$, $Q > 0.5$, and when $Q > 0.5$, $P > 0.5$, this shows that both P and Q are >0.5 at the same time, and also implying that both <0.5 at the same time (because there are only two cases in truth-degree range ($[1 - \beta, \beta] - \{0.5\})^n$), and thus, by Definition 11.11′, it follows that $P \Leftrightarrow {}_{>0.5}Q$. ∎

Theorem 11.4′ *Let P and Q be two flexible-propositional formulas. If $P \Leftrightarrow {}_tQ$, then $P \Rightarrow {}_{>0.5}Q$ and $Q \Rightarrow {}_{>0.5}P$.*

Proof Suppose $P \Leftrightarrow {}_tQ$. From Definition 11.7, the truth-degrees of P and Q are always equal, so any $(\varepsilon_1, \varepsilon_2, \ldots, \varepsilon_n) \in ([1 - \beta, \beta] - \{0.5\})^n$ which can make $P(\varepsilon_1, \varepsilon_2, \ldots, \varepsilon_n) > 0.5$ also can make $Q(\varepsilon_1, \varepsilon_2, \ldots, \varepsilon_n) > 0.5$ necessarily, and thus it follows that $P \Rightarrow {}_{>0.5}Q$; and, any $(\varepsilon_1, \varepsilon_2, \ldots, \varepsilon_n) \in ([1 - \beta, \beta] - \{0.5\})^n$ which can make $Q(\varepsilon_1, \varepsilon_2, \ldots, \varepsilon_n) > 0.5$ also can make $P(\varepsilon_1, \varepsilon_2, \ldots, \varepsilon_n) > 0.5$ necessarily, and thus it follows that $Q \Rightarrow {}_{>0.5}P$. ∎

Theorem 11.5′ *Let P and Q be two flexible-propositional formulas. If $P \Leftrightarrow {}_tQ$, then $P \Leftrightarrow {}_{>0.5}Q$.*

Proof Suppose $P \Leftrightarrow {}_tQ$. From Definition 11.7, the truth-degrees of P and Q are always equal, then, in truth-degree range ($[1 - \beta, \beta] - \{0.5\})^n$, both P and Q are >0.5 at the same time and also <0.5 at the same time, and thus, by Definition 11.11, it follows that $P \Leftrightarrow {}_{>0.5}Q$. ∎

11.6.3   Rules of Near-True Inference and Near-True Inference

1. Rules of near-true inference

As stated above, the near-true logical implication to which an always near-true implication, i.e., an implicational near-true tautology, corresponds is actually a rule of inference in truth-degreed logic without middle; conversely, such a rule of inference in truth-degreed logic without middle can also be written as a near-true logical implication. On the other hand, a near-true tautology in truth-degreed logic and a tautology in traditional two-valued logic have the same logical expression. Thus, in the sense of near-true, logical implications in traditional two-valued logic are just the near-true logical implications in truth-degreed logic without middle, and rules of inference in traditional two-valued logic are also, or can be treated as, the rules of inference in truth-degreed logic without middle. But as a rule of inference in truth-degreed logic without middle, its premises and conclusion are both near-true, namely, their truth-degrees are all >0.5. So, for definite, the corresponding argument forms should give clear indication of this characteristic; for example, now, the universal modus ponens in truth-degreed logic without middle is

$$
\begin{array}{c}
A(x) \to B(y), t(A(x) \to B(y)) > 0.5 \\
\dfrac{A(x_0), t(A(x_0)) > 0.5}{\therefore B(y_0), t(B(y_0)) > 0.5}
\end{array} \tag{11.78'}
$$

We call the rules of inference in the sense of near-true to be the **rules of near-true inference**. Thus, the rule of inference shown by expression (11.78') above is **near-true-universal modus ponens (near-true-UMP** for short). With rules of near-true inference, the inference in truth-degreed logic without middle would be done. We call this kind of inference following rules of near-true inference to be the **near-true inference**.

The near-true-UMP demands that major premise $A(x) \to B(y)$ satisfies $t(A(x) \to B(y)) > 0.5$. But it can be proven that "if $t(A(x)) > 0.5$ then $t(B(y)) > 0.5$)" implies "$t(A(x) \to B(y)) > 0.5$" (the proof is similar to previous degree-true-UMP, so here omitted). Therefore, we can use the former in place of the latter to formulate near-true-UMP as

$$
\begin{array}{c}
(A(x), t(A(x)) > 0.5) \to (B(y), t(B(y) > 0.5) \\
\dfrac{A(x_0), t(A(x_0)) > 0.5}{\therefore B(y_0), t(B(y_0)) > 0.5}
\end{array} \tag{11.79'}
$$

Thus, we do not need to judge whether $t(A(x) \to B(y) > 0.5$ actually, while only we need to know $A(x) \to B(y)$ satisfying correspondence relation $(A(x), t(A(x)) > 0.5) \to (B(y), t(B(y) > 0.5)$.

From the relation between the truth-degree of a flexible proposition and the consistency-degree of object in the proposition with the corresponding flexible linguistic value, the above near-true-UMP is also tantamount to the following universal modus ponens with consistency-degrees shown by expression (11.80′), and vice versa.

$$(A(x), c_A(x) > 0.5) \rightarrow (B(y), c_B(y) > 0.5)$$
$$\frac{A(x_0), t(A(x_0)) > 0.5}{\therefore B(y_0), c_B(y_0) > 0.5} \qquad (11.80′)$$

Example 11.2′ Suppose there is a proposition: if the degree of fat of a person is > 0.5, then the degree of heavy of the person is >0.5; also known: The degree of fat of Zhang is 0.85. Since $0.85 > 0.5$, we have the conclusion: The degree of heavy of Zhang is >0.5. This argument is also a near-true inference, and its formal version is

$$(\text{Fat}(x), c_{\text{Fat}}(x) > 0.5) \rightarrow (\text{Heavy}(y), c_{\text{Heavy}}(y) > 0.5)$$
$$\frac{\text{Fat}(x_{\text{Zhang}}), c_{\text{Fat}}(x_{\text{Zhang}}) = 0.85 > 0.5}{\therefore \text{Heavy}(y_{\text{Zhang}}), c_{\text{Heavy}}(y_{\text{Zhang}}) > 0.5}$$

The natural language version of this argument is:

$$\frac{\text{If a person is fat, then the person is heavy}}{\therefore \text{Zhang is heay}}$$
$$\text{Zhang is fat}$$

Conversely, the argument in natural language similar to this expression way is just a near-true inference following near-true-UMP, or in other words, the arguments of this kind of expression way are the deductive reasoning suitable for near-true inference.

From expression (11.80′) and the example we see that, similar to degree-true inference, when conducting a near-true inference, the computing and judging of the truth-degrees (which are really consistency-degrees) of minor premise would be involved, and in outcome of the argument, only a scope of $t(B(y_0))$ (i.e., $c_B(y_0)$) is pointed out, but the specific number is not given; as to which number being the number object y_0 is also unknown. In fact, the meaning of the conclusion is only: $\exists \, y_0 \in \text{core}(B)^+$ such that $c_B(y_0) > 0.5$, that is, $t(B(y_0)) > 0.5$ (near-true). Therefore, near-true inference also is only an inference about scope of truth-degrees, which is not the exact reasoning at the level of truth-degree.

Similarly, only by the macro-correspondence relation between $c_A(x)$ and $c_B(y)$, the $c_B(y_0)$ and y_0 cannot be obtained. But, now the scopes of $c_B(y_0)$ and y_0 are reduced.

Finally, it would note that the near-true inference above is the reasoning on the range of truth-degrees without middle (0.5), and then, in practical reasoning, when occurs the evidence fact of truth-degree = 0.5, corresponding reasoning would be not done. However, a special treatment can also be done according to specific problems, that is, including 0.5 into near-true to do near-true inference.

11.7 Pure Formal Symbol Deduction in Truth-Degreed Logic

It can be seen that except for the characteristic of premise's and conclusion's truth-degrees >0, or, >0.5, there is no difference between the symbolic deducing processes of degree-true inference and near-true inference and that of usual inference in two-valued logic. Therefore, pure formal symbol deduction (including propositional and predicate symbol deductions) can be done in truth-degreed logic and truth-degreed logic without middle, and corresponding formal systems can also be established. But, note that the pure formal symbol deduction in truth-degreed logic can be done only in the sense of degree-true (so this kind of symbol deduction can be called the degree-true symbol deduction, the corresponding formal systems can be called the degree-true formal systems); and the pure formal symbol deduction in truth-degreed logic without middle can be done only in the sense of near-true (so this kind of symbol deduction can be called the near-true symbol deduction, the corresponding formal systems can be called the near-true formal systems); otherwise, the corresponding symbol deduction and formal systems would be senseless.

11.8 Summary

In this chapter, we proposed the concept of flexible propositions and founded the corresponding truth-degreed logic and inference theories. First, we considered the representations and truth values of flexible propositions; next, we defined the numerical truth value; that is, the truth-degree, of a flexible proposition, introduced the formal representations of flexible propositions and the flexible propositions with composite linguistic values, particularly, introduced the algebraic compound flexible proposition and the flexible proposition with a synthetic linguistic value, then analyzed and expounded the computation principles and methods of truth-degrees of the various flexible propositions, and then founded the corresponding truth-degreed logic algebras; and then we introduced flexible-propositional formulas and flexible-predicate formulas, extended true and false and proposed the concepts of degree-true, degree-false, near-true, and near-false, extended the validity of argument forms and proposed the concepts of degree-valid argument form and near-valid argument form, and extended the concepts of tautology and logical implication and proposed the concepts of degree-true tautology, degree-true logical implication, near-true tautology, and near-true logical implication, thus obtaining rules of degree-true inference and rules of near-true inference, and establishing the principles and methods called degree-true inference and near-true inference in truth-degreed logic.

The main points and results of the chapter are as follows:

- Flexible propositions are the propositions that contain flexible linguistic values. A flexible proposition can be represented in two forms of possession and membership.
- A compound flexible proposition can be written as a flexible proposition with a composite linguistic value that includes flexible proposition with a combined linguistic value and flexible proposition with a synthetic linguistic value, and vice versa.
- The numerical truth value of a flexible proposition, that is, the truth-degree, is the degree of the corresponding object having the corresponding flexible linguistic value or belonging to the corresponding flexible set, which are, respectively, called the connotative-truth-degree and denotative-truth-degree of the flexible proposition. The truth-degree of a compound flexible proposition also can be indirectly obtained by using the truth-degrees of the corresponding component propositions through certain computation.
- The ranges of degrees of flexible propositions are all symmetrical about 0.5. The sum of truth-degrees of relatively negative flexible propositions is 1, and the relation is called the complement law of truth-degrees.
- On the basis of the truth-degree computation models of compound flexible propositions, the corresponding truth-degreed logic algebras can be founded.
- In truth-degreed logic, there are no the valid argument forms in the traditional sense and no tautology in traditional sense and exist no universally usable rules of inference at the level of truth-degree, and exact reasoning at the level of truth-degree cannot be done in general.
- In truth-degreed logic, we can define the valid argument forms, the tautologies, and the rules of inference in the sense of degree-true (i.e., truth-degree >0), that is, so-called degree-valid argument forms, degree-true tautologies, and rules of degree-true inference, and thus in truth-degreed logic, the inference in the sense of degree-true, i.e., so-called degree-true inference, can be done, but which can only apply to unconventional propositions. In truth-degreed logic without middle, we can define the valid argument forms, the tautologies, and the rules of inference in the sense of near-true (i.e., truth-degree >0.5), that is, so-called near-valid argument forms, near-true tautologies, and rules of near-true inference, and thus in truth-degreed logic without middle, the inference in the sense of near-true, i.e., so-called near-true inference, can be done, which can apply to usual flexible propositions. Degree-true inference and near-true inference are both the inference at the level of truth-degree scope.

- The degree-true tautologies in truth-degreed logic, the near-true tautologies in truth-degreed logic without middle, and the near-true tautologies in traditional two-valued logic share the same logical expressions.
- In the sense of degree-true or near-true, the pure formal symbol deduction on propositional formulas and predicate formulas can also be done, and corresponding formal systems can also be established.

Reference

1. Lian S (2009) Principles of imprecise-information processing. Science Press, Beijing

Chapter 12
Flexible-Linguistic-Truth-Valued Logic and Corresponding Inference

Abstract This chapter founds the fundamental theory of flexible-linguistic-truth-valued logic. First, it introduces two flexible linguistic truth values of flexible propositions, rough-true and rough-false, and then it deduces the computation models for the flexible linguistic truth values of compound flexible propositions according to the relation between numerical truth values and flexible linguistic truth values and then founds the flexible-two-valued logic algebraic system. Next it examines flexible propositional logic and flexible predicate logic in the flexible-two-valued logic as well as corresponding inference, proposes the concepts of rough-valid argument form, rough-true tautology, rough-true logical implication, and rule of rough-true inference, and establishes the inference principles and methods called rough-true inference. In addition, it also discusses the approximate reasoning based on rough-true-UMP. In particular, it proposes the terminology of logical semantics of a proposition and defines anew the linguistic truth values of compound flexible propositions based on the logical semantics. Lastly, it sketches flexible command logic, negation-type logic, and opposite-type logic.

Keywords Flexible-linguistic-truth-valued logic · Rough-true inference · Logical semantics

Besides numerical truth values, the flexible propositions also have linguistic truth values. In this chapter, we consider the linguistic truth values of flexible propositions and the corresponding logic and inference.

12.1 Linguistic Truth Values of Flexible Propositions—Rough-True and Rough-False

We know that "true" and "false" are two linguistic truth values in the traditional two-valued logic. Then, are these two linguistic truth values appropriate for the flexible propositions? Obviously, if "true" and "false" are rigid, that is, "rigid true"

© Springer Science+Business Media Singapore 2016   291
S. Lian, *Principles of Imprecise-Information Processing*,
DOI 10.1007/978-981-10-1549-6_12

and "rigid false," then they are not suitable, but if they can also be flexible, then they are also applicable to the flexible propositions. Actually, for the flexible propositions, "true" and "false" are originally flexible, that is, "flexible true" and "flexible false." In fact, since the linguistic values in flexible propositions are flexible concepts, so there is no a rigid standard and boundary whether object(s) in a flexible proposition have corresponding flexible linguistic value or whether which belongs to corresponding flexible set. For example, since "tall" is a flexible linguistic value, so the truth value of proposition "Jack is a tall" can only be flexible true or flexible false. This is to say, flexible propositions have linguistic truth values "true" and "false" as well, but this kind of true and false is actually flexible. In order to distinguish, we may as well refer to this kind of "flexible true" and "flexible false" of flexible propositions as **rough-true** and **rough-false**, and denote them separately by italics T and F [1].

Actually, linguistic truth values "rough-true" and "rough-false" are also the summarization of numerical truth values, that is, truth-degrees, of flexible propositions, they are also two relatively negative flexible linguistic values on both range $[0, 1]$ of denotation-truth-degrees and range $[1 - \beta, \beta]$ $(\beta \geq 1)$ of connotation-truth-degrees, and they, respectively, denote two relatively complemented flexible classes in $[0, 1]$ and $[1 - \beta, \beta]$. The consistency functions of "rough-true" and "rough-false" on range $[0, 1]$ of truth-degrees are, respectively:

$$c_T(t) = t, \quad 0 \leq t \leq 1 \tag{12.1}$$

$$c_F(t) = 1 - t, \quad 0 \leq t \leq 1 \tag{12.2}$$

whose graphs are shown in Fig. 12.1; the support set, core, and extended core of "rough-true" are separately $(0, 1]$, $\{1\}$, and $(0.5, 1]$, and the support set, core, and extended core of "rough-false" are separately $[0, 1)$, $\{0\}$, and $[0, 0.5)$. The consistency functions of "rough-true" and "rough-false" on range $[1 - \beta, \beta]$ $(\beta \geq 1)$ of truth-degrees are, respectively

$$c_T(t) = t, \quad 1 - \beta \leq t \leq \beta \tag{12.3}$$

$$c_F(t) = 1 - t, \quad 1 - \beta \leq t \leq \beta \tag{12.4}$$

whose graphs are shown in Fig. 12.2; the support set, core, and extended core of "rough-true" are separately $(0, \beta]$, $[1, \beta]$, and $(0.5, \beta]$; and the support set, core, and extended core of "rough-false" are separately $[1 - \beta, 1)$, $[1 - \beta, 0]$, and $[1 - \beta, 0.5)$.

It can be seen that for any truth-degree $t \in [0, 1]$ or $[1 - \beta, \beta]$ $(\beta \geq 1)$, the degree of it being rough-true is t itself, and the degree of it being rough-false is $1 - t$. It shows that the truth-degree originally is with respect to rough-true, that is, the truth-degree as numerical truth value is actually the degree of a proposition being "true" or "rough-true." From this, we see that the true and false in two-valued logic are not always rigid true and rigid false, and they can also be flexible true and flexible false. In fact, when the linguistic values in a proposition are rigid linguistic values, that is, this proposition is a rigid proposition, its "true" or "false" is rigid true or rigid

Fig. 12.1 Rough-true and
rough-false on truth-degree
range [0, 1]

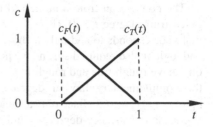

Fig. 12.2 Rough-true and
rough-false on truth-degree
range $[1 - \beta, \beta]$

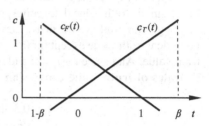

false; while when the linguistic values in a proposition are flexible linguistic values,
that is, this proposition is flexible proposition, its "true" or "false" is flexible true or
flexible false. For example, the truth value of proposition "Jack is an Olympic
champion" is rigid true or rigid false since the "Olympic champion" is a rigid
linguistic value, while the truth value of proposition "Jack is an excellent sportsman"
is flexible true or flexible false since the "excellent sportsman" is a flexible linguistic
value. Actually, in daily logic thinking, we sometimes use rigid true and rigid false,
while sometimes use flexible true and flexible false, only we do not note that.

Note that because flexible propositional clusters have respective truth-degree
range, so β in truth-degree range $[1 - \beta, \beta]$ is actually a variable; thus, different
truth-degree ranges have respective different flexible linguistic truth values
T (rough-true) and F (rough-false). This is to say, flexible linguistic truth values T and
F are not corresponding to a fixed truth-degree range, but corresponding to multiple,
even infinite, truth-degree ranges. But from a logical point of view, these different Ts
and Fs have no distinction and also need not to be distinguished. Therefore, when
using T and F later, we do not consider those specific truth-degree ranges they belong
to, while only discuss the T and F on the general truth-degree range $[1 - \beta, \beta]$.

12.2   Conversion from Numerical Truth Values to Flexible Linguistic Truth Values

A flexible proposition has both numerical truth value and linguistic truth value, then
for one and the same flexible proposition p, what is the linguistic truth value that its
numerical truth value corresponds to? That will involve the conversion from a
numerical truth value to a linguistic truth value.

The conversion from a numerical truth value to a linguistic truth value is for a given truth-degree $t \in [0, 1]$ or $[1 - \beta, \beta]$ ($\beta \geq 1$) to find the linguistic truth value that t corresponds to and only a choice or outcome can be given (in usual language and logic it also required so, and we just do so). Now, the linguistic truth values here only have rough-true and rough-false, and the two are relatively negative, then from the complementary law of degrees, necessarily $c_T(t) > 0.5$ or $c_F(t) > 0.5$ or $c_T(t) = c_F(t) = 0.5$. Since only a choice is asked, then the linguistic truth value with a greater consistency-degree should be chosen as the linguistic truth value that t corresponds to. Thus, for truth-degree $t \in [0, 1]$ or $[1 - \beta, \beta)$, if $c_T(t) > 0.5$, then the linguistic truth value that t corresponds to is "rough-true" (T); if $c_F(t) > 0.5$, then the linguistic truth value that t corresponds to is "rough-false" (F). Thus, any numerical truth value except for 0.5 can be converted into a corresponding linguistic truth value. And since $c_T(t) = t$ and $c_F(t) = 1 - t$, so we have the following rules.

Rules of truth value conversion: for truth-degree $t \in [1 - \beta, \beta]$ ($\beta \geq 1$),

$$\text{if } t > 0.5, \text{then convert } t \text{ into } T;$$
$$\text{if } t < 0.5, \text{then convert } t \text{ into } F;$$
$$\text{if } t = 0.5, \text{then do not convert}.$$

It can be seen from the above conversion rules for truth values that

(1) For any $t \in [0, 0.5)$ or $[1 - \beta, 0.5)$, $c_F(t) > 0.5$, and $c_T(t) < 0.5$, that is, t is rough-false; conversely, if t is rough-false, then necessarily $t \in [0, 0.5)$ or $[1 - \beta, 0.5)$.
(2) For any $t \in (0.5, 1]$ or $(0.5, \beta]$, $c_T(t) > 0.5$, and $c_F(t) < 0.5$, that is, t is rough-true; conversely, if t is rough-true, then necessarily $t \in (0.5, 1]$ or $(0.5, \beta]$.

As thus, the two linguistic truth values rough-true and rough-false actually denote, respectively, 2 subregions $[0, 0.5)$ and $(0.5, 1]$ on truth-degree range $[0, 1]$, or 2 subregions $[1 - \beta, 0.5)$ and $(0.5, \beta]$ on truth-degree range $[1 - \beta, \beta]$ ($\beta \geq 1$). That means rough-false and rough-true actually become 2 rigid linguistic truth values on truth-degree ranges, and the medium truth-degree 0.5 is excluded from $[0, 1]$ and $[1 - \beta, \beta]$. Thus, their consistency functions are shown in Fig. 12.3.

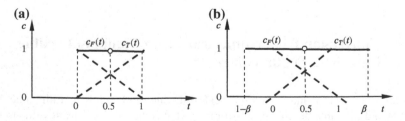

Fig. 12.3 Rigid-ended rough-true and rough-false. **a** Rigid-ened rough-true and rough-false on $[0, 1]$, **b** Rigid-ened rough-true and rough-false on $[1 - \beta, \beta]$

This is to say, speaking from the concept alone, rough-true and rough-false are two flexible linguistic truth values, but when judging the truth of a proposition, they are hardened into two rigid linguistic truth values. Or in other words, rough-true and rough-false are originally two flexible linguistic truth values defined on truth-degree ranges [0, 1] and [1 − β, β] (1 ≤ β), but in effect, they become two rigid linguistic truth values defined on truth-degree ranges [0, 1]–{0.5} and [1 − β, β]–{0.5}. Reviewing the "near-true" and "near-false" in Sect. 11.6, we can see that the "rough- true" and "rough-false" after harden here are just the "near-true" and "near-false" said there. This is to say, in conceptual, the "rough-true" and "rough-false" are two flexible linguistic truth values, but in practical, they are just rigid linguistic truth values "near-true" and "near-false."

12.3 Operations of Flexible-Linguistic-Truth-Values and Flexible-Two-Valued Logic Algebra

In the following, we define the computation models of truth values of compound flexible propositions based on the two flexible linguistic truth values of rough-true and rough-false.

Firstly, from the conversion rules of truth values above, we have immediately the following theorem.

Theorem 12.1 *The flexible proposition p is rough-true if and only if $t(p) > 0.5$; and p is rough-false if and only if $t(p) < 0.5$.*

And from the above-defined linguistic truth values of rough-true and rough-false and the computation formulas of truth-degrees of compound propositions, we have the following theorem.

Theorem 12.2 *Let p and q be two propositions, then*

(1) $p \wedge q$ *is rough-true if and only if p is rough-true and q is rough-true.*
(2) $p \vee q$ *is rough-true if and only if p is rough-true or q is rough-true.*
(3) $\neg p$ *is rough-true if and only if p is rough-false.*
(4) $p \rightarrow q$ *is rough-true if and only if p is rough-false or q is rough-true.*
(5) $p \longleftrightarrow q$ *is rough-true if and only if p and q are both rough-true or both rough-false.*

Proof We only prove (1) and (4).

Let $p \wedge q$ be rough-true. From this, we have $t(p \wedge q) > 0.5$; while $t(p \wedge q) = \min\{t(p), t(q)\}$, so $\min\{t(p), t(q)\} > 0.5$, which shows $t(p) > 0.5$ and $t(q) > 0.5$, thus p is rough-true and q is rough-true. Conversely, let p be rough-true and q be rough-true. Then $t(p) > 0.5$ and $t(q) > 0.5$, thus $\min\{t(p), t(q)\} > 0.5$; while $\min\{t(p), t(q)\} = t(p \wedge q)$, so $t(p \wedge q) > 0.5$, therefore $p \wedge q$ is rough-true.

Let $p \rightarrow q$ be rough-true. From this, we have $t(p \rightarrow q) > 0.5$; while $t(p \rightarrow q) = \max(1 - t(p), q)$, so $\max\{1 - t(p), q\} > 0.5$, which shows $1 - t(p) > 0.5$ or $t(q) > 0.5$; then, if $1 - t(p) > 0.5$, then $t(p) < 0.5$, that is, p is rough-false, in this case, no matter what values $t(q)$ takes, always p is rough-false; similarly, for $t(q) > 0.5$, that is, q is rough-true, right now what values $t(p)$ takes does not matter either. Conversely, let p be rough-false or q be rough-true. Then, if p is rough-false, then $t(p) < 0.5$, thus, $t(\neg p) > 0.5$, that is, $\neg p$ is rough-true, thus, $1 - t(p) > 0.5$, right now necessarily $\max\{1 - t(p), t(q)\} > 0.5$, so $p \rightarrow q$ is rough-true; if q is rough- true, then just the same, $\max\{1 - t(p), t(q)\} > 0.5$, thus $p \rightarrow q$ is also rough-true.∎

It can be seen that Theorem 12.2 is equivalent to the definitions of linguistic truth values of compound flexible propositions $p \wedge q$, $p \vee q$, $\neg p$, $p \rightarrow q$, and $p \leftarrow \rightarrow q$, meanwhile, 5 kinds of operations of two flexible linguistic truth values of rough-true and rough-false are given.

Definition 12.1 Let T and F denote separately flexible linguistic truth values of rough-true and rough-false, and we define 5 kinds of operations $\wedge$, $\vee$, $\neg$, $\rightarrow$ and $\leftarrow \rightarrow$ on $\{T, F\}$ as follows:

$\wedge$	T	F
T	T	F
F	F	F

$\vee$	T	F
T	T	T
F	T	F

p	$\neg p$
T	F
F	T

$\rightarrow$	T	F
T	T	F
F	T	T

$\leftarrow \rightarrow$	T	F
T	T	F
F	F	T

Which are in order called flexible logical multiplication, flexible logical addition, flexible logical negation, flexible logical implication, and flexible logical equivalence. Where T and F, respectively, denote rough-true and rough-false.

Thus, conceptually we have founded a **flexible linguistic truth-valued logic** on the basis of truth-degreed logic, that is, truth-degree range $[1 - \beta, \beta]$ ($\beta \geq 1$) ([0, 1] is included, same below); however, in effect, this flexible linguistic truth-valued logic is a rigid linguistic truth-valued logic on the truth-degree range $[1 - \beta, \beta]$– $\{0.5\}$ ($\beta \geq 1$). Since it is a two-valued logic, so we call it simply **flexible-two-valued logic**. It can be verified that flexible linguistic truth values T and F for operations $\wedge$, $\vee$, and $\neg$ satisfy all the laws that rigid linguistic truth values T and F for operations $\wedge$, $\vee$, and $\neg$. Therefore, $\langle\{T, F\}, \wedge, \vee, \neg\rangle$ also forms a Boolean algebra, we call this Boolean algebra the 2-element flexible propositional algebra, or 2-element flexible truth-valued algebra.

12.4   Logical Semantics of Propositions

We discover that a usual simple proposition always implicates itself being true (or rough-true) regardless of whether it is really true (or rough-true) or not. For instance, the proposition "Jack is tall" always implicates which is rough-true regardless of how tall Jack really is and what is the truth-degree of this proposition. For another instance, the proposition "Snow is black" implicates which is true if only viewed from the literal expression, although in fact it is false. Generally, the literal meaning of simple proposition "x_0 is A" is "x_0 has A" or "x_0 belongs to A," but actually, x_0 only has A to a degree or belongs to A to a degree. Therefore, proposition "x_0 is A" itself in fact implicates it is true (or rough-true, this point can also be seen from the conversion from numerical values to pure flexible linguistic values in Sect. 7.3.1).

Further, we discover that a compound proposition not only implicates itself being true but also implicates the logical relation between the truths of its component propositions, and the two are consistent. For instance, conjunctive proposition $p \wedge q$ implicates "$p \wedge q$ true" and "p true and q true," while implicational proposition $p \rightarrow q$ implicates "$p \rightarrow q$ true" and "if p true, then q true."

Since the truth of a proposition can be determined only through practice test or logical inference, so the implicating of a proposition for the truth of itself or the logical relation between the truths of its component propositions is only a assumption or agreement for the truth of itself and the logical relation between the truths of its component propositions.

We called the assumption and agreement of a proposition for the truth of itself and the logical relation between the truths of its component propositions to be the **logical semantics of the proposition**.

It can be seen that the truth values in the logical semantics of a proposition usually are always "true" or "rough-true." We call the logical semantics that only contains "true" or "rough-true" to be the natural logical semantics or standard logical semantics of a proposition.

The natural logical semantics of a proposition is embodied in its literal description. In two-valued logic, the natural logical semantics of 5 common basic compound propositions are shown in Table 12.1.

Table 12.1 Natural logical semantics of 5 basic compound propositions

Compound propositions	Logical semantics in traditional two-valued logic	Logical semantics in flexible-two-valued logic
$p \wedge q$	p true and q true	p rough-true and q rough-true
$p \vee q$	p true or q true	p rough-true or q rough- true
$\neg p$	not p true	not p rough-true
$p \rightarrow q$	If p true then q true	If p rough-true then q rough-true
$p \leftarrow \rightarrow q$	p true if and only if q true	p rough-true if and only if q rough-true

Table 12.2 The standard logical semantics of 5 basic compound propositions in truth-degreed logic

Compound propositions	Standard logical semantics
$p \wedge q$	p near-true and q near-true
$p \vee q$	p near-true or q near-true
$\neg p$	not p near-true
$p \rightarrow q$	If p near-true, then q near-true
$p \leftarrow \rightarrow q$	p near-true if and only if q near-true

Now we further see that in the usual mathematical logic, the logical semantics of propositional formulas are actually all natural logical semantics or, in other words, standard logical semantics. Actually, it is also the case in the more traditional formal logic. For instance, equalities $\neg p \wedge p = F$ and $\neg p \vee p = T$ are the symbol representations of the famous law of contradiction (a proposition and its negation cannot be true at the same time) and law of excluded middle (a proposition and its negation cannot be false at the same time) in formal logic. From these two equalities, it can be seen that the logical semantics of propositional formulas $\neg p \wedge p$ and $\neg p \vee p$ are just such natural logical semantics as shown in Table 12.1.

What the above stated is the logical semantics of propositions in two-valued logic; then, in multivalued truth-degreed logic, the standard logical semantics of simple proposition p is "$t(p) > 0.5$," i.e., "p near-true," and the standard logical semantics of 5 basic compound flexible propositions are shown in Table 12.2.

We now understand that the literal meaning of a proposition is also the embodiment of its logical semantics. We used "literal meanings" in Sect. 11.2 to deduce the computation formula of the truth-degree of a compound proposition, but now we can use logical semantics to deduce again the computation formulas of the truth-degrees of compound propositions. It is not difficult to verify that the results are exactly the same as the formulas in Sect. 11.2. Therefore, the logical semantics of a compound proposition decides the truth computation model of the compound proposition, the logical semantics of compound propositions are the basis of logical operations, and the truth-degree computations are in essence based on the logical semantics of compound propositions. As a matter of fact, in Sect. 11.2.3, the basic reason why the two groups of indirect computation formulas of the truth-degree of compound propositions are the same is because the logical semantics of corresponding compound propositions are the same.

Now that the natural logical semantics of proposition "$x_0 \in A$" always refers itself being rough-true or near-true, and then membership relation "$x_0 \in A$" is tantamount to "proposition '$x_0 \in A$' rough-true," that is, $t(x_0 \in A) > 0.5$. While t $(x_0 \in A) = m_A(x_0)$, so $x_0 \in A \Rightarrow m_A(x_0) > 0.5$. Conversely, if $m_A(x_0) > 0.5$, while $m_A(x_0) = t(x_0 \in A)$, so $t(x_0 \in A) > 0.5$, that is, proposition "$x_0 \in A$" rough-true, that is also, $x_0 \in A$; consequently, it follows that $m_A(x_0) > 0.5 \Rightarrow x_0 \in A$. Thus, to sum up, we have $x_0 \in A \Leftrightarrow m_A(x_0) > 0.5$. On the other hand, $m_A(x_0) > 0.5 \Leftrightarrow x_0 \in \text{core}(A)^+$, so we also have $x_0 \in A \Leftrightarrow x_0 \in \text{core}(A)^+$.

Table 12.3 The extended logical semantics of 5 basic compound flexible propositions in truth-degreed logic

Compound proposition	Extended logical semantics
$p \wedge q$	p degree-true and q degree-true
$p \vee q$	p degree-true or q degree-true
$\neg p$	not p degree-true
$p \rightarrow q$	If p degree-true, then q degree-true
$p \leftarrow \rightarrow q$	p degree-true if and only if q degree-true

Similarly, we can also have $A(x_0) \Leftrightarrow c_A(x_0) > 0.5$ and $A(x_0) \Leftrightarrow x_0 \in \mathrm{core}(A)^+$. Therefore, by the arbitrariness of x_0, we have the following theorem.

Theorem 12.3 *Let A be a flexible linguistic value on measurement space U, then for any $x \in U$, we have*

$$x \in A \Leftrightarrow m_A(x) > 0.5 \tag{12.5}$$

$$x \in A \Leftrightarrow x \in \mathrm{core}(A)^+ \tag{12.6}$$

$$A(x) \Leftrightarrow c_A(x) > 0.5 \tag{12.7}$$

$$A(x) \Leftrightarrow x \in \mathrm{core}(A)^+ \tag{12.8}$$

Extending "near-true" or "$t(p) > 0.5$" in standard logical semantics of a flexible proposition as "degree-true" or "$t(p) > 0$," the corresponding logical semantics is called the extended logical semantics of the flexible proposition. Shown in Table 12.3 are the extended logical semantics of 5 basic compound flexible propositions.

Note that since natural logical semantics is a tacit logical semantics of propositions, so for distinguishing, if we treat a certain proposition p: x_0 is A, with its extended logical semantics, then it should be rewritten as p': x_0 is A to a certain degree.

Finally, it should be noted that:

① The compound flexible propositions stated above refer to the logic compound flexible propositions, while the algebraic compound flexible propositions have no this wording of logical semantics.
② The logical semantics of a flexible proposition with a composite linguistic value is similar to the logical semantics of a simple flexible proposition.
③ In the situation without special specifying, the logical semantics mentioned later all refer to standard logical semantics.

12.5 Definitions and Computation Models of Linguistic Truth Values of Compound Flexible Propositions Based on Logical Semantics

With the logical semantics described by rough-true and rough-false of compound propositions, we then can use the logical semantics in place of "literal meaning" of a proposition to directly deduce the definition and computation model of linguistic truth value of a compound flexible proposition.

(1) The natural logical semantics of conjunctive compound flexible proposition $p \wedge q$ is: p rough-true and q rough-true. That is to say, viewed from the form, $p \wedge q$ represents the conjunction of propositions p and q, but viewed from the sense of truth, the meaning of $p \wedge q$ actually and always is: p rough-true and q rough-true. Then, according to this logical semantics, if the real situations of p and q are:

① p and q are both rough-true, then which just accords with this logical semantics, so in this case $p \wedge q$ is rough-true;

② At least one of p and q is not rough-true, that is, rough-false, then which goes contrary to this logical semantics, so in this case $p \wedge q$ is rough-false.

Thus, the correspondence relation between truth values of $p \wedge q$ and those of p and q is shown in Table 12.4.

(2) The natural logical semantics of disjunctive compound flexible proposition $p \vee q$ is: p rough-true or q rough-true. That is to say, viewed from the form, $p \vee q$ represents the disjunction of propositions p and q, but viewed from the sense of truth, the meaning of $p \vee q$ actually and always is: p rough-true or q rough-true. Then, according to this logical semantics, if the real situations of p and q are:

① At least one of p and q is rough-true, then which just accords with this logical semantics, so in this case $p \vee q$ is rough-true;

② p and q are both rough-false, then which goes contrary to this logical semantics, so in this case $p \vee q$ is rough-false.

Thus, the correspondence relation between the truth values of $p \vee q$ and those of p and q is shown in Table 12.5.

Table 12.4 Truth values of $p \wedge q$ based on the logical semantics

$t(p)$	$t(q)$	$t(p \wedge q)$
F	F	F
F	T	F
T	F	F
T	T	T

Table 12.5 Truth values of $p \lor q$ based on the logical semantics

$t(p)$	$t(q)$	$t(p \lor q)$
F	F	F
F	T	T
T	F	T
T	T	T

Table 12.6 Truth values of $\neg p$ based on logical semantics

$t(p)$	$t(\neg p)$
F	T
T	F

(3)   The natural logical semantics of negative compound flexible proposition $\neg p$ is: $\neg p$ rough true. Then, according to this logical semantics, if the real situations of p are:

① p is rough-false, that is, $\neg p$ is rough-true, then which just accords with this logical semantics, so in this case $\neg p$ is rough-true;

② p is rough-true, that is, $\neg p$ is rough-false, then which goes contrary to this logical semantics, so in this case $\neg p$ is rough-false.

Thus, the correspondence relation between truth values of $\neg p$ and those of p is shown in Table 12.6.

(4)   The natural logical semantics of implicational compound flexible proposition $p \rightarrow q$ is: if p rough-true, then q rough-true. Then, according to this logical semantics, if the real situations of p and q are:

① p rough-true and q rough-true, obviously, which is completely in accordance with the logical semantics of $p \rightarrow q$, so in this case, $p \rightarrow q$ is rough-true;

② p rough-false but q rough-true, which seems not to tally with the logical semantics of $p \rightarrow q$, but it does not go contrary to the logical semantics of $p \rightarrow q$ either, so in this case, $p \rightarrow q$ can also be treated as rough-true;

③ p rough-false and q rough-false, this case is also neither tallying with nor going contrary to the logical semantics of $p \rightarrow q$, so in this case $p \rightarrow q$ is also rough-true;

④ p rough-true but q rough-false, obviously, which is completely contrary to the logical semantics of proposition $p \rightarrow q$, so in this case $p \rightarrow q$ is rough-false.

Consequently, the correspondence relation between truth values of $p \rightarrow q$ and those of p and q is shown in Table 12.7.

It can be seen that the truth values of $p \rightarrow q$ can also be summed up as: $p \rightarrow q$ is rough-true when and only when p is rough-false or q is rough-true. Thus, we have

Table 12.7 Truth values of
$p \rightarrow q$ based on logical
semantics

$t(p)$	$t(q)$	$t(p \rightarrow q)$
F	F	T
F	T	T
T	F	F
T	T	T

$$p \rightarrow q \Leftrightarrow \neg p \vee q$$

Actually, p rough-false is also $\neg p$ rough-true, q rough-false is also $\neg q$ rough-true. Therefore, the above ② is also equivalent to $\neg p$ rough-true and q rough- true; the ③ is also equivalent to $\neg p$ rough-true and $\neg q$ rough-true. Thus, the logical semantics of compound proposition $p \rightarrow q$, "if p rough-true, then q rough-true," is also equivalent to "p is a sufficient condition for q."

(5) Similarly, from the natural logical semantics "q true when and only when p true" of equivalent compound proposition $p \leftarrow \rightarrow q$, we can have the correspondence relation between truth values of $p \leftarrow \rightarrow q$ and those of p and q are shown in Table 12.8.

Comparing Tables 12.4, 12.5, 12.6, 12.7, and 12.8 with Theorem 12.2, it can be seen that the two are completely the same. That is to say, linguistic truth values obtained based on the truth-degree computation formulas of compound propositions are the same as the linguistic values obtained based on their logical semantics. Just the same, these 5 tables define 5 operations on set $\{F, T\}$. Obviously, the truth operations defined by them are also the truth operations given by Definition 12.1, and the 5 operations are also the computation models of linguistic truth vales of the corresponding compound propositions.

The above discussions show that logical semantics is also a criterion to judge the truth of a proposition. If the real truth values of component propositions of a compound proposition accord with this criterion, then this compound proposition is rough-true, else it would be rough-false.

It can be seen that the definitions and computation models of linguistic truth values of compound flexible propositions above we obtain based on the logical semantics of propositions are completely consistent with the truth definitions and computation models of the corresponding compound propositions in traditional two-valued logic. Actually, the logical semantics of propositions is very important and crucial. It is the basis of logical operations and logical reasoning. With regard to the logical semantics of the compound proposition, we will further discuss it in another work.

Table 12.8 Truth values of
$p \leftarrow \rightarrow q$ based on logical
semantics

$t(p)$	$t(q)$	$t(p \leftarrow \rightarrow q)$
F	F	T
F	T	F
T	F	F
T	T	T

12.6 Rough-True Inference

12.6.1 Rough-True Tautologies and Rough-True Logical Implication

Definition 12.2 Let $P(p_1, p_2, \ldots, p_n)$ and $Q(p_1, p_2, \ldots, p_n)$ be two flexible propositional formulas, if for arbitrary $\varepsilon_1, \varepsilon_2, \ldots, \varepsilon_n \in \{T, F\}$, always

$$P(\varepsilon_1, \varepsilon_2, \ldots, \varepsilon_n) = Q(\varepsilon_1, \varepsilon_2, \ldots, \varepsilon_n)$$

then we call $P(p_1, p_2, \ldots, p_n)$ is logically equivalent to $Q(p_1, p_2, \ldots, p_n)$ in the sense of rough-true, symbolically,

$$P(p_1, p_2, \ldots, p_n) \Leftrightarrow_T Q(p_1, p_2, \ldots, p_n)$$

or

$$P(p_1, p_2, \ldots, p_n) =_T Q(p_1, p_2, \ldots, p_n)$$

The two expressions are called the rough-true logical equivalence.

Those expressions in Table 12.9 are some important rough-true logical equivalences.

Table 12.9 Some important rough-true logical equivalences

E_1	$\neg\neg P \Leftrightarrow_T P$	(Double negative law)
E_2	$P \wedge P \Leftrightarrow_T P,$ $P \vee P \Leftrightarrow_T P$	(Idempotent laws)
E_3	$P \wedge Q \Leftrightarrow_T Q \wedge P,$ $P \vee Q \Leftrightarrow_T P \vee Q$	(Commutative laws)
E_4	$(P \wedge Q) \wedge R \Leftrightarrow_T P \wedge (Q \wedge R),$ $(P \vee Q) \vee R \Leftrightarrow_T P \vee (Q \vee R)$	(Associative laws)
E_5	$P \wedge (Q \vee R) \Leftrightarrow_T P \wedge Q \vee P \wedge R,$ $P \vee (Q \wedge R) \Leftrightarrow_T (P \vee Q) \wedge (P \vee R)$	(Distributive laws)
E_6	$P \wedge (P \vee Q) \Leftrightarrow_T P,$ $P \vee (P \wedge Q) \Leftrightarrow_T P$	(Absorption laws)
E_7	$\neg(P \wedge Q) \Leftrightarrow_T \neg P \vee \neg Q,$ $\neg(P \vee Q) \Leftrightarrow_T \neg P \wedge \neg Q$	(De Morgan's laws)
E_8	$P \vee F \Leftrightarrow_T P,$ $P \wedge T \Leftrightarrow_T P$	(Identity laws)
E_9	$P \vee T \Leftrightarrow_T T,$ $P \wedge F \Leftrightarrow_T F$	(0–1 laws)
E_{10}	$P \vee \neg P \Leftrightarrow_T T$ $P \wedge \neg P \Leftrightarrow_T F$	(Negation laws)
E_{11}	$P \rightarrow Q \Leftrightarrow_T \neg P \vee Q$	(Implicational expression)
E_{12}	$P \leftarrow \rightarrow Q \Leftrightarrow_T (P \rightarrow Q) \wedge (Q \rightarrow P)$	(Equivalent expression)

It is not hard to prove these logical equivalences in Table 12.9 with truth tables, so here the proof is omitted.

Definition 12.3 Let $P(p_1, p_2, \ldots, p_n)$ and $Q(p_1, p_2, \ldots, p_n)$ be two flexible propositional formulas, for arbitrary $\varepsilon_1, \varepsilon_2, \ldots, \varepsilon_n \in \{T, F\}$, if $P(\varepsilon_1, \varepsilon_2, \ldots, \varepsilon_n) = T$, then $Q(\varepsilon_1, \varepsilon_2, \ldots, \varepsilon_n) = T$, and then we call that $P(p_1, p_2, \ldots, p_n)$ logically implies $Q(p_1, p_2, \ldots, p_n)$ in the sense of rough-true, symbolically,

$$P(p_1, p_2, \ldots, p_n) \Rightarrow_T Q(p_1, p_2, \ldots, p_n)$$

This expression is called the rough-true logical implication.

Those expressions in the Table 12.10 are some important rough-true logical implications.

We only prove I_3, and the rest are left for the readers.

Proof Suppose $(P \rightarrow Q) \wedge P = T$, then from the definition of $\wedge$, necessarily $P \rightarrow Q = T$ and $P = T$; whereas when $P \rightarrow Q = T$, then necessarily $P = F$ or $Q = T$; now we know $P = T$, so it can only be $Q = T$. Thus, when $P = T$ and $Q = T$, left-hand side of $I_3 = T$, and also right-hand side of $I_3 = T$. ■

Theorem 12.4 *Let P and Q be two flexible propositional formulas, then $P \Leftrightarrow_T Q$, if and only if $P \Rightarrow_T Q$ and $Q \Rightarrow_T P$.*

Proof Suppose $P \Leftrightarrow_T Q$, then by the definition of "$\Leftrightarrow_T$" (Definition 12.2), it follows that the truth values of P and Q are always equal, then, any $\varepsilon_1, \varepsilon_2, \ldots, \varepsilon_n \in \{T, F\}$ which can make $P(\varepsilon_1, \varepsilon_2, \ldots, \varepsilon_n)$ rough-true then also can make $Q(\varepsilon_1, \varepsilon_2, \ldots, \varepsilon_n)$ rough-true, thus, we have $P \Rightarrow_T Q$. Conversely, any $\varepsilon_1, \varepsilon_2, \ldots, \varepsilon_n \in \{T, F\}$ which can make $Q(\varepsilon_1, \varepsilon_2, \ldots, \varepsilon_n)$ rough-true then also can make $P(\varepsilon_1, \varepsilon_2, \ldots, \varepsilon_n)$ rough-true; thus, we have $Q \Rightarrow_T P$.

Suppose $P \Rightarrow_T Q$ and $Q \Rightarrow_T P$, then by the definition of "$\Rightarrow_T$" (Definition 12.3), it follows that P and Q rough-true at the same time; additionally, since P and Q only can take two truth values T and F, so they have also necessarily rough-false at the same time. So the truth values of them are always equal. Thus, by the definition of "$\Leftrightarrow_T$" (Definition 12.2), it follows that $P \Leftrightarrow_T Q$. ■

Table 12.10 Some important rough-true logical implications

I_1	$P \Rightarrow_T P \vee Q$	(Law of addition)
I_2	$P \wedge Q \Rightarrow_T P,$ $P \wedge Q \Rightarrow_T Q$	(Law of reduce)
I_3	$(P \rightarrow Q) \wedge P \Rightarrow_T Q$	(Modus ponens)
I_4	$(P \rightarrow Q) \wedge \neg Q \Rightarrow_T \neg P$	(Modus tollens)
I_5	$(P \vee Q) \wedge \neg P \Rightarrow_T Q$	(Disjunctive syllogism)
I_6	$(P \rightarrow Q) \wedge (Q \rightarrow R) \Rightarrow_T P \rightarrow R$	(Hypothetical syllogism)
I_7	$(P \leftarrow \rightarrow Q) \wedge (Q \leftarrow \rightarrow R) \Rightarrow_T P \leftarrow \rightarrow R$	

Definition 12.4 Let $P(p_1, p_2, ..., p_n)$ be a flexible propositional formula.

(1) If for arbitrary $\varepsilon_1, \varepsilon_2, ..., \varepsilon_n \in \{T, F\}$, always $P(\varepsilon_1, \varepsilon_2, ..., \varepsilon_n) = T$, then $P(p_1, p_2, ..., p_n)$ is called a tautology in the sense of rough-true on set $\{T, F\}$ of truth values, or a rough-true tautology for short.
(2) If for arbitrary $\varepsilon_1, \varepsilon_2, ..., \varepsilon_n \in \{T, F\}$, always $P(\varepsilon_1, \varepsilon_2, ..., \varepsilon_n) = F$, then $P(p_1, p_2, ..., p_n)$ is called a contradiction in the sense of rough-false on set $\{T, F\}$, or a rough-false contradiction for short.

When a flexible propositional formula is a rough-true tautology, the propositional formula also is said to be the **logically rough-valid**.

Example 12.1 Given the flexible propositional formulas:

(1) $(p \rightarrow q) \wedge p \rightarrow q$
(2) $(p \rightarrow q) \wedge \neg q \rightarrow \neg p$
(3) $(p \rightarrow q) \wedge \neg p \rightarrow \neg q$

It can be proved that expressions (1) and (2) are both rough-true tautology, but expression (3) is not.

Proof We only prove (1) is always rough-true.

1) Suppose $(p \rightarrow q) \wedge p$ rough-true. Then, necessarily $p \rightarrow q$ and p are both rough-true. And by the definition of operation $\rightarrow$, in the case of $p \rightarrow q$ rough-true and antecedent p rough-true, consequent q can only be rough-true. That is to say, there is no the situation of antecedent $(p \rightarrow q) \wedge p$ rough-true while consequent q rough-false. Therefore, $(p \rightarrow q) \wedge p \rightarrow q$ is a rough-true tautology. By the definition of operation $\rightarrow$, then $(p \rightarrow q) \wedge p \rightarrow q$ rough-true.
(2) Suppose $(p \rightarrow q) \wedge p$ rough-false. By the definition of operation $\rightarrow$, then $(p \rightarrow q) \wedge p \rightarrow q$ rough-true.

Synthesizing (1) and (2), $(p \rightarrow q) \wedge p \rightarrow q$ is a rough-true tautology.

Definition 12.5 Let P be a flexible predicate formula, and U be its domain of individuals.

For interpretation I in U,

(1) If P is rough-true, then P is called rough-true on interpretation I.
(2) If P is rough-false, then P is called rough-false on interpretation I.

For any interpretation I in U,

(1) If P is always rough-true, then P is always rough-true on U, or a rough-true tautology on U.
(2) If P is always rough-false, then P is always rough-false on U, or a rough-false contradiction on U.

Definition 12.6 Let P be a flexible predicate formula. For any domain of individuals

(1) If P is always rough-true, then P is called a rough-true tautology.
(2) If P is always rough-false, then P is called a rough-false contradiction.

Theorem 12.5 *Let P and Q be two flexible propositional formulas. $P \Rightarrow_T Q$, if and only if $P \to Q - T$ (i.e., $P \to Q$ is a implicational rough-true tautology).*

Proof Let $P \Rightarrow_T Q$. Then, when P rough-true, Q necessarily is rough-true; thus, by the definition of truth values of $P \to Q$, $P \to Q$ is rough-true; When P rough-false, for arbitrary Q, by the definition of truth values of $P \to Q$, always $P \to Q$ rough-true.

Conversely, let $P \to Q$ be always rough-true, i.e., $P \to Q$ is a implicational rough-true tautology, by the definition of truth-degree connective "$\to$," then there would not occur the situation of P rough-true but Q rough-false. Therefore, P logically implies Q in the sense of rough-true, that is, $P \Rightarrow_T Q$.   ∎

Theorem 12.4 is to say that just the same as the relation between implicational tautology and logical implication in traditional two-valued logic, in flexible-two-valued logic, one implicational rough-true tautology also corresponds to one rough-true logical implication and vice versa.

12.6.2   Rules of Rough-True Inference and Rough-True Inference

1. Rough-valid argument form

Definition 12.7 In flexible linguistic truth-valued logic, an argument form is rough-valid means that which can always guarantee that a rough-true conclusion follows from rough-true premise(s), that is, when premises rough-true, conclusion also rough-true.

Theorem 12.6 *In flexible linguistic truth-valued logic, an argument form is rough-valid if and only if its corresponding implication is always rough true, that is, which is a rough-true tautology.*

Proof Let $P(p_1, p_2, …, p_n)$ and $Q(p_1, p_2, …, p_n)$ be two flexible propositional formulas, $P \Rightarrow Q$ be an argument form in flexible-two-valued logic, and $P \to Q$ be corresponding implication.

Suppose $P \Rightarrow Q$ is rough-valid. Then, when P rough-true, necessarily Q rough-true. This is to say, in the condition of $P \Rightarrow Q$ being rough-valid, the case of P rough-true and Q rough-false would not occur. While by the truth definition of $P \to Q$, when P rough-false, always $P \to Q$ rough-true no matter what value of Q. Thus, for any values of propositional variables $p_1, p_2, …, p_n$, $P \to Q$ is always rough-true, or logically rough-valid.

Conversely, suppose that $P \rightarrow Q$ is a rough-true tautology, that is, for any values of propositional variables $p_1, p_2, ..., p_n$, $P \rightarrow Q$ is always rough-true. By the truth definition of $P \rightarrow Q$, then there are three cases: ① P rough-true and Q rough-true; ② P rough-false and Q rough-true; and ③ P rough-false and Q rough-false. It can be seen that in the condition of $P \rightarrow Q$ being a rough-true tautology, the case of P rough-true but Q rough-false would not occur. That is to say, when P rough-true, necessarily Q rough-true. Thus, $P \Rightarrow Q$ is a rough-valid argument form. ∎

From Theorem 12.6, we can obtain a rough-valid argument form through an implicational rough-true tautology; or conversely, by judging whether the corresponding implication is a rough-true tautology, we can judge the rough validity of an argument form.

2. Rules of rough-true inference

The same situation as traditional two-valued logic, in flexible-two-valued logic, an rough-true logical implication just is a rule of inference. Thus, those rough-true logical implications in Table 12.8 are all rules of inference in flexible-two-valued logic. Actually, it can be verified that the tautologies in traditional two-valued logic are all the rough-true tautologies on truth value set $\{T, F\}$. Thus, in the sense of rough-true, logical implications in traditional two-valued logic are just the logical implications in the flexible-two-valued logic, and rules of inference in traditional two-valued logic are also, or can be treated as, the rules of inference in flexible-two-valued logic. But the rules of inference in flexible-two-valued logic are the rules of inference in the sense of rough-true, so, for definite, the corresponding argument form should give clear indication of this characteristic. For example, in flexible-two-valued logic, the universal modus ponens is:

$$\frac{\begin{array}{l} A(x) \rightarrow B(y) \text{ rough-true} \\ A(x_0) \text{ rough-true} \end{array}}{\therefore B(y_0) \text{ rough-true}} \qquad (12.9)$$

We call the rules of inference in the sense of rough-true to be the **rules of rough-true inference**. Thus, the rule of inference shown by expression (12.9) above is **rough-true-universal modus ponens (rough-true-UMP** for short). We call this kind of inference following rules of rough-true inference to be the **rough-true inference**.

A rule of rough-true inference is the argument form that can guarantee that rough-true conclusion follows from rough-true premise(s). Actually, the rough-true contains true really. Viewing from the level of truth-degree, rough-true denotes really subrange $(0.5, \beta]$ $(1 \le \beta)$ of truth-degrees, so a rule of rough-true inference is just an argument form that can guarantee that conclusion whose truth-degree $\in (0.5, \beta]$ follows from premises whose truth-degrees $\in (0.5, \beta]$. That shows that rules of rough-true inference are more general rules of inference, and the rough-true inference is more general inference, while usual rules of inference and inference in traditional two-valued logic are only the special cases of the rules of rough-true inference and rough-true inference.

3. Rough-true deduce reasoning

Here, we only consider the rough-true deduce reasoning following rough-true-UMP.

The rough-true-UMP demands that major premise $A(x) \rightarrow B(y)$ must be rough-true. But we also cannot directly judge its truth, while using "$A(x)$ rough true $\rightarrow B$ (y) rough-true" in place of "$A(x) \rightarrow B(y)$ rough-true." Because by the definition of implication connective $\rightarrow$ as an operator, "$A(x)$ rough-true $\rightarrow B(y)$ rough-true" implies "$A(x) \rightarrow B(y)$ rough-true." Thus, the rough-true-UMP can be rewritten as

$$
\begin{array}{c}
A(x)\ \text{rough−true} \rightarrow B(y)\ \text{rough-true} \\
\underline{A(x_0)\ \text{rough-true}} \\
\therefore B(y_0)\ \text{rough-true}
\end{array}
\qquad (12.10)
$$

As thus, when reasoning, we need to consider whether major premises $A(x) \rightarrow B(y)$ satisfies "$A(x)$ rough−true $\rightarrow B(y)$ rough−true." Actually, according to the logical semantics of propositions, here major premises $A(x) \rightarrow B(y)$ satisfies "$A(x)$ rough−true $\rightarrow B(y)$ rough−true " originally (this is tantamount to say, those major premises in practical reasoning are always rough-true. Of course, which may be real rough-true, but also can be supposed rough-true, otherwise, corresponding reasoning would be senseless). Thus, when reasoning, we only need to consider whether minor premise $A(x_0)$ is rough true.

We see that the conclusion in expression (12.10) is $B(y_0)$ rough-true, but which number is the number object y_0 in which it is not pointed out. In fact, the meaning of the conclusions is only: $\exists y_0 \in \text{core}(B)^+$ such that $B(y_0)$ rough-true. This is like the situation of near-true-universal modus ponens in Sect. 11.6.3, that is, we only know existing y_0 but not get it. Of course, through L-N converting, the approximate value of y_0 can be obtained from linguistic value B.

Example 12.2 Suppose there is the following argument:

$$
\begin{array}{c}
\text{If a person is tall and fat, then the person is heavy} \\
\underline{\text{Zhang is tall and fat}} \\
\therefore \text{Zhang is heavy}
\end{array}
$$

Since "tall," "fact," and "heavy" are all flexible linguistic values, so here premise and conclusion are all flexible propositions; thus, their truth values are all rough-true. This is to say, this argument is actually a rough-true deduce reasoning following rough-true-UMP, whose symbolic version is

$$
\begin{array}{c}
\text{Tall}(x) \wedge \text{Fat}(y)\ \text{rough-true} \rightarrow \text{Heavy}(z)\ \text{rough-true} \\
\underline{\text{Tall}(x_{\text{Zhang}}) \wedge \text{Fat}(y_{\text{Zhang}})\ \text{rough-true}} \\
\therefore \text{Heavy}(z_{\text{Zhang}})\ \text{rough-true}
\end{array}
$$

From expression (12.10) and Example 12.2, it can be seen that usual reasoning with flexible propositions is really rough-true inference we said here. In other words, rough-true inference is applied to the deduce reasoning on flexible propositions.

Further, we see that because the conceptual flexible linguistic truth value "rough-true" is hardened into rigid linguistic truth value "near-true" in practical application, therefore the conceptual rules of rough-true inference are actually the corresponding rules of near-true inference in reasoning. Thus, the conceptual rough-true inference is really the near-true inference.

4. Pure formal rough-true symbol deduction

It can be seen that except for the characteristic of premises and conclusion being both rough-true, the symbol deduction processes of rough-true inference and usual inference in two-valued logic have no difference. Therefore, pure formal symbol deduction (including propositional and predicate symbol deductions) can be done in flexible-two-valued logic, and various formal systems (including propositional calculus formal systems and predicate calculus formal systems) can also be established. Viewed from the syntactic structure, there is no any difference between these formal inference and formal systems and those in traditional two-valued logic. Of course, the formal systems in flexible-two-valued logic should be called flexible propositional calculus formal systems or flexible predicate calculus formal systems.

12.7   Approximate Reasoning Following Rough-True-UMP

Examining carefully the rough-true-UMP stated in last section, it can be seen that where minor premise $A(x_0)$ can match the antecedent $A(x)$ of major premise $A(x) \to B(y)$, and the two are both rough-true, but the truth-degrees of the two are not necessarily the same. The truth-degree of a proposition equals to the consistency-degree of object having corresponding flexible linguistic value in the proposition. It can be seen from this that the rough-true-UMP may be used to do approximate reasoning.

Here, approximate reasoning is also the fuzzy inference in the fuzzy set theory, which is a kind of reasoning with non-exact matching flexible predicate, whose basic scheme is

If x is A, then y is B (this is a simplified formulation of corresponding universal proposition)
x_0 is A'
_____
What is y_0?

$$(12.11)$$

Or simply

$$
\begin{array}{c}
A \to B \\
A' \\
\hline
B'?
\end{array} \qquad (12.12)
$$

where x, $x_0 \in U$, y, $y_0 \in V$, A, and A' are flexible linguistic values of feature $\mathcal{F}$, A' is approximate to A semantiually; B and B' are flexible linguistic values of feature $\mathcal{G}$; A, A' and B, B' are, respectively, defined on one-dimensional measurement spaces U and V.

Next we give an example.

Example 12.3 Suppose there is the flexible proposition: If x is small, then y is large, which represents the correspondence relation between x in space U and y in space V; and its known $x_0 \in U$ is relatively small. Question: How about the corresponding y_0?

Solution Let A and B, respectively, denote "small" and "large," then the original flexible proposition can be expressed as $A(x) \to B(y)$, and the known fact "x_0 is relatively small" can expressed as $A'(x_0)$; thus, the problem becomes the following argument:

$$
\begin{array}{c}
A(x)\ (\text{rough}-\text{true}) \to B(y)\ (\text{rough}-\text{true}) \\
A'(x_0)\ (\text{rough}-\text{true}) \\
\hline
y_0?
\end{array} \qquad (A)
$$

Since here linguistic value A' in minor premises only is an approximate value of antecedent linguistic value A in major premise, the two cannot completely match, so rough-true-UMP cannot be directly used for the argument. But it can be seen that if minor premise $A'(x_0)$ can be transformed into $A(x_0)$, and $A(x_0)$ rough true yet, that is, $c_A(x_0) > 0.5$, then we can use rough-true-UMP to do the reasoning. That is,

$$
\begin{array}{c}
A(x)\ (\text{rough-true}) \to B(y)\ (\text{rough-true}) \\
A(x_0)\ (\text{rough-true}) \\
\hline
\therefore B(y_0)\ (\text{rough-true})
\end{array}
$$

Thus, this outcome $B(y_0)$ of the reasoning is also the result of expression (A) above, that is, the result of doing approximate reasoning with $A(x) \to B(y)$ and $A'(x_0)$.

Then, how does $A'(x_0)$ be transformed into $A(x_0)$? The method is: First, find consistency-degree $c_A(x_0)$ and then judge whether $c_A(x_0) > 0.5$, if $c_A(x_0) > 0.5$, then $A(x_0)$ rough true, and thus we can change $A'(x_0)$ to $A(x_0)$. Finding consistency-degree $c_A(x_0)$ has two situation: ① When x_0 is known, find $c_A(x_0)$ directly; ② when only A' is known but x_0 is not explicit, take peak-value point $\xi_{A'}$ of A' as x_0, then find $c_A(\xi_{A'})$.

Similarly, the y_0 in reasoning outcome above is not known in general. If require, we then can obtain an approximate value of y_0 from flexible linguistic value B by L-N conversion.

The above analysis shows that we can utilize rough-true-UMP to realize the approximate reasoning with flexible propositions. But it can be seen that the linguistic value in conclusion of this kind of approximate reasoning is always the conclusion (linguistic value B) in major premise. This makes that the obtained approximate linguistic value is not enough appropriate. This is a limitation of this approximate reasoning method. Therefore, we will further research and discuss approximate reasoning in Part V of the book.

12.8 Relations Between Truth-Degreed Logic, Flexible-Two-Valued Logic, and Traditional Two-Valued Logic

Previously, we founded truth-degreed logic and flexible-two-valued logic, and then, what relations are there between these two kinds of logic and between them and the traditional two-valued logic? Firstly, we would consider the relation between truth-degree, rough-true, and rough-false, as well as true and false in the traditional two-valued logic.

(1) **Relation between truth-degree and "true" and "false"**

Comparing the truth-degree with the "true" and "false" in the traditional two-valued logic, it can be seen that for rigid propositions, truth values "true" and "false" in traditional two-valued logic are tantamount to truth-degrees 1 and 0 in truth-degreed logic; and for flexible propositions, "true" and "false" in traditional two-valued logic are tantamount to "near-true" and "near-false" in truth-degreed logic, that is, subranges (0.5, 1] and [0, 0.5) in denotation-truth-degree range [0, 1], or subranges $(0.5, \beta]$ and $(1 - \beta, 0.5]$ in connotation-truth-degree range $[1 - \beta, \beta]$. That shows "true" and "false" in traditional two-valued logic are linguistic truth values, but not numerical truth values. Although usually "true" and "false" are denoted by 1 and 0, there 1 and 0 are only symbols representing linguistic truth values but not the real numerical values 1 and 0.

(2) **Relation between truth-degree and "rough-true" and "rough-false"**

It can be seen from the definition that the "rough-true" and "rough-false" are linguistic truth values, while the truth-degree is numerical truth values. "rough-true" and "rough false" are the summarizations of truth-degrees, while the truth-degrees are the instances of "rough-true" and "rough-false." But a truth-degree itself is also the degree of "rough-true."

(3) **Relation between the "true" and "false" and the "rough-true" and "rough-false"**

Comparing the "rough-true" and "rough-false" with the "true" and "false" in traditional two-valued logic, it can be seen that

① For the rigid propositions, "rough-true" is the extension of "true," while "true" is the contraction of "rough-true"; just the same, "rough-false" is the extension of "false" and "false" is the contraction of "rough-false";
② For the flexible propositions, "true" and "false" are also "rough-true" and "rough-false."

It can be seen from the relation between truth values that truth-degreed logic, flexible-two-valued logic, and traditional two-valued logic have the following relations:

(1) Traditional two-valued logic and flexible-two-valued logic are actually both linguistic truth-valued logic founded on truth-degree range $[1 - \beta, \beta]$ $(1 \le \beta)$. From the relations between the truth-degree and the "true" and "false" as well as their operations, and from the relations between the truth-degree and the "rough-true" and "rough-false" as well as their operations, we can say that traditional two-valued logic and flexible-two-valued logic are both founded on the basis of truth-degreed logic.
(2) For the rigid propositions, traditional two-valued logic is just rigid-two-valued logic, which is tantamount to special truth-degreed logic that only takes two values of 0 and 1; for the flexible propositions, traditional two-valued logic is tantamount to two-valued logic based on hardened "rough-true" and "rough-false."
(3) Viewed from truth values, flexible-two-valued logic includes traditional two-valued logic, or in other words, flexible-two-valued logic is the extension of rigid-two-valued logic.

12.9 Flexible Command Logic

Command logic is also called imperative logic, which is a logic branch that studies the logic characteristics and the inference of commands expressed by imperative sentences. The command in command logic is generally consists of command receivers and the command tasks, but the command receivers that can be known from context can be omitted in the command. For instance, the following are just several commands:

① Jack, get a pen!
② Silence, please!
③ John, don't go!

④   Jack and John, please listen carefully!
⑤   Let us go to the library or the sport stadium!
⑥   If the furnace temperature is a bit low, turn the air door bigger!
⑦   If and only if I receive your phone, I will remit to you!

Like propositions, the commands also have positive and negative commands, simple and compound commands, and compound commands can be separated as conjunctive, disjunctive, implicational, and equivalent commands. In fact, the above ③ is just a negative command, the rest are positive commands, ④ is a conjunctive command, ⑤ is a disjunctive command, ⑥ is an implicational command, and ⑦ is an equivalent command.

It can be seen that the task of a command actually is a certain action or behavior, while the action and behavior have corresponding states. The words that describe states are adverbs, which are called adverbial adjunct in the grammar. Many of linguistic values that describe the action states of objects are flexible linguistic values. For example,

Run quickly!
Please do not make loud noise!
Hit them hard!
We should study hard!

The "quickly," "loud," "hard," and "hard" in these commands are all flexible linguistic values.

We call a command that contains flexible linguistic values to be the flexible command.

It can be seen that flexible commands are often encountered and used in our daily life. Actually, in the production rules of usual control and planning, the consequents of many rules are all flexible commands. For example, the command in the consequent of the above ⑥ is just a flexible command. Therefore, the study and processing of flexible commands cannot be shunned away in imprecision information processing.

We call the command logic that studies flexible commands to be the flexible command logic. Similar to the previous flexible proposition logic, flexible command logic studies the logic theory about flexible commands, especially approximate reasoning with flexible commands.

12.10   Negative-Type Logic and Opposite-Type Logic

Examining the true and false in traditional two-valued logic, the 0 and 1 in truth-degreed logic, and the T (rough-true) and F (rough-false) in flexible-two-valued logic, it can be found that the opposite, or in other words, symmetrical, truth values in these logics are all relatively negative, and these logics have a common characteristics, that is, they all have negation operation. Having negation operation is due to these logics are based on the **propositions with negation** (i.e., the proposition whose linguistic values are the linguistic values with negation), while the flexible linguistic values in propositions with negation have negation operation. A flexible linguistic value and its negation are relatively negative, which causes the propositions are relatively negative, which in turn causes the corresponding truth values are relatively negative.

In view of the characteristics of containing negation operation and symmetrical truth values being relatively negative, we call this type of logic to be **negation-type logic**. Negation-type logic is also the logic with relatively negative propositions, or in other words, relatively negative linguistic values, as background. Thus, the truth-degreed logic previously discussed in Chap. 11 and the flexible linguistic truth-valued logic in this chapter and the traditional two-valued logic are all negation-type logic.

We know that there are also relatively opposite linguistic values except for relatively negative linguistic values. And relatively opposite linguistic values would result in another type of propositions—**proposition with opposite**, that is, the proposition whose linguistic values are the linguistic values with opposite and then forming another type of logic—**opposite-type logic**, that is, the logic with relatively opposite linguistic values and relatively opposite propositions as background and containing opposite operation. For opposite-type logic, we will make detailed discussion in another work.

12.11 Summary

In this chapter, we founded the fundamental theory of flexible-linguistic-truth-valued logic. First, we introduced two flexible linguistic truth values of flexible propositions, rough-true and rough-false, and then we deduced the computation models for the flexible linguistic truth values of compound flexible propositions according to the relation between numerical truth values and flexible linguistic truth values and then founded the flexible-two-valued logic algebraic system. Next we examined flexible propositional logic and flexible predicate logic in the flexible-two-valued logic as well as corresponding inference, proposed the concepts of rough-valid argument form, rough-true tautology, rough-true logical implication, and rule of rough-true inference, and established the inference principles and methods called rough-true inference. In addition, we also discussed the approximate reasoning based on rough-true-UMP. In particular, we proposed the terminology of logical semantics of a proposition and defined anew the linguistic truth values of compound flexible propositions based on the logical semantics. Lastly, we sketched flexible command logic, negation-type logic, and opposite-type logic.

The main points and results of the chapter are:

- Flexible propositions have two relatively negative flexible linguistic truth values —rough-true and rough-false, which are also the result of the relatively negative flexible partition of the truth-degree range. These two flexible linguistic truth values have similar computation models to those of the usual linguistic truth values "true" and "false," so there occur the flexible-two-valued logic algebraic systems based on the rough-true and rough-false.
- On the basis of flexible-two-valued logic operations, flexible propositional logic and flexible predicate logic can be founded. Similar to the traditional two-valued logic, there are the concepts of rough-true tautologies, rough-false contradictions, rough-true logical implication, rough-true logical equivalence, and so on in flexible-two-valued logic. Further, there are also the corresponding rules of rough-true inference such as rough-true-universal modus ponens (rough-true-UMP).
- Inference in flexible-two-valued logic is still exact inference at the level of linguistic truth value, in which the pure formal symbolic deduce and usual deduce reasoning are in formal not essential different from those in the traditional two-valued logic, but to do the former must be in the essence of "rough-true." Besides, in flexible-two-valued logic, some approximate reasoning can be indirectly realized by utilizing rough-true-UMP.
- The relations between truth-degreed logic, flexible-two-valued logic, and traditional two-valued logic are:
  Traditional two-valued logic and flexible-two-valued logic are actually both linguistic truth-valued logics founded on truth-degree range $[1 - \beta, \beta]$ $(1 \leq \beta)$. For rigid propositions, traditional two-valued logic is tantamount to special truth-degreed logic that only takes the two values of 0 and 1; for flexible propositions, traditional two-valued logic is tantamount to the linguistic truth-valued logic based on hardened two values of rough-true and rough-false. Therefore, flexible-two-valued logic includes traditional two-valued logic, or in other words, it is the extension of traditional two-valued logic.
- The assumption or agreement for the truth of a proposition itself or the logic relation between the truths of its component propositions is the logical semantics of the proposition. The logical semantics that always refers to "true," "rough-true," or "near-true" is the natural logical semantics or standard logical semantics of a proposition, and that referring to "degree-true" is the extended logical semantics of a proposition. In the natural logical semantics,

$$A(x) \Leftrightarrow c_A(x) > 0.5 \Leftrightarrow x \in \text{core}(A)^+ \Leftrightarrow m_A(x) > 0.5 \Leftrightarrow x \in A$$

Logical semantics is the criterion to judge the truth of a proposition and also a basis for logical operations and logical inference.
- Rough-true and rough-false and their operations are all conceptual; in the practical application, they then are tantamount to the corresponding near-true and near-false and their operations.

Reference

1. Lian S (2009) Principles of imprecise-information processing. Science Press, Beijing

Part V
Approximate Reasoning with Flexible Linguistic Rules and Approximate Evaluation of Flexible Linguistic Functions: Reasoning and Computation with Imprecise Information and Knowledge

Chapter 13
Flexible Linguistic Rules and Their Numerical Model

Abstract This chapter introduces and makes an all-round examination of flexible linguistic rules. First, it expounds the concept and types of flexible linguistic rules; discusses the transformation and reduction of them; analyzes the truth domains and logical semantics of flexible linguistic rules; analyzes the mathematical essences and mathematical backgrounds of flexible linguistic rules; and then presents a numerical model of a flexible linguistic rule, which provides a condition for using mathematical approach to realize reasoning and computation with flexible linguistic rules. Besides, it discusses the relationship between the flexible linguistic rules and the flexible linguistic functions and correlations, and gives a query on the practice of treating a fuzzy rule as a fuzzy relation in the traditional fuzzy set theory.

Keywords Flexible linguistic rules · Flexible linguistic functions · Numerical-model representative

Flexible linguistic rules are a kind of common and important representation form of imprecise information and knowledge, which play an indispensable and important role in imprecise-information processing. This chapter introduces flexible linguistic rules and their type and transformation; analyzes their truth domains, logical semantics, mathematical background, and mathematical essence; and then presents a numerical model of flexible linguistic rules.

13.1 What Is a Flexible Linguistic Rule?

Production rule, or simply rules, is a kind of knowledge representation technique used widely in artificial intelligence. The general form of a rule is as follows:

$$\text{If} \quad \langle \text{Antecedent} \rangle \quad \text{then} \quad \langle \text{Consequent} \rangle$$

or symbolically,

© Springer Science+Business Media Singapore 2016
S. Lian, *Principles of Imprecise-Information Processing*,
DOI 10.1007/978-981-10-1549-6_13

$$A \to C$$

where *Antecedent* (A) is a simple proposition or compound proposition and *Consequent* (C) is a simple proposition or command, which can also be a compound proposition or command. The antecedent is premise or condition, and the consequent is conclusion or action. When the premise in antecedent occurs or the condition is satisfied, then the conclusion in consequent is produced or the action is executed.

For the production rule, there are multiple different wordings in the literature, such as if-then rule, conditional statement, inference sentence, and implicational. The production rules represent relations of cause–result, result–cause, correspondence, response, etc., between things. Viewed from the angle of logic, a production rule is an implicational proposition or an implicational command, which represents an implication relation. But viewed abstractly from the angle of mathematics, a production rule always represents the correspondence relation between its antecedent and consequent [1].

Definition 13.1 We call a production rule whose antecedent or consequent contains flexible linguistic values to be the flexible linguistic rule or flexible rule for short.

The following are some examples of flexible rules (expressed in natural language):

(1) If temperature is high and sunshine, and water and fertilizer are sufficient, then plants grow fast.
(2) If the voltage is high or the electric resistance is small, then the electric current is large.
(3) If one is gifted plus who is studious, then he/she can achieve good results.
(4) If the furnace temperature is too high, then close the air door smaller or reduce the speed of the fan.
(5) If the body becomes fat, then do more sports and diet properly.
(6) If an apple has a big size, symmetrical shape, bright color, and smooth skin, then this apple is a superior apple.
(7) If the road becomes bad or visibility is down or an obstacle appears at not far ahead, then reduce speed appropriately.
(8) If bank notes are issued excessively plus supplies are not enough plus there exists severe monopoly, then the price will rise sharply.
(9) If one has a high fever, headache, and poor appetite, then he/she probably has caught a bad cold.
(10) If the weather is especially sultry and without any wind, then there highly probably will be a heavy storm.
(11) If x is close to 0, then $y = 1$.
(12) If $x = 0$, then y is far greater than 10.

It can be seen that viewed from the level of linguistic values, a flexible rule is also an implicational compound flexible proposition or flexible command.

Note that here the "flexible" in "flexible linguistic rule" and "flexible rule" is with respect to the linguistic values in antecedent and consequent of a rule, but not with respect to the correspondence relation between antecedent and consequent of the rule. The "flexible" rule with respect to the correspondence relation between antecedent and consequent of a rule will be discussed in Chap. 25.

Since a flexible linguistic value is the summarization and a general term of a batch of numerical values, so viewed deep in the level of numerical values, or in other words, substituting the objects in a proposition by corresponding measurements, that is, numerical values, a flexible rule is also an implicational compound possessive relation or membership relation made of possessive relations or membership relations, for instance, $A(x) \rightarrow B(y)$ and $x \in A \rightarrow y \in B$. Also, if representing strictly in the first-order predicate, then a flexible rule is an universal implicational compound flexible proposition or flexible command, as $\forall x A(x) \rightarrow \exists y B(y)$. (This is just the major premise of universal modus ponens.)

13.2 Types of Flexible Linguistic Rules

Actually, flexible linguistic rules widely exist in our knowledge especially experiential knowledge and common sense, or in other words, much of our knowledge are all represented by the form of flexible rules. In order to understand flexible rules more comprehensively and more deeply, next we classify flexible rules from different angles.

1. **Classification according to the structures and the linguistic values**

According to the structures of antecedent and consequent and the nature of flexible linguistic values, we present the following several types of flexible rules.

We call the rule of the form

$$A(x) \rightarrow B(y) \tag{13.1}$$

to be the flexible rule with single condition and single conclusion; call the rules of the forms

$$\left.\begin{array}{l} A_1(x_1) \wedge A_2(x_2) \wedge \cdots \wedge A_n(x_n) \rightarrow B(y) \\ A_1(x_1) \vee A_2(x_2) \vee \cdots \vee A_n(x_n) \rightarrow B(y) \\ A_1(x_1) \oplus A_2(x_2) \oplus \cdots \oplus A_n(x_n) \rightarrow B(y) \end{array}\right\} \tag{13.2}$$

to be the flexible rule with multiple conditions and single conclusion; and then call them in order to be the conjunction-type rule, disjunction-type rule (the two are collectively called the combination-type rule), and synthesis-type rule; call the rules of the forms

$$\left.\begin{array}{l} A(x) \rightarrow B_1(y_1) \wedge B_2(y_2) \wedge \cdots \wedge B_m(y_m) \\ A(x) \rightarrow B_1(y_1) \vee B_2(y_2) \vee \cdots \vee B_m(y_m) \end{array}\right\} \qquad (13.3)$$

to be the flexible rule with single condition and multiple conclusions; and call the rules of the forms

$$\left.\begin{array}{l} A_1(x_1) \wedge A_2(x_2) \wedge \cdots \wedge A_n(x_n) \rightarrow B_1(y_1) \wedge B_2(y_2) \wedge \cdots \wedge B_m(y_m) \\ A_1(x_1) \wedge A_2(x_2) \wedge \cdots \wedge A_n(x_n) \rightarrow B_1(y_1) \vee B_2(y_2) \vee \cdots \vee B_m(y_m) \\ A_1(x_1) \vee A_2(x_2) \vee \cdots \vee A_n(x_n) \rightarrow B_1(y_1) \wedge B_2(y_2) \wedge \cdots \wedge B_m(y_m) \\ A_1(x_1) \vee A_2(x_2) \vee \cdots \vee A_n(x_n) \rightarrow B_1(y_1) \vee B_2(y_2) \vee \cdots \vee B_m(y_m) \\ A_1(x_1) \oplus A_2(x_2) \oplus \cdots \oplus A_n(x_n) \rightarrow B_1(y_1) \wedge B_2(y_2) \wedge \cdots \wedge B_m(y_m) \\ A_1(x_1) \oplus A_2(x_2) \oplus \cdots \oplus A_n(x_n) \rightarrow B_1(y_1) \vee B_2(y_2) \vee \cdots \vee B_m(y_m) \end{array}\right\} \quad (13.4)$$

to be the flexible rule with multiple conditions and multiple conclusions.

Where A, B, A_i, and B_j ($i = 1, 2, \ldots, n; j = 1, 2, \ldots, m$) in the expressions (13.1)–(13.4) above are all atomic linguistic values, and x, y, x_i, and y_j are one-dimensional or multidimensional numerical variables.

Additionally, we call the rules with a composite linguistic value of the forms

$$\left.\begin{array}{l} A_1 \wedge A_2 \wedge \cdots \wedge A_n(x_1, x_2, \ldots, x_n) \rightarrow B(y) \\ A_1 \vee A_2 \vee \cdots \vee A_n(x_1, x_2, \ldots, x_n) \rightarrow B(y) \end{array}\right\} \qquad (13.5)$$

to be the flexible rule with a combined linguistic value; and call the rule with a composite linguistic value of the form

$$A_1 \oplus A_2 \oplus \cdots \oplus A_n(x_1, x_2, \ldots, x_n) \rightarrow B(y) \qquad (13.6)$$

to be the flexible rule with a synthetic linguistic value.

Where A_i ($i = 1, 2, \ldots, n$) and B in three expressions above are all atomic linguistic values, and x_i and y are one-dimensional or multidimensional numerical variables.

From the representation of possessive relation form of compound flexible propositions (see Sect. 11.2.1), it can be seen that a flexible rule with multiple conditions and single conclusion can be rewritten as a rule with a composite linguistic value; conversely, a flexible rule with a composite linguistic value can also be rewritten as a flexible rule with multiple conditions, so a flexible rule with multiple conditions and single conclusion is equivalent to the corresponding flexible rule with a composite linguistic value.

Further, we call the flexible rule with multiple conditions and single conclusion to be the standard flexible rule, and call the standard flexible rule in which the numerical variables are one-dimensional variables to be typical flexible rule.

It is shown the rules (1), (6), (9), and (10) in the examples in last section are conjunction-type rules; the rules (2) and (7) are disjunction-type rules; and rules (3) and (8) are synthesis-type rules; and except that rules (4) and (5) are the rule with multiple conclusions, the remaining rules are all rules with single conclusion, which are standard flexible rules and typical flexible rules.

Finally, we point out that since, generally speaking, a flexible rule whose antecedent is a compound proposition and consequent is also a compound proposition or command can always be split or transformed logically into multiple flexible rules with multiple conditions and single conclusions, that is, standard flexible rules, and the standard flexible rules are more convenient for reasoning and use, we discuss mainly standard flexible rules, especially, typical flexible rules, in the chapters and sections later.

2. Classification according to the characteristics of feature values of the antecedent and consequent

The rules discussed above are all such rules in which the feature values in the antecedent and consequent are all linguistic values, but there are also some special rules in which the feature values in antecedent or consequent are numerical values.

We refer to the rule whose feature values in the antecedent and consequent are both linguistic values as a L-L (short for language–language) rule, refer to the rule whose feature values are the linguistic values in the antecedent but the numerical value in the consequent as a the L-N (short for language–number) rule, and refer to the rule whose feature values are the numerical values in the antecedent while the linguistic values in the consequent as a N-L (short for number-language) rule. Thus, according to the characteristics of the feature values in the antecedent and consequent, flexible rules can be separated into three types of L-L rules, L-N rules, and N-L rules. For instance, the rule (11) in last section is an L-N rule, and rule (12) is an N-L rule, while the rest are all L-L rules.

3. Classification according to the nature of flexible linguistic values

According to the nature of flexible linguistic values, flexible rules can be separated into property–property rules, property–relation rules, relation–property rules, relation–relation rules, and hybrid–property rules.

(1) Property–property rules

A property–property rule is the rule whose linguistic values in the antecedent and consequent are both property-type flexible linguistic values. For example,

$$\text{If } x \text{ is large, then } y \text{ is small}$$

which is a property–property rule.

(2) Property–relation rule

A property–relation rule is the rule that the linguistic values in the antecedent are property-type flexible linguistic value, while the linguistic values in consequent are relation-type flexible linguistic value. For instance,

If x is large, then y is far greater than z

which is just a property-relation rule.

(3) Relation–property rules

A relation–property rule is the rule that the linguistic values in the antecedent are relation-type flexible linguistic values, while the linguistic values in the consequent are the property-type flexible linguistic values. For example,

If x and y are approximately equal, then z is near z'

which is just a relation–property rule.

(4) Relation–relation rules

A relation–relation rule is the rule that the linguistic values in both the antecedent and consequent are relation-type flexible linguistic values. For example,

If x and y are approximately equal, then u is far greater than v

which is a relation–relation rule.

(5) Hybrid-property rules

A hybrid-property rule is the rule that the linguistic values in the antecedent are relation-type flexible linguistic values and property-type flexible linguistic values, while the linguistic values in the consequent are property flexible linguistic values. For instance,

If x and y are approximately equal and x is smaller, then y is also smaller

which is a hybrid-property rule.

4. Classification according to the variables

We call the flexible rule in which the variables are all numerical variable to be a numerical variable rule. Those variables (as x, y, z, x_i, y_i, $\boldsymbol{x}$, $\boldsymbol{y}$, etc.) in the previous expressions of various rules are all numerical variables; therefore, those rules above are all numerical variable rules. But, there is also a kind of rules, in which the variables are linguistic variables. We call the flexible rule in which the variables are all linguistic variable to be a linguistic variable rule. For example,

If X is A, then Y is B

which just is a linguistic variable rule, which is really

$$\text{If } X = A, \text{then } Y = B$$

and can also be formalized as

$$A(x) \rightarrow B(y)$$

even

$$\forall x A(x) \rightarrow \exists y B(y)$$

Actually, numerical variable rule and linguistic variable rule are only two deferent expression forms of one and same rule, the semantics of the two are really the same. For example, we use linguistic variables V, R, and I to represent separately voltage (quantity), electric resistance (quantity), and electric current (quantity), and then, the above rule (2) can be expressed as

$$\text{If } V \text{ is high or } R \text{ is small, then } I \text{ is large}$$

that is

$$\text{If } V = \text{high or } R = \text{small, then } I = \text{large}$$

This is a linguistic variable rule. Of course, it can also be formalized as

$$\text{High } (V) \vee \text{Small } (R) \rightarrow \text{Large } (I)$$

or

$$\forall V \text{ High } (V) \vee \forall R \text{ Small } (R) \rightarrow \exists I \text{ Large } (I)$$

But if we use numerical variables v, r, and i to represent separately voltage (quantity), electric resistance (quantity), and electric current (quantity), then the above rule (2) can also be expressed as

$$\text{High } (v) \vee \text{Small } (r) \rightarrow \text{Large } (i) \quad \text{or} \quad v \in \text{High} \vee r \in \text{Small} \rightarrow i \in \text{Large}$$

This is two numerical variable rules. Similarly, they can also be expressed by quantified predicate formula as

$$\forall v \text{ High } (v) \vee \forall r \text{Small } (r) \rightarrow \exists i \text{ Large } (i)$$

Since a numerical variable rule is equivalent to the corresponding linguistic variable rule, that is, both represent the correspondence (relation) between antecedent linguistic values and consequent linguistic values, we would not distinguish them later.

5. Classification according to the natures of correspondence relations

As previous stated, viewed abstractly, a production rule represents the correspondence relation between its antecedent and consequent. Since the correspondence relations between things have the distinction of certainty and uncertainty, flexible rules can be separated into certain rules and uncertain rules.

Certain flexible rules are the flexible rules modified by words "necessary," "certain," and so on. For example, the component proposition "if p then q" in proposition "'if p then q' is necessary" is a certain flexible rule. Uncertain flexible rules are then the type of flexible rules modified by words "possible," "very likely," and so forth. For example, the component proposition "if p then q" in proposition "'if p then q' is possible" just is an uncertain flexible rule.

The statement describing a certain or uncertain flexible rule is also a proposition, and the kind of proposition can also be represented as the form of rule. For instance, "'if p then q' is necessary" can be represented as "'if p then necessarily q,'" and "'if p then q' is possible" can be represented as "'if p then possibly q.'" Obviously, the two expression forms are just the expression forms in daily language. But, in daily language, the modifiers "necessary," "certain," and so on describing certain rules are often omitted, thus making that a rule describing a certain rule and the certain rule described by which become one and the same rule.

In logic, the propositions containing words "necessarily," "possibly," and so on are called the modal propositions. Therefore, this kind of flexible rules containing words "necessarily," "possibly," and so on can also be called the modal flexible rules. Further, according to modal words, we then separate modal flexible rules into necessarily-type modal flexible rules, possibly type modal flexible rules, and flexible possibly-type modal flexible rules. For example, in the preceding rules in last section, rule (9) is a possibly-type modal flexible rule and rule (10) is a flexible possibly-type modal flexible rule; the rest are simplified necessarily-type modal flexible rules, that is, certain flexible rules.

For the uncertain rules and modal flexible rules, we will further discuss in Chaps. 24 and 25.

13.3   Transformation and Reduction of Flexible Linguistic Rules

From the finding methods of the membership-consistency functions of multidimensional linguistic values (see Sect. 4.2), it is known that a multidimensional linguistic value can be transformed into a one-dimensional linguistic value through space transformation; that means that through space transformation, a rule with multidimensional linguistic values can be transformed into a rule with one-dimensional linguistic values, or more generally, a non-typical flexible rule be transformed into a typical flexible rule. Next, we analyze it through some instances.

Example 13.1 Let $A(x_1, x_2, \ldots, x_n) \rightarrow B(y)$ be a rule with multidimensional linguistic values whose antecedent linguistic value is property-type linguistic value, $(x_1, x_2, \ldots, x_n) \in U \subset \mathbf{R}^n$, and $y \in V$. where U is not the direct measurement space that A belongs to, and the direct measurement space A belongs to is really a one-dimensional space $[a, b]$, but there exists mapping $\varphi \colon U \rightarrow [a, b]$, for $(x_1, x_2, \ldots, x_n) \in U, \varphi(x_1, x_2, \ldots, x_n) = r \in [a, b]$. Thus, flexible linguistic value A on U can be transformed into flexible linguistic value A on $[a, b]$, and then, the flexible proposition $A(x_1, x_2, \ldots, x_n)$ is transformed into flexible proposition $A(r)$. Thus, the original rule with multidimensional linguistic value, $A(x_1, x_2, \ldots, x_n) \rightarrow B(y)$, also is transformed into rule with one-dimensional linguistic value, $A(r) \rightarrow B(y)$.

Example 13.2 Let $R(x_1, x_2, \ldots, x_n) \rightarrow B(y)$ be a rule with multidimensional linguistic value whose antecedent linguistic value is a relation-type linguistic value, $(x_1, x_2, \ldots, x_n) \in U \subset \mathbf{R}^n$. Since R is a relation flexible linguistic value on U, so U is not the direct measurement space that A belongs to. But if there exists mapping $\varphi \colon U \rightarrow [a, b]$, for $(x_1, x_2, \ldots, x_n) \in U, \varphi(x_1, x_2, \ldots, x_n) = r \in [a, b]$, then flexible linguistic value A on U can be transformed into flexible linguistic value A on $[a, b]$; further, flexible proposition $A(x_1, x_2, \ldots, x_n)$ is transformed into flexible proposition $A(r)$. Thus, the original rule with multidimensional linguistic value, $R(x_1, x_2, \ldots, x_n) \rightarrow B(y)$, is also transformed into rule with one-dimensional linguistic value, $R(r) \rightarrow B(y)$.

Example 13.3 Let $A(x) \rightarrow R(y_1, y_2, \ldots, y_m)$ be a rule with multidimensional linguistic value whose consequent linguistic value is a relation-type flexible linguistic value, $x \in U \subset \mathbf{R}$, and $(y_1, y_2, \ldots, y_m) \in V$. V is not the direct measurement space that R belongs to. Yet there exists function $r = \psi(y_1, y_2, \ldots, y_m)$, flexible linguistic value R on m-dimensional space V can be transformed into flexible linguistic value R on one-dimensional space $[a, b]$. Thus, flexible proposition $R(y_1, y_2, \ldots, y_m)$ is transformed into $R(r)$, which is tantamount to that rule with multidimensional linguistic value, $A(x) \rightarrow R(y_1, y_2, \ldots, y_m)$, is transformed into rule with one-dimensional linguistic value, $A(x) \rightarrow R(r)$.

Example 13.4 Let $R_n(x_1, x_2, \ldots, x_n) \rightarrow R_m(y_1, y_2, \ldots, y_m)$ be a relation–relation rule, $(x_1, x_2, \ldots, x_n) \in U \subset \mathbf{R}^n$, and $(y_1, y_2, \ldots, y_m) \in V \subset \mathbf{R}^m$. U and V are not the direct measurement spaces that R_n and R_m belong to, respectively, but there exists functions $r_n = \varphi(x_1, x_2, \ldots, x_n)$ and $r_m = \psi(y_1, y_2, \ldots, y_m)$, which can transform flexible linguistic value R_n on n-dimensional space U into flexible linguistic value R_n on one-dimensional space $[a_n, b_n]$, and can transform flexible linguistic value R_m on m-dimensional space V into flexible linguistic value R_m on one-dimensional space $[a_m, b_m]$. Thus, flexible proposition $R_n(x_1, x_2, \ldots, x_n)$ is transformed into $R_n(r_n)$; $R_m(y_1, y_2, \ldots, y_m)$ is transformed into $R_m(r_m)$, which is tantamount to rule with multidimensional linguistic value; and $R_n(x_1, x_2, \ldots, x_n) \rightarrow R_m(y_1, y_2, \ldots, y_m)$ is transformed into rule with one-dimensional linguistic value, $R_n(r_n) \rightarrow R_m(r_m)$.

Example   13.5 Let $A_1(x_{1_1}, x_{1_2}, \ldots, x_{1_r}) \wedge A_2(x_{2_1}, x_{2_2}, \ldots, x_{2_s}) \wedge \cdots \wedge A_n(x_{n_1},$ $x_{n_2}, \ldots, x_{n_t}) \rightarrow B(y_1, y_2, \ldots, y_m)$ be a conjunction non-typical flexible rule, where $(x_{i_1}, x_{i_2}, \ldots, x_{i_j}) \in U_i (i = 1, 2, \ldots, n), (y_1, y_2, \ldots, y_m) \in V, A_1, A_2, \ldots, A_n$ and B are multidimensional property-type or relation-type flexible linguistic values. U_i and V are not the direct measurement space that A_i and B belong to, respectively, but there exists mappings $r_i = \varphi(x_{i_1}, x_{i_2}, \ldots, x_{i_j})$ and $r = (y_1, y_2, \ldots, y_m)$ which can transform flexible linguistic value A_i on multidimensional space U_i into flexible linguistic value A_i on one-dimensional space $[a_i, b_i]$ and transform flexible linguistic value B on multidimensional space V into flexible linguistic value B on one-dimensional space $[a, b]$; thus, flexible proposition $A_i(x_{i_1}, x_{i_2}, \ldots, x_{i_j})$ is transformed into $A_i(r_i)$ and $B(y_1, y_2, \ldots, y_m)$ is transformed into $B(r)$. Consequently, the non-typical flexible rule $A_1(x_{1_1}, x_{1_2}, \ldots, x_{1_r}) \wedge A_2(x_{2_1}, x_{2_2}, \ldots, x_{2_s}) \wedge \cdots \wedge A_n(x_{n_1}, x_{n_2}, \ldots, x_{n_t}) \rightarrow B(y_1, y_2, \ldots, y_m)$ is transformed into typical flexible rule $A_1(r_1) \wedge A_2(r_2) \wedge \cdots \wedge A_n(r_n) \rightarrow B(r)$.

It can be seen that from these examples that as long as there exists such mathematical transformations, that is, mappings, the rule with multidimensional linguistic value can be transformed into rule with one-dimensional linguistic value. Thus, an approach is provided for solution of relevant problems of rules with multidimensional linguistic value and non-typical flexible rules. The fact of rule transformation also shows that rules with multidimensional linguistic value and non-typical flexible rules can all be reduced to rules with one-dimensional linguistic value and typical flexible rules.

13.4   Truth Domains and Logical Semantics of Flexible Linguistic Rules

1. Truth domains of flexible propositional forms

Let A be a flexible linguistic value on n-dimensional measurement space U. Then, for arbitrary $x_0 \in U$, a flexible proposition $A(x_0)$ is always formed. Therefore, when x is a variable, possessive relation $A(x)$ actually is a flexible propositional form, which represents a propositional cluster $\{A(x_0)|x_0 \in U\}$ with flexible linguistic value A.

We know that for any $x_0 \in U$, when and only when $c_A(x_0) > 0$, that is, $x_0 \in \text{supp}(A)$, $t(A(x_0)) > 0$, that is, proposition $A(x_0)$ is true in a certain degree or, in other words, $A(x_0)$ is rough-true. Thus, the corresponding flexible set A is also a kind of truth set or truth domain—we can call it the **rough-truth domain**. Since $t(A(x_0)) > 0.5 \Leftrightarrow c_A(x_0) > 0.5 \Leftrightarrow x_0 \in \text{core}(A)^+$, the extended core $\text{core}(A)^+$ of flexible set A is also the **near-truth domain** of flexible propositional form $A(x)$. Rough-truth set is the **conceptual truth domain** of flexible propositional form $A(x)$, and near-truth set is the **practical truth domain** of $A(x)$.

Similarly, membership relation $x \in A$ is also a kind of flexible propositional form, of which the conceptual truth domain and practical truth domain are also separately flexible set A and extended core $\text{core}(A)^+$.

From Sect. 11.2, we know that the possessive relation forms of 5 basic compound propositions are separately $A \wedge B(x, y), A \vee B(x, y), \neg A(x), \neg A \vee B(x, y)$ and $(\neg A \vee B) \wedge (A \vee \neg B) (x, y)$ and the membership relation forms are separately $(x, y) \in A \times V \cap U \times B, (x, y) \in A \times V \cup U \times B, x \in A^c \times V \cup U \times B$ and $(x, y) \in (A^c(x, y) \in A^c \times V \cup U \times B) \cap (A \times V \cup U \times B^c)$. Then, the conceptual truth domains and practical truth domains of the flexible propositional forms with the 5 pairs of compound possessive relations and membership relations are separately one by one the corresponding flexible sets $A \times V \cap U \times B, A \times V \cup U \times B$, $A^c, A^c \times V \cup U \times B$, and $(A^c \times V \cup U \times B) \cap (A \times V \cup U \times B^c)$ and extended cores $\text{core}(A)^+ \times V \cap U \times \text{core}(B)^+$, $\text{core}(A)^+ \times V \cup U \times \text{core}(B)^+$, $\text{core}(A^c)^+, \text{core}(A^c)^+ \times V \cup U \times \text{core}(B)^+$, and $(\text{core}(A^c)^+ \times V \cup U \times \text{core}(B)^+) \cap (\text{core}(A)^+ \times V \cup U \times \text{core}(B^c)^+$ (as shown in Figs. 13.1, 13.2, 13.3, and 13.4, where the gray parts are practical truth domains).

2. Truth domain of a flexible rule

We know that the $A(x) \rightarrow B(y)$ as a flexible rule refers to universal implicational compound flexible proposition $\forall x A(x) \rightarrow \exists y B(y)$. Considering that the antecedent $A(x)$ is a sufficient condition for the consequent $B(y)$, and speaking in term of the corresponding flexible sets, the complete meaning of flexible rule $A(x) \rightarrow B(y)$ is as follows: For any x in measurement space U, if x belongs to flexible set A with a certain degree, then there exists at least one y in measurement space to belong to flexible set B with a certain degree; or else, the corresponding y belongs to flexible sets B or B^c with a certain degree. It can be seen that the x here is arbitrary, but y is dependent on x, that is, which is bound by x. Therefore, the truth-degree of rule $A(x) \rightarrow B(y)$ cannot be decided completely by the mutually independent truth-degrees of antecedent and consequent like proposition $A(x_0) \rightarrow B(y_0)$, but according to specific problem to analyze specifically that for x_0 in U which can make $t(A(x_0)) > 0$ $t(A(x_0)) > 0.5$, which y_0 in the corresponding V can make $t(B(y_0)) > 0$ $(t(B(y_0)) > 0.5)$. We denote set $\{y_0 | y_0 \in V, \text{when } t(A(x_0)) > 0, t(B(y_0)) > 0\}$ as B^s and set $\{y_0 | y_0 \in V, \text{when } t(A(x_0)) > 0.5, t(B(y_0)) > 0.5\}$ as $(\text{core}(B)^+)^s$. Thus, the conceptual truth domain of flexible rule $A(x) \rightarrow B(y)$ is

$$A \times V \cap U \times B^s \cup A^c \times V \cap U \times (B \cup B^c)$$
$$= A \times V \cap U \times B^s \cup A^c \times V \cap U \times V$$
$$= A \times V \cap U \times B^s \cup A^c \times V$$
$$= A^c \times V \cup U \times B^s$$

namely

Fig. 13.1 The truth domain
of flexible propositional form
$A \wedge B(x, y)$

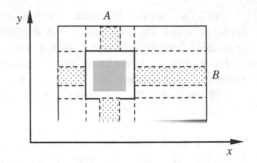

Fig. 13.2 The truth domain
of flexible propositional form
$A \vee B(x, y)$

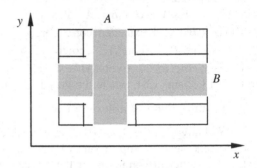

Fig. 13.3 The truth domain
of flexible propositional form
$A(x) \longrightarrow B(y)$

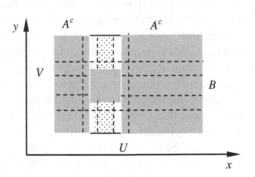

Fig. 13.4 The truth domain
of flexible propositional form
$A(x) \longleftarrow \longrightarrow B(y)$

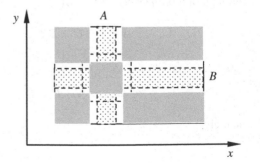

$$A^c \times V \cup U \times B^s \ (B^s \subseteq B) \tag{13.7}$$

and the practical truth domain is

$$\text{core}(A^c)^+ \times V \cup U \times (\text{core}(B)^+)^s \ ((\text{core}(B)^+)^s \subseteq \text{core}(B)^+) \tag{13.8}$$

Actually, in consideration of the possessive relation form and membership relation form of rule $\forall x A(x) \rightarrow \exists y B(y)$ are separately $\forall x \exists y \neg A \vee B(x,y)$ and $\forall x \exists y (x,y) \in A^c \times V \cup U \times B$, and in consideration of y being dependent on x and $(\text{core}(B)^+)^s \subseteq \text{core}(B)^+$, then the above conceptual truth domain and practical truth domain of rule $\forall x A(x) \rightarrow \exists y B(y)$ can also be derived from the conceptual truth domain $A^c \times V \cup U \times B$ and practical truth domain $\text{core}(A^c)^+ \times V \cup U \times \text{core}(B)^+$ of propositional form $A(x) \rightarrow B(y)$.

It can be seen that for arbitrary B^s and $(\text{core}(B)^+)^s$,

$$A^c \times V \cup U \times B^s \subseteq A^c \times V \cup U \times B \tag{13.9}$$

$$\text{core}(A^c)^+ \times V \cup U \times (\text{core}(B)^+)^s \subseteq \text{core}(A^c)^+ \times V \cup U \times \text{core}(B)^+ \tag{13.10}$$

This shows that regions $A^c \times V \cup U \times B$ and $\text{core}(A^c)^+ \times V \cup U \times \text{core}(B)^+$ are separately the most one, that is, least upper bound, of all possible conceptual truth domain and practical truth domain of rule $\forall x A(x) \rightarrow \exists y B(y)$. Thus, though we cannot give definitely the truth domains of flexible rule $\forall x A(x) \rightarrow \exists y B(y)$, we can know their scopes. On the other hand, the $A^c \times V \cup U \times B$ and $\text{core}(A^c)^+ \times V \cup U \times \text{core}(B)^+$ at right-hand side of the above two expressions are just separately the conceptual truth domain and practical truth domain of flexible propositional form $A(x) \rightarrow B(y)$ (as shown in Fig. 13.3). That is to say, the truth domain of flexible propositional form $A(x) \rightarrow B(y)$ is the least upper bound of the truth domain of flexible rule $\forall x A(x) \rightarrow \exists y B(y)$. This is the relation between the truth domain of flexible rule $\forall x A(x) \rightarrow \exists y B(y)$ and the truth domain of flexible propositional form $A(x) \rightarrow B(y)$.

From the truth domains of rule $A(x) \rightarrow B(y)$, it is not hard to further deduce the truth domains of other rules.

3. Logical semantics of a flexible rule

Since flexible rules are a kind of universal proposition, they should have logical semantics. Next, we first analyze what is the logical semantics of rule $A(x) \rightarrow B(y)$. Using predicate to represent rule $A(x) \rightarrow B(y)$, it is the universal proposition:

$$\forall x A(x) \rightarrow \exists y B(y)$$

Its literal meaning is for $\forall x \in U$, if x is A, then there exists $y \in V$, y is B. According to the literal meaning of this rule, the truth relation between its antecedent and consequent is as follows:

For $\forall x \in U$, if "x is A" is near-true, then there exists $y \in V$, "y is B" is near-true.

Obviously, this is just the logical semantics of rule $A(x) \to B(y)$. Its formal representation is

$$t(A(x)) > 0.5 \to t(B(y)) > 0.5 \tag{13.11}$$

or

$$t(x \in A) > 0.5 \to t(y \in B) > 0.5 \tag{13.12}$$

It can be seen that the logical semantics of rule $A(x) \to B(y)$ is consistent with the logical semantics of simple implicational proposition $A(x_0) \to B(y_0)$ it covers. As a matter of fact, the logical semantics of a rule is also the abstraction of the logical semantics of the simple implicational proposition it covers.

From the logical semantics of rule $A(x) \to B(y)$, it is not hard to further deduce the logical semantics of other rules. For example,

① The logical semantics of conjunctive rule $A_1(x_1) \wedge A_2(x_2) \wedge \cdots \wedge A_n(x_n) \to B(y)$ is as follows:

If $A_1(x_1)$ is near-true and $A_2(x_2)$ is near-true and ... and $A_n(x_n)$ is near-true, then $B(y)$ is near-true

That is,

$$t(A_1(x_1)) > 0.5 \wedge t(A_2(x_2)) > 0.5 \wedge \cdots \wedge t(A_n(x_n)) > 0.5 \to t(B(y)) > 0.5 \tag{13.13}$$

② The logical semantics of disjunctive rule $A_1(x_1) \vee A_2(x_2) \vee \cdots \vee A_n(x_n) \to B(y)$ is as follows:

If $A_1(x_1)$ is near-true or $A_2(x_2)$ is near-true or... or $A_n(x_n)$ is near-true, then $B(y)$ is near-true

That is,

$$t(A_1(x_1)) > 0.5 \vee t(A_2(x_2)) > 0.5 \vee \cdots \vee t(A_n(x_n)) > 0.5 \to t(B(y)) > 0.5 \tag{13.14}$$

And the logical semantics of rules with a combined linguistic value, $A_1 \wedge A_2 \wedge \cdots \wedge A_n(x_1, x_2, \ldots, x_n) \to B(y)$ and $A_1 \vee A_2 \vee \cdots \vee A_n(x_1, x_2, \ldots, x_n) \to B(y)$, are separately

if $A_1 \wedge A_2 \wedge \cdots \wedge A_n(x_1, x_2, \ldots, x_n)$ is near-true, then $B(y)$ is near-true

and

if $A_1 \vee A_2 \vee \cdots \vee A_n(x_1, x_2, \ldots, x_n)$ is near-true, then $B(y)$ is near-true

That is,

$$t(A_1 \wedge A_2 \wedge \cdots \wedge A_n(x_1, x_2, \ldots, x_n)) > 0.5 \rightarrow t(B(y)) > 0.5 \qquad (13.15)$$

and

$$t(A_1 \vee A_2 \vee \cdots \vee A_n(x_1, x_2, \ldots, x_n)) > 0.5 \rightarrow t(B(y)) > 0.5 \qquad (13.16)$$

③ The logical semantics of rule with a synthetic linguistic value, $A_1 \oplus A_2 \oplus \cdots \oplus A_n(x_1, x_2, \ldots, x_n) \rightarrow B(y)$, is as follows:

If $A_1 \oplus A_2 \oplus \cdots \oplus A_n(x_1, x_2, \ldots, x_n)$ is near-true, then $B(y)$ is near-true

That is,

$$t(A_1 \oplus A_2 \oplus \cdots \oplus A_n(x_1, x_2, \ldots, x_n)) > 0.5 \rightarrow t(B(y)) > 0.5 \qquad (13.17)$$

4. The extended logical semantics of flexible rules

The logical semantics stated above is the natural (or standard) logical semantics of flexible rules, while the extended logical semantics of flexible rules, like those of flexible propositions, is described by using "degree-true," that is, if the antecedent degree-true then the consequent degree-true. For example, the extended logical semantics of rule $A(x) \rightarrow B(y)$ is as follows: if $A(x)$ degree-true then $B(y)$ degree-true, or formally and quantitatively, $t(A(x_0)) > 0 \rightarrow t(B(y)) > 0$. The extended logical semantics of other rules is similar.

5. Truth domain and logical semantics of a rule with linguistic values with opposite

The above-stated truth domains and logical semantics are both for the usual rules with linguistic values with negation. A rule with linguistic values with opposite also has its truth domain and logical semantics. Since the support set of a linguistic value with opposite is tantamount to the extended core of a linguistic value with negation, so the conceptual truth domain and practical truth domain of rule with linguistic values with opposite, $A(x) \rightarrow B(y)$, are as follows:

$$A^c \times V \cup U \times B^s \quad (B^s \subseteq B) \qquad (13.18)$$

$$\mathrm{supp}(A^c) \times V \cup U \times \mathrm{supp}(B^s) \quad (\mathrm{supp}(B^s) \subseteq \mathrm{supp}(B)) \qquad (13.19)$$

From this, it is not hard to further deduce the truth domains of other rules with linguistic values with opposite.

If we use linguistic truth values to describe the logical semantics of rules, then the logical semantics of rules with linguistic values with opposite is no different from those of the rules with linguistic values with negation. But if we use truth-degrees, then "near-true" in the logical semantics of the rule with linguistic values with opposite is "truth-degree >0," but not "truth-degree >0.5." For instance, for rule with linguistic values with opposite, $A(x) \rightarrow B(y)$, its logical semantics described by linguistic truth values is as follows:

For $\forall x \in U$, if $A(x)$ is near-true, then there exists $y \in V$, and $B(y)$ is near-true, while its logical semantics described by truth-degree is

$$t(A(x)) > 0 \rightarrow t(B(y)) > 0 \tag{13.20}$$

From that, it is not hard to further deduce the logical semantics of other rules with linguistic values with opposite.

13.5　Mathematical Essence, Mathematical Background, and Numerical Model of Flexible Linguistic Rules

1. Implication and correspondence

We know that viewed from the angle of logic, a flexible rule is just an implicational compound flexible proposition. So a flexible rule represents the implication relation between its antecedent and consequent.

The so-called implication is that the antecedent is a sufficient condition for the consequent. Therefore, the complete expression of rule $A(x) \rightarrow B(y)$ should be as follows:

$$\text{if } A(x) \text{ then } B(y), \text{else } B(y) \vee \neg B(y)$$

while separately speaking, the sentence is as follows:

$$\begin{cases} A(x) \mapsto B(y) \\ \neg A(x) \mapsto B(y) \vee \neg B(y) = V(y)(V = B \vee \neg B) \end{cases} \tag{13.21}$$

(Note that here the arrow symbol "$\mapsto$" represents "correspondence" but not "implication"). That is to say, completely speaking, the implication includes the two propositional correspondences.

We examine again the following argument:

$$\frac{A(x) \rightarrow B(y)}{A(x_0)}$$
$$\therefore B(x_0)$$

It can be seen that although the implication includes the correspondences between the antecedent and consequent in the two cases of the rule's antecedent being true and false, rule $A(x) \rightarrow B(y)$ is used in usual on condition that fact $A(x_0)$ is known, so only the correspondence $A(x) \mapsto B(y)$ whose antecedent is true is used in the usual logical inference actually, while the correspondence $\neg A(x) \mapsto B(y) \vee \neg B(y)$ whose antecedent is false is not used at all. That means in reasoning the rule that represents the implication relation can completely be treated as a correspondence between antecedent and consequent to be used. In the following, we further consider the mathematical background and mathematical essence of a flexible rule as a correspondence.

2. Mathematical essence and mathematical background of a flexible rule

Firstly, let us examine the simplest L-L rule $A(x) \rightarrow B(y)$.

We know that the complete representation of rule $A(x) \rightarrow B(y)$ is $\forall x A(x) \rightarrow \exists y B(y)$, which means for any $x \in U$, if x has A, then there exists $y \in V$ to have B. Thus, viewed as a whole, rule $\forall x A(x) \rightarrow \exists y B(y)$ represents the correspondence $A \mapsto B$ from flexible linguistic value A to B. Actually, if rewriting $A(x) \rightarrow B(y)$ into form $A(X) \rightarrow B(Y)$ of linguistic variables, then the flexible-linguistic-valued correspondence $A \mapsto B$ can be obtained immediately. Then viewed from the level of flexible linguistic functions, flexible rule $A(x) \rightarrow B(y)$ is also tantamount to a pair (A, B) of values of a certain flexible linguistic function $Y = f(X)$ on the corresponding universe of discourse. Thus, the correspondence $A \mapsto B$ follows still. In a word, viewed from the angle of mathematics, flexible rule $A(x) \rightarrow B(y)$ is essentially flexible-linguistic-valued correspondence $A \mapsto B$.

And from the relation between flexible linguistic values and flexible sets, the flexible-linguistic-valued correspondence $A \mapsto B$ just is the flexible set correspondence $A \mapsto B$. Thus, the **mathematical essence** of flexible rule $A(x) \rightarrow B(y)$ is the flexible set correspondence $A \mapsto B$. Actually, if rewriting the rule $A(x) \rightarrow B(y)$ of possessive relation into rule $x \in A \rightarrow y \in B$ of membership relation, then the flexible set correspondence $A \mapsto B$ can be more directly obtained.

To sum up, flexible rule $A(x) \rightarrow B(y)$ can be simply written as flexible-linguistic-valued correspondence, i.e., flexible set correspondence, $A \mapsto B$.

From Sect. 9.2, we know that the mathematical background of flexible-linguistic-valued correspondence $A \mapsto B$ is a function or correlation from A to B, which is called the background function or background correlation of correspondence $A \mapsto B$. Thus, the mathematical background of flexible rule $A(x) \rightarrow B(y)$ is also a function or correlation from A to B, which is also the background function or background correlation of rule $A(x) \rightarrow B(y)$.

3. Numerical model of a flexible rule

From Sect. 9.2, we known that in theory, flexible-linguistic-valued correspondence $A \mapsto B$ can be represented as its background function or background correlation. Thus, flexible rule $A(x) \rightarrow B(y)$ can be translated into a binary relation. From Sect. 9.3.3, we known that the binary relation is the numerical model of flexible-linguistic-valued correspondence $A \mapsto B$, and thus, it is also the numerical model of the flexible rule $A(x) \rightarrow B(y)$. However, in practical problems, the background function or background correlation of a flexible rule is often unknown. From Sects. 9.3.3 and 9.4, we can take universal relation $\text{core}(A)^+ \times \text{core}(B)^+$ as a numerical-model representative of flexible-linguistic-valued correspondence $A \mapsto B$. Thus, universal relation $\text{core}(A)^+ \times \text{core}(B)^+$ is also a numerical model (representative) of flexible rule $A(x) \rightarrow B(y)$.

Actually, taking $\text{core}(A)^+ \times \text{core}(B)^+$ as the numerical model of rule $A(x) \rightarrow B(y)$ is in accord with the logical semantics $t(A(x)) > 0.5 \rightarrow t(B(y)) > 0.5$ of rule $A(x) \rightarrow B(y)$. In fact, $t(A(x)) > 0.5 \Leftrightarrow x \in \text{core}(A)^+$ and $t(B(y)) > 0.5 \Leftrightarrow y \in \text{core}(B)^+$.

With universal relation as the numerical model, we can use mathematical method to examine and apply flexible rules, and by which, we also can estimate the practical background function or background correlation of a rule. In fact, region $\text{core}(A)^+ \times \text{core}(B)^+$ is a block point in corresponding product space $U \times V$ (as shown in Fig. 13.5a), which gives obviously the scope of graph of background function or background correlation of rule $A(x) \rightarrow B(y)$. Of course, $\text{core}(A)^+ \times \text{core}(B)^+$ is the practical representative of rule $A(x) \rightarrow B(y)$, while the conceptual representative of the rule is then $\text{supp}(A) \times \text{supp}(B)$ (as shown in Fig. 13.5b).

According to the above analysis and result about rule $A(x) \rightarrow B(y)$, in general, the rules with n-conditions are as follows:

$$A_1(x_1) \wedge A_2(x_2) \wedge \cdots \wedge A_n(x_n) \rightarrow B(y)$$
$$A_1(x_1) \vee A_2(x_2) \vee \cdots \vee A_n(x_n) \rightarrow B(y)$$
$$A_1(x_1) \oplus A_2(x_2) \oplus \cdots \oplus A_n(x_n) \rightarrow B(y)$$

that is, rules with a *n*-ary composite linguistic value

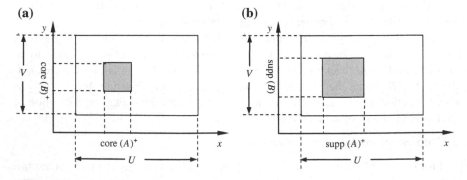

Fig. 13.5 Graph of the numerical-model representative of rule $A(x) \rightarrow B(y)$

$$A_1 \wedge A_2 \wedge \cdots \wedge A_n(x_1, x_2, \ldots, x_n) \rightarrow B(y)$$
$$A_1 \vee A_2 \vee \cdots \vee A_n(x_1, x_2, \ldots, x_n) \rightarrow B(y)$$
$$A_1 \oplus A_2 \cdots \oplus A_n(x_1, x_2, \ldots, x_n) \rightarrow B(y)$$

can be separately written briefly as flexible-linguistic-valued correspondences

$$A_1 \wedge A_2 \wedge \cdots \wedge A_n \mapsto B \tag{13.22}$$

$$A_1 \vee A_2 \vee \cdots \vee A_n \mapsto B \tag{13.23}$$

$$A_1 \oplus A_2 \oplus \cdots \oplus A_n \mapsto B \tag{13.24}$$

and the mathematical essence of them are separately flexible set correspondences

$$A_1 \cap A_2 \cap \cdots \cap A_n \mapsto B \tag{13.25}$$

$$A_1 \cup A_2 \cup \cdots \cup A_n \mapsto B \tag{13.26}$$

$$A_1 \times A_2 \times \cdots \times A_n \mapsto B \tag{13.27}$$

and the numerical-model representatives are separately local universal relations

$$\mathrm{core}(A_1 \cap A_2 \cap \cdots \cap A_n)^+ \times \mathrm{core}(B)^+ \tag{13.28}$$

$$\mathrm{core}(A_1 \cup A_2 \cup \cdots \cup A_n)^+ \times \mathrm{core}(B)^+ \tag{13.29}$$

$$\mathrm{core}(A_1 \times A_2 \times \cdots \times A_n)^+ \times \mathrm{core}(B)^+ \tag{13.30}$$

Of them, the two-dimensional block points $\mathrm{core}(A_1 \cap A_2)^+ \times \mathrm{core}(B)^+$ and $\mathrm{core}(A_1 \cup A_2)^+ \times \mathrm{core}(B)^+$ are shown in Figs. 13.6 and 13.7.

Fig. 13.6 The geometry of numerical-model representative of rule $A_1 \wedge A_2 \rightarrow B$

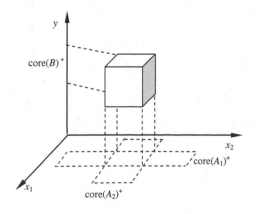

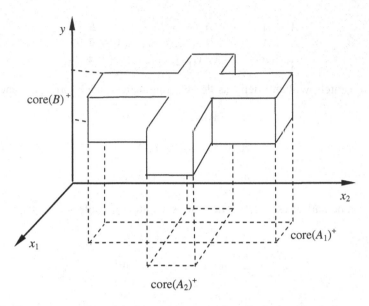

Fig. 13.7 The geometry of numerical-model representative of rule $A_1 \vee A_2 \to B$

Note that if we split rule $A_1 \vee A_2 \to B$ into $A_1 \to B$ and $A_2 \to B$, then the corresponding numerical-model representatives are $\mathrm{core}(A_1)^+ \times \mathrm{core}(B)^+$ and $\mathrm{core}(A_2)^+ \times \mathrm{core}(B)^+$ (their geometries are shown in Fig. 13.8).

Fig. 13.8 The geometry of numerical-model representative of rules $A_1 \to B$ and $A_2 \to B$

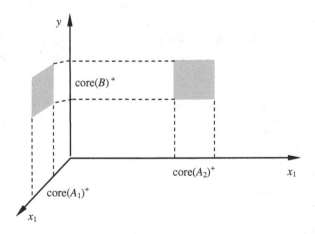

13.6   A Query on the Fuzzy Relational Representation of a Fuzzy Rule

In the last section, we analyzed flexible rules in the two aspects of logic and mathematics, and on the basis of their mathematical essence and mathematical background, we obtained naturally the numerical model $\text{core}(A)^+ \times \text{core}(B)^+$ of flexible rule $A(x) \to B(y)$. While the universal relation $\text{core}(A)^+ \times \text{core}(B)^+$ is obviously a kind of rigid relation.

However, in the fuzzy set theory, a flexible rule (called fuzzy rule in fuzzy set theory) is treated as a binary flexible relation (called fuzzy relation in fuzzy set theory) and also be treated according to implication relation. In fact, in fuzzy set theory, rule $A(x) \to B(y)$ is represented into fuzzy relation

$$R = \int_{U \times V} \mu_R(x,y)/(x,y)$$

where membership function $\mu_R(x,y)$, called implication operator, is a certain operation of membership functions $\mu_A(x)$ and $\mu_B(y)$. For example, $\mu_R(x,y) = \max\{1 - \mu_A(x), \mu_B(y)\}$ is just a basic implication operator, and the fuzzy set R which corresponds to is $A^c \times V \cup U \times B$.

The question now is that in fuzzy set theory, it does not be explained that, and we also are unable to perceive that, what is the theory basis of representing a fuzzy rule $(A(x) \to B(y))$ into a binary flexible relation $(A^c \times V \cup U \times B)$? In fact, no matter viewed from the level of flexible linguistic values, that is, flexible sets, or viewed from the level of the elements, that is, numerical values, of flexible sets, the correspondence between flexible linguistic values of antecedent and consequent of a flexible rule and the binary relation—function or correlation summarized by the rule are all rigid relation. Even if a flexible rule is treated as an implication relation, which would still be represented as a binary rigid relation. From Sect. 13.4, we know that fuzzy set $A^c \times V \cup U \times B$ should be the conceptual truth domain of rule $A(x) \to B(y)$ as implication relation.

13.7   Relationship Between the Flexible Linguistic Rules and the Flexible Linguistic Functions and Correlations

Now, we see that a flexible rule, viewed from the angle of mathematics, is essentially a flexible-linguistic-valued correspondence, and a flexible linguistic function (or flexible linguistic correlation) is then a set of flexible-linguistic-valued correspondences. That is to say, the flexible rule (set) and the flexible linguistic function (or flexible linguistic correlation) are actually two kinds of equivalent representation forms of the correspondence relation between flexible linguistic values. From the forms of expression, we can also so say that the flexible rules are a logical representation of flexible linguistic function, and the flexible linguistic function is a mathematical representation of the flexible rules. In fact, a set of flexible rules represents a function (or correlation) or a subfunction (or sub correlation) from the range of linguistic values in antecedent to the range of linguistic value in consequent; conversely, a flexible linguistic function (or flexible linguistic correlation) can also be represented as a set of flexible rules.

Here also in need of special note is:

① Since a disjunction-type flexible rule with multiple conditions, $A_1 \vee A_2 \vee \cdots \vee A_n \rightarrow B$, can be split into multiple rules with single condition, $A_1 \rightarrow B, A_2 \rightarrow B, \ldots, A_n \rightarrow B$, thus, from the set of disjunction-type rules

$$\{(A_1 \vee A_2 \vee \cdots \vee A_n \rightarrow B)|A_i \subset U_i \quad (i = 1, 2, \ldots, n.), B \subset V\}$$

multiple flexible linguistic functions of a variable, i.e., a set of flexible linguistic functions of a variable

$$\{(A_i, B)|A_i \subset U_i), B \subset V\}, \quad i = 1, 2, \ldots, n.$$

can follows; conversely, from a set of flexible linguistic functions of a variable:

$$\{(A_i, B)|A_i \subset U_i), B \subset V\}, \quad i = 1, 2, \ldots, n.$$

We can also have a set of disjunction-type flexible rules with multiconditions:

$$\{(A_1 \vee A_2 \vee \cdots \vee A_n \rightarrow B)|A_i \subset U_i \quad (i = 1, 2, \ldots, n.), B \subset V\}$$

② In a set of flexible rules which represents a flexible linguistic correlation, there are the rules whose antecedents are the same but consequents are different. Of course, we can also combine them into a rule, of which the consequent is the disjunction of multiple flexible linguistic values.

13.8   Summary

In this chapter, we introduced and made an all-round examination of flexible linguistic rules. First, we expounded the concept and types of flexible linguistic rules and discussed the transformation and reduction of them; then we analyzed the truth domains and logical semantics of flexible linguistic rules; and then we analyzed the mathematical essences and mathematical backgrounds of flexible linguistic rules and presented a numerical model of a flexible linguistic rule, which provides a condition for using mathematical approach to realize reasoning and computation with flexible linguistic rules. Besides, we also discussed the relationship between the flexible linguistic rules and the flexible linguistic functions and correlations, and gave a query on the practice of treating a fuzzy rule as a fuzzy relation in the traditional fuzzy set theory.

The main points and results of the chapter are as follows:

- There are many types of flexible rules, among which the most frequently mentioned are as follows: conjunction-type rules, disjunction-type rules, and synthesis-type rules got from the structure, and corresponding rule with a combined linguistic value and rule with a synthetic linguistic value got from the linguistic values, as well as typical flexible rules.
- Through space transformation, a flexible rule with multidimensional linguistic values can be transformed into a flexible rule with one-dimensional linguistic values, or more generally, a non-typical flexible rule can be transformed to a typical flexible rule. Thus, non-typical flexible rules can be reduced to typical flexible rules.
- Speaking from the angle of concept or logic, a flexible rule represents the implication relation between its antecedent and consequent; however, viewed from the practical or mathematical angle, a flexible rule represents the correspondence between its antecedent and consequent, also the correspondence between its linguistic values, that is, corresponding flexible sets, in antecedent and consequent.
- The mathematical essence of a flexible rule is the corresponding flexible set correspondence, which is the summarization of the local function or correlation between corresponding measurement spaces, and the function or correlation summarized is the mathematical background of the flexible rule.
- The local universal relation formed by the Cartesian product of the extended cores of antecedent and consequent linguistic values of a flexible rule summarizes background function or background correlation of the rule, which can be treated as a numerical-model representative, whose geometrical interpretation is a block point in corresponding measurement space.
- In logic, the implication relation $A \rightarrow B$ covers the correspondence $A \mapsto B$, while in mathematics, the corresponding sets of the former contains that of the latter; treating a rule $(A(x) \rightarrow B(y))$ only as a correspondence $(A \mapsto B)$ can completely satisfy the require of reasoning, which is closer to the mathematical background of rule; treating a rule as an implication relation does not agree with the pattern and require of usual logical reasoning and go far away from the mathematical background of rules, and there would occur large information redundancy.
- The truth domains of a flexible rule have the separation of the conceptual truth domain and practical truth domain. The relation between the truth domains of a flexible rule and the corresponding flexible propositional form is that the former is the least upper bound of the latter.

342   13 Flexible Linguistic Rules and Their Numerical Model

- The flexible linguistic rules and the flexible linguistic functions and flexible linguistic correlations are two kinds of equivalent representation forms of the correspondence relation between flexible linguistic values.

Reference

1. Lian S (2009) Principles of imprecise-information processing. Science Press, Beijing

Chapter 14
Adjoint Functions of a Flexible Linguistic Rule

Abstract This chapter first reveals the triple functional relations implicated by a flexible rule and proposes the term of adjoint function of a rule and gives some specific methods and reference models for the construction of rules' adjoint functions.

Keywords Adjoint functions · Adjoint measured functions · Adjoint degreed functions · Adjoint truth-degreed functions

Further analyzing the correspondence relations between the antecedent and consequent of a flexible linguistic rule at three levels of measure, degree, and truth-degree, we find that a flexible linguistic rule implicates really triple functions. This chapter will analyze and discuss the characteristics of these functional relations and the ideas and methods of obtaining them based on the mathematical background, mathematical essence, numerical model, and logical semantics of flexible linguistic rules.

14.1 Functional Relations Implicated by a Flexible Linguistic Rule

From Sect. 13.5 we know that flexible linguistic rule $A(x) \rightarrow B(y)$ represents essentially the correspondence (relation) between flexible linguistic values A and B. Further viewed from mathematical background, this rule also implicates the correspondence relation between measures x and y, that is, $x \mapsto y$. And because x and y are variables, so when y varies with x, consistency-degree $c_B(y)$ also varies with $c_A(x)$. Therefore, rule $A(x) \rightarrow B(y)$ actually also implicates the correspondence relation between consistency-degrees $c_A(x)$ and $c_B(y)$, that is, $c_A(x) \mapsto c_B(y)$. And from $c_A(x) = t(A(x))$ and $c_B(y) = t(B(y))$; thus, rule $A(x) \rightarrow B(y)$ actually also implicates the correspondence relation between truth-degrees $t(A(x))$ and $t(B(y))$, that is, $t(A(x)) \mapsto t(B(y))$ [1].

Actually, the logical semantics "if $t(A(x)) > 0.5$ then $t(B(y)) > 0.5$" of the rule implicates that $t(p_x)$ and $t(q_y)$ have a certain correspondence relation, and then consistency-degrees $c_A(x)$ and $c_B(y)$, and measures x and y also have a certain correspondence relations.

These correspondence relations $x \mapsto y$, $c_A(x) \mapsto c_B(y)$ and $t(p_x) \mapsto t(q_y)$ may be usual function, also may be correlation. While correlations can also be viewed as a kind of multiple-valued function. Thus, we can say that rule $A(x) \rightarrow B(y)$ implicates functional relations of three different levels of measure, degree, and truth-degree.

Of course, the above is only the analysis of the simplest property–property rule with single condition. But it is not hard to see from it that the other types of flexible rules all implicate such triple functional relations.

In nature, the triple functional relations implicated by a flexible rule are in order **measured function**, **degreed function**, and **truth-degreed function** from inside to outside. If they are explicitly expressed, they are 3 functions accompanying a rule. We call them three to be **adjoint functions** of a rule.

It can be seen that if the adjoint degreed function of corresponding flexible rule can be known, then the problem occurred in approximate reasoning based on rough-true universal modus ponens in Sect. 12.7 would not be hard to be solved (actually, the degreed function here is also the quantitative model of flexible linguistic functions, see Sect. 9.3.3). Therefore, these adjoint functions of rules are of great importance to reasoning based on flexible rules. And the measured function actually links logical reasoning with numerical computation.

It need also to be pointed out that rules involved in this chapter only refer to the rules in usual language and professional fields, but not include implicational tautologies in logic, that is, rules of inference, such as $(p \rightarrow q) \wedge p \Rightarrow q$, $(p \rightarrow q) \wedge \neg q \Rightarrow \neg p$, and $(p \rightarrow q) \wedge (q \rightarrow r) \Rightarrow p \rightarrow r$. Though viewed from the form, this type of inference rules also belongs to production rules, they are production rules in logic, that is, logic rules. Logic rules are different from the rules in usual language and professional fields, which are the abstract models and basic frames to do reasoning with the latter. Compared with the usual field rules, logic rules are the rules of a higher level, or in other words, a kind of meta-rules. Therefore, these logic rules have no corresponding truth-degreed functions, nor degreed functions, even less measured functions. In fact, a logic rule actually includes infinite specific arguments, while the truth-degreed functions, degreed functions, and measured functions in these arguments all differ in thousands of ways, so they cannot have a common functional relation. On the other hand, the arguments in imprecise-information processing are all arguments with specific field knowledge. Though these arguments need inference rules, does not need and involve the trueness degreed function, degreed function, and measured function of inference rules themselves.

14.2   Analysis of the Graph Spaces of Flexible Rules' Adjoint Functions

Although a flexible rule implicates triple functional relations, to obtain the accurate expressions of the corresponding adjoint functions is difficult or even impossible. So we can only consider the approximate expressions of these functions. To this end, we first make an analysis of the graph spaces of these adjoint functions.

Since the truth-degree $t(A(x_0))$ of proposition $A(x_0)$ equal numerically to consistency-degree $c_A(x_0)$, or membership degree $m_A(x_0)$, so the adjoint truth-degreed function of a rule is in essence the same to its adjoint degreed function. Therefore, we only consider the adjoint degreed function and adjoint measured function of rules. Also because a degreed function based on consistency-degree covers really a degreed function based on membership degree, so we only consider the former below.

1. The graph spaces of degreed functions

We first examine the graph space of adjoint degreed function of flexible rule with single condition, $A \to B$. Since consistency-degrees $c_A(x)$ and $c_B(y)$ are also truth-degrees $t(A(x))$ and $t(B(y))$ in number, so the logical semantics "if $t(A(x)) > 0.5$ then $t(B(y)) > 0.5$" of the rule is embodied as "if $c_A(x) > 0.5$ then $c_B(y) > 0.5$" at the level of consistency-degree. In other words, in terms of consistency-degree, the logical semantics of the rule is "if $c_A(x) > 0.5$, then $c_B(y) > 0.5$." Thus, the logical semantics of the rule also gives the basic characteristic of the functional relation between degree d_A of antecedent and degree d_B of consequent of the rule, that is, when $d_A \in (0.5, \beta_A]$ then $d_B \in (0.5, \beta_B]$. On the other hand, in argument, the requirement "$t(A(x_0)) > 0.5$" of rough-true inference for minor premise $A(x_0)$ reflected at the level of consistency-degree is just "$c_A(x_0) > 0.5$." Thus, for the degreed function $d_B = f_d(d_A)$, actually only the function section on interval $(0.5, \beta_A]$ needs to be considered. Thus, although the domain and the range of the degreed function are separately $[\alpha_A, \beta_A]$ and $[\alpha_B, \beta_B]$ in concept, they are then $(0.5, \beta_A]$ and $(0.5, \beta_B]$ in practice. Thus, the graph space of degreed function of the rule with single condition, $A \to B$, is

$$(0.5, \beta_A] \times (0.5, \beta_B] \tag{14.1}$$

(as shown by the gray region in Fig. 14.1). It can be seen that the space is much smaller than the original spaces $[\alpha_A, \beta_A] \times [\alpha_B, \beta_B]$.

We then examine the graph space of the adjoint degreed function of a rule with multiple conditions/a composite linguistic value.

It can be seen that the (indirect) domain of definition of degrees of antecedent $A_1 \wedge A_2$ of conjunction-type rule $A_1 \wedge A_2 \to B$ is $(\alpha_1, \beta_1] \times (\alpha_2, \beta_2]$, and the range of values of the degrees of its consequent B is $(\alpha, \beta]$. Thus, in conceptual, the domain and the range of degreed function of rule $A_1 \wedge A_2 \to B$ are separately $(\alpha_1, \beta_1] \times (\alpha_2, \beta_2]$ and $(\alpha, \beta]$. But under the constraints of requirement "$c_{A_1 \wedge A_2}(x_1, x_2) > 0.5$" of

Fig. 14.1 Graphic
representation of graph space
of adjoint degreed function of
the rule with single condition,
$A \rightarrow B$

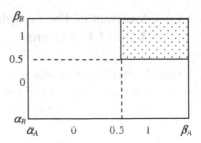

rough-true inference and logical semantics "if $c_{A_1 \wedge A_2}(x_1, x_2) > 0.5$, then $c_B(y) > 0.5$" of the rule, we really only need to consider the function section on subregion $(0.5, \beta_1] \times (0.5, \beta_2]$. Therefore, the domain and the range of the degreed function are separately $(0.5, \beta_1] \cap (0.5, \beta_2]$ and $(0.5, \beta]$ in practice. Thus, the graph space of adjoint degreed function of rule $A_1 \wedge A_2 \rightarrow B$ is

$$((0.5, \beta_1] \times (0.5, \beta_2]) \times (0.5, \beta] \tag{14.2}$$

(as shown in Fig. 14.2a).

Likewise, from logical semantics "if $c_{A_1 \vee A_2}(x_1, x_2) > 0.5$, then $c_B(y) > 0.5$" and requirement "$c_{A_1 \vee A_2}(x_1, x_2) > 0.5$" of rough-true inference, the graph space of adjoint degreed function of disjunction rule $A_1 \vee A_2 \rightarrow B$ is then

$$((0.5, \beta_1] \times (a_2, \beta_2]) \cup ((a_1, \beta_1] \times (0.5, \beta_2]) \times (0.5, \beta] \tag{14.3}$$

(as shown in Fig. 14.2b).

Note: because rule $A_1 \vee A_2 \rightarrow B$ is actually a union of $A_1 \rightarrow B$ and $A_2 \rightarrow B$, so its adjoint degreed function should be originally two discrete two-dimensional subspaces $(0.5, \beta_1] \times (0.5, \beta]$ and $(0.5, \beta_2] \times (0.5, \beta]$ (as shown in Fig. 14.3), while three-dimensional subspace $((0.5, \beta_1] \times (a_2, \beta_2]) \cup ((a_1, \beta_1] \times (0.5, \beta_2]) \times (0.5, \beta]$ is then a extension of the two subspaces in corresponding three-dimensional

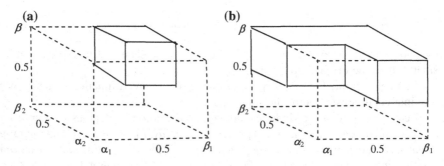

Fig. 14.2 Graphic representation of graph spaces of adjoint degreed functions of rules with 2 conditions/2-ary composite linguistic values, $A_1 \wedge A_2 \rightarrow B$ and $A_1 \vee A_2 \rightarrow B$

Fig. 14.3 Graphic
representation of graph spaces
of adjoint degreed functions
of rules $A_1 \rightarrow B$ and $A_2 \rightarrow B$

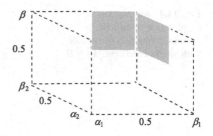

subspace $(\alpha_1, \beta_1] \times (\alpha_2, \beta_2] \times (\alpha, \beta]$. We will see later that to do so does not affect the design of corresponding adjoint degreed function.

And the graph space of adjoint degreed function of synthesis rule $A_1 \oplus A_2 \rightarrow B$ is

$$R \times (0.5, \beta] \tag{14.4}$$

where $R = \{(d_1, d_2)|(d_1, d_2) \in [\alpha_1, \beta_1] \times [\alpha_2, \beta_2]$ and $w_1d_1 + w_2d_2 > 0.5$, $w_1 + w_2 = 1\}$

From the above analysis, it is not hard to imagine that the graph spaces of adjoint degreed functions of general rules with n-conditional/n-ary composite linguistic value, $A_1 \wedge A_2 \wedge \cdots \wedge A_n \rightarrow B, A_1 \vee A_2 \vee \cdots \vee A_n \rightarrow B$ and $A_1 \oplus A_2 \oplus \cdots \oplus A_n \rightarrow B$.

2. The graph spaces of measured functions

Actually, from the numerical-model representative, i.e., universal relation core $(A)^+ \times \text{core}(B)^+$, of flexible rule $A \rightarrow B$ (see Sect. 13.5), the graph space of adjoint measured function of a flexible rule can be obtained immediately—this universal relation itself just is the graph space of adjoint measured function of corresponding rule. But in the following, we consider a new the graph space of adjoint measured function of a rule from logical semantics of the rule and requirement of rough-true inference.

Suppose A and B arc of full-peak, of which the support sets are separately (s_A^-, s_A^+) and (s_B^-, s_B^+), and extended cores are separately (m_A^-, m_A^+) and (m_B^-, m_B^+). It can be seen that $t(A(x)) > 0.5$ and $t(B(y)) > 0.5$ embodied on the measures x and y are just the $x \in \text{core}(A)^+ = (m_A^-, m_A^+)$ and $y \in \text{core}(B)^+ = (m_B^-, m_B^+)$. Thus, by the logical semantics of rule and the requirement of rough-true inference, the domain and range of adjoint measured function of rule $A \rightarrow B$ in conceptual are separately (s_A^-, s_A^+) and (s_B^-, s_B^+), but in practical are separately (m_A^-, m_A^+) and (m_B^-, m_B^+). Therefore, the graph space of adjoint measured function of this rule is rectangular region.

$$\left(m_A^-, m_A^+\right) \times \left(m_B^-, m_B^+\right) \tag{14.5}$$

(see Fig. 14.4). It can be observed that this rectangular region $(m_A^-, m_A^+) \times (m_B^-, m_B^+)$ just is $\text{core}(A)^+ \times \text{core}(B)^+$. That is to say, the graph space of the adjoint

Fig. 14.4 An example of
graph space of adjoint
measured function of the rule
with single condition, $A \rightarrow B$

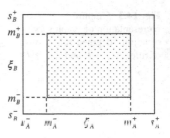

measured function of a rule obtained from logical semantics of the rule and
requirement of rough-true inference is completely consistent with the
numerical-model representative of the rule. Obviously, this space is greatly reduced
from the conceptual function space $(s_A^-, s_A^+) \times (s_B^-, s_B^+)$. From Fig. 14.4, we can
also see the shapes of graph spaces of adjoint measured function of corresponding
rules when flexible linguistic values A or B are of semi-peak.

Similarly, from the requirement of rough-true inference and rule's logical
semantics or the numerical-model representative of the rule, the graph spaces of
adjoint measured functions of rules with 2 conditions/2-ary composite linguistic
values are one by one.

$$\left[\left(m_{A_1}^-, m_{A_1}^+\right) \times \left(m_{A_2}^-, m_{A_2}^+\right)\right] \times (m_B^-, m_B^+) \tag{14.6}$$

$$\left[\left(\left(m_{A_1}^-, m_{A_1}^+\right) \times \left(s_{A_2}^-, s_{A_2}^+\right)\right) \cup \left(\left(m_{A_2}^-, m_{A_2}^+\right) \times \left(s_{A_1}^-, s_{A_1}^+\right)\right)\right] \times (m_B^-, m_B^+) \tag{14.7}$$

$$R \times (m_B^-, m_B^+) \tag{14.8}$$

Here $R = \{(x_1, \ x_2) | (x_1, \ x_2) \in (s_{A_1}^-, \ s_{A_1}^+) \times (s_{A_2}^-, \ s_{A_2}^+)$, and $w_1 m_{A_1}(x_1) +
w_2 m_{A_2}(x_2) > 0.5, w_1 + w_2 = 1\}$. Of which, the graph spaces of adjoint measured
functions of the conjunction-type and the disjunction-type rules
$A_1(x_1) \wedge A_2(x_2) \rightarrow B(y)$ and $A_1(x_1) \vee A_2(x_2) \rightarrow B(y)$ are shown in Fig. 14.5a, b,
but the graph space of synthesis rule $A_1 \oplus A_2(x_1, x_2) \rightarrow B(y)$ cannot be draw
concretely due to the values of rights w_1 and w_2 are not fixed. But from

$$R \supseteq \left(m_{A_1}^-, m_{A_1}^+\right) \times \left(m_{A_2}^-, m_{A_2}^+\right)$$

we have

$$R \times (m_B^-, m_B^+) \supseteq \left[\left(m_{A_1}^-, m_{A_1}^+\right) \times \left(m_{A_2}^-, m_{A_2}^+\right)\right] \times (m_B^-, m_B^+)$$

Thus, from Fig. 14.5, we can also imagine the shapes of graph spaces of adjoint
measured functions of corresponding rules when flexible linguistic values A_1, A_2, or
B are of semi-peak.

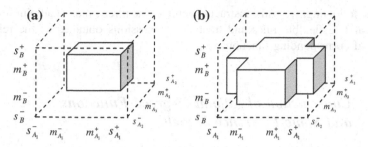

Fig. 14.5 Examples of graph spaces of adjoint measured functions of a rule with 2-conditions/a 2-ary composite linguistic value

Note: because rule $A_1 \vee A_2 \rightarrow B$ is actually a union of $A_1 \rightarrow B$ and $A_2 \rightarrow B$, so its adjoint measured function should be originally two discrete two-dimensional subspaces $(m_{A_1}^-, m_{A_1}^+) \times (m_B^-, m_B^+)$ and $(m_{A_2}^-, m_{A_2}^+) \times (m_B^-, m_B^+)$, while three-dimensional subspace $[((m_{A_1}^-, m_{A_1}^+) \times (s_{A_2}^-, s_{A_2}^+)) \cup ((m_{A_2}^-, m_{A_2}^+) \times (s_{A_1}^-, s_{A_1}^+))] \times (m_B^-, m_B^+)$ is then a extension of the two subspaces in corresponding three-dimensional subspace $(s_{A_1}^-, s_{A_1}^+) \times (s_{A_2}^-, s_{A_2}^+) \times (s_B^-, s_B^+)$. We will see later that to do so does not affect the design of corresponding adjoint measured function.

From the above analysis, it is not hard to imagine that the graph spaces of adjoint measured functions of general rules with n conditions/an n-ary composite linguistic values, $A_1 \wedge A_2 \wedge \cdots \wedge A_n \rightarrow B$, $A_1 \vee A_2 \vee \cdots \vee A_n \rightarrow B$ and $A_1 \oplus A_2 \oplus \cdots \oplus A_n \rightarrow B$.

14.3   Construction of Adjoint Functions of Flexible Rules and Some Reference Models

The graph spaces of the adjoint functions of rules provide the theoretical basis for constructing corresponding functions. In principle, the corresponding approximate adjoint functions can be induced and deduced from known sample data in the graph spaces of the adjoint functions of rules by using certain kind of mathematical methods (such as curve fitting, regression analysis, interpolation) or machine leaning approaches (such as neural network learning). But we see from the above analyses that under the constraints of the mathematical essence, logical semantics of rules and rough-true inference, the valid graph space of an adjoint function of a flexible rule is relatively small. The smaller space means that the function is simpler, say, it can be monotonic or even linear (in fact, theoretically, if the space is sufficiently smaller, then any function in the space can all be treated as an approximate adjoint function of rule). Therefore, we take directly linear function, that is, straight line, plane, or hyperplane, as an approximate adjoint function of corresponding correlation or functional relation. Of course, if there exist sample data, which can be used to guide the construction of corresponding approximate functions.

In the following, we will construct concretely the three kinds of adjoint functions of typical L-L flexible rules and treat the expressions obtained as the reference models of corresponding functions.

14.3.1 Construction of Adjoint Degreed Functions and Some Reference Models

1. Adjoint degreed function of a rule with single condition

Let $A \rightarrow B$ be a rule with single condition.

(1) Suppose that in region $(0.5, \beta_A] \times (0.5, \beta_B]$ consistency-degrees $c_A(x)$ and $c_B(y)$ are of correlation. Then make the median line of the rectangular region (as shown in Fig. 14.6a), we have function

$$d_B = \frac{\beta_B - 0.5}{2}, \quad 0.5 < d_A \leq \beta_A \qquad (14.9)$$

It can be seen that the maximum error between the degree d_B obtained from this function and the expected value d_B' does not exceed $\frac{1}{2}(\beta_B - 0.5)$. Then, we can take this function as an adjoint degreed function of rule $A \rightarrow B$.

(2) Suppose that consistency-degrees $c_A(x)$ and $c_B(y)$ are of a functional relation in region $(0.5, \beta_A] \times (0.5, \beta_B]$. Then make a diagonal of rectangular region (as shown in Fig. 14.6b), we have function

$$d_B = \frac{\beta_B - 0.5}{\beta_A - 0.5}(d_A - 0.5) + 0.5, \quad 0.5 < d_A \leq \beta_A \qquad (14.10)$$

It can be seen that the maximum error between the degree d_B obtained from this function and the expected value d_B' does not exceed $\beta_B - 0.5$, and the average error does not exceed $\frac{1}{2}(\beta_B - 0.5)$. Then, we can take this function as an adjoint degreed function of rule $A \rightarrow B$.

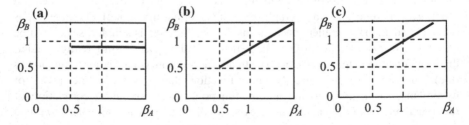

Fig. 14.6 Examples of adjoint degreed functions of a rule with single condition

(3)   Suppose that there are sample data (d_{A_i}, d_{B_i}) $(i = 1, 2, ..., n)$. Then, we can construct the following adjoint degreed function (an example of its graph is shown in Fig. 14.6c),

$$d_B = \kappa d_A + \lambda, \quad \kappa \geq 1, \lambda \geq 0, 0.5 < d_A \leq \beta_A \tag{14.11}$$

where κ and λ are two adjustable parameters, whose values can be determined by the distributing characteristic of the sample points (d_{A_i}, d_{B_i}) $(i = 1, 2, ..., n)$.

2. Adjoint degreed function of a rule with multiple conditions/a composite linguistic value

It can be seen that the adjoint degreed function of a rule with composite linguistic value should be the functional relation between the overall degree of rule's antecedent and the degree of rule's consequent. Thus, for a rule with multiple conditions/a composite linguistic value should find the overall degree of its antecedent firstly and then consider the relation between the overall degree and the degree of consequent of the rule. We know already that subregions $(0.5, \beta_1] \times (0.5, \beta_2]$ and $((0.5, \beta_1] \times (\alpha_2, \beta_2]) \cup ((\alpha_1, \beta_1] \times (0.5, \beta_2])$ are separately the domains of definition of degrees of $A_1 \wedge A_2$ and $A_1 \vee A_2$, and the corresponding degree computation formulas are

$$d_{A_1 \wedge A_2} = \min\{d_{A_1}, d_{A_2}\}$$
$$d_{A_1 \vee A_2} = \max\{d_{A_1}, d_{A_2}\}$$

From this, it is not hard to see that on the region $(0.5, \beta_1] \times (0.5, \beta_2]$, the minimum of $d_{A_1 \wedge A_2}$ is $\min\{0.5, 0.5\} = 0.5$, the maximum is $\max\{\beta_1, \beta_2\} = \beta_\wedge$. Thus, the operation min maps two-dimensional region $(0.5, \beta_1] \times (0.5, \beta_2]$ to one-dimensional region $(0.5, \beta_\wedge]$. Thus, we construct an function from region $(0.5, \beta_\wedge]$ to $(0.5, \beta_B]$:

$$d_B = \frac{\beta_B - 0.5}{\beta_\wedge - 0.5} (d_{A_1 \wedge A_2} - 0.5) + 0.5 \quad d_{A_1 \wedge A_2} \in (0.5, \beta_\wedge] \, (\beta_\wedge = \min\{\beta_1, \beta_2\})$$

This is a straight line passing points $(0.5, 0.5)$ and $(\beta_\wedge, \beta_B)$ in space $(0.5, \beta_\wedge] \times (0.5, \beta_B]$. Considering that $d_{A_1 \wedge A_2} = \min\{d_{A_1}, d_{A_2}\}$, so the adjoint degreed function of conjunction-type rule $A_1 \wedge A_2 \to B$ should be

$$d_B = \frac{\beta_B - 0.5}{\beta_\wedge - 0.5} (d_{A_1 \wedge A_2} - 0.5) + 0.5 \tag{14.12}$$

where $\beta_\wedge = \min\{\beta_1, \beta_2\}$, $d_{A_1 \wedge A_2} = \min\{d_{A_1}, d_{A_2}\}$, $(d_{A_1}, d_{A_2}) \in (0.5, \beta_1] \times (0.5, \beta_2]$. This is a plane in space $((0.5, \beta_1] \times (0.5, \beta_2]) \times (0.5, \beta_B]$.

Note: we do not substitute the expression of $d_{A_1 \wedge A_2}$ into Eq. (14.12), but write them separately, which is because in reasoning whether $d_{A_1 \wedge A_2} > 0.5$ needs to be judged first.

Likewise, we can also construct an adjoint degreed function of the disjunction-type rule $A_1 \vee A_2 \rightarrow B$ as follows:

$$d_B = \frac{\beta_B - 0.5}{\beta_\vee - 0.5}(d_{A_1 \vee A_2} - 0.5) + 0.5 \tag{14.13}$$

where $\beta_\vee = \max\{\beta_1, \beta_2\}$, $d_{A_1 \vee A_2} = \max\{d_{A_1}, d_{A_2}\}$, $(d_{A_1}, d_{A_2}) \in ((0.5, \beta_1] \times (\alpha_2, \beta_2]) \cup ((\alpha_1, \beta_1] \times (0.5, \beta_2])$. This is a plane in space $((0.5, \beta_1] \times (\alpha_2, \beta_2]) \cup ((\alpha_1, \beta_1] \times (0.5, \beta_2]) \times (0.5, \beta_B]$.

It can be seen that although this function, in form, is regard as a function in space $((0.5, \beta_1] \times (\alpha_2, \beta_2]) \cup ((\alpha_1, \beta_1] \times (0.5, \beta_2]) \times (0.5, \beta]$, in effect, which are two functions located separately in two-dimensional spaces $(0.5, \beta_1] \times (0.5, \beta]$ and $(0.5, \beta_2] \times (0.5, \beta]$.

Also, the overall degree of the antecedent $A_1 \oplus A_2$ of synthesis rule $A_1 \oplus A_2 \rightarrow B$ is

$$d_{A_1 \oplus A_2} = w_1 d_{A_1} + w_2 d_{A_2}, \quad w_1 + w_2 = 1, w_1, w_2 \in [0, 1]$$

and the domain of this degree is region $[\alpha_1, \beta_1] \times [\alpha_2, \beta_2]$ ($\alpha_1 \leq 0, 1 \leq \beta_1, \alpha_2 \leq 0, 1 \leq \beta_2$). It is not hard to see that in region $[\alpha_1, \beta_1] \times [\alpha_2, \beta_2]$, the maximum of $d_{A_1 \oplus A_2}$ is $w_1\beta_1 + w_2\beta_2$, denote $\beta_\oplus$. Thus, we construct adjoint degreed function of rule: $A_1 \oplus A_2 \rightarrow B$ as

$$d_B = \frac{\beta_B - 0.5}{\beta_\oplus - 0.5}(d_{A_1 \oplus A_2} - 0.5) + 0.5 \tag{14.14}$$

where $\beta_\oplus = w_1\beta_1 + w_2\beta_2$, $d_{A_1 \oplus A_2} = w_1 d_{A_1} + w_2 d_{A_2}$, $(d_{A_1}, d_{A_2}) \in [\alpha_1, \beta_1] \times [\alpha_2, \beta_2]$. This is a plane in space $([\alpha_1, \beta_1] \times [\alpha_2, \beta_2]) \times (0.5, \beta_B]$ (but not necessarily continuous).

It can be see that the three adjoint degreed functions in general are suitable to the rules that between degrees of antecedent and consequent of it is functional relation, and they cannot further be optimized. We can also use methods similar to those previous for constructing adjoint degreed functions of a rule with single condition to construct the adjoint functions of multiple variables as the expressions in Eqs. (14.9) and (14.11) suitable for correlations or can being optimized.

Generalizing the adjoint degreed functions of a rule with 2 conditions/a 2-ary composite linguistic value, we could obtain the more general adjoint degreed functions of a rule with n conditions/an n-ary composite linguistic value.

14.3.2   Construction of Adjoint Measured Functions and Some Reference Models

Notice first: in order to reduce the graph space of rule's measured function and make the corresponding reasoning easier, for the measured function, in what follows we only discuss those rules whose antecedents are semi-peak values. It is conceivable that the consequent of a rule whose antecedents are semi-peak values may also be semi-peak value, but may also be full-peak value. Because the rule whose antecedents are full-peak values can be spilt into the rules whose antecedents are semi-peak values (see Sect. 16.2 for the split principle). Therefore, it would be enough for us to only discuss the measured functions of the rules whose antecedents are semi-peak values.

1. Adjoint measured functions of a rule with single condition

Let $A \to B$ be a rule with single condition, A and B be all negative semi-peak value, and their extended core be separately $(m_A, \xi_A]$ and $(m_B, \xi_B]$.

(1) Suppose that measures x and y are of correlation in region $(m_A, \xi_A] \times (m_B, \xi_B]$. Then, make the median line of the rectangular region (as shown in Fig. 14.7a), we have function

$$y = \frac{\xi_B - m_B}{2}, \quad m_A < x \le \xi_A \tag{14.15}$$

Then, it can be treated as an adjoint measured function of rule $A \to B$.

(2) Suppose that measures x and y are of functional relation in region $(m_A, \xi_A] \times (m_B, \xi_B]$. Then, make the diagonal of the rectangular region (as shown in Fig. 14.7b), we have function

$$y = \frac{\xi_B - m_B}{\xi_A - m_A}(x - m_A) + m_B, \quad m_A < x \le \xi_A \tag{14.16}$$

Then, it can be treated as an adjoint measured function of rule $A \to B$.

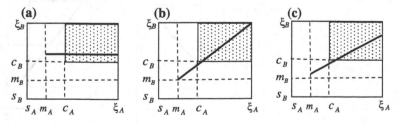

Fig. 14.7 Examples of adjoint measured functions of a rule with single condition whose consequent is semi-peak value

(3) Suppose that there are sample data $(x_i, y_i) \in (m_A, \xi_A] \times (m_B, \xi_B]$ $(i = 1, 2, ..., n)$. Then, we can construct the following adjoint measured function (its example is as shown in Fig. 14.7c):

$$y = k\frac{\xi_B - m_B}{\xi_A - m_A}(x - m_A) + m_B + \lambda, \quad m_A < x \le \xi_A \qquad (14.17)$$

where κ and λ are two adjustable parameters, $\kappa > 1$, $\lambda > 0$, whose values can be determined by the distributing characteristic of the sample points (x_i, y_i) $(i = 1, 2, ..., n)$.

The above three functions are all for the rules whose antecedent linguistic value A and consequent linguistic value B are both negative semi-peak value. Similarly, we can also give the adjoint measured functions of rule $A \to B$ whose A or B is positive semi-peak value. In the following, we then consider the adjoint measured functions of rule $A \to B$ whose A is semi-peak value but whose B is full-peak value. Since the extended core of a full-peak value is also a rectangular region, so we can also use the above-stated method to construct the corresponding measured function. For instance, generally, the following measured function can be constructed:

$$y = k\frac{m_B^+ - m_B^-}{\xi_A - m_A^-}(x - m_A^-) + m_B^- + \lambda, \quad m_A^- < x \le \xi_A \qquad (14.18)$$

and

$$y = k\frac{m_B^+ - m_B^-}{m_A^+ - \xi_A}(x - \xi_A) + m_B^- + \lambda, \quad \xi_A \le x < m_A^+ \qquad (14.19)$$

where $\kappa \ge 1, \lambda \ge 0$ whose graphs are separately shown in Fig. 14.8a, b.

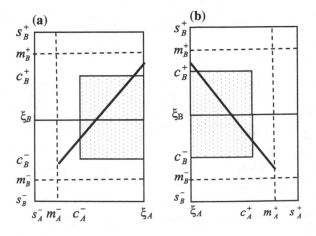

Fig. 14.8 Examples of adjoint measured function of a rule with single condition whose consequent is a full-peak value

2. **Adjoint measured functions of a rule with multiple conditions/a composite linguistic value**

We firstly still consider the adjoint measured functions of a rule with 2 conditions/a 2-ary composite linguistic value. Here, flexible linguistic values are all negative semi-peak values.

From Sect. 14.2, it can be known that the graph space of adjoint measured function of the conjunction-type rules $A_1 \wedge A_2 \rightarrow B$ is $((m_{A_1}, \ \xi_{A_1}] \cap (m_{A_2}, \xi_{A_2}]) \times (m_B, \xi_B]$. Ordinarily, any function in the space can be taken as an adjoint measured function of rule $A_1 \wedge A_2 \rightarrow B$. In consideration of the logical semantics, "if $c_{A_1 \wedge A_2}(x_1, x_2) > 0.5$, then $c_B(y) > 0.5$" of the rule, and $c_{A_1 \wedge A_2}(x_1, x_2) = \min \{c_{A_1}(x_1), \ c_{A_2}(x_2)\}$, also $\min\{c_{A_1}(x_1), \ c_{A_2}(x_2)\} = c_{A_1}(x_1)$ or $\min\{c_{A_1}(x_1), c_{A_2}(x_2)\} = c_{A_2}(x_2)$; therefore, we construct 2 measured functions which separately satisfy "$c_B(y) > 0.5$ when $c_{A_1}(x_1) > 0.5$" and "$c_B(y) > 0.5$ when $c_{A_2}(x_2) > 0.5$":

$$
\begin{cases}
y = \frac{\xi_B - m_B}{\xi_{A_1} - m_{A_1}}(x_1 - m_{A_1}) + m_B; & m_{A_1} < x_1 \le \xi_{A_1}, m_{A_2} < x_2 \le \xi_{A_2} & \text{as } \min\{c_{A_1}(x_1), c_{A_2}(x_2)\} = c_{A_1}(x_1) \\
y = \frac{\xi_B - m_B}{\xi_{A_2} - m_{A_2}}(x_2 - m_{A_2}) + m_B & m_{A_1} < x_1 \le \xi_{A_1}, \ m_{A_2} < x_2 \le \xi_{A_2} & \text{as } \min\{c_{A_1}(x_1), c_{A_2}(x_2)\} = c_{A_2}(x_2)
\end{cases}
$$

$$(14.20)$$

As shown in Fig. 14.9a, the two functions together are adjoint measured functions of the rule $A_1 \wedge A_2 \rightarrow B$.

Based on the graph space $[((m_{A_1}^-, \ m_{A_1}^+) \times (s_{A_2}^-, \ s_{A_2}^+)) \cup ((m_{A_2}^-, \ m_{A_2}^+) \times (s_{A_1}^-, \ s_{A_1}^+))] \times (m_B^-, \ m_B^+)$ of adjoint measured function of the disjunction-type rule $A_1(x_1) \vee A_2(x_2) \rightarrow B(y)$, we construct the two function expressions as follows:

$$
\begin{cases}
y = \frac{\xi_B - m_B}{\xi_{A_1} - m_{A_1}}(x_1 - m_{A_1}) + m_B; & m_{A_1} < x_1 \le \xi_{A_1}, s_{A_2} < x_2 \le \xi_{A_2} & \text{as } \min\{c_{A_1}(x_1), c_{A_2}(x_2)\} = c_{A_1}(x_1) \\
y = \frac{\xi_B - m_B}{\xi_{A_2} - m_{A_2}}(x_2 - m_{A_2}) + m_B & s_{A_1} < x_1 \le \xi_{A_1}, m_{A_2} < x_2 \le \xi_{A_2} & \text{as } \min\{c_{A_1}(x_1), c_{A_2}(x_2)\} = c_{A_2}(x_2)
\end{cases}
$$

$$(14.21)$$

Its graphs are shown in Fig. 14.9b.

It can be seen that although the two functions, in form, are regarded as functions in space $[((m_{A_1}^-, \ m_{A_1}^+) \times (s_{A_2}^-, \ s_{A_2}^+)) \cup ((m_{A_2}^-, \ m_{A_2}^+) \times (s_{A_1}^-, \ s_{A_1}^+))] \times (m_B^-, \ m_B^+)$, in effect, which are two functions located separately in two-dimensional spaces $(m_{A_1}^-, m_{A_1}^+) \times (m_B^-, m_B^+)$ and $(m_{A_2}^-, m_{A_2}^+) \times (m_B^-, m_B^+)$.

Based on the graph space $R \times (m_B, \xi_B]$ (where $R = \{(x_1, x_2) \mid (x_1, x_2) \in (s_{A_1}, \xi_{A_1}] \times (s_{A_2}, \xi_{A_2}]$ and $w_1 c_{A_1}(x_1) + w_2 c_{A_2}(x_2) > 0.5, w_1 + w_2 = 1\})$ of adjoint measured function of the synthesis-type rule $A_1 \oplus A_2 (x_1, x_2) \rightarrow B(y)$, we construct the following adjoint measured function, whose graph is shown in Fig. 14.9c.

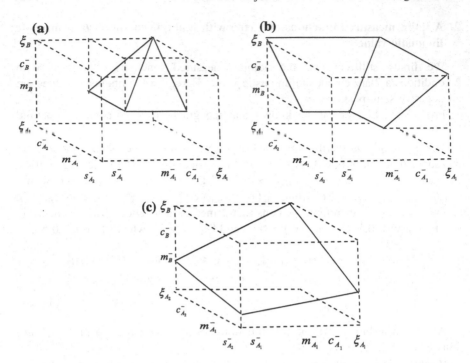

Fig. 14.9 Examples of adjoint measured functions of a rule with 2 conditions/a 2-ary composite linguistic value

$$
\begin{vmatrix}
x_1 & x_2 & y & 1 \\
\xi_{A_1} & \xi_{A_2} & \xi_B & 1 \\
s_{A_1}^- & \xi_{A_2} & m_B^- & 1 \\
\xi_{A_1} & s_{A_2}^- & m_B^- & 1
\end{vmatrix} = 0
\tag{14.22}
$$

where $s_{A_1} < x_1 \leq \xi_{A_1}$, $s_{A_2} < x_2 \leq \xi_{A_2}$, $w_1 c_{A_1}(x_1) + w_2 c_{A_2}(x_2) > 0.5$, $w_1 + w_2 = 1$.

It can be seen that this is a plane passing points $(\xi_{A_1}, \xi_{A_2}, \xi_B)$, $(s_{A_1}^-, \xi_{A_2}, m_B^-)$ and $(\xi_{A_1}, s_{A_2}^-, m_B^-)$.

Here, it should be noted that although the graph space of adjoint measured function of rule $A_1 \oplus A_2 \rightarrow B$, in theory, is $\text{core}(A_1 \times A_2)^+ \times \text{core}(B)^+$, extended core $\text{core}(A_1 \times A_2)^+$, in general, cannot explicitly given; therefore, here $R \times (m_B, \xi_B]$ representing space $\text{core}(A_1 \times A_2)^+$ becomes already $\text{supp}(A_1 \times A_2) \times \text{core}(B)^+$ actually.

From the above adjoint measured functions of a rule with single condition and a rule with 2 conditional/a 2-ary composite linguistic value, we then can have the more general adjoint measured functions of a rule with n conditional/a n-ary composite linguistic value.

14.4   Adjoint Functions of Non-typical Flexible Linguistic Rules

Obtaining directly the adjoint functions of non-typical flexible rules would encounter difficulty. But from Sect. 13.3, it is known that a non-typical flexible rule can be transformed into a typical flexible rule through space transformation. And for the latter, we have already had methods to obtain adjoint functions. Based on this way of thinking, we discuss the approaches of obtaining the adjoint functions of non-typical flexible rules below.

1. **Adjoint functions of a property–property rule with the antecedent of multidimensional linguistic value**

Let $A(x_1, x_2, \ldots, x_n) \to B(y)$ be a property–property rule whose antecedent is a multidimensional atomic linguistic value, $(x_1, x_2, \ldots, x_n) \in U \subset \mathbf{R}^n$, and $y \in V$. Suppose that there exits mapping $\varphi: U \to [a, b]$, $r = \varphi(x_1, x_2, \ldots, x_n)$, flexible linguistic value A on U can be transformed into flexible linguistic value A on $[a, b]$. Thus, multidimensional flexible rule $A(x_1, x_2, \ldots, x_n) \to B(y)$ can be transformed into one-dimensional flexible rule $A(r) \to B(y)$. Obviously, for this one-dimensional flexible rule, we are able to obtain its triple adjoint functions. Therefore, the triple adjoint functions of original rule $A(x_1, x_2, \ldots, x_n) \to B(y)$ can be indirectly obtained.

(1) Let the adjoint degreed function of rule $A(r) \to B(y)$ be

$$d_B = \frac{\beta_B - 0.5}{\beta_A - 0.5}(d_A - 0.5) + 0.5, \quad 0.5 < d_A \le b \tag{14.23}$$

Since transforming φ only changes the space and dimensions of flexible linguistic value A, it does not change the correspondence relation between the linguistic values of antecedent and consequent at the level of degree. Therefore, this degreed function should also be an adjoint degreed function of the original rule $A(x_1, x_2, \ldots, x_n) \to B(y)$.

(2) Let the adjoint measured function of rule $A(r) \to B(y)$ be

$$y = \frac{\xi_B - m_B}{\xi_A - m_A}(r - m_A) + m_B; \quad m_A < r \le \xi_A \text{ or } \xi_A \le r < m_A$$

This measured function is the functional relation between r and y, while $r = \varphi(x_1, x_2, \ldots, x_n)$, then substitute it into the above expression, and then we have

$$y = \frac{\xi_B - m_B}{\xi_A - m_A}(\varphi(x_1, x_2, \ldots, x_n) - m_A) + m_B \tag{14.24}$$

where $m_A < \varphi(x_1, x_2, \ldots, x_n) \le \xi_A$ or $\xi_A \le \varphi(x_1, x_2, \ldots, x_n) < m_A$. This is just an adjoint measured function of the original rule $A(x_1, x_2, \ldots, x_n) \to B(y)$.

Example 14.1 Suppose there is rule r_1: if (x, y) is near point $(5, 6)$, then u is large. Using A and B to separately denote "near" and "large," we define their consistency functions as follows:

$$c_A(z, 0, 1, 2) = 2 - z, \quad 0 < z \le 2$$
$$c_B(u, 8, 9, 10) = u - 8, \quad 8 < u \le 10$$

Using adjoint measured function (14.16), then it follows that

$$u = \frac{\xi_B - m_B}{\xi_A - m_A}(z - m_A) + m_B$$

Take $z = \varphi(x, y) = \sqrt{(x-5)^2 + (y-6)^2}$, $m_A = 1.5$, $\xi_A = 0$, $m_B = 8.5$, $\xi_B = 10$, then we have measured function

$$u = 10 - \sqrt{(x-5)^2 + (y-6)^2}, \quad 0 \le \sqrt{(x-5)^2 + (y-6)^2} < 1.5$$

The graph of this function is a conical surface (as shown in Fig. 14.10). Now viewed conversely, this measured function can also be interpreted as rule r_1. This shows that it is appropriate to treat this function as the mathematical model of rule r_1.

2. Adjoint functions of a property–property rule with multiple conditions and multidimensional linguistic values

Let $A_1(x_1, x_2, \ldots, x_n) \wedge A_2(y_1, y_2, \ldots, y_m) \to B(z)$ be a conjunction-type property–property rule with multidimensional linguistic values, where $(x_1, x_2, \ldots,$

Fig. 14.10 An example of the adjoint measured function of a property–property rule with the antecedent of multidimensional linguistic value

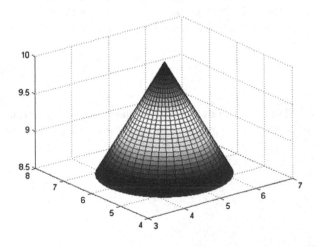

$x_n) \in U \subset \mathbf{R}^n$, and $(y_1, y_2, ..., y_m) \in V \subset \mathbf{R}^m$. Suppose that there exit mappings φ: $U \to [a_1, b_1]$, $r_1 = \varphi(x_1, x_2, ..., x_n)$ and $\psi: V \to [a_2, b_2]$, $r_2 = \psi(y_1, y_2, ..., y_m)$ such that flexible linguistic value A_1 on U can be transformed into flexible linguistic value A_1 on $[a_1, b_1]$, and flexible linguistic value A_2 on V can be transformed into flexible linguistic value A_2 on $[a_2, b_2]$. Consequently, original flexible rule that contains multidimensional linguistic values

$$A_1(x_1, x_2, ..., x_n) \wedge A_2(y_1, y_2, ..., y_m) \to B(z)$$

is transformed into the flexible rule that contains only one-dimensional linguistic values

$$A_1(r_1) \wedge A_2(r_2) \to B(z)$$

From this, we find the triple adjoint functions of the original rule $A_1(x_1, x_2, ..., x_n) \wedge A_2(y_1, y_2, ..., y_m) \to B(z)$.

(1) Let the adjoint degreed function of rule $A_1(r_1) \wedge A_2(r_2) \to B(z)$ be

$$d_B = \begin{cases} \frac{\beta_B - 0.5}{\beta_{A_1} - 0.5}(d_{A_1} - 0.5) + 0.5, & 0.5 < d_{A_1} \le \beta_{A_1}, 0.5 < d_{A_2} \le \beta_{A_2} \\ \frac{\beta_B - 0.5}{\beta_{A_2} - 0.5}(d_{A_2} - 0.5) + 0.5, & 0.5 < d_{A_1} \le \beta_{A_1}, 0.5 < d_{A_2} \le \beta_{A_2} \end{cases} \quad (14.25)$$

Then, this is also an adjoint degreed function of the original rule.

(2) Let the adjoint measured function of rule $A_1(r_1) \wedge A_2(r_2) \to B(z)$ be

$$\begin{cases} y = \frac{\xi_B - m_B}{\xi_{A_1} - m_{A_1}}(r_1 - m_{A_1}) + m_B, & m_{A_1} < r_1 \le \xi_{A_1}, \quad m_{A_2} < r_2 \le \xi_{A_2} \\ y = \frac{\xi_B - m_B}{\xi_{A_2} - m_{A_2}}(r_2 - m_{A_2}) + m_B, & m_{A_1} < r_1 \le \xi_{A_1}, \quad m_{A_2} < r_2 \le \xi_{A_2} \end{cases}$$

while $r_1 = \varphi(x_1, x_2, ..., x_n)$ and $r_2 = \psi(y_1, y_2, ..., y_m)$, substitute them into the expressions above, then we have

$$\begin{cases} y = \frac{\xi_B - m_B}{\xi_{A_1} - m_{A_1}}(\varphi(x_1, x_2, ..., x_n) - m_{A_1}) + m_B \\ y = \frac{\xi_B - m_B}{\xi_{A_2} - m_{A_2}}(\psi(y_1, y_2, ..., y_m) - m_{A_2}) + m_B \end{cases} \quad (14.26)$$

where $m_{A_1} < \varphi(x_1, ..., x_n) \le \xi_{A_1}$, $m_{A_2} < \psi(y_1, ..., y_m) \le \xi_{A_2}$. This is just an adjoint measured function of the original rule.

3. Adjoint function of a relation–property rule

Let $R(x_1, x_2, ..., x_n) \to B(y)$ be a relation–property rule whose antecedent is a relational linguistic value, $(x_1, x_2, ..., x_n) \in U \subset \mathbf{R}^n$. Suppose that there exists mapping $\varphi: U \to [a, b]$, $r = \varphi(x_1, x_2, ..., x_n)$ such that flexible linguistic value A on U can be transformed into flexible linguistic value A on $[a, b]$. Thus, the original relation–property rule $R(x_1, x_2, \ldots, x_n) \rightarrow B(y)$ is transformed into $R(r) \rightarrow B(y)$. From this, we find the triple adjoint functions of the original rule $R(x_1, x_2, \ldots, x_n) \rightarrow B(y)$.

(1) Let the adjoint degreed function of rule $R(r) \rightarrow B(y)$ be

$$d_B = \frac{\beta_B - 0.5}{\beta_R - 0.5}(d_R - 0.5) + 0.5, \quad 0.5 < d_R \leq \beta_R \tag{14.27}$$

Since transforming φ only changes the space, dimensions, and property of flexible linguistic value R, it does not change the correspondence relation between antecedent and consequent linguistic values at the level of degree. Therefore, this degreed function should also be an adjoint degreed function of the original rule $R(x_1, x_2, \ldots, x_n) \rightarrow B(y)$.

(2) Let the adjoint measured function of rule $R(r) \rightarrow B(y)$ be

$$y = \frac{\xi_B - m_B}{\xi_A - m_R}(r - m_R) + m_B; \quad m_R < r \leq \xi_R \text{ or } \xi_R \leq r < m_R$$

This measured function is the functional relation between r and y, while $r = \varphi(x_1, x_2, \ldots, x_n)$, then substitute it into the expression above, we have

$$y = \frac{\xi_B - m_B}{\xi_R - m_R}(\varphi(x_1, x_2, \ldots, x_n) - m_R) + m_B \tag{14.28}$$

where $m_R < \varphi(x_1, x_2, \ldots, x_n) \leq \xi_R$ or $\xi_R \leq \varphi(x_1, x_2, \ldots, x_n) < m_R$. This is just an adjoint measured function of the original rule $R(x_1, x_2, \ldots, x_n) \rightarrow B(y)$.

Example 14.2 Suppose that there is rule r_2: if x is close to y, then u is small. Using R and B to separately denote "close to" and "small," we define its consistency function as follows:

$$c_R(z, 0.8, 0.9, 1, 1.1, 1.2) = \begin{cases} 10z - 8, & 0.8 < z \leq 1 \\ 17 - 10z, & 1 \leq z < 1.2 \end{cases}$$

$$c_B(u, 0, 1, 2) = 2 - u, \quad 0 \leq u < 2$$

and employing adjoint measured function (14.35), then it follows that

$$u = \frac{\xi_B - m_B}{\xi_R - m_R}(z - m_R) + m_B$$

Taking $z = \varphi(x, y) = \frac{x}{y}$, then when $0.85 < \frac{x}{y} \leq 1$, $m_R = 0.85$, $\xi_R = 1$, $m_B = 1.5$ and $\xi_B = 0$; when $1 \leq \frac{x}{y} < 1.15$, $m_R = 1.15$, $\xi_R = 1$, $m_B = 1.5$, and $\xi_B = 0$. Thus, we have measured function

Fig. 14.11 An example of the approximate measured function of a relation–property rule

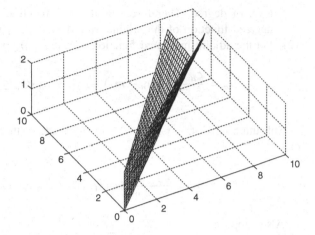

$$u = \begin{cases} 10 - 10\frac{x}{y}, & 0.85 < \frac{x}{y} \le 1 \\ 10\frac{x}{y} - 10, & 1 \le \frac{x}{y} < 1.15 \end{cases}$$

This is just an adjoint measured function of the original rule, its graph is shown in Fig. 14.11. Now viewed conversely, this measured function can also be interpreted as rule r_2.

4. **Adjoint functions of a flexible rule with the antecedent and consequent of multidimensional linguistic value**

Let $A(x_1, x_2, \ldots, x_n) \to B(y_1, y_2, \ldots, y_m)$ be a rule whose antecedent and consequent are all multidimensional linguistic values, $(x_1, x_2, \ldots, x_n) \in U$, $(y_1, y_2, \ldots, y_m) \in V$; A, B can be property-type linguistic values and relation-type linguistic values. Suppose there exist transformations $r_n = \varphi(x_1, x_2, \ldots, x_n)$ and $r_m = \psi(y_1, y_2, \ldots, y_m)$ such that flexible linguistic value A on n-dimensional space U can be transformed into flexible linguistic value A on one-dimensional space $[a, b]$ and flexible linguistic value B on m-dimensional space V can be transformed into flexible linguistic value B on one-dimensional space $[c, d]$. Thus, non-typical flexible rule $A(x_1, x_2, \ldots, x_n) \to B(y_1, y_2, \ldots, y_m)$ is transformed into typical flexible rule $A(r_n) \to B(r_m)$. Based on this, we find the triple adjoint functions of original rule $A(x_1, x_2, \ldots, x_n) \to B(y_1, y_2, \ldots, y_m)$.

(1) Let the adjoint degreed function of rule $A(r_n) \to B(r_m)$ be

$$d_B = \frac{\beta_B - 0.5}{\beta_A - 0.5}(d_A - 0.5) + 0.5, \quad 0.5 < d_A \le \beta_A \tag{14.29}$$

Since transformations φ and ψ only change the spaces, dimensions, and properties of flexible linguistic value A and B, they do not change the correspondence relation between antecedent and consequent linguistic values at the level of degree. Therefore, this degreed function should also be an adjoint degreed function of the original rule $A(x_1, x_2, \ldots, x_n) \rightarrow R(y_1, y_2, \ldots, y_m)$.

(2) Let the adjoint measured function of rule $A(r_n) \rightarrow B(r_m)$ be

$$r_m = \frac{\xi_B - m_B}{\xi_A - m_A}(r_n - m_A) + m_B; \quad m_A < r_n \leq \xi_A \text{ or } \xi_A \leq r_n < m_A$$

Substitute $r_n = \varphi(x_1, x_2, \ldots, x_n)$ into right side of the equation above, it follows that

$$r_m = \frac{\xi_B - m_B}{\xi_A - m_A}(\varphi(x_1, x_2, \ldots, x_n) - m_A) + m_R$$

Also suppose

$$(y_1, y_2, \ldots, y_m) = \psi^{-1}(r_m)$$

Thus, from the above two equations we have

$$(y_1, y_2, \ldots, y_m) = \psi^{-1}\left(\frac{\xi_B - m_R}{\xi_A - m_A}(\varphi(x_1, x_2, \ldots, x_n) - m_A) + m_R\right) \quad (14.30)$$

This is an adjoint measured function of the original rule $A(x_1, x_2, \ldots, x_n) \rightarrow B$ $(y_1, y_2, \ldots, y_m)$.

It can be seen that Eq. (14.30) is really a vector-valued function. This vector-valued function is the mathematical background and background function of the flexible rule whose antecedent and consequent are all multidimensional linguistic values—the most typical non-typical flexible rule. Actually, it is not hard to see that the measured functions of all rules whose consequents are multidimensional flexible linguistic values are certain kind of vector-valued function.

We see from the above-stated that for non-typical flexible rules, we can transform them into typical flexible rules through certain kinds of mathematical transformations, that is, mappings, then find the adjoint functions of the latter; further derive the adjoint functions of original rule. The measured function instances in Examples 14.1 and 14.2 above just verify the correctness of this method.

5. Adjoint functions of a rule with multiple conclusions

From the logical relation between antecedent and consequent as well as the logical relation between various terms in consequent, it is not hard to see that the rules with multiple conclusions, $A \rightarrow B_1 \wedge B_2 \wedge \cdots \wedge B_m$ and $A \rightarrow B_1 \vee B_2 \vee \cdots \vee B_m$) (where A is either a simple proposition or a compound proposition $A_1 \wedge A_2 \wedge \cdots \wedge A_n$, $A_1 \vee A_2 \vee \cdots \vee A_n$, or $A_1 \oplus A_2 \oplus \cdots \oplus A_n$) can be split into m rules with single conclusion, $A \rightarrow B_1, A \rightarrow B_2, \ldots, A \rightarrow B_m$, which are separately either conjunctive or disjunctive; Or conversely speaking, the conjunction and the disjunction of rules with single conclusion, $A \rightarrow B_1, A \rightarrow B_2, \ldots, A \rightarrow B_m$ are separately

$A \to B_1 \wedge B_2 \wedge \cdots \wedge B_m$ and $A \to B_1 \vee B_2 \vee \cdots \vee B_m$. Consequently, the conjunction and disjunction of the adjoint functions of the m rules with single conclusion are separately the adjoint functions of the two rules with multiple conclusions. Thus, to obtain the adjoint functions of the rules with multiple conclusions, $A \to B_1 \wedge B_2 \wedge \cdots \wedge B_m$ and $A \to B_1 \vee B_2 \vee \cdots \vee B_m$, we need to firstly find the adjoint functions of rules with single conclusion, $A \to B_1, A \to B_2, \ldots, A \to B_m$, respectively, then join them separately by conjunction ($\wedge$) and disjunction ($\vee$). For example, the adjoint degreed function of rule $A \to B_1 \wedge B_2 \wedge \cdots \wedge B_m$ would be

$$f_{d_1}(d_A) \wedge f_{d_2}(d_A) \wedge \cdots \wedge f_{d_m}(d_A)$$

and the adjoint degreed function of rule $A \to B_1 \vee B_2 \vee \cdots \vee B_m$ would be

$$f_{d_1}(d_A) \vee f_{d_2}(d_A) \vee \cdots \vee f_{d_m}(d_A)$$

where $f_{d_i}(d_A)$ is the adjoint degreed function of rule $A \to B_i$ ($i = 1, 2, \ldots, m$).

6. Adjoint functions of a rule with linguistic values with opposite

All types of adjoint functions above are all for the flexible rules in negation-type logic. But the rule with linguistic values with opposite in opposite-type logic should also have similar adjoint functions. Therefore, it is also necessary to find the adjoint functions of a rule with linguistic values with opposite.

Since the near-true in the logical semantics of a rule with linguistic values with opposite means truth-degree >0, rather than >0.5. Therefore, the domains and ranges of triple adjoint functions of a rule with linguistic values with opposite should be somewhat different from those of the adjoint functions of a rule with linguistic values with negation. To be specific, the domains and ranges of the truth-degreed function and degreed function of a rule with single condition should take 0 as the lower bound, and those of measured function should take critical points as the lower bound or upper bound. For example, for the rule with linguistic values with opposite, $A(x) \to B(y)$, the domains of its truth-degreed function and degreed function are $[0, \beta_A]$, and the ranges are $[0, \beta_B]$. Thus, the graph spaces of the functions are $[0, \beta_A] \times [0, \beta_B]$, correspondingly, the graph space of its measured function is $[s_A, \xi_A] \times [s_B, \xi_B]$ or $[\xi_A, s_A] \times [s_B, \xi_B]$.

As thus, the thinking and method of finding the adjoint functions of a rule with linguistic values with opposite are completely the same as the previous; only the corresponding functional expressions and domains are somewhat different. And the difference is then: the location of 0.5 of the original adjoint truth-degreed function and degreed function is just 0 in the adjoint truth-degreed function and degreed function of a rule with linguistic values with opposite; the location of the median point of the original adjoint measured function is the critical point of the adjoint truth-degreed function and degreed function of a rule with linguistic values with opposite. The same is true for the differences between domains. Since it is so, we only need to make a little modifications of the original adjoint functions, that is, changing 0.5 into 0, and changing the median point (as m_A and m_B) into critical

point (as s_A and s_B), then we can have the corresponding adjoint functions of a rule
with linguistic values with opposite.

14.5   Adjoint Functions of Flexible Linguistic Rules in Extended Logical Semantics

The adjoint functions discussed above are based on the natural (standard) logical
semantics of rules, in the following we consider the adjoint functions of flexible
rules in extended logical semantics.

The extended logical semantics of a flexible rule is as follows: if antecedent
degree is true, then consequent degree would be true (for details, see Sect. 13.4). It
can be seen that according to the extended logical semantics, the graph space of the
adjoint degreed function of a flexible rule is the product of corresponding subranges
with 0 as infimum of range of consistency functions of flexible linguistic values of
antecedent and consequent of the rule, and the graph space of the adjoint measured
function of a flexible rule is the product of corresponding support sets of flexible
linguistic values of antecedent and consequent of the rule. For example, the
extended logical semantics of rule with single condition, $A(x) \rightarrow B(y)$, is if A
(x) degree-true then $B(y)$ degree-true, that is, $t(A(x)) > 0 \rightarrow t(B(y)) > 0$. Then,
according to the extended logical semantics, the graph space of the adjoint degreed
function of the rule is $(0, \beta_A] \times (0, \beta_B]$ (as shown in Fig. 14.12) and the graph space
of the adjoint measured function of the rule is $(s_A^-, s_A^+) \times (s_B^-, s_B^+)$ (as shown in
Fig. 14.13).

Similarly, based on such graph spaces, using the methods similar to preceding
methods, we can also construct the approximate adjoint degreed function and
adjoint measured function of a flexible rule in extended logical semantics. But, by
comparing the graph spaces of adjoint degreed functions and adjoint measured
functions of flexible rules in extended logical semantics with the corresponding
graph spaces of adjoint degreed functions and adjoint measured functions of flexible
rules in standard logical semantics in Sect. 14.2, it can be seen that the former are
much bigger than the latter. Therefore, the errors of adjoint functions constructed
for flexible rules in extended logical semantics are much greater than the errors of
corresponding adjoint functions constructed in standard logical semantics.

Fig. 14.12 Example of graph
spaces adjoint degreed
functions of a flexible rule in
extended logical semantics

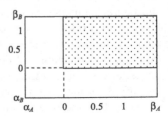

Fig. 14.13 Example of graph spaces adjoint measured functions of a flexible rule in extended logical semantics

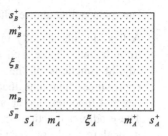

14.6   Optimization of Flexible Linguistic Rules' Adjoint Functions

In the above, we analyzed and discussed the adjoint functions of all types of flexible rules and gave some reference models. It can be seen that these reference models are actually similar to piecewise or blocking linear interpolation functions. Because of the constraints of rough-true inference and logical semantics of rules, the graph spaces of the adjoint functions of rules have be reduced greatly, and for the measured function, we also used the technique of semi-peak valued rules making that the graph spaces of corresponding functions are further reduced. So it should say that these reference models given can satisfy the requirements of some practical problems.

Nonetheless, it still cannot be guaranteed that the above-stated methods and models can satisfy the requirements of all practical problems. Then, to further improve the accuracy of the adjoint functions of rules, on the basis of the above-stated methods and models, we can use machine learning or the optimization methods to optimize further these functions. In fact, in principle, as long as there is certain amount of sample data, and by machine learning, then the approximate degree of the adjoint functions can be effectively improved.

Besides, there is also other way to raise the approximate degree of the adjoint functions, that is, narrowing the support sets of the flexible linguistic values in rules because in theory, the method we use to construct adjoint functions is the function approximation, that is, using local linear functions to approximate a global non-linear function. Obviously, the narrower the domains of the local functions are, the better the approximation effects would be and the higher approximate degree would be. Of course, narrowing support sets would involve the adjustments of rules and rule sets.

14.7   Summary

In this chapter, we first revealed the triple functional relations implicated by a flexible rule, proposed the term of adjoint function of a rule, and gave some specific methods and reference models for the construction of rules' adjoint functions.

The main points and results of the chapter are as follows:

- A flexible rule actually implicates functions or correlations at three levels of truth-degree, degree, and measure whose concrete expressions are then three adjoint functions of the rule; we can only try to find their corresponding approximate functions, that is, approximate truth-degreed function, approximate degreed function, and approximate measured function.
- These adjoint functions of rules can be obtained in principle by using mathematical methods or machine learning methods; but under the constraints of rough (near)-true inference and natural (standard) logical semantics of rules, the graph spaces of adjoint functions of rules are already smaller relatively, so for typical rules we can also linear functions as the corresponding approximate adjoint functions; for non-typical rules we then can use space transforming method to obtain their adjoint functions. In extended logical semantics, the graph spaces of adjoint functions of rules are the biggest; thus, the errors of corresponding approximate adjoint functions are greater.
- Combined with sample data, the given reference models of adjoint functions can be further optimized, and the approximate degree of the adjoint functions can also raised by narrowing the support sets of the corresponding flexible linguistic values.

Reference

1. Lian S (2009) Principles of imprecise-information processing. Science Press, Beijing

Chapter 15
Reasoning and Computation with Flexible Linguistic Rules

Abstract This chapter expounds, in the frame of logic, the principles, and methods of reasoning with flexible linguistic rules. First, it presents natural inference with data conversion(s) according to the relation between numerical values and flexible linguistic values. Then, it presents the principles and methods of reasoning with truth-degrees and reasoning with degrees based on rule's adjoint functions, on the basis of rough-true (near-true) inference. And it discusses reasoning with degrees with data conversion(s) and approximate reasoning and computation based on reasoning with degrees. In addition, it discusses parallel reasoning with degrees.

Keywords Natural inference · Reasoning with truth-degrees · Reasoning with degrees · Approximate reasoning · Approximate computation

The reasoning with flexible linguistic rules is a central technique of imprecise-information processing. On the basis of the previous Chaps. 6, 7, 11–14, this chapter will discuss the (approximate) reasoning and computation with flexible linguistic rules.

15.1 Natural Inference with Data Conversion(S)

The so-called natural inference is just usual reasoning. Although the symbols (linguistic values) in flexible rules are flexible, the reasoning with flexible rules is not different from that with general rules when the evidence fact matches completely with the premise of corresponding rule. However, since flexible linguistic values and numerical values can be converted mutually, the natural inference with flexible rules has its unique points. In fact, with data conversion(s), natural inference with flexible rules can be used in three ways:

① To do N-L conversion firstly then to do reasoning (as shown in Fig. 15.1);
② To do L-N conversion after reasoning (as shown in Fig. 15.2); and

© Springer Science+Business Media Singapore 2016   367
S. Lian, *Principles of Imprecise-Information Processing*,
DOI 10.1007/978-981-10-1549-6_15

Fig. 15.1 The diagram of
natural inference with N-L
conversion

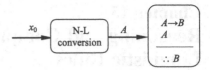

Fig. 15.2 The diagram of
natural inference with L-N
conversion

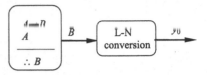

③ To do N-L conversion firstly then to do reasoning and to do L-N conversion
after reasoning (as shown in Fig. 15.3).

(In these figures, x_0 and y_0 are numerical values, and A and B are flexible linguistic
values.) In the following, we give some examples.

Example 15.1 Suppose there is a flexible rule of classifying apples set for a robot:

```
If an apple has a big size, symmetrical shape, bright color, and
smooth skin, then this apple is a superior apple.
```

And it is known that the observed values of size, shape, color, and skin of an apple
from the robot are separately a_1, a_2, a_3, and a_4. Question: Will the robot classify the
apple into the superior apple?

Solution It can be seen that because the linguistic values in antecedent of this
classifying rule are flexible linguistic values, while these observed values of the
apple from robot are numerical values, the observed values are needed to be converted into flexible linguistic values firstly before robot classifying the apple
according to this rule.

We denote flexible linguistic values "big," "symmetrical," "bright," "smooth,"
and "superior" one by one as A_1, A_2, A_3, A_4, and B, and then, the original rule can be
symbolized as $A_1 \wedge A_2 \wedge A_3 \wedge A_4 \rightarrow B$.

Suppose the consistency-degrees of numerical values a_1, a_2, a_3, and a_4 with the
corresponding flexible linguistic values "big," "symmetrical," "bright," and
"smooth" are one by one 1.0, 0.86, 1.05, and 0.98, that is, $c_{A_1}(a_1) = 1.0$,
$c_{A_2}(a_2) = 0.86$, $c_{A_3}(a_3) = 1.05$,   and $c_{A_4}(a_4) = 0.98$.   Clearly,   the   4

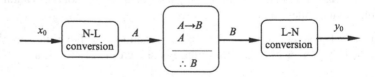

Fig. 15.3 The diagram of natural inference with both N-L and L-N conversions

consistency-degrees are all greater than 0.5. Thus, numerical values a_1, a_2, a_3, and a_4 can be converted separately into flexible linguistic values A_1, A_2, A_3, and A_4. And then, from the logic relation between A_1, A_2, A_3, and A_4, conjunctive flexible linguistic value $A_1 \wedge A_2 \wedge A_3 \wedge A_4$ follows. Of course, from

$$c_{A_1 \wedge A_2 \wedge A_3 \wedge A_4}(a_1, a_2, a_3, a_4) = \min\{c_{A_1}(a_1), c_{A_2}(a_2), c_{A_3}(a_3), c_{A_4}(a_4)\}$$
$$= \min\{1.0, 0.86, 1.05, 0.98\} = 0.86 > 0.5$$

the flexible linguistic value $A_1 \wedge A_2 \wedge A_3 \wedge A_4$ can also be obtained directly from numerical values a_1, a_2, a_3, and a_4.

Now, from fact $A_1 \wedge A_2 \wedge A_3 \wedge A_4$ and rule $A_1 \wedge A_2 \wedge A_3 \wedge A_4 \to B$, and according to *modus ponens*, the conclusion B follows. Thus, the robot will put the apple into the class of superior apples.

It can be seen that in this example, there is an N-L conversion before reasoning.

Example 15.2 Suppose there is a flexible rule used for robot driving:

    If the road becomes bad or visibility is down or an obstacle appears
    at not far ahead, then reduce speed appropriately.

And it is known that the sighted values of road condition, visibility, and distance ahead from the robot's eyes in car running separately are a_1, a_2, and a_3. Question: How should the robot now specifically operate?

Solution It can be seen that since the linguistic values in antecedent of the rule are flexible linguistic values, the sighted values from the robot's eyes should be converted into flexible linguistic values firstly before reasoning with this rule. Also since the linguistic value in the consequent of the rule is a flexible linguistic value, while the robot operating the car requires a precise quantity, the outcome (flexible linguistic value) would be needed to be converted into a numerical value after reasoning.

We denote flexible linguistic values "bad," "low," "not far," and "appropriately" one by one as A_1, A_2, A_3, and B, and then, the original rule is symbolized as $A_1 \vee A_2 \vee A_3 \to B$.

Suppose the consistency-degrees of numerical values a_1, a_2, and a_3 with the corresponding flexible linguistic values "bad," "low," and "not far" are one by one 0.15, 0.95, and 0, that is, $c_{A_1}(a_1) = 0.15$, $c_{A_2}(a_2) = 0.95$, and $c_{A_3}(a_3) = 0$. Obviously, only consistency-degree $c_{A_2}(a_2)$ is greater than 0.5. Thus, numerical value a_2 can be converted into flexible linguistic value A_2. But since A_2 implies $A_1 \vee A_2 \vee A_3$, disjunctive flexible linguistic value (fact) $A_1 \vee A_2 \vee A_3$ can still follow. Of course, from

$$c_{A_1 \vee A_2 \vee A_3}(a_1, a_2, a_3) = \max\{c_{A_1}(a_1), c_{A_2}(a_2), c_{A_3}(a_3)\}$$
$$= \max\{0.15, 0.95, 0\} = 0.95 > 0.5$$

the flexible linguistic value $A_1 \vee A_2 \vee A_3$ can also be obtained directly from numerical values a_1, a_2, and a_3.

Thus, from fact $A_1 \vee A_2 \vee A_3$ and rule $A_1 \vee A_2 \vee A_3 \rightarrow B$, and according to *modus ponens*, the conclusion B follows. That is, the robot should make the car slow down appropriately. But, how much is the specific operating quantity?

To obtain the specific operating quantity, the flexible linguistic value B must be converted into a number b. According to the conversion methods given in Sect. 7.3.1, theoretically speaking, any number in the extended core of flexible linguistic value B can be taken as the numerical value b. Thus, the robot specifically operates to reduce speed with this operating quantity b.

It can be seen that in this example, there is an N-L conversion before reasoning and an L-N conversion after reasoning.

Example 15.3 Suppose there is a flexible rule describing economic phenomena:

```
If bank notes are issued excessively plus supplies are not enough
plus there exists severe monopoly, then the price will rise
sharply.
```

And it is known that the amounts of banknote issuance, supplies, and assessment of management in a certain period separately are a_1, a_2, and a_3. Question: Will the price rise sharply?

Solution Similarly, the linguistic values in the antecedent of this rule are flexible linguistic values, while the relevant statistical data given in the problem are numerical values, so we should firstly convert these numerical values into flexible linguistic values.

We denote flexible linguistic values "(issued) excessively," "(supplies) not enough," "severe (monopoly)," and "(rising) sharply" one by one as A_1, A_2, A_3, and B, and then, the original rule can be symbolized as $A_1 \oplus A_2 \oplus A_3 \rightarrow B$.

Suppose the consistency-degrees of a_1, a_2, and a_3 with the corresponding flexible linguistic values "(issued) excessively," "(supplies) not enough," and "severe (monopoly)" are one by one 1.25, 0.78, and 1.36, that is, $c_{A_1}(a_1) = 1.25$, $c_{A_2}(a_2) = 0.78$, and $c_{A_3}(a_3) = 1.36$. And suppose the weights of the three consistency-degrees are separately 0.35, 0.25, and 0.40. Since

$$c_{A_1 \oplus A_2 \oplus A_3}(a_1, a_2, a_3) = 1.25 \times 0.35 + 0.78 \times 0.25 + 1.36 \times 0.40 = 1.1765 > 0.5$$

numerical values a_1, a_2, and a_3 can be converted into synthetic flexible linguistic value $A_1 \oplus A_2 \oplus A_3$.

Thus, from fact $A_1 \oplus A_2 \oplus A_3$ and rule $A_1 \oplus A_2 \oplus A_3 \rightarrow B$, and according to *modus ponens*, the conclusion B follows. That is, the price will rise sharply.

Similarly, there is also an N-L conversion before reasoning in the example. But there is no L-N conversion after reasoning. Of course, if required, an L-N conversion can also be done after reasoning. In fact, "(rise) sharply" can be converted into a real interval indicating the rising range, $[b_1, b_2]$ (i.e., the extended core of "sharply") or a number indicating specific degree, b (a number in the extended core).

All the reasoning in the above is for rules with single conclusion, while for rules with multiple conclusions, $A \rightarrow B_1 \wedge B_2 \wedge \cdots \wedge B_m$ and $A \rightarrow B_1 \vee B_2 \vee \cdots \vee B_m$ (where A is a simple proposition or a compound proposition $A_1 \wedge A_2 \wedge \cdots \wedge A_n$, $A_1 \vee A_2 \vee \cdots \vee A_n$, or $A_1 \oplus A_2 \oplus \cdots \oplus A_n$), the general form of corresponding natural inference is as follows:

$$\frac{\begin{array}{l} A \rightarrow B_1 \wedge B_2 \wedge \cdots \wedge B_m \\ A \end{array}}{\therefore B_1 \wedge B_2 \wedge \cdots \wedge B_m} \tag{15.1}$$

and

$$\frac{\begin{array}{l} A \rightarrow B_1 \vee B_2 \vee \cdots \vee B_m \\ A \end{array}}{\therefore B_1 \vee B_2 \vee \cdots \vee B_m} \tag{15.2}$$

Now, if it is required to convert conclusion of flexible linguistic value form $B_1 \wedge B_2 \wedge \cdots \wedge B_m$ or $B_1 \vee B_2 \vee \cdots \vee B_m$ into numerical value form, then firstly convert B_i into number $x_i (i = 1, 2, \ldots, m)$, respectively, and then join them by conjunction ($\wedge$) or disjunction ($\vee$) and express as

$$x_1 \wedge x_2 \wedge \cdots \wedge x_m$$

or

$$x_1 \vee x_2 \vee \cdots \vee x_m$$

From what is stated above, it can be seen that with data conversion(s), the natural inference with flexible rules can be applied to problem solving such as classifying, forecasting, decision, control, and so forth. And as long as the granule sizes of relevant flexible linguistic values are suitable, and the rules are proper, then the outcomes obtained will meet the requirements. Further, we see that a natural inference system with interfaces of both N-L and L-N conversions realizes actually a mapping from the measurement space of antecedent linguistic values of rules to the measurement space of consequent linguistic values. Therefore, the natural inference utilizing this approach is effectively the approximate evaluation of the background function of corresponding flexible rules. It is easy to see that the smaller the granule sizes of corresponding flexible linguistic values are, the higher is the precision of the approximate value obtained. Thus, this method can be used to construct a function approximator (as shown in Fig. 15.4). And in theory, such a function approximator can approximate any (non-chaotic) continuous function on a measurement space if the rules are many enough and the granule sizes of corresponding flexible linguistic values are small enough.

Fig. 15.4 The diagram of a function approximator formed by a natural inference system with data conversion interfaces

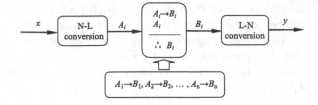

From the above, we can see that the natural inference with data conversion(s) can completely be applied for solving the relevant practical problems, so it should have a good application prospect. However, we find that it has also some shortcomings, mainly,

① The obtained reasoning result (flexible language value) is not further portrayed and subdivided.
② It is difficult to further improve the accuracy of the output numerical value in the case of the granule sizes of flexible linguistic values being invariant.
③ It is hard to realize the approximate reasoning when facts do not match completely with the premises of a rule.

For this reason, in the following, we will introduce the methods of reasoning and computation utilizing adjoint functions of a rule.

15.2   Reasoning with Truth-Degrees

We know that the near-true-UMP (short for Universal Modus Ponens) in truth-degreed logic (see Sect. 11.6.3) is

$$\frac{(A(x), t(A(x)) > 0.5) \rightarrow (B(y), t(B(y) > 0.5)}{A(x_0), \quad t(A(x_0)) > 0.5}}{\therefore B(y_0), t(B(y_0)) > 0.5} \tag{15.3}$$

that is

$$\frac{A(x)(\text{near-true}) \rightarrow B(y)(\text{near-true})}{A(x_0)(\text{near-true})}}{\therefore B(y_0)(\text{near-true})} \tag{15.3'}$$

and the rough-true-UMP in flexible-two-valued logic (see Sect. 12.6.2) is

$$\frac{A(x)(\text{rough-true}) \rightarrow B(y)(\text{rough-true})}{A(x_0)(\text{rough-true})}}{\therefore B(y_0)(\text{rough-true})} \tag{15.4}$$

But when reasoning, rough-true-UMP is really near-true-UMP. That is to say, when reasoning, "$A(x)$ rough-true $\rightarrow B(y)$ rough-true" is really "$A(x)$ near-true $\rightarrow B$ (y) near-true," i.e., "$(A(x), t(A(x)) > 0.5) \rightarrow (B(y)), t(B(y) > 0.5)$," while this expression implicates the correspondence relation or functional relation between truth-degrees $t(A(x))$ and $t(B(y))$. In Chap. 14, we examined specially this kind of relations and gave some reference models of approximate functions [1].

Let p_x: $A(x)$ $(x \in U)$ and q_y: $B(y)$ $(y \in V)$. Now, we consider flexible rule $p_x \rightarrow q_y$ at the level of proposition. Since there exists a truth-degreed functional relation between antecedent and consequent of rule $p_x \rightarrow q_y$, its adjoint truth-degreed function $t_{q_y} = f_t(t_{p_x})$ satisfies: when $t_{p_x} > 0.5$, $f_t(t_{p_x}) = t_{q_y} > 0.5$. Thus, from expression (15.3), we have the following argument form:

$$\frac{\begin{array}{ll} \text{(rule)} & (p_x, t_{p_x}) \rightarrow (q_y, f_t(t_{p_x})) \\ \text{(fact)} & (p_{x_0}, t_{p_0}) \end{array}}{\begin{array}{ll} \text{(result)} & (q_{y_0}, t_{q_0}) \end{array}} \qquad (15.5)$$

where $t_{p_x} = t(p_x)$, $f_t(t_{p_x}) = t_{q_y}$ is the adjoint truth-degreed function of rule $p_x \rightarrow q_y$, $t_{p_0} = t(p_{x_0})$ is the truth-degree of proposition p_{x_0} (fact), $t_{q_0} = f_t(t_{p_0})$ is the truth-degree of proposition q_{y_0} (conclusion).

Because the adjoint truth-degreed function of a rule is constrained by the logical semantics of the rule, when $t_{p_x} > 0.5$ necessarily $f_t(t_{p_x}) = t_{q_y} > 0.5$ (for details, see the analysis of graph space of rule's adjoint truth-degreed function in Sect. 14.2). Thus, from this, on the one hand, it can follow that $t(p_x \rightarrow q_y) > 0.5$, i.e., major premise $p_x \rightarrow q_y$ near-true; on the other hand, when, truth-degree of minor premise, $t_{p_0} > 0.5$, necessarily $t_{q_0} = f_t(t_{p_0}) > 0.5$. This shows that the argument form shown by the above expression (15.5) is a valid argument form in the sense of near-true. Thus, it can be taken as a rule of near-true inference. In consideration that the characteristic of this rule of inference is that computing of truth-degrees is added on the basis of near-true-UMP, and the t_{q_0} in reasoning result is specific truth-degree (unlike degree-true inference, near-true inference, and rough-true inference only knowing a scope of truth-degrees, that is, $t_{q_0} > 0, t_{q_0} > 0.5$, or q_{y_0} rough-true), so we call it to be truth-degree-level universal modus ponens, or **truth-degree-level-UMP** for short. And then, we call the reasoning following truth-degree-level-UMP to be **reasoning with truth-degrees**.

It can be seen that besides doing symbolic matching of the evidence fact to the antecedent of a rule, reasoning with truth-degrees also needs to judge whether $t_{p_0} > 0.5$, and if yes, then substituting t_{p_0} into $f_t(t_{p_x})$ and $f_t(t_{p_0}) = t_{q_0}$ follows; further, the result (q_{y_0}, t_{q_0}) of the argument is obtained. This is to say, in the process of reasoning with truth-degrees, judging and computing of truth-degrees accompanies judging and deducing of symbols. Therefore, the basic principle of reasoning with truth-degrees can also be simply expressed as follows:

deducing of propositional symbols + computing of truth − degrees

For simplicity, we rewrite the above expression (15.5), namely truth-degree-level-UMP, as the following form

$$\frac{\begin{array}{c}(p_x \rightarrow q_y; f_t(t_{p_x})) \\ (p_{x_0}, t_{p_0})\end{array}}{(q_{y_0}, t_{q_0})} \qquad (15.6)$$

The following reasoning with truth-degrees also all uses the representation form disjointing a rule and its truth-degreed function.

From the above statement, we see that reasoning with truth-degrees is logical. In fact, the truth-degree-level-UMP as basis comes down in one continuous line with rough-true-UMP, near-true-UMP, UMP, and MP, which is the further quantification of rough-true-UMP and near-true-UMP and can obtain the specific truth-degree of conclusion (this solves in a certain sense the problem that truth-degree-level exact inference cannot be done in truth-degreed logic), on the other hand, which also can be viewed as a generalization of traditional UMP. In fact, the truth-degree-level-UMP with truth-degreed function $t_{q_y} = t_{p_x}$, and when $t_{p_0} = 1$, is tantamount to traditional UMP.

Generally, the general scheme of reasoning with truth-degrees with the rule with multiple conditions is as follows:

$$\frac{\begin{array}{c}(\boldsymbol{p_x} \rightarrow q_y; f_t(t_{p_x})) \\ (\boldsymbol{p_{x0}}, t_{p_0})\end{array}}{(q_{y_0}, t_{q_0})} \qquad (15.7)$$

where $\boldsymbol{p_x} = p_{x_1} \wedge p_{x_2} \wedge \cdots \wedge p_{x_n}$, $p_{x_1} \vee p_{x_2} \vee \cdots \vee p_{x_n}$, or $p_{x_1} \oplus p_{x_2} \oplus \cdots \oplus p_{x_n}$, $f_t(t_{p_x})$ is the adjoint truth-degreed function of rule $\boldsymbol{p_x} \rightarrow q_y$, $\boldsymbol{p_{x0}} = p_{x_{10}} \wedge p_{x_{20}} \wedge \cdots \wedge p_{x_{n0}}$, $p_{x_{10}} \vee p_{x_{20}} \vee \cdots \vee p_{x_{n0}}$, or $p_{x_{10}} \oplus p_{x_{20}} \oplus \cdots \oplus p_{x_{n0}}$, $t_{p_0} = t(\boldsymbol{p_{x0}}) = \min\{t(p_{x_{10}}), t(p_{x_{20}}), \ldots, t(p_{x_{n0}})\}$, $\max\{t(p_{x_{10}}), t(p_{x_{20}}), \ldots, t(p_{x_{n0}})\}$, or $\sum_{i=1}^{n} w_i t(p_{x_{i0}})$ (from Sects. 11.2 and 11.3) and $t_{p_0} > 0.5$, and $t_{q_0} = t(q_{y_0}) = f_t(t_{p_0}) > 0.5$.

The argument following this scheme still judges firstly whether $t_{p_{a_0}} > 0.5$ if yes, then substitute $t_{p_{a_0}}$ into $f_t(t_{p_x})$, $f_t(t_{p_0}) = t_{q_0}$ follows; further, result (q_{y_0}, t_{q_0}) follows. But t_{p_0} is the overall truth-degree of facts $p_{x_{10}}, p_{x_{20}}, \ldots, p_{x_{n0}}$, so before judging whether $t_{p_0} > 0.5$, if t_{p_0} is not given, then we should find t_{p_0} firstly from the truth-degrees $t(p_{x_{10}}), t(p_{x_{20}}), \ldots, t(p_{x_{n0}})$ of component propositions.

Note: The above reasoning with truth-degrees is actually oriented to the flexible rules in negation-type logic, for the rules with linguistic values with opposite in opposite-type logic, and then, we need to judge whether $t_{p_0} > 0$ when reasoning and in the obtained result (q_{y_0}, t_{q_0}), and $t_{q_0} > 0$.

Note that the reasoning with truth-degrees here is not to compute the truth-degree of conclusion q_{y_0} by the truth-degrees of premises $p_x \rightarrow q_y$ and p_{x_0}. That is to say, the truth-degree of conclusion q_{y_0} is actually not directly related to that of premise $p_x \to q_y$. The reason is that as stated in 14.1, though inference rule $(p \to q) \wedge p \Rightarrow q$ is also a kind of production rule, it is actually a kind of meta-rules, whose semantics is that conclusion q is near-true when both premises $p \to q$ and p are near-true, but there is no corresponding computation formula or function of truth-degrees. While in actual argument, premise $p \to q$ is always assumed relatively true, that is, its truth-degree is bigger than 0.5; otherwise, the argument cannot be done.

15.3 Reasoning with Degrees

Expanding propositional forms p_x and q_y in rule $p_x \to q_y$ into possessive relation forms, we then have rule $A(x) \to B(y)$ $(x \in U, y \in V)$. Correspondingly, from the relation between adjoint truth-degreed function and adjoint degreed function of a rule (see Sect. 14.1), the adjoint truth-degreed function of rule $p_x \to q_y$ just is the adjoint degreed function of rule $A(x) \to B(y)$ numerically (and vice versa). Thus, from the truth-degree-level-UMP in the last section, we have the following inference form with degreed function and degreed computation:

$$
\begin{array}{ll}
(\text{rule}) & (A(x) \to B(y); f_d(d_A)) \\
(\text{fact}) & (A(x_0), d_{A_0}) \\
\hline
(\text{result}) & (B(y_0), d_{B_0})
\end{array}
\tag{15.8}
$$

where $f_d(d_A)$ is the adjoint degreed function of rule $A(x) \to B(y)$, d_{A_0} (i.e. $c_A(x_0)) > 0.5$ is the degree of linguistic value A in proposition $A(x_0)$ as evidence fact, d_{B_0} (i.e. $c_B(y_0)) = f_d(d_A) > 0.5$ is the degree of linguistic value B in proposition $B(y_0)$ as the result of reasoning.

It can be seen that this form of inference only is a version of the truth-degree-level-UMP. So it is equivalent to truth-degree-level-UMP. Since the numerical computation here is the computation about degrees, therefore, we call this inference form to be the degree-level universal modus ponens, or **degree-level-UMP** for short, and call the reasoning following degree-level-UMP to be the **reasoning with degrees**.

The semantics of UMP with degrees is as follows: if the degree d_A of x having A is >0.5, then degree of y having B is $f_d(d_A) > 0.5$; now, it is already known that the degree of some x_0 having A is d_{A_0} and $d_{A_0} > 0.5$, so there exits y_0, the degree of which having B is $d_{B_0} = f_d(d_A)$ and $d_{B_0} > 0.5$. Thus, The process of reasoning with degrees is firstly to do matching of symbolic patterns of the linguistic value as evidence fact to the linguistic value of rule's antecedent and then judge whether $d_{A_0} > 0.5$; if yes, then substitute d_{A_0} into $f_d(d_A)$, $d_{B_0} = f_d(d_{A_0})$ follows; further result (B, d_{B_0}) follows; from the range of degreed function (see Sect. 14.2), it is known that necessarily $d_{B_0} > 0.5$.

It can be seen that reasoning with degrees is actually a kind of inference at the level of the predicate. So, simply speaking, the basic principle of reasoning with degrees is **deducing of predicate symbols + computing of degrees**.

From Sect. 13.5, it is known that a flexible rule is essentially the correspondence between the linguistic values of antecedent and consequent. Thus, rule $A(x) \rightarrow B$ (y) is reduced as $A \rightarrow B$. Therefore, correspondingly, propositions $A(x_0)$ and $B(y_0)$ as evidence fact and inference result can be briefly denoted as linguistic values A and B, while two-tuples $(A(x_0), d_{A_0})$ and $(B(y_0), d_{B_0})$ be briefly denoted as linguistic values with degrees (A, d_{A_0}) and (B, d_{B_0}). Thus, the above expression (15.8) can be simplified as the following form:

$$\frac{(A \rightarrow B; \ f_d(d_A))}{\frac{(A, d_{A_0})}{(B, d_{B_0})}} \tag{15.8'}$$

From the statement above, we see that reasoning with degrees is actually a version of reasoning with truth-degrees, so it also logical. In fact, reasoning with degrees is the quantification of flexible predicate inference, which is a kind of logical inference with predicates that is both qualitative and quantitative. On the other hand, reasoning with degrees can also be viewed as a generalization of traditional inference with predicates. As a matter of fact, the reasoning with degrees with the degreed function $d_B = d_A$, and when degree $d_{A_0} = 1$, just is tantamount to traditional inference with predicate.

Generally, the general scheme of reasoning with degrees with the rule with multiple conditions/a composite linguistic value is as follows:

$$\frac{(A \rightarrow B; \ f_d(d_A))}{\frac{(A, d_{A_0})}{(B, d_{B_0})}} \tag{15.9}$$

where $A = A_1 \wedge A_2 \wedge \cdots \wedge A_n$, $A_1 \vee A_2 \vee \cdots \vee A_n$, or $A_1 \oplus A_2 \oplus \cdots \oplus A_n$, $f_d(d_A)$ is the degreed function of rule $A \rightarrow B$, $d_{A_0} = \min\{d_{A_{10}}, d_{A_{20}}, \ldots, d_{A_{n0}}\}$, $\max\{d_{A_{10}}, d_{A_{20}}, \ldots, d_{A_{n0}}\}$, or $\sum_{i=1}^{n} w_i d_{A_{i0}}$ (from t_{p_0} in expression (15.7)) and $d_{A_0} > 0.5$, $d_{A_{i0}}$ is the degree of $A_i (i = 1, 2, \ldots, n)$, and $d_{B_0} = f_d(d_{A_0}) > 0.5$.

The reasoning following this scheme is still firstly to do symbolic matching of the fact to the antecedent of a rule and then judges whether $d_{A_a} > 0.5$; if yes, then substitute d_{A_a} into $f_d(d_A)$, $d_{B_0} = f_d(d_{A_0})$ follows; further result (B, d_{B_0}) follows. But d_{A_0} is the overall degree of composite linguistic value A as a fact, so before judging whether $d_{A_0} > 0.5$, if d_{A_0} is not given, then we should find firstly the corresponding d_{A_0} from the degrees $d_{A_{10}}, d_{A_{20}}, \ldots, d_{A_{n0}}$ of component linguistic values.

Example 15.4 Let A_1, A_2, A_3, and B denote separately flexible linguistic values "(issued) excessively," "(supplies) not enough," "severe (monopoly)," and "(rise) sharply," and there are known facts: $(A_1, 1.25)$, $(A_2, 0.78)$, and $(A_3, 1.36)$. Try to do reasoning with degrees with flexible rule $A_1 \oplus A_2 \oplus A_3 \rightarrow B$.

Solution According to the general scheme of reasoning with degrees above, the reasoning with degrees to be done is

$$\frac{(A_1 \oplus A_2 \oplus A_3 \to B; \ f_d(d_A))}{(A_1 \oplus A_2 \oplus A_3, d_{A_0})}$$
$$\frac{}{(B, d_{B_0})}$$

It can be seen that to do this reasoning with degrees, it is needed to construct the adjoint degreed function $f_d(d_A)$ of rule $A_1 \oplus A_2 \oplus A_3 \to B$ and compose original evidence facts $(A_1, 1.25)$, $(A_2, 0.78)$, and $(A_3, 1.36)$ into $(A_1 \oplus A_2 \oplus A_3, d_{A_0})$.

For simplicity, we directly reference to the Eq. (14.14) in Sect. 14.3.1; let the adjoint degreed function $f_d(d_A)$ of rule $A_1 \oplus A_2 \oplus A_3 \to B$ be

$$d_B = \frac{\beta_B - 0.5}{\beta_\oplus - 0.5}(d_A - 0.5) + 0.5$$

where $\beta_\oplus = w_1 \beta_{A_1} + w_2 \beta_{A_2} + w_3 \beta_{A_3}$, $d_A = \sum_{i=1}^{3} w_i d_{A_i}$, $0 < d_{A_j} \le \beta_{A_j}$ $(j = 1, 2, 3)$.

Suppose the maximum of consistency-degrees of "(issued) excessively" is $\beta_{A_1} = 2.5$, the maximum of consistency-degrees of "(supplies) not enough" is $\beta_{A_2} = 1.8$, and the maximum of consistency-degrees of "severe (monopoly)" is $\beta_{A_3} = 2.0$, and from the weights given in Example 15.3, $w_1 = 0.35$, $w_2 = 0.25$, and $w_3 = 0.4$ and then $\beta_\oplus = \sum_{i=1}^{3} w_i \beta_{A_i} = 0.35 \times 2.5 + 0.25 \times 1.8 + 0.4 \times 2.0 \approx 2.1$; Also, suppose then the maximum of consistency-degrees of "(rise) sharply" is $\beta_B = 2.0$. Substituting the two numbers above into the above function expression, we have a actual adjoint degreed function $f_d(d_A)$ of rule $A_1 \oplus A_2 \oplus A_3 \to B$:

$$d_B = \frac{15}{16} d_A + \frac{1}{32}$$

And from the operation rules of flexible linguistic values with degrees (see Sect. 7.2),

$$(A_1, 1.25) \oplus (A_2, 0.78) \oplus (A_3, 1.36)$$
$$= (A_1 \oplus A_2 \oplus A_3, 1.25 \times 0.35 + 0.78 \times 0.25 + 1.36 \times 0.40)$$
$$= (A_1 \oplus A_2 \oplus A_3, 1.2)$$

Now, the major and minor premises of the reasoning with degrees separately are as follows:

$$\left(A_1 \oplus A_2 \oplus A_3 \to B; \ \frac{15}{16} d_A + \frac{1}{32} \right)$$

and

$$(A_1 \oplus A_2 \oplus A_3, 1.2)$$

It is clear that the composite linguistic value $A_1 \oplus A_2 \oplus A_3$ in the minor premise matches completely with the antecedent linguistic value in the major premise, and $d_{A_0} = 1.2 > 0.5$. Therefore, the corresponding reasoning with degrees can be done.

Substituting $d_A = 1.2$ into function $d_R = \frac{15}{16}d_A + \frac{1}{32}$, $d_B \approx 1.16$ follows. Consequently, the result of the reasoning with degrees is: $(B, 1.16)$. That is, the price will rise sharply, and the strength is 1.16.

Actually, for evidence facts $(A_1, 1.25)$, $(A_2, 0.78)$, and $(A_3, 1.36)$, if flexible rule is $A_1 \wedge A_2 \wedge A_3 \rightarrow B$ or $A_1 \vee A_2 \vee A_3 \rightarrow B$, the corresponding reasoning with degrees can also be done.

Finally, it should be noted that:

① The above reasoning with degrees is actually oriented to the flexible rules in negation-type logic, while for the rules with linguistic values with opposite in opposite-type logic, we need to judge whether $d_{A_0} > 0$ in reasoning, and d_{B_0} in reasoning result is also >0.

② Although the reasoning with degrees has its specific representation schemes and methods, if applied flexibly, it can be used to achieve reasoning in a variety of situations and of a variety of requirements, such as reasoning with data conversion(s), multistep reasoning, multipath reasoning, parallel reasoning, or even approximate reasoning and computation. These will be expounded in details in the following sections. The examples of the engineering applications of reasoning with degrees see "Sect. 18.6 Principle of flexible control with examples."

15.4 Reasoning with Degrees with Data Conversion(S)

Reasoning with degrees requires the evidence facts in premise to be flexible linguistic values with degrees, but in practical problems, some known evidence facts are numerical values. In such a case, we can convert numerical values into flexible linguistic values with degrees firstly and then do reasoning with degrees (as shown in Fig. 15.5). It can be seen that this is also the reasoning with degrees with N-Ld conversion.

For the method converting a numerical value x_0 into a linguistic value with degree, see Sect. 7.3.2. While for a group of numerical values as evidence facts, x_1, x_2, ..., x_n, it is needed to convert firstly x_1, x_2, ..., x_n into linguistic values with degrees (A_{1_i}, d_1), (A_{2_j}, d_2),..., (A_{n_k}, d_n), respectively, and then, according to the logic relation between them (that is, the structure type of the antecedent linguistic

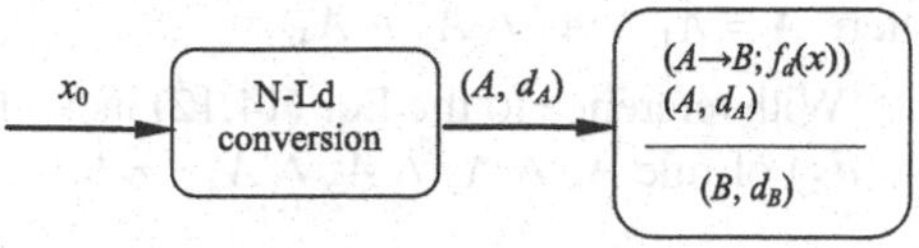

Fig. 15.5 Diagram of reasoning with degrees with N-Ld conversion

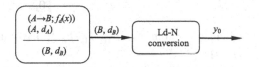

Fig. 15.6 Diagram of reasoning with degrees with Ld-N conversion

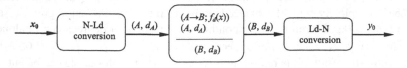

value of the corresponding rule), compose these linguistic values with degrees into a composite linguistic value with degree (for the composing rules, see Sect. 7.2).

The result of reasoning with degrees is a linguistic value with degree (B, d_B), but sometimes, we need to know what is the number (y_0) it corresponds to. In this case, we can convert linguistic value with degree (B, d_B) into a numerical value (as shown in Fig. 15.6). It can be seen that this is also the reasoning with degrees with Ld-N conversion.

For the specific method converting linguistic value with degree (B, d_B) into a numerical value, see Sect. 7.3.2.

The third case is that known evidence facts are numerical values and the final result after reasoning is also required to be a numerical value. In this situation, it is needed to convert relevant numerical values into flexible linguistic values with degrees before reasoning and to convert flexible linguistic values with degrees as reasoning result into a numerical value (as shown in Fig. 15.7). It can be seen that this is also the reasoning with degrees with both N-Ld and Ld-N conversions.

Actually, the Figs. 15.5, 15.6, and 15.7 above are additional three usages of reasoning with degrees (which are similar to the three usages of natural inference with flexible rule in previous Sect. 15.1).

Example 15.5 We use reasoning with degrees to solve the problem in Example 15.1 above.

Solution It can be seen that the corresponding reasoning with degrees is as follows:

$$(A_1 \wedge A_2 \wedge A_3 \wedge A_4 \rightarrow B; f_d(d_A))$$
$$\frac{(A_1 \wedge A_2 \wedge A_3 \wedge A_4, d_{A_0})}{(B, d_{B_0})}$$

the images show:

appears as part of Fig. 15.5 (x₀ → N-Ld conversion)
appears as part of Fig. 15.5

Fig. 15.7 Diagram of reasoning with degrees with both N-Ld and Ld-N conversions

here $A = A_1 \wedge A_2 \wedge A_3 \wedge A_4$.

With reference to the Eq. (14.12) in Sect. 14.3.1, let the adjoint degreed function $f_d(d_A)$ of rule $A_1 \wedge A_2 \wedge A_3 \wedge A_4 \to B$ be

$$d_B = \frac{\beta_B - 0.5}{\beta_\wedge - 0.5}(d_A - 0.5) + 0.5$$

where $\beta_\wedge = \min\{\beta_{A_1}, \beta_{A_2}, \beta_{A_3}, \beta_{A_4}\}$, $d_A = \min\{d_{A_1}, d_{A_2}, d_{A_3}, d_{A_4}\}$, $0.5 < d_{A_j} \le \beta_{A_j}$ $(j = 1, 2, 3, 4)$.

Suppose the maximum of consistency-degrees of "big" is $\beta_{A_1} = 1.5$, the maximum of consistency-degrees of "symmetrical" is $\beta_{A_2} = 1.1$, the maximum of consistency-degrees of "bright" is $\beta_{A_3} = 1.35$, and the maximum of consistency-degrees of "smooth" is $\beta_{A_4} = 1.15$. Then, $\beta_\wedge = \min\{\beta_{A_1}, \beta_{A_2}, \beta_{A_3} \beta_{A_4}\}$ $= \{1.5, 1.1, 1.35, 1.15\} = 1.1$. Suppose then the maximum of consistency-degrees of "superior" is $\beta_B = 1.2$. Substituting the above two numbers into the above function expression, we have an actual adjoint degreed function of rule $A_1 \wedge A_2 \wedge A_3 \wedge A_4 \to B$:

$$d_B = 1\frac{1}{6}d_A - \frac{1}{12}$$

And from $c_{A_1}(a_1) = 1.0, c_{A_2}(a_2) = 0.86, c_{A_3}(a_3) = 1.05$ and $c_{A_4}(a_4) = 0.98$ given in Example 15.1, we convert separately numerical values a_1, a_2, a_3, and a_4 into flexible linguistic values with degrees $(A_1, 1.0)$, $(A_2, 0.86)$, $(A_3, 1.05)$, and $(A_4, 0.98)$. And then, compose them according to the logic relation among them, and thus, it follows that conjunctive flexible linguistic value with degree:

$$(A_1 \wedge A_2 \wedge A_3 \wedge A_4, \min\{1.0, 0.86, 1.05, 0.98\}) = (A_1 \wedge A_2 \wedge A_3 \wedge A_4, 0.86)$$

Now, the major and minor premises of the reasoning with degrees separately are as follows:

$$\left(A_1 \wedge A_2 \wedge A_3 \wedge A_4 \to B; \ 1\frac{1}{6}d_A - \frac{1}{12} \right)$$

and

$$(A_1 \wedge A_2 \wedge A_3 \wedge A_4, 0.86)$$

Obviously, the composite linguistic value $A_1 \wedge A_2 \wedge A_3 \wedge A_4$ in minor premise matches completely with the antecedent linguistic value in major premise, and $d_{A_0} = 0.86 > 0.5$ Therefore, corresponding reasoning with degrees can be done.

Substituting $d_A = 0.86$ into degreed function $d_B = 1\frac{1}{6}d_A - \frac{1}{12}$, $d_B \approx 0.75$ follows. Thus, the result of the reasoning is $(B, 0.75)$. That is, this apple can be put into the class of superior apples, but its membership-degree is only 0.75.

Example 15.6 We use reasoning with degrees to solve the problem in Example 15.2 above.

Solution It can be seen that the corresponding reasoning with degrees is as follows:

$$\frac{(A_1 \vee A_2 \vee A_3 \rightarrow B; \ f_d(d_A))}{(A_1 \vee A_2 \vee A_3, d_{A_0})}$$
$$\overline{(B, d_{B_0})}$$

here $A = A_1 \vee A_2 \vee A_3$.

With reference to the Eq. (14.13) in Sect. 14.3.1, let the adjoint degreed function $f_d(d_A)$ of rule $A_1 \vee A_2 \vee A_3 \rightarrow B$ be

$$d_B = \frac{\beta_B - 0.5}{\beta_\vee - 0.5}(d_A - 0.5) + 0.5$$

where $\beta_\vee = \max\{\beta_{A_1}, \beta_{A_2}, \beta_{A_3}\}$, $d_A = \max\{d_{A_1}, d_{A_2}, d_{A_3}\}$, $0.5 < d_{A_j} \le \beta_{A_j}$ $(j = 1, 2, 3)$.

Suppose the maximum of consistency-degrees of "(road) bad" is $\beta_{A_1} = 1.5$, the maximum of consistency-degrees of "(visibility) low" is $\beta_{A_2} = 1.0$, and the maximum of consistency-degrees of "not far (ahead)" is $\beta_{A_3} = 1.2$. Then, $\beta_\vee = \max\{\beta_{A_1}, \beta_{A_2}\beta_{A_3}\} = \{1.5, 1.0, 1.2\} = 1.5$. Suppose then the maximum of consistency-degrees of "appropriately (reduce speed)" is $\beta_B = 1.25$. Substituting the above two numbers into the above function expression, we have an actual adjoint degreed function of rule $A_1 \vee A_2 \vee A_3 \rightarrow B$:

$$d_B = 1.5d_A - 0.25$$

And from $c_{A_1}(a_1) = 0.15$, $c_{A_2}(a_2) = 0.95$, and $c_{A_3}(a_3) = 0$ given in Example 15.2, we convert separately numerical values a_1, a_2, and a_3 into flexible linguistic values with degrees $(A_1, 0.15)$, $(A_2, 0.95)$, and $(A_3, 0)$. And then, compose them according to the logic relation, and thus, it follows that disjunctive flexible linguistic value with degree

$$(A_1 \vee A_2 \vee A_3, \max\{0.15, 0.95, 0\}) = (A_1 \vee A_2 \vee A_3, 0.95)$$

Now, the major and minor premises of the reasoning with degrees separately are as follows:

$$(A_1 \vee A_2 \vee A_3 \rightarrow B; \ 1.5d_A - 0.25)$$

and

$$(A_1 \vee A_2 \vee A_3, 0.95)$$

Clearly, the composite linguistic value $A_1 \vee A_2 \vee A_3$ in minor premise matches completely with the antecedent linguistic value in major premise, and $d_{A_0} = 0.95 > 0.5$. Therefore, corresponding reasoning with degrees can be done.

Substituting $d_A = 0.95$ into degreed function $d_B = 1.5\, d_A - 0.25$, $d_B \approx 1.18$ follows. Thus, the result of the reasoning is $(D, 1.18)$. That is, promptly reduce speed appropriately in strength 1.15.

And the specific operating quantity to which degree 1.18 corresponds should be obtained by converting flexible linguistic value with degree $(B, 1.18)$ (the converting method, see Sect. 7.3.2). In theory, the numerical value obtained by converting flexible linguistic value with degree $(B, 1.18)$ is more accurate as the numerical value obtained by converting pure flexible linguistic value B.

Example 15.7 We use reasoning with degrees to solve the problem in Example 15.3 above.

Solution It can be seen that the corresponding reasoning with degrees is as follows:

$$\frac{(A_1 \oplus A_2 \oplus A_3 \to B;\ f_d(d_A))}{(A_1 \oplus A_2 \oplus A_3, d_{A_0})}{(B, d_{B_0})}$$

here $A = A_1 \oplus A_2 \oplus A_3$.

But, this reasoning with degrees has been done in Example 15.4 above; here, we need only to convert numerical values from relevant statistical data, a_1, a_2, and a_3, into flexible linguistic values with degrees $(A_1, 1.25)$, $(A_2, 0.78)$, and $(A_3, 1.36)$. Besides, if required, the flexible linguistic value with degree as reasoning result, $(B, 1.16)$, can also be converted into a number indicating specific degree, b.

All the reasoning with degrees in the above is for rules with single conclusion, while for rules with multiple conclusions, $A \to B_1 \wedge B_2 \wedge \cdots \wedge B_m$ and $A \to B_1 \vee B_2 \vee \cdots \vee B_m$ (where A is a simple proposition or a compound proposition $A_1 \wedge A_2 \wedge \cdots \wedge A_n$, $A_1 \vee A_2 \vee \cdots \vee A_n$, or $A_1 \oplus A_2 \oplus \cdots \oplus A_n$); the general form of corresponding reasoning with degrees is as follows

$$\frac{(A \to B_1 \wedge B_2 \wedge \cdots \wedge B_m;\ f_{d_1}(d_A) \wedge f_{d_2}(d_A) \wedge \cdots \wedge f_{d_m}(d_A))}{(A, d_{A_0})}{(B_1, d_{B_1}) \wedge (B_2, d_{B_2}) \wedge \cdots \wedge (B_m, d_{B_m})} \tag{15.10}$$

and

$$\frac{(A \to B_1 \vee B_2 \vee \cdots \vee B_m; f_{d_1}(d_A) \vee f_{d_2}(d_A) \vee \cdots \vee f_{d_m}(d_A))}{(A, d_{A_0})}{(B_1, d_{B_1}) \vee (B_2, d_{B_2}) \vee \cdots \vee (B_m, d_{B_m})} \tag{15.11}$$

where $f_{d_i}(d_A)$ is the adjoint degreed function of rule $A \rightarrow B_i (i = 1, 2, ..., m)$.

Now, if it is required to convert the conclusion of flexible linguistic value with degree $(B_1, d_{B_1}) \wedge (B_2, d_{B_2}) \wedge \cdots \wedge (B_m, d_{B_m})$ or $(B_1, d_{B_1}) \vee (B_2, d_{B_2}) \vee \cdots \vee (B_m, d_{B_m})$ into numbers, we firstly convert flexible linguistic value with degree (B_i, d_{B_i}) into number $x_i (i = 1, 2, ..., m)$, respectively and then join them by conjunction ($\wedge$) or disjunction ($\vee$) and express as follows:

$$x_1 \wedge x_2 \wedge \cdots \wedge x_m$$

or

$$x_1 \vee x_2 \vee \cdots \vee x_m$$

From what is stated above, we can see that like natural inference with data conversion(s), the reasoning with degrees with data conversion(s) can also be applied to problem solving such as classifying, forecasting, decision, control, and so forth. And as long as the granule sizes of relevant flexible linguistic values are suitable, and the rules are proper, especially when the corresponding adjoint degreed functions of rules are also suitable, the outcomes obtained will meet the requirements.

15.5 Utilizing Reasoning with Degrees to Realize Approximate Reasoning and Approximate Computation

Reasoning with degrees is a kind of inference of exact symbol matching and degree computing, but utilizing reasoning with degrees, we can also realize the approximate reasoning of inexact symbol matching and the approximate computation (of course, if considering that rule's adjoint degreed function itself is approximate, then reasoning with degrees is also a kind of approximate reasoning).

15.5.1 Utilizing Reasoning with Degrees to Realize Approximate Reasoning

1. Basic principle

Here, the approximate reasoning refers to the inexact symbol matching inference of the following form:

$$\frac{\begin{array}{c} A \rightarrow B \\ A' \end{array}}{?} \tag{15.12}$$

Here, A' is an approximate value of A (as the superposed value of A). Viewed from appearance, this is not reasoning with degrees, but it is not hard to see that as long as A' can be converted into (A, d_{A_0}), reasoning with degrees can be used to realize this kind of approximate inference. As a matter of fact, after converting A' to (A, d_{A_0}), the above approximate reasoning can be changed to the reasoning with degrees of the form

$$\frac{\begin{array}{c}(A \rightarrow B;\ f_d(d_A)) \\ (A, d_{A_0})\end{array}}{(B, ?)} \tag{15.13}$$

As to the conversion from A' to (A, d_{A_0}), see the method of conversion from $A'(x_0)$ to $A(x_0)$ in Sect. 12.7. The conversion from pure linguistic value A' to linguistic value with degree (A, d_{A_0}) can be called L'-Ld conversion.

From reasoning with degrees, the deduced result of the above expression (15.13) should be (B, d_{B_0}). However, this kind of result generally does not conform to the requirement of approximate reasoning. Since viewed from given fact A', its deduced result should also be a B' that is approximate to B. Therefore, after reasoning with degrees, we still need to convert (B, d_{B_0}) into appropriate $(B', d_{B'_0})$. The method of this conversion is to find out firstly corresponding y_0 from (B, d_{B_0}) and then take y_0 as the peak value point to construct a flexible linguistic value with reference to the "size" of B; then, this flexible linguistic value can be treated as B' obtained. The conversion from linguistic value with degree (B, d_{B_0}) to pure linguistic value B' can be called Ld-L' conversion.

Thus, utilizing reasoning with degrees, we have realized inexact symbol matching approximate reasoning, whose principle and process are illustrated in Fig. 15.8.

2. **Approximate reasoning with the rule with multiple** conditions/a **composite linguistic value**

The above approximate reasoning is the approximate reasoning with the rule with single condition and atomic linguistic value, whose basic principles are also applicable to the approximate reasoning with the rule with multiple conditions/a composite linguistic value. However, from Sect. 6.8, it is known that the approximation relation between composite linguistic values is not that direct and simple like the approximation relation between atomic linguistic values, so when to

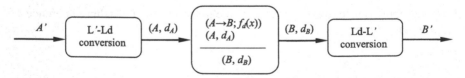

Fig. 15.8 Diagram of the principle of approximate reasoning realized by utilizing reasoning with degrees

conduct approximate reasoning with the rule with a composite linguistic value, the conversion from composite linguistic value A' as an evidence fact to the corresponding linguistic value with degree (A, d_{A_0}) is also relatively complex and troublesome. In fact, to convert composite linguistic value A' into the corresponding linguistic value with degree (A, d_{A_0}), we need firstly to convert its component linguistic values or ingredient values $A_1', A_2', \ldots, A_n'$ separately into the corresponding linguistic values with degrees $(A_1, d_{A_{10}})$ $(A_2, d_{A_{20}})$, ..., $(A_n, d_{A_{n0}})$ (since these linguistic values are already atomic linguistic values, the conversion method is the same as that the previous) and then implement corresponding composition of these linguistic values with degrees according to the operation types in the original composite linguistic value A'; the obtained composite linguistic value with degree is just the corresponding (A, d_{A_0}).

Specifically speaking, let $A_a' = A_1' \wedge A_2' \wedge \cdots \wedge A_n'$, whose conversion method is firstly to convert its component linguistic values $A_1', A_2', \ldots, A_n'$ separately into linguistic values with degrees $(A_1, d_{A_{10}}), (A_2, d_{A_{20}}), \ldots, (A_n, d_{A_{n0}})$ and then conduct conjunctive composition of the latter. By the composition rules in Sect. 7.2,

$$(A_1, d_{A_{10}}) \wedge (A_2, d_{A_{20}}) \wedge \cdots \wedge (A_n, d_{A_{n0}}) = (A_1 \wedge A_2 \wedge \cdots \\ \wedge A_n, \min\{d_{A_{10}}, d_{A_{20}}, \ldots, d_{A_{n0}}\})$$

Thus, the original conjunctive linguistic value $A_a' = A_1' \wedge A_2' \wedge \cdots \wedge A_n'$ is converted into

$$(A_1 \wedge A_2 \wedge \ldots \wedge A_n, \min\{d_{A_{10}}, d_{A_{20}}, \ldots, d_{A_{n0}}\})$$

Similarly, disjunctive linguistic value $A_o' = A_1' \vee A_2' \vee \cdots \vee A_n'$ is converted into

$$(A_1 \vee A_2 \vee \cdots \vee A_n, \max\{d_{A_{10}}, d_{A_{20}}, \ldots, d_{A_{n0}}\})$$

and synthetic linguistic value $A_s' = A_1' \oplus A_2' \oplus \cdots \oplus A_n'$ is converted into

$$\left(A_1 \oplus A_2 \oplus \ldots \oplus A_n, \sum_{i=1}^{n} w_i d_{A_{i0}} \right), \quad \sum_{i=1}^{n} w_i = 1$$

After composite linguistic value A' as an evidence fact is converted into corresponding linguistic value with degree (A, d_{A_0}), the approximate reasoning with the rule with a composite linguistic value also becomes the reasoning with degrees with the rule with a composite linguistic value in previous Sect. 15.3. As to the treatment of deduced result, it is the same as that of the previous approximate reasoning with the rule with single condition and atomic linguistic value.

3. Approximate reasoning with multirules

Actually, approximate reasoning in practical problems is often the following approximate reasoning with multirules:

$$A_1 \rightarrow B_1$$
$$A_2 \rightarrow B_2$$
$$\ldots$$
$$A_m \rightarrow B_m$$
$$\frac{A}{?} \tag{15.14}$$

where A_1, A_2, ..., A_m and A are flexible linguistic values on one-dimensional measurement space U, and B_1, B_2, ..., B_m, and B are flexible linguistic values on one-dimensional measurement space V (here, B_i and B_j ($i, j \in \{1, 2, ..., n\}$, $i \neq j$) may be the same), or more general approximate reasoning:

$$A_1 \rightarrow B_1$$
$$A_2 \rightarrow B_2$$
$$\ldots$$
$$A_m \rightarrow B_m$$
$$\frac{A}{?} \tag{15.15}$$

where $A_i = A_{i_1} \wedge A_{i_2} \wedge \cdots \wedge A_{i_n}$, $A_{i_1} \vee A_{i_2} \vee \cdots \vee A_{i_n}$, or $A_{i_1} \oplus A_{i_2} \oplus \cdots \oplus A_{i_n}$, A_i ($i = 1, 2, ..., m$), and A are composite flexible linguistic values on n-dimensional product space $U = U_1 \times U_2 \times \cdots \times U_n$.

For the approximate reasoning with multirules, we need firstly to examine which rule's antecedent linguistic value that the evidence fact A is closer to (actually, if A_1, A_2,..., A_n are all basic flexible linguistic values or composite linguistic values composed by basic flexible linguistic values, then A can only be approximate to one flexible linguistic value among them) and then choose the rule to do approximate reasoning. Suppose the chosen rule is $A_k \rightarrow B_k$, $k \in \{1, 2, ..., n\}$, and then, approximate reasoning with multirules just is reduced to the following approximate reasoning with single rule:

$$A_k \rightarrow B_k$$
$$\frac{A}{?} \tag{15.16}$$

The method to examine which rule's antecedent linguistic value that the evidence fact A being closer to is to compute the approximate degrees of A and the antecedent linguistic values of all rules (see Sect. 6.8), and then, one with the maximum approximate degree is the closest.

15.5.2 *Utilizing Reasoning with Degrees to Realize Approximate Computation*

From Sect. 15.4 above, we see that on the one hand, for the numerical evidence fact, we can convert it into a linguistic value with degree and then conduct reasoning with degrees; on the other hand, we can also convert the deduced result of a linguistic value with degree into a number. Then, couldn't the combination of the two just realize the approximate evaluation of the background function of a rule? That is to say, the approximate computation of numerical values can also be realized by utilizing reasoning with degrees. Actually, what is shown in Fig. 15.7 above is just the principle and process of this kind of approximate evaluation of a function. The principles of utilizing reasoning with degrees to realize approximate evaluation of a function can also be more specifically illustrated in Fig. 15.9.

Further, we see that a system of reasoning with degrees with interfaces of both N-L and L-N conversions realizes really a function approximator (as shown in

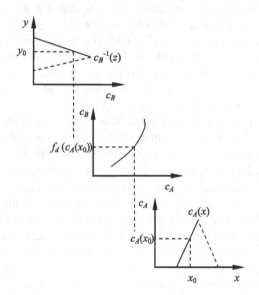

Fig. 15.9 Illustration of the principle of approximate evaluation utilizing reasoning with degrees

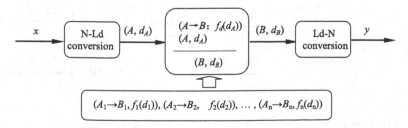

Fig. 15.10 The diagram of a function approximator formed by a system of reasoning with degrees with interfaces of data conversions

Fig. 15.10). And in theory, such a function approximator can also approximate any (non-chaotic) continuous function on a measurement space.

15.6 Parallel Reasoning with Degrees in the Sense of Degree-True

In Sect. 11.5.4, we discussed the deductive reasoning in the sense of "degree-true," i.e., degree-true inference. Now, we further consider the reasoning with degrees in the sense of degree-true.

Actually, speaking from inference scheme, the reasoning with degrees in the sense of degree-true has no difference with the reasoning with degrees in the sense of near-true above. However, in the reasoning with degrees in the sense of degree-true, the graph space of rule's adjoint degreed function is $[\alpha_A, \beta_A] \times [\alpha_B, \beta_B]$ (but not former $(0.5, \beta_A] \times (0.5, \beta_B]$ in near-true inference), and it requires degree d_{A_0} in evidence fact (A, d_{A_0}) to satisfy $d_{A_0} > 0$ (but not $d_{A_0} > 0.5$ in near-true inference). And then, if to convert flexible linguistic value (B, d_{B_0}) into number y_0, then the $y_0 \in [\alpha_B, \beta_B]$ (but not that $y_0 \in (0.5, \beta_B]$ in near-true inference).

That the degree d_{A_0} in the evidence fact (A, d_{A_0}) is >0 but not >0.5 in the reasoning with degrees in the sense of degree-true is such that numerical evidence facts x_0 or $x_1, x_2, \ldots, x_n$ may cause multiple flexible rules to be triggered simultaneously and form a parallel reasoning with degrees.

In fact, let $A_1, A_2, \ldots, A_n$ be a group of basic flexible linguistic values, it can be known from Sect. 7.3.2 that for the numerical value $x_0 \in \text{supp}(A_i) \cap \text{supp}(A_{i+1}) = (s_{A_{i+1}}^-, s_{A_i}^-)$ $(i \in \{1, 2, \ldots, n-1\})$, the corresponding degrees d_{A_i} and $d_{A_{i+1}}$ in flexible linguistic values with degree (A_i, d_{A_i}) and $(A_{i+1}, d_{A_{i+1}})$ obtained from converting x_0 are all >0. Thus, according to rule of degree-true inference, the corresponding flexible rules $A_i \to B_j$ and $A_{i+1} \to B_k$ will be triggered at the same time, thus forming a kind of parallel reasoning with degrees (as shown in Fig. 15.11).

Further, let U and V be two one-dimensional measurement spaces, $A_1, A_2, \ldots, A_n$ be a group of basic flexible linguistic values on U, and $B_1, B_2, \ldots, B_m$ be a group of basic flexible linguistic values on V, and let $x \in U$ and $y \in V$. Suppose that x can be converted into (A_i, d_{A_i}) and $(A_{i+1}, d_{A_{i+1}})$, in which d_{A_i} and $d_{A_{i+1}}$ satisfy separately

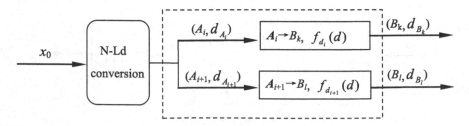

Fig. 15.11 Diagram of parallel reasoning with degrees in the sense of degree-true

$0 < d_{A_i} < 1$ and $0 < d_{A_{i+1}} < 1$, and y can be converted into (B_j, d_{B_j}) and $(B_{j+1}, d_{B_{j+1}})$, in which d_{B_j} and $d_{B_{j+1}}$ satisfy separately $0 < d_{B_j} < 1$ and $0 < d_{B_{j+1}} < 1$. Thus, when x and y as evidences are conjunctive relation, there can occur $2 \times 2 = 4$ conjunctive flexible linguistic values with degree: $(A_i, d_{A_i}) \wedge (B_j, d_{B_j})$, $(A_i, d_{A_i}) \wedge (B_{j+1}, d_{B_{j+1}})$, $(A_{i+1}, d_{A_{i+1}}) \wedge (B_j, d_{B_j})$ and $(A_{i+1}, d_{A_{i+1}}) \wedge (B_{j+1}, d_{B_{j+1}})$. While $(A_i, d_{A_i}) \wedge (B_j, d_{B_j}) = (A_i \wedge B_j, \min\{d_{A_i}, d_{B_j}\})$, $(A_i, d_{A_i}) \wedge (B_{j+1}, d_{B_{j+1}}) = (A_i \wedge B_{j+1}, \min\{d_{A_i}, d_{B_{j+1}}\})$, $(A_{i+1}, d_{A_{i+1}}) \wedge (B_j, d_{B_j}) = (A_{i+1} \wedge B_j, \min\{d_{A_{i+1}}, d_{B_j}\})$ and $(A_{i+1}, d_{A_{i+1}}) \wedge (B_{j+1}, d_{B_{j+1}}) = (A_{i+1} \wedge B_{j+1}, \min\{d_{A_{i+1}}, d_{B_{j+1}}\})$. Thus, flexible rules $A_i \wedge B_j \rightarrow C_{k_1}$, $A_i \wedge B_{j+1} \rightarrow C_{k_2}$, $A_{i+1} \wedge B_j \rightarrow C_{k_3}$, and $A_{i+1} \wedge B_{j+1} \rightarrow C_{k_4}$ can be triggered at the same time, thus realizing a parallel reasoning with degrees in which 4 rules are simultaneously executed.

Similarly, when x and y are disjunction or synthesis relation, there can also occur 4 composite flexible linguistic values with degree, respectively, and then, they can trigger simultaneously 4 corresponding flexible rules, respectively, forming parallel reasoning with degrees.

Generally, let U_1, U_2, ..., U_s be s one-dimensional measurement spaces. Suppose $x_i \in U_i$ ($i = 1, 2, ..., s$) can be converted into a pair of two flexible linguistic values with degree (A_{ij}, d_{ij}) and (A_{ij+1}, d_{ij+1}), in which the degrees d_{ij} and d_{ij+1} are all greater than 0. Thus, when x_1, x_2, ..., x_s are conjunction or disjunction or synthesis relation, there will occur $2 \times 2 \times \cdots \times 2 = 2^s$ composite flexible linguistic values with degree of the form

$$\left(A_{1_j} \wedge A_{2_k} \wedge \cdots \wedge A_{s_l}, \min\{d_{1_j} d_{2_k}, \ldots, d_{s_l}\}\right)$$

or

$$\left(A_{1_j} \vee A_{2_k} \vee \cdots \vee A_{s_l}, \max\{d_{1_j} d_{2_k}, \ldots, d_{s_l}\}\right)$$

or

$$\left(A_{1_j} \oplus A_{2_k} \oplus \cdots \oplus A_{s_l}, w_1 d_{1_j} + w_2 d_{2_k} + \ldots + w_s d_{s_l}\right), \quad \left(\sum_{i=1}^{s} w_i = 1\right)$$

and then, 2^s flexible rules of the form

$$A_{1_j} \wedge A_{2_k} \wedge \cdots \wedge A_{s_l} \rightarrow B_{w_1}$$

or

$$A_{1_j} \vee A_{2_k} \vee \cdots \vee A_{s_l} \rightarrow B_{w_2}$$

or

$$A_{1_j} \oplus A_{2_k} \oplus \cdots \oplus A_{s_l} \rightarrow B_{w_3}$$

would be triggered, forming parallel reasoning with degrees.

From the above statement, we see that for some numerical values, through N-L conversion, a parallel reasoning with degrees in the sense of degree-true can be realized. However, the parallel reasoning with degrees in the sense of degree-true has some shortcomings and limitations.

① Speaking from the principle, the parallel reasoning cannot always be performed, but only performed for those numerical values in the flexible boundary regions between corresponding flexible linguistic values;
② The logic basis of this parallel reasoning is degree-true inference, and the graph space of the adjoint degreed function of the rule used is the product of corresponding support sets, so the error of corresponding adjoint degreed function is larger. Consequently, the error of numerical value (y_0) obtained by conversion from the result of the reasoning is also larger.
③ The relationship between flexible linguistic values in various reasoning conclusions is relatively complex. They may be the same flexible linguistic value, but can also be different flexible linguistic values, and may be adjacent flexible linguistic values, but can also be non-adjacent flexible linguistic values. Therefore, to synthesize these conclusions and then convert it into a numerical value (y_0), we can merely make specific analyses to specific problems but not give a unified method.

15.7 Approximate Computation with the Adjoint Measured Functions of Rules

Although utilizing reasoning with degrees the approximate computation of the value of the background function of a rule can be realized, the adjoint measured function of a rule just is an approximate function of background function of the rule. If the adjoint measured function of a rule can be constructed or is known, then we can directly use it to conduct the approximate computation of the value of the background function of the rule. Since the adjoint measured function of a rule is only a local approximate function of relevant global background function on the universe of discourse, the first step of this kind of approximate computation is to choose the adjoint measured function of a corresponding rule according to known number x_0. After the function is determined, substitute x_0 into it and conduct computation. Thus, generally speaking, the approximate computing process with the adjoint measured functions of rules is **function choosing + computation** (as shown in Fig. 15.12).

Fig. 15.12 Diagram of approximate computing with the adjoint measured functions of rules

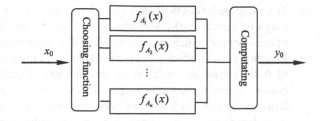

Since a measured function is dependent on a rule, here function choosing is closely connected to the rules. The specific choosing method is for a known number x_0, firstly judge which rule's antecedent linguistic value is more compatible with it [i.e., which consistency-degree is more greater; this is also tantamount to converting x_0 into a corresponding linguistic value A (for the conversion method, see Sect. 7.3.1)] and then take the adjoint measured function of the rule as the chosen measured function.

Of course, the corresponding measured function can also be directly determined from the region where x_0 locates. That is, the measured function is chosen whose domain is where x_0 locates. Since the domain of the adjoint measured function of a rule is the extended core of the corresponding flexible linguistic value, by Theorem 5.2, this is actually tantamount to still choosing measured function from rules.

Since the function choosing is always decided by the corresponding flexible rule, this kind of approximate computation with the adjoint measured functions of rules is actually "navigated" by rules and reasoning with rules. Or in other words, this kind of approximate computation with the adjoint measured functions of rules is actually a kind of numerical computation based on logical inference.

For the application examples of approximate computation with the adjoint measured functions of rules, see "Sect. 18.6 Principle of flexible control with examples."

15.8 Summary

In this chapter, we expounded, in the frame of logic, the principles and methods of reasoning with flexible linguistic rules. First, we presented natural inference with data conversion(s) according to the relation between numerical values and flexible linguistic values. Then, we presented the principles and methods of reasoning with truth-degrees and reasoning with degrees based on rule's adjoint functions, on the basis of rough-true (near-true) inference. And we discussed reasoning with degrees with data conversion(s) and approximate reasoning and computation based on reasoning with degrees. In addition, we discussed parallel reasoning with degrees.

The main points and results of the chapter are as follows:

- Natural inference with data conversion(s) is a simple and easy method of reasoning and approximate computation with flexible rules about imprecise information.

- Truth-degree-level-UMP as a rule of inference is the further quantification of rough-true (near-true)-UMP, which is the rough-true (near-true)-UMP at the level of truth-degree; which also can be viewed as a generalization of traditional UMP.
- Reasoning with truth-degrees is a kind of exact propositional inference at the level of truth-degree and realized by the method of deducing propositional symbols + computing of truth-degrees in the framework of truth-degree-level-UMP. Reasoning with truth-degrees solves at least theoretically the problem that truth-degree-level exact inference cannot be realized in truth degreed logic.
- Degree-level-UMP is a version of truth-degree-level-UMP, which is the UMP at the level of degree (consistency-degree), which can also be viewed as a kind of generalization of traditional UMP.
- Reasoning with degrees is a version of reasoning with truth-degrees, which is a kind of exact predicate inference at the level of degree and realized by the method of deducing predicate symbols + computing of degrees in the framework of degree-level-UMP. Reasoning with degrees further quantifies traditional qualitative predicate inference, or say it is a combination of qualitative and quantitative reasoning. From the performing process, it is the numerical computation under guidance of logical inference and the logical inference supported by numerical computation.
- Reasoning with degrees has many usage ways; it can be used for multistep reasoning and multipath reasoning, and combined with data conversion(s), it can be adapted to various practical problems. In particular, utilizing reasoning with degrees, we can realize approximate reasoning and approximate computing.
- Natural inference with data conversion(s) and reasoning with degrees simulate "percepts-thinking-response" process of human brain; they have a good prospect of application in the fields of intelligent robots, intelligent agents, expert systems, knowledge-based systems, decision support systems, and natural language processing.
- For some numerical values, through N-L conversion, a parallel reasoning with degrees in the sense of degree-true can be realized. However, the parallel reasoning with degrees has shortcomings and limitations.

Reference

1. Lian S (2009) Principles of imprecise-Information processing. Science Press, Beijing

Chapter 16
Approximate Evaluation of Flexible Linguistic Functions

Abstract This chapter analyzes and reveals the approximate evaluation principles of flexible linguistic functions from the angle of mathematics and gives the corresponding methods. Besides, it discusses approximate computation utilizing the approximate evaluation of flexible linguistic functions and comments on the traditional fuzzy logic system.

Keywords Flexible linguistic functions · Approximate evaluation · Interpolation of linguistic function

The so-called approximate evaluation of flexible linguistic functions is to find the value of a flexible linguistic function in the situation of unknowing its total expressing. This is also often encountered in practical problems just like the approximate evaluation of numerical functions. We know that there are already many theories and methods for the approximate evaluation of the numerical functions. Then, what are the theories and methods for the approximate evaluation of the flexible linguistic functions? This chapter will discuss this problem.

Firstly a statement:

① The approximate relation between flexible linguistic values discussed in this chapter is the approximate relation defined in Sect. 6.8.
② Since the correlation is actually also a kind of multivalued function, so for the convenience of narrating, in the following we call uniformly the background function and background correlation of a flexible linguistic function to be the background function and denote them by one and the same symbol.

16.1 Approximate Evaluations in Two Cases

There are approximate evaluations of the flexible linguistic functions in two cases.

© Springer Science+Business Media Singapore 2016
S. Lian, *Principles of Imprecise-Information Processing*,
DOI 10.1007/978-981-10-1549-6_16

1. Approximate evaluation with single pair of corresponding values

That is, known a pair of corresponding values, (A, B), satisfying certain flexible linguistic function $Y = f(X)$ and approximate value A' of the assumed value A of independent variable to find corresponding function value $B' = f(A')$. The geometrical representation of this kind of approximate evaluation (with respect to the typical linguistic functions) is by a known point (X_A, Y_B) on block point curve (surface or hypersurface) $Y = f(X)$ (X_A and Y_B are the extended cores of A and B) and $X_{A'}$ closing to horizontal coordinate X_A of the point to determine the vertical coordinate $Y_{B'}$ of another point $(X_{A'}, Y_{B'})$ on this curve (surface or hypersurface).

2. Approximate evaluation with multiple pairs of corresponding values

That is, known a group of pairs of corresponding values, (A_1, B_1), (A_2, B_2), ..., (A_m, B_m), satisfying certain flexible linguistic function $Y = f(X)$ and an assumed value A of independent variable to find the corresponding function value $B = f(A)$. The geometrical representation of this kind of approximate evaluation (with respect to typical flexible linguistic functions) is by a group of known points (X_{A1}, Y_{B1}), (X_{A2}, Y_{B2}), ..., $X_{Am}, Y_{Bm})$ on block point curve (surface or hypersurface) $Y = f(X)$ and horizontal coordinate X_A of point (X_A, Y_B) to determine vertical coordinate Y_B of the point.

16.2 Approximate Evaluation of a Monovariate Flexible Linguistic Function with Single Pair of Corresponding Values

16.2.1 Basic Issues and Ideas

Firstly let us consider the approximate evaluation of the monovariate typical flexible linguistic function.

Suppose a pair of corresponding values, (A, B), satisfying flexible linguistic function $Y = f(X)$ on one-dimensional measurement space U is known, and the flexible linguistic value $A' \subset U$ is approximate to A, now it is asked to find function value $B' = f(A')$. From the mathematical essence and numerical model of a flexible linguistic function, that is tantamount to say, known a point (X_A, Y_B) (X_A and Y_B are the extended cores of A and B) on block point curve $Y = f(X)$ in two-dimensional space $U \times V$, and $X_{A'}$ closing to horizontal coordinate X_A of this point, ask to determine vertical coordinate $Y_{B'}$ of another point $(X_{A'}, Y_{B'})$ on this curve.

Actually, this kind of approximate evaluation problem of the linguistic functions is completely similar to the problem finding function value $y'_0 = f(x'_0)$ from a known pair of corresponding values. (x_0, y_0), satisfying numerical function $y = f(x)$ and x'_0 closing to assumed value x_0 of independent variable.

From this it is not hard to see that since block point $(X_{A'} = Y_{B'})$ is also on curve $Y = f(X)$, and its horizontal coordinate $X_{A'}$ is close to horizontal coordinate X_A of block point (X_A, Y_B) on curve $Y = f(X)$, when the background function of function $Y = f(X)$ is not discontinuous or chaos, vertical coordinate $Y_{B'}$ of block point $(X_{A'},$ $Y_{B'})$ should be also close to vertical coordinate Y_B of (X_A, Y_B). Thus, to find $Y_{B'}$, such problems as the orientation of $Y_{B'}$ relatively to Y_B, distance of $Y_{B'}$ to Y_B and the size of $Y_{B'}$ would be involved. In other words, if the orientation of $Y_{B'}$ relatively to Y_B, the distance of $Y_{B'}$ to Y_B and the size of $Y_{B'}$ are all known, then $Y_{B'}$ would be completely determined. Since flexible linguistic value B' should be approximate to B, so B' actually is the approximate value of B. Thus, from our definition of the approximate relation of flexible linguistic values, the sizes of $Y_{B'}$ and Y_B are totally the same. Next, we analyze and discuss how the problem of the orientation and distance of $Y_{B'}$ is to be solved [1].

1. Orientation of $Y_{B'}$ and refining of a linguistic function

It can be seen that similar to the situation of numerical functions, the orientation of $Y_{B'}$ relatively to Y_B is decided by the orientation of $X_{A'}$ relatively to X_A and the property of linguistic function $Y = f(X)$. In fact,

① If $Y = f(X)$ is monotone increasing at point X_A, then when $X_{A'}$ is located at the negative side of X_A, then $Y_{B'}$ is located at the negative side of Y_B; when $X_{A'}$ is located at the positive side of X_A, then $Y_{B'}$ is located at the positive side of Y_B.

② If $Y = f(X)$ is monotone decreasing at point X_A, then $X_{A'}$ is located at the negative side of X_A, then $Y_{B'}$ is located at the positive side of Y_B; when $X_{A'}$ is located at the positive side of X_A, then $Y_{B'}$ is located at the negative side of Y_B.

③ If $Y = f(X)$ is convex at point X_A, then no matter what side $X_{A'}$ is located at X_A, $Y_{B'}$ is always located at the negative side of Y_B; If $Y = f(X)$ is concave at point X_A, then no matter what side $X_{A'}$ is located at X_A, $Y_{B'}$ is always located at the positive side of Y_B.

Since the orientation between flexible linguistic values is defined by the orientation between their peak-value points, so the orientation of $X_{A'}$ at X_A can be determined by the orientation of peak-value point $\xi_{A'}$ at ξ_A. Thus, orientation of $X_{A'}$ at X_A would be easy to determine. However, how can we know the property of flexible linguistic function $Y = f(X)$ at X_A? Obviously, if the whole block point curve $Y = f(X)$ is known, or at least the adjacent block points of block point (X_A, Y_B) are known, then the property of function $Y = f(X)$ at X_A can be determined. But there is only one known block point of (X_A, Y_B) now. Therefore, the property of function $Y = f(X)$ at point X_A cannot be determined directly at the level of the block point curve. Then, we penetrate deep within block point (X_A, Y_B) to analyze the property of corresponding background function $f_{AB}(x)$. Since although block point (X_A, Y_B) is small, it is still a region space, so curve $y = f_{AB}(x)$ should have a certain run in (X_A, Y_B). Obviously, based on background function $f_{AB}(x)$ having different properties, when $X_{A'}$ closes to X_A, there would be $Y_{B'}$s from different orientations to correspond to Y_B. As shown in Fig. 16.1, for one and the same horizontal coordinate $X_{A'}$, with respect to background function $f_1(x)$, the corresponding vertical

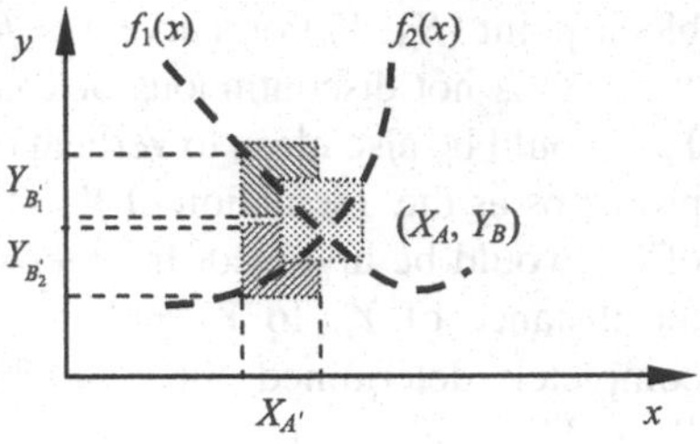

Fig. 16.1 An illustration of
the relation between
background functions and
orientations of $Y_{B'}$

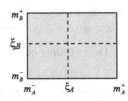

Fig. 16.2 Image of block
point (X_A, Y_B)

coordinate is $Y_{B'_1}$, but with respect to background function $f_2(x)$, the corresponding
vertical coordinate is $Y_{B'_2}$.

Next we take block point (X_A, Y_B) to which the pair of corresponding values,
(A, B), corresponds whose two values are both full-peak values (as shown in
Fig. 16.2) as an example to analyze the possible properties of background function
$f_{AB}(x)$ in this small region.

From the figure, it can be seen that the peak-value points of full-peak values
A and B divide block point $X_A \times Y_B$ into four parts. For the convenience of narrating, we refer to the semi-region at the negative side of peak-value point ζ_A in X_A
as the negative semi-region of X_A, denote X_A^-, and refer to the semi-region at the
positive side of peak-value point ζ_A in X_A as the positive semi-region of X_A, denote
X_A^+. Similarly, we refer to the semi-region at the negative side of peak-value point
ζ_B in Y_B as the negative semi-region of Y_B, denote Y_B^-, and refer to the semi-region
at the positive side of peak-value point ζ_B in Y_B as the positive semi-region of Y_B,
denote Y_B^+.

Now suppose for background function $f_{AB}(x)$, there is a group of data points
$(x_i, y_i) \in X_A \times Y_B (i = 1, 2, \ldots n)$ which can be treated as sample data [which can be
the original sample data of generating valued pair (A, B)]. Then, if for all $x_k \in X_A^-$,
always $y_k = f_{AB}(x_k) \in Y_B^+$, and for all $x_l \in X_A^+$, always $y_l = f_{AB}(x_l) \in Y_B^-$, then the
distributive scope of data points (i.e., valued pairs) of background function $f_{AB}(x)$ in
region $X_A \times Y_B$ is roughly shown in Fig. 16.3a (the gray region is the distributive
scope of the data points, which is the same below). If for all $x_k \in X_A^-$, always
$y_k = f_{AB}(x_k) \in Y_B^-$, and for all $x_l \in X_A^+$, always $y_l = f_{AB}(x_l) \in Y_B^+$, then the distributive scope of data points of background function $f_{AB}(x)$ in region $X_A \times Y_B$ is
roughly shown in Fig. 16.3b. The rest can be reasoned by analogy. Thus, as shown
in Fig. 16.3, the data points of background function $f_{AB}(x)$ have totally 9 distribution situations in region $X_A \times Y_B$.

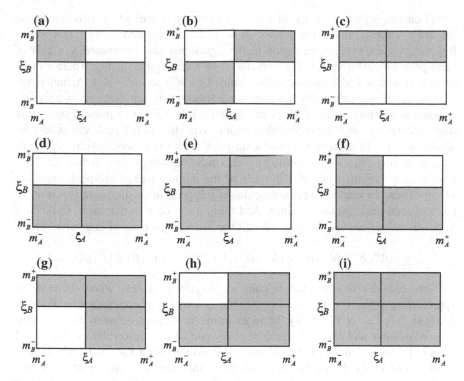

Fig. 16.3 An analysis of the distribution situation of data points of background function $f_{AB}(x)$ of a rule

From the figure, we can visually see all the possible distribution situations of data points of background function $f_{AB}(x)$ in region $X_A \times Y_B$, and which implicates all possible runs of the function curve of single-valued background function in region $X_A \times Y_B$. For instance, of them, (a) is obviously decreasing, (b) is increasing, (d) implicates concave, (e) implicates convex, etc. It can be seen that these properties of background function $f_{AB}(x)$ are helpful for determining the orientation of $Y_{B'}$. In fact, for Fig. 16.3a, when $X_{A'}$ is located at the negative side of X_A, $Y_{B'}$ can only be located at the positive side of Y_B; when $X_{A'}$ is located at the positive side of X_A, $Y_{B'}$ can only be located at the negative side of Y_B. For Fig. 16.3b; when $X_{A'}$ is located at the negative side of X_A, then $Y_{B'}$ can only be located at the negative side of Y_B; when $X_{A'}$ is located at the positive side of X_A, then $Y_{B'}$ can only be located at the positive side of Y_B. For Fig. 16.3c, no matter which side $X_{A'}$ is located at X_A, $Y_{B'}$ is always located at the negative side of Y_B. For Fig. 16.3d, no matter which side $X_{A'}$ is located at X_A, $Y_{B'}$ is always located at the positive side of Y_B. Then, the problem now is how to indicate all kinds of properties of background function $f_{AB}(x)$ so that the orientation of $Y_{B'}$ can be easily determined in evaluation or inference.

From Fig. 16.3, it can be further seen that if only semi-regions $X_A^- \times Y_B$ and $X_A^+ \times Y_B$ are considered, then the 9 kinds of possible background functions

$f_{AB}(x)$ on full-region $X_A \times Y_B$ all can be separately described by two semi-peak linguistic valued pairs. As a matter of fact, from the mathematical essence of linguistic values, every semi-region in the figure just can be denoted as a pair of semi-peak linguistic values. For instance, two semi-regions in Fig. 16.3a can be separately denoted as linguistic valued pairs (A^-, B^+) and (A^+, B^-). Actually, the way of describing summarily a background function by semi-regions is also tantamount to splitting a pair of full-peak linguistic values into two pairs of semi-peak linguistic values. From the reducibility of approximation of full-peak values and the single direction of approximation of semi-peak values (see Sect. 6.8), two pairs of semi-peak linguistic values formed by decomposing single pair of full-peak linguistic values together just play the role of the original pair of full-peak linguistic values. Thus, the original 9 possible pairs of full-peak linguistic values become 18 pairs of semi-peak linguistic values. And then, it can be seen that these 18 pairs of linguistic values can also be reduced into the following 6 pairs of linguistic values:

$$(A^-, B^-), (A^-, B^+), (A^+, B^-), (A^+, B^+), (A^-, B), (A^+, B)$$

Now consider that for the 6 pairs of linguistic values, when there is an approximate value A' close to A^- or A^+, the orientation of corresponding $Y_{B'}$ relatively to Y_{B^-}, Y_{B^+} or Y_B. Firstly, let us examine the relation between the two signs (that is, positive and negative signs) of every pair of linguistic values in the first 4 pairs of linguistic values above and the orientation between $Y_{B'}$ and Y_B when $X_{A'}$ is approximate to X_A. It can be seen that since the signs of former and latter two semi-peak values in a pair of semi-peak linguistic values are just the signs of semi-region Y_B^- or Y_B^+ where the two semi-peak values locate, and the approximation to a semi-peak value is from single direction, so for each one of these 4 pairs of linguistic values, the sign of first linguistic value is just consistent with the sign of orientation of approximate linguistic value A', and the sign of second linguistic value is just consistent with the sign of orientation of $Y_{B'}$ needing to be determined. That is to say, the two signs of former and latter linguistic values of a pair of semi-peak linguistic values also indicate the orientations of $X_{A'}$ and $Y_{B'}$ meanwhile.

For the pair of linguistic values whose second linguistic value is full-peak value, we can set $Y_{B'}$ and Y_B to overlap, that is, $Y_{B'} = Y_B$. In fact, as shown in Fig. 16.4 (take pair of linguistic values, (A^-, B), as an instance), since the range of background function is full-region Y_B, no matter what side $Y_{B'}$ is located at Y_B, $Y_{B'}$ cannot totally cover the range Y_B of background function $f_{AB}(x)$ on X_A^-, which means that corresponding block point $(X_{A'}, Y_{B'})$ cannot cover the curve of background function $y = f_{AB}(x)$. That obviously would cause a very large error, so it's undesirable. And to make $Y_{B'}$ cover completely the range of $f_{AB}(x)$ on X_A^-, we then can only make $Y_{B'}$ and Y_B overlap.

Thus, for the above-stated 6 pairs of linguistic values, the determination and indication of the orientation of $Y_{B'}$ are both solved.

In the above, we solve the determination and the indication of the orientation of $Y_{B'}$ by splitting a pair of corresponding values of a linguistic function (i.e., a rule). Actually, since approximating to a semi-peak value is single direction, so for the

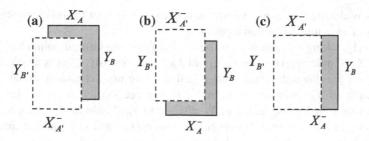

Fig. 16.4 An illustration of the relation between orientations of $Y_{B'}$ and Y_B

linguistic valued pair in which second linguistic value B is a semi-peak value, the orientation of $Y_{B'}$ relatively to Y_B is consistent with the sign of this semi-peak value B. That is to say, we can do not split those pairs of linguistic values whose first linguistic value is a full-peak value while second is a semi-peak value.

The split of a pair of linguistic values is also the refining of the corresponding linguistic function, which is also the more meticulous and exact summarization and approximation of the corresponding background function. In fact, from Fig. 16.4, it can be seen that except for the last case (i), the graph scopes of the background functions in the previous 8 cases all reduce a little.

2. Distance from $Y_{B'}$ to Y_B

Now we consider how to find the distance from $Y_{B'}$ to Y_B. For this problem, we firstly examine the relation between distances $d_{BB'}$ and $d_{AA'}$.

Since block points (X_A, Y_B) and $(X_{A'}, Y_{B'})$ are on one and the same block point curve, so in the situation of the background function of function $Y = F(X)$ being not discontinuous and chaos, $Y_{B'}$ should also be close to Y_B when $X_{A'}$ is close to X_A. That is to say, distances $d_{BB'}$ and $d_{AA'}$ are related in a way. Thus, as long as distance $d_{AA'}$ is known, distance $d_{BB'}$ can be estimated. However, distance is an absolute difference after all, and since $d_{BB'}$ and $d_{AA'}$ belong to different universe of discourse, so it is relatively difficult to directly assess the specific functional relation between the two. In consideration of

$$D_{W'W} = \frac{d_{W'W}}{r_W}, \quad s_{W'W} = 1 - D_{W'W}$$

(here $D_{W'W}$ and $s_{W'W}$ are separately the difference-degree and approximation-degree between linguistic values W' and W, $r_W = \xi_W - m_W^-$ or $m_W^+ - \xi_W$ is the approximate radius (see Definition 6.19)); therefore, the relation between difference-degrees $D_{B'B}$ and $D_{A'A}$ and the relation between approximation-degrees $s_{B'B}$ and $s_{A'A}$ are consistent with the relation between distances $d_{BB'}$ and $d_{A'A}$. Thus, we can turn to consider the relation between $s_{B'B}$ and $s_{A'A}$. Then, what relation should be there between $s_{B'B}$ and $s_{A'A}$? Of course, it can be various correspondence relations, such as linear function, nonlinear function. But in consideration of that the size of region

$X_A \times Y_B$ is already very small, so we can approximately treat the relation between s_B '$_B$ and $s_{A'A}$ as a linear functional relation.

Actually, taking a dynamic view, point (X_A, Y_B) is the limit of point $(X_{A'}, Y_{B'})$, so $Y_{B'}$ and $X_{A'}$ should approximate to Y_B and X_A synchronously. That is to say, for any pair $(X_{A'}, Y_{B'})$, although distances $d_{BB'}$ and $d_{AA'}$ are not necessarily equal and the step length of $Y_{B'}$ approximating to Y_B is not necessarily equal to that of $X_{A'}$ approximating to X_A, the degree of $Y_{B'}$ closing to Y_B should be the close to that of $X_{A'}$ closing to X_A. And from the relation between $D_{Y'Y}$ and $s_{Y'Y}$, and for simplicity, we directly consider the relation between $D_{B'B}$ and $D_{A'A}$. Let

$$D_{B'B} = kD_{A'A}$$

Further, from $D_{B'B} = \frac{d_{B'B}}{r_B}$ and $D_{A'A} = \frac{d_{A'A}}{r_A}$, we have

$$\frac{d_{B'B}}{r_B} = k\frac{d_{A'A}}{r_A}$$

Consequently,

$$d_{B'B} = k\frac{d_{A'A}}{r_A} \cdot r_B \tag{16.1}$$

This is the computation formula of distance $d_{B'B}$, where k is an adjustable parameter.

Particularly, when take $k = 1$, we have

$$d_{B'B} = d_{A'A}\frac{r_B}{r_A} \tag{16.2}$$

This equation is also tantamount to taking

$$D_{B'B} = D_{A'A}$$

that is, the computation formula of $d_{B'B}$ when $s_{B'B} = s_{A'A}$.

$$d_{B^{-\prime}B^-} = d_{A^{-\prime}A^-}\frac{r_{B^-}}{r_{A^-}}$$

Of course, this distance formula is a universal formula, which includes all 4 kinds of correspondence relations of linguistic values A, B, A' and B' as semi-peak values. For instance, for the pair of corresponding values, (A^-, B^-), and approximate value $A^{-\prime}$, the computation formula of corresponding $d_{B^-\prime B^-}$ (using formula 16.2) is

$$d_{B^{-1}B^-} = d_{A^{-1}A^-} \frac{r_{B^-}}{r_{A^-}}$$

As to when function values B and B' are full-peak value, since we set directly $Y_{B'} = Y_B$, so there is no need to compute distance $d_{B'B}$.

16.2.2   AT Method of Approximate Evaluation

In the above, we analyzed and discussed three problems of the orientation, distance, and size of $Y_{B'}$ and gave corresponding thinking and method for the solution. Further, it is not hard to see that when the orientation and distance of $Y_{B'}$ are obtained, actually we do not need to find out $Y_{B'}$, the support set and core of linguistic value B' can be deduced, certainly its consistency function can also be obtained.

In fact, let

$$\text{supp}(B') = \left(s_{B'}^-, s_{B'}^+\right)$$
$$\text{core}(B') = \left[c_{B'}^-, c_{B'}^+\right]$$

where s_B^- and s_B^+ are two critical points of flexible linguistic value B, c_B^- and c_B^+ are two core–boundary points of B, $s_{B'}^-$ and $s_{B'}^+$ are two critical points of flexible linguistic value B', $c_{B'}^-$ and $c_{B'}^+$ are two core–boundary points of B'. Suppose that distance $d_{B'B}$ has already been obtained by formula (16.1). Since the size of supp (B') and the size of supp(B) are the same, the size of core(B') and the size of core (B) are the same.

When B is a negative semi-peak value, that is, $Y_{B'}$ is located at the negative side of Y_B, we have

$$s_{B'}^- = s_B^- - d_{B'B}, \quad s_{B'}^+ = s_B^+ - d_{B'B} \tag{16.3}$$

$$c_{B'}^- = c_B^- - d_{B'B}, \quad c_{B'}^+ = c_B^+ - d_{B'B} \tag{16.4}$$

When B is a positive semi-peak value, that is, $Y_{B'}$ is located at the positive side of Y_B, we have

$$s_{B'}^- = s_B^- + d_{B'B}, \quad s_{B'}^+ = s_B^+ + d_{B'B} \tag{16.5}$$

$$c_{B'}^- = c_B^- + d_{B'B}, \quad c_{B'}^+ = c_B^+ + d_{B'B} \tag{16.6}$$

When B is a full-peak value, that is, $Y_{B'} = Y_B$, we have

$$s_{B'}^- = s_B^-, ; \; s_{B'}^+ = s_B^+ \tag{16.7}$$

$$c_{B'}^- = c_B^-, \;\; c_{B'}^+ = c_B^+ \tag{16.8}$$

With the critical points and core–boundary points of B', its consistency function $c_{B'}(y)$ and membership function $m_{B'}(y)$ can be directly written out, but $c_{B'}(y)$ and $m_{B'}(y)$ of B' can also be indirectly obtained through doing translation transformation of consistency function $c_B(y)$ and membership function $m_B(y)$ of B. The concrete procedure is as follows:

When B is a negative semi-peak value, that is, $Y_{B'}$ is located at the negative side of Y_B, set

$$c_{B'}(y) = c_B(y + d_{B'B}) \tag{16.9}$$

When B is a positive semi-peak value, that is, $Y_{B'}$ is located at the positive side of Y_B, set

$$c_{B'}(y) = c_B(y - d_{B'B}) \tag{16.10}$$

When B is a full-peak value, that is, $Y_{B'} = Y_B$, set

$$c_{B'}(y) = c_B(y) \tag{16.11}$$

With consistency function $c_{B'}(y)$, peak-value point $\xi_{B'}$ is implied in them. But we can also directly find peak-value point $\xi_{B'}$ from distance $d_{B'B}$. In fact,

When B is a negative semi-peak value, that is, $Y_{B'}$ is located at the negative side of Y_B,

$$\xi_{B'} = \xi_B - d_{B'B} \tag{16.12}$$

When B is a positive semi-peak value, that is, $Y_{B'}$ is located at the positive side of Y_B,

$$\xi_{B'} = \xi_B + d_{B'B} \tag{16.13}$$

When B is a full-peak value, that is, $Y_{B'} = Y_B$,

$$\xi_{B'} = \xi_B \tag{16.14}$$

It can be seen that the above several groups of equations for finding the critical points, core–boundary points, consistency function and peak-value points of flexible linguistic value B' are tantamount to do translation transformation separately of the corresponding elements of flexible linguistic value B. Therefore, we call these translation transformations as doing translation transformation of linguistic value B.

To sum up the above results, from a pair of corresponding values, (A, B), satisfying flexible linguistic function $Y = f(X)$ and A' approximating to A (A and A' are both semi-peak value), the corresponding function value $f(A') = B'$ can be obtained by using the following steps and method:

(1) Find out peak-value points ξ_A and $\xi_{A'}$ of flexible linguistic values A and A', find distance $d_{A'A}$ between extended cores $X_{A'}$ and X_A by ξ_A and $\xi_{A'}$;
(2) Compute distance $d_{B'B}$ by using formula (16.1) or (16.2);
(3) According to the correspondence relation of orientations decided by valued pair (A, B), determine the orientation of $Y_{B'}$ relatively to Y_B, then choose appropriate formulae from Eqs. (16.3)–(16.8) to find critical points $s_{B'}^-$ and $s_{B'}^+$, and core–boundary points $c_{B'}^-$ and $c_{B'}^-$ of linguistic value B', and obtaining the extended core of B', and then write out consistency function $c_{B'}(y)$ of B' (if requiring); or doing directly translation transforming of consistency function $c_B(y)$ of linguistic value B with Eqs. (16.9)–(16.11) to obtain $c_{B'}(y)$ of B' Of course, here B'.

Of obtained is only an approximate value of $F(A')$.

Thus, we have solved the approximation evaluation problem of flexible linguistic functions with one pair of corresponding values.

It can be seen that in this approximate evaluation method there are two key techniques: the one is finding approximate-degree $s_{B'B}$ from approximate-degree $s_{A'A}$, which is tantamount to "transmitting" the approximate-degree of linguistic values A' and A to linguistic values B and B' through correspondence relation (A, B); another key technique is translation transformation. Therefore, we call this method to be the approximate evaluation method using approximate-degree transmission and translation transformation, or AT method for short.

Note: although this method is named AT method, actually only difference-degree is used in it rather than approximate-degree. Such naming is because the approximate-degree and the difference-degree can be derived from one another, and approximate-degree the formulation is more appropriate with approximate evaluation.

16.2.3 An Analysis of Rationality and Effectiveness of AT Method

Actually, since the AT method is obtained based on the numerical model (representative) of a flexible linguistic function, so it is naturally reasonable. In the following we do further analysis.

Firstly, we use the relation between distances $d_{B'B}$ and $d_{A'A}$ to derive the approximate function of background function summarized by corresponding pair of corresponding values.

As stated above, when B is a negative semi-peak value, that is, $Y_{B'}$ is located at the negative side of Y_B,

$$\xi_{B'} = \xi_B - d_{B'B}$$

while

$$d_{B'B} = d_{A'A} \frac{r_B}{r_A} \quad (\text{here take } k = 1)$$

From the above two equations, it follows that

$$\xi_{B'} = \xi_B - d_{A'A} \frac{r_B}{r_A} \tag{16.15}$$

Thus, for valued pair (A^-, B^-), then $d_{A'A} = \xi_A - \xi_{A'}$ and $\frac{r_B}{r_A} = \frac{\xi_B - m_B^-}{\xi_A - m_A^-}$; thus,

$$\xi_{B'} = \xi_B - (\xi_A - \xi_{A'}) \frac{\xi_B - m_B^-}{\xi_A - m_A^-}$$

It can be seen that ξ_A, ξ_B, m_A^-, and m_B^- at the right-hand side of the above equation are all known. Thus, by using this equation, $\xi_{B'}$ can be directly obtained from $\xi_{A'}$. Then, when peak-value point $\xi_{A'}$ varies in X_A^-, peak-value point $\xi_{B'}$ also follows to vary in Y_B^-. That is to say, when peak-value points $\xi_{A'}$ and $\xi_{B'}$ are viewed as variables, the above equation is just the expression of function between $\xi_{A'}$ and $\xi_{B'}$. Thus, using variables x and y to replace $\xi_{A'}$ and $\xi_{B'}$ in the above equation, we have

$$y = \xi_B - (\xi_A - x) \frac{\xi_B - m_B^-}{\xi_A - m_A^-}, \quad m_A^- \le x \le \xi_A \tag{16.16}$$

Thus, we obtain a function defined on the X_A^-. It can be seen that this is a linear function, whose graph is a line segment connecting points (ξ_A, ξ_B) and $(m_A^- m_B^-)$. And this line segment is just a diagonal of rectangular region $X_A^- \times Y_B^-$ (as shown in Fig. 16.7a). That shows that for any $x' \in X_A^-$, substitute which into expression (16.16), necessarily the found $y' \in Y_B^-$.

On the other hand, region $X_A^- \times Y_B^-$ is also the graph space of background function $f_{A-B-}(x)$ of valued pair (A^-, B^-); thus, necessarily $f_{A-B-}(x') \in Y_B^-$. Therefore, $|y' - f_{A-B-}(x')| < |Y_B^-|$, which shows that y' and $f_{A-B-}(x')$ should be approximate. Consequently, function (16.16) is the approximate function of background function $f_{A-B-}(x)$ of valued pair (A^-, B^-).

Similarly, we can also derive the corresponding approximate functions of valued pairs (A^-, B^+), (A^+, B^-), (A^+, B^+), (A^-, B) and (A^+, B) as follows, their graphs are shown in Fig. 16.5.

$$y = \xi_B + (\xi_A - x) \frac{m_B^+ - \xi_B}{\xi_A - m_A^-}, \quad m_A^- \le x \le \xi_A \tag{16.17}$$

$$y = \xi_B - (x - \xi_A) \frac{\xi_B - m_B^=}{m_A^+ - \xi_A}, \quad \xi_A \le x \le m_A^+ \tag{16.18}$$

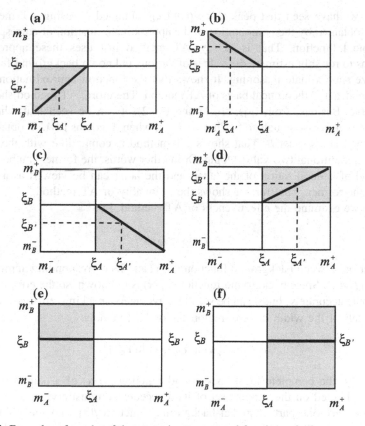

Fig. 16.5 Examples of graphs of the approximate measured functions of rules

$$y = \xi_B + (x - \xi_A)\frac{m_B^+ - \xi_B}{m_A^+ - \xi_A}, \quad \xi_A \leq x \leq m_A^+ \qquad (16.19)$$

$$y = \xi_B, \quad m_A^- \leq x \leq \xi_A \qquad (16.20)$$

$$y = \xi_B, \quad \xi_A \leq x \leq m_A^+ \qquad (16.21)$$

Readers may have found that here these approximate measured functions are also the kind of function similar to the expression (14.16) with only some difference in the expression form. Thus, by using these approximate measured functions, we can perform approximate computation of the background functions of rules.

Now we have seen that peak-value point $\xi_{B'}$ obtained by using AT method is actually obtained by the computation of the approximate function of corresponding background function. That is to say, AT method just uses these approximate functions to roughly estimate the values of various unknown background functions to realize approximate reasoning. It shows that the function approximation is just the basic idea and theoretical basis of AT method. Therefore, AT method should be reasonable. Besides, from Eqs. (16.1) and (16.2), it can be seen that when $d_{A'A}$ equals to 0, $d_{B'B}$ also equals to 0. That is to say, when A' equals to A, B' obtained by AT method is also just B. That shows AT method is compatible with the modus ponens in traditional two-valued logic. Or in other words, the former can be viewed as a kind of generalization of the latter, and the latter can be viewed as a special case of the former. That further shows the rationality of AT method.

Now we examine the effectiveness of AT method. Let

$$e = \left| f_{AB}(x) - f_{AB}^*(x) \right|$$

be the error between background function $f_{AB}(x)$ and its corresponding approximate function $f_{AB}^*(x)$. Since background function $f_{AB}(x)$ is unknown, so the error e is hard to compute accurately, but it can be seen that averagely speaking, error e should not exceed half of the width of semi-region Y_B^- or Y_B^+, namely

$$e \leq \max\left\{ widt(Y_B^-), widt(Y_B^+) \right\} \tag{16.22}$$

Actually, the complete local background function $f_{AB}(x)$ of a pair of full-peak values is defined on the support set of its antecedent linguistic value A, which is originally only one part of global background function $f(x)$, and the AT method splits a pair of full-peak values into two pairs of semi-peak values, which is tantamount to dividing local function $f_{AB}(x)$ into two sections, and then further reduces these two function sections to on two semi-extended cores of A. That is to say, background function $f_{AB}(x)$ and its approximate function $f_{AB}^*(x)$ we now talk and discuss about are only one small part of a complete background function, whose domain is already relatively narrow. Therefore, function $f_{AB}(x)$ itself is very likely monotonous and even linear. Then, for such function $f_{AB}(x)$, we use linear function $f_{AB}^*(x)$ to approximate it, then error e would not be very large. Even if function $f_{AB}(x)$ is still nonlinear and non-monotonous, but because parameter k in Eq. (16.1) is adjustable, additionally the limitation of the scope of values, the error should also be limited and controllable. In fact, even if formula (16.2) is used to obtain $d_{B'B}$, theoretically speaking, for any error requirement $\varepsilon > 0$, as long as $widt()$ is sufficiently small, we can make $e < \varepsilon$. Of course, to make $widt()$ sufficiently small, then extended cores X_A and Y_B must be both sufficiently small.

16.3 Approximate Evaluation of a Multivariate Flexible Linguistic Function with Single Pair of Corresponding Values

Now on the basis of approximate evaluation principle of monovariate typical flexible linguistic functions, we further study the approximate evaluation of multivariate typical flexible linguistic functions.

16.3.1 Issues and Ideas of the Approximate Evaluation of Multivariate Flexible Linguistic Functions

The problem of the approximate evaluation of a multivariate flexible linguistic function is also how to determine the orientation and distance of $Y_{B'}$. Since independent variable's assumed value A in the pair of corresponding values, (A, B), of a multivariate flexible linguistic function is a conjunctive composite flexible linguistic value, so the problem about the orientation and distance of corresponding $Y_{B'}$ is relatively complex.

1. Splitting of the valued pairs and the orientation of $Y_{B'}$

Firstly, we split all pairs of corresponding values of a flexible linguistic function into the valued pairs in which independent variables' assumed values are all the conjunctions of semi-peak values. For instances, $(A_1^- \wedge A_2^+ \wedge \cdots \wedge A_n^-, B^+)$, $(A_1^- \wedge A_2^+ \wedge \cdots \wedge A_n^+, B^-), \ldots, (A_1^+ \wedge A_2^+ \wedge \cdots \wedge A_n^-, B)$. Thus, the multivariate flexible linguistic functions in the following are just the ones whose independent variables' assumed values are all semi-peak linguistic values and whose function values are semi-peak values or full-peak values.

Since a linguistic value has two semi-peak values: a positive one and a negative one, the valued pair in which independent variable's value is the conjunctive value of n full-peak values can be split into 2^n valued pairs in which independent variables' assumed values are the conjunctive values of semi-peak values, and a total of all these possible valued pairs is 3×2^n. Of them, every valued pair summarizes the background function on the corresponding subregion in n-dimensional extended core X_A. The correspondence relation of signs of linguistic values in every valued pair just indicates the correspondence relation of orientations of the corresponding $X_{A'}$ and $Y_{B'}$. That is, when $X_{A'}$ is located at the Θ side of X_A, (Θ is a n-ary repeatable arrangement of *, * $\in \{+, -\}$), the corresponding $Y_{B'}$ is located at the negative side, positive side of Y_B or overlaps with Y_B. That is to say, the orientation of $Y_{B'}$ relatively to Y_B is completely decided by the corresponding valued pair directly.

Extended core X_A of the full-peak conjunctive value A is an n-dimensional region, from the relation between a linguistic value and its extended core, the splitting of the valued pair stated above is tantamount to divide extended core X_A

Fig. 16.6 An illustration of the division of two-dimensional extended core $X_{A_1} \times X_{A_2}$ and their orientations

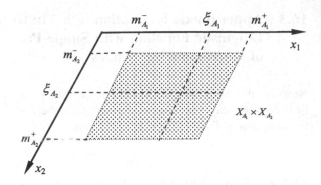

into 2^n subregions according to the peak-value points (two-dimensional extended core and its division of subregions are shown in Fig. 16.6), every subregion is the extended core of a semi-peak conjunctive value. That means the extended core of a full-peak conjunctive value has 2^n orientations altogether, every orientation is a repeatable arrangement of the corresponding n signs. For instance, two-dimensional extended core X_A as shown in Fig. 16.6 is divided into 4 subregions, so there occur 4 orientations, namely ++, −+, +−, and −−.

It needs to be explained that although extended core $X_{A'}$ of an approximate value can approach X_A from any direction around X_A, the descriptions of orientations of $X_{A'}$ relatively to X_A can only be limited 2^n. That is because a composite linguistic value on a multidimensional space is different from an atomic linguistic value on a multidimensional space. As thus, for composite linguistic value A and atomic flexible linguistic value A on an n-dimensional space, though corresponding approximate values can approach separately them from any direction around them, for composite linguistic value A, the corresponding approaching can only described by orientations, while for atomic linguistic value A, the corresponding approaching can be described by directions. Obviously, the former description is rough, while the latter description is exact.

2. Computation of distance

Since here independent variable's value A is a conjunctive linguistic value composed by n semi-peak linguistic values, whose extended core is the intersection of n orthogonal extended cores, and the (description of) orientation of the extended core is an arrangement of orientations of n one-dimensional extended cores, the distance, difference-degree, and approximate-degree of the extended core of A can only be, respectively, defined by the distance, difference-degree, and approximate-degree of n one-dimensional extended cores, or in other words, which can only be defined in the sense of its coordinate components, but not by the whole of the extended core, that is, the peak-value point vector. And from the definitions of the approximate relation and approximate-degree of composite linguistic values, the approximate relation and approximate-degree between two composite linguistic values are just the approximate relation and approximate-degree between their extended cores.

Let $A = A_1 \wedge A_2 \wedge \cdots \wedge A_n$, and $A' = A'_1 \wedge A'_2 \wedge \cdots \wedge A'_n$, where $A, A' \subset U = U_1 \times U_2 \times \cdots \times U_n$, A_i and $A'_i (i = 1, 2, \ldots, n)$ are both semi-peak value, X_A and $X_{A'}$ are separately the extended cores of A and A'. Similarly, here also take

$$s_{B'B} = k \cdot s_{A'A}$$

that is,

$$D_{B'B} = k \cdot D_{A'A}$$

and by Eq. (6.23), it should follows that

$$D_{A'A} = \max\left\{ D_{A'_1 A_1}, D_{A'_2 A_2}, \ldots, D_{A'_n A_n} \right\}$$

On the other hand, $D_{B'B} = \dfrac{d_{B'B}}{r_B}$. Thus,

$$\frac{d_{B'B}}{r_B} = k \cdot \max\left\{ D_{A'_1 A_1}, D_{A'_2 A_2}, \ldots, D_{A'_n A_n} \right\}$$

Consequently,

$$d_{B'B} = k \cdot \max\{ D_{A'_1 A_1}, D_{A'_2 A_2}, \ldots, D_{A'_n A_n} \} \cdot r_B$$
$$= k \cdot \max\{ \frac{d_{A'_1 A_1}}{r_{A_1}}, \frac{d_{A'_2 A_2}}{r_{A_2}}, \ldots, \frac{d_{A'_n A_n}}{r_{A_n}} \} \cdot r_B \tag{16.23}$$

16.3.2 AT Method of Approximate Evaluation of Multivariate Flexible Linguistic Functions

Based on the above analysis on the approximate evaluation problem of the multivariate flexible linguistic functions, we can have the AT method of the corresponding approximate evaluation.

Let $A = A_1 \wedge A_2 \wedge \cdots \wedge A_n$, and $A' = A'_1 \wedge A'_2 \wedge \cdots \wedge A'_n$, where $A, A' \subset U = U_1 \times U_2 \times \cdots \times U_n$,, $B \subset V$, $A_i, A'_i (i = 1, 2, \ldots, n)$ and B are all one-dimensional semi-peak values. Then, the AT method of approximate evaluation of a multivariate flexible linguistic function $Y = F(X)$ is as follows:

(1) Find out distances $d_{A_1 A'_1}, d_{A_2 A'_2}, \ldots, d_{A_n A'_n}$ between the corresponding components of linguistic valued vectors A and A';
(2) Find distance $d_{BB'}$ by formula (16.23);

(3) Determine the orientation of $Y_{B'}$ relatively to Y_B according to the correspondence relation between orientations decided by valued pair (A, B), then choose
appropriate formulas from Eqs. (16.3)–(16.8) do corresponding translation
transformation of critical points s_B^- and s_B^+ and core–boundary points c_B^- and
c_B^+ of linguistic value B to obtain critical points $s_{B'}^-$ and $s_{B'}^+$ and core–boundary
points $c_{B'}^-$ and $c_{B'}^+$ of linguistic value B', and obtaining the extended core of B',
and then write out consistency function $c_{B'}(y)$ of linguistic value B'
(if requiring); or do directly translation transformation of consistency function
$c_B(y)$ of linguistic value B with Eqs. (16.9) (16.11) to obtain consistency
function $c_{B'}(y)$ of linguistic value B'.

Next, we make a simple analysis of the rationality and effectiveness of this AT
method of approximate evaluation of multivariate flexible linguistic functions.
Firstly, taking the pair of corresponding values of a bivariate flexible linguistic
function as an example, we examine the approximate function of its background
function.

As stated previously, when B is a negative semi-peak value, that is, $Y_{B'}$ is located
at the negative side of Y_B,

$$\xi_{B'} = \xi_B - d_{B'B}$$

while

$$d_{B'B} = d_{A_k'A_k}\frac{r_B}{r_{A_k}} \quad \text{(here take } k = 1)$$

From the above two equations, we have

$$\xi_{B'} = \xi_B - d_{A_k'A_k}\frac{r_B}{r_{A_k}} \tag{16.24}$$

For the valued pair $(A_1^+ \wedge A_2^+, B^-)$, suppose that $\max\{D_{A_1'A_1}, D_{A_2'A_2}\} = D_{A_2'A_2}$,
then,

$$d_{A_k'A_k}\frac{r_B}{r_{A_k}} = \left(\xi_{A_2}' - \xi_{A_2}\right)\frac{\xi_B - m_B^-}{m_{A_2}^+ - \xi_{A_2}}$$

Thus,

$$\xi_{B'} = \xi_B - \left(\xi_{A_2}' - \xi_{A_2}\right)\frac{\xi_B - m_B^-}{m_{A_2}^+ - \xi_{A_2}}$$

It can be seen that ξ_B, ξ_{A_2}, $m_{A_2}^+$ and m_B^- on the right-hand side of the above
equation are all known. Thus, when peak-value point $\left(\xi_{A_1'}, \xi_{A_2'}\right)$ varies in

positive-positive subregion of X_A, peak-value point $\xi_{B'}$ follows to vary in the negative subregion of Y_B. That is to say, when $\xi_{A'_2}$ and $\xi_{B'}$ are viewed as variables, this equation is also a formula or function of finding $\xi_{B'}$ directly by $\xi_{A'_2}$. Thus, using variables x_1, x_2 and y to replace $\xi_{A'_1}$, $\xi_{A'_2}$ and $\xi_{B'}$ in the equation, we have

$$y = \xi_B - (x_2 - \xi_{A_2})\frac{\xi_B - m_B^-}{m_{A_2}^+ - \xi_{A_2}} \tag{16.25}$$

where $\xi_{A_2} \leq x_2 \leq m_{A_2}^+$, $\xi_{A_1} \leq x_1 \leq m_{A_1}^+$. Thus, we obtain a function on the positive-positive subregion of X_A. It can be seen that this is a linear function, whose graph is a plane through points $\left(\xi_{A_1}, \xi_{A_2}, \xi_B\right)$, $\left(m_{A_1}^+, \xi_{A_2}, \xi_B\right)$, $\left(m_{A_1}^+, m_{A_2}^+, m_B^-\right)$ and $\left(\xi_{A_1}, m_{A_2}^+, m_B^-\right)$ (as shown in Fig. 16.7a). This function is an approximate function of background function $f_{AB}(x, y)$ of valued pair $\left(A_1^+ \wedge A_2^+, B^-\right)$. However, if $\max\left\{D_{A'_1 A_1}, D_{A'_2 A_2}\right\} = D_{A_1 A'_1}$, then the corresponding approximate function is

$$y = \xi_B - (x_1 - \xi_{A_1})\frac{\xi_B - m_B^-}{m_{A_1}^+ - \xi_{A_1}} \tag{16.26}$$

where $\xi_{A_1} \leq x_1 \leq m_{A_1}^+, \xi_{A_2} \leq x_2 \leq m_{A_2}^+$. The graph is a plane through points $\left(\xi_{A_1}, \xi_{A_2}, \xi_B\right)$, $\left(m_{A_1}^+, \xi_{A_2}, m_B^-\right)$, $\left(m_{A_1}^+, m_{A_2}^+ m_B^-\right)$ and $\left(\xi_{A_1}, m_{A_2}^+, \xi_B\right)$ (as shown in Fig. 16.7b).

Similarly, we can also derive the approximate functions of background function $f_{AB}(x)$ of other valued pairs. Particularly, for the valued pairs in which B is full-peak value, we take the approximate functions of their background functions all as

$$y = \xi_B$$

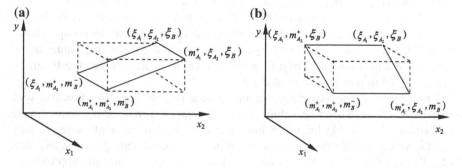

Fig. 16.7 Illustrations of bivariate approximate functions

Generalizing the above approximate functions of the pairs of corresponding values of a bivariate flexible linguistic function, the general expression of the approximate functions of the pairs of corresponding values of n-variable flexible linguistic functions is

$$y = \xi_B + \left(x_k - \xi_{A_k}\right) \frac{\xi_B - m_B}{m_{A_2} - \xi_{A_k}}, \quad k \in \{1, 2, \ldots, n\} \tag{16.27}$$

where $\quad m_{A_j} < x_j \leq \xi_{A_k} \quad$ or $\quad \xi_{A_k} \leq r_j < m_{A_j} (j = 1, 2, \ldots, n),$

$\max\left\{D_{A'_1 A_1}, D_{A'_2 A_2}, \ldots, D_{A'_n A_n}\right\} = D_{A_k A'_k};$ and

$$y = \xi_B \tag{16.28}$$

here $m_{A_j} < x_j \leq \xi_{A_k}$ or $\xi_{A_k} \leq x_j < m_{A_j} (j = 1, 2, \ldots, n)$, B is full-peak value.

These approximate functions are all linear functions. AT method just is using these linear functions to approximate various kinds of unknown background function to realize approximate evaluation. That is to say, AT method of approximate evaluation of multivariate flexible linguistic function is the generalization of AT method of approximate evaluation of unary flexible linguistic functions; thus, the rationality and effectiveness of AT method are further shown.

16.4   Approximate Evaluation of a Flexible Linguistic Function with Multiple Pairs of Corresponding Values

16.4.1   Basic Ideas and Approaches

The problem of the approximate evaluation of a flexible linguistic function with multiple pairs of corresponding values is known several pairs of corresponding values, $(A_1, B_1), (A_2, B_2), \ldots, (A_m, B_m)$, satisfying certain flexible linguistic function $Y = f(X)$ on universe of discourse U ($A_i \in U, B_i \in V$, here we take typical flexible linguistic function as a representative), which form a function on the group of basic flexible linguistic values on U or only a part of pairs of corresponding values of the function, but the expression of $f(X)$ is not known, then, for any flexible linguistic value $A \in U$, to find function value $f(A)$.

From the above approximate evaluation of a flexible linguistic function with single pair of corresponding values, it can be supposed that for the approximate evaluation of a flexible linguistic function with multiple pairs of corresponding values, we can firstly select an A_k from all pairs of corresponding values, (A_1, B_1), $(A_2, B_2), \ldots, (A_m, B_m)$, which is mostly close to A, then to find the approximate value B'_k by using corresponding (A_k, B_k) and A, then, this B'_k is just the approximate value of $f(A)$ found.

This is an approach of approximate evaluation of a flexible linguistic function with multiple pairs of corresponding values. It can be seen that this approach is really reducing the approximate evaluation with multiple pairs of corresponding values to that with single pair of corresponding values. For the latter we can use AT method. However, note that the prerequisite of using AT method is linguistic value A must be approximate to A_k. That is to say, linguistic value A not only must be relatively most close to A_k, but it also must be absolutely approximate to A_k. That means if the known valued pairs are not "dense" enough, then maybe for certain fact A, the AT method cannot be used to do approximate evaluation due to the corresponding valued pair (A_k, B_k) unable to be found. Such is a limitation of this approach.

We know that for the usual numerical function, when several pairs of corresponding values, that is, sample data (a_1, b_1), (a_2, b_2), ..., (a_m, b_m) of certain function $y = f(x)$ are known but not the function (expression) itself, there are usually the following approaches to evaluate the function value $f(a)$:

(1) If the sample data show the unknown function $y = f(x)$ is a mono-valued function, that is, a usual function, then, firstly construct interpolation function $g(x)$ by using certain kind of interpolation methods with the these samples, then find $g(a) = b$ as the approximate value of $f(a)$; or, using certain kind of function fitting technique to obtain the corresponding approximate function $f^*(x)$, then find $f^*(a) = b$ as an approximate value of $f(a)$;

(2) If the sample data show that the unknown function $y = f(x)$ is a multivalued function, that is, a usual correlation, then, we can use certain kind of regression analysis method to obtain the corresponding approximate function $\hat{f}(x)$, then find $\hat{f}(a) = b$ as an approximate value of $f(a)$.

Enlightened by the approximate evaluation methods of numerical functions, for the approximate evaluation of linguistic functions, we have similar ideas and approaches.

(1) With samples (A_1, B_1), (A_2, B_2), ..., (A_m, B_m), use certain kind of linguistic interpolation methods to construct linguistic interpolation function $g(X)$, then find $g(A) = B$ as an approximate value of $f(A)$. We call the approach to be the **interpolation method**.

(2) With samples (A_1, B_1), (A_2, B_2), ..., (A_m, B_m), use certain kind of linguistic function fitting techniques to obtain a corresponding approximate global linguistic function $f^*(X)$, then find $f^*(A) = B$ as an approximate value of $f(A)$. We call the approach to be the **approximate global function method**.

The two approaches are both utilizing the exact evaluation of the corresponding approximate linguistic function to realize the approximate evaluation of the original linguistic function, where the key is to get a suitable approximate linguistic function. This is actually raising an acquisition problem of linguistic functions. Since the acquisition of the linguistic functions already belongs to the category of knowledge discovery, about which we will make special discussions in Sect. 19.6.

16.4.2 Interpolation of Linguistic Function

Below we give an interpolation method of linguistic function based on the interpolation formula of corresponding background function. The main steps are as follows:

① Convert pairs of linguistic values, (A_1, B_1), (A_2, B_2), ..., (A_m, B_m), into pairs of numerical values, (a_1, b_1), (a_2, b_2), ..., (a_m, b_m), and convert linguistic value A into numerical value a;

② Select and use a certain existent interpolation formula $y = f^*(x)$ to find out $f^*(a) = b$;

③ Convert numerical value b into flexible linguistic value B.

This method is translating firstly the approximate evaluation of a flexible linguistic function into the approximate evaluation of a numerical function and doing approximate evaluation of the numerical function, then converting the obtained value of the approximate function into a flexible linguistic value.

It can be seen that the key of this method is the conversion from linguistic valued pairs (A_1, B_1), (A_2, B_2), ..., (A_m, B_m) to numerical valued pairs (a_1, b_1), (a_2, b_2),..., (a_m, b_m) and the selection of interpolation formula. The former is actually the conversion from A_i to a_i and from B_i to b_i ($i = 1, 2, ..., m$), that is, the conversion from a flexible linguistic value to a numerical value. The usual method of this conversion has already been given in Sect. 7.3.1, while to convert a flexible linguistic value into its peak-value point can be the first choice. Here, we just convert A_i to its peak-value point ξ_{A_i} and convert B_i to its peak-value point ξ_B. The selection of the interpolation formula should be decided by the characteristics of practical problems. As to the conversion from result b of the approximate computation to flexible linguistic value, B is comparatively easy. In fact, we can take b as the peak-value point to define a suitable flexible linguistic value, or convert b into corresponding flexible linguistic value B_k according to the consistency-degrees of b with all flexible linguistic values $B_1, B_2, ... B_m$ on universe of discourse V (see for reference the conversion methods given in Sect. 7.3.1). Nonetheless, for some problems (e.g., control), numerical value b can be directly used, thus the conversion from b to B is not needed.

Since the computation of a flexible function is actually reduced to the computation of a usual numerical function, this kind of linguistic function interpolation should be mostly suitable for flexible functions.

16.4.3 Interpolation of Linguistic Function and Numerical Approximate Computation, and to Comment Concurrently Fuzzy Logic System

It can be seen that in the above interpolation of flexible linguistic function, if what is known is not flexible linguistic value A but numerical value a, meanwhile the conversion of resulting numerical value b to linguistic value B is not needed, then such interpolation of linguistic function just becomes the interpolation of numerical function. That means the interpolation of the linguistic functions can be used to realize the approximate evaluation computation of corresponding background numerical functions.

Actually, there are already scholars who have proved that fuzzy controller in essence is a certain interpolators and that the commonly used fuzzy control algorithms can all be reduced to certain interpolation methods [2]. Besides, literature [3] also points out that the fuzzy reasoning method proposed by Zadeh effectively expresses the proximate and interpolative reasoning used by humans.

We know that the key link in fuzzy control is fuzzy inference, while fuzzy inference is a kind of approximate reasoning at the level of linguistic values and with rules with linguistic values. Since a rule with linguistic values actually covers multiple or even infinite number of special functions or correlations, hence the numerical interpolation functions realized by fuzzy control algorithms are actually a kind of general interpolation function, or in other words, it is a kind of interpolation function that is not optimized for a specific problem. Thus, the approximate effect of this kind of interpolation function cannot be guaranteed and can only depend on luck. This is just an important cause why fuzzy inference and fuzzy control are good for some problems but not so for some other problems.

An interpolation function represented by fuzzy controller is a so-called fuzzy logic system. In order to raising the accuracy of a fuzzy logic system, people employ generally the method of machine learning to adjust appropriately some parameters (such as the orientation of the peak-value point and the width of the support set) of corresponding membership function. That is tantamount to an approach of the optimization of the interpolation function. However, people find that although adjusting some parameters of the corresponding membership function according to the requirement of practical problems can indeed make the corresponding fuzzy logic system optimized and can even make the fuzzy logic system, that is, the corresponding fuzzy interpolation function to approximate to any continuous function on a compact set $U \subset \mathbf{R}^n$ [4]; however, at this time the membership functions cannot coincide with the original semantics of the corresponding flexible linguistic values, and they even are greatly mismatched. Thus, there occurs "the dilemma between precision and interpretability" [4]. In fact, in this situation, except for the name of "membership function," the so-called fuzzy logic system has already not any connections with the fuzzy logic in the real sense, but becomes a neural network system.

Fuzzy controllers, the kind of interpolation function, is not proposed by people from the angle of the linguistic function interpolation, but which is discovered after the analysis of principle of fuzzy controllers. Then, we now consider this problem from the interpolation of linguistic function; better results will surely be gained

16.5 Summary

In this chapter, we analyzed and revealed the approximate evaluation principles of flexible linguistic functions from the angle of mathematics and gave the corresponding methods. Besides, we also discussed approximate computation utilizing the approximate evaluation of flexible linguistic functions and commented on the traditional fuzzy logic system.

The main points and results of the chapter are as follows:

- The approximate evaluation of a flexible linguistic function is for a certain assumed value of independent variables to find the corresponding function value in the situation that several pairs of corresponding values satisfying the function are already known but the expression of the function is unknown. Its principle is similar to the approximate evaluation of a numerical function.
- For the approximate evaluation of a flexible linguistic function with single pair of corresponding values, the method of approximate-degree transmission and translation transformation, that is, the so-called AT method, can be used.
- For the approximate evaluation of a flexible linguistic function with multiple pairs of corresponding values, there are the following three approaches:

  ① By the approximate relation between the given assumed values of independent variables and the corresponding linguistic values in the samples, translate the approximate evaluation of flexible linguistic function with multiple pairs of corresponding values into that of flexible linguistic function with single pair of corresponding values.
  ② With the samples given, construct a linguistic interpolation function by using certain linguistic interpolation method, then find the corresponding function value as an approximate value found.
  ③ With the samples given, obtain the corresponding approximate flexible linguistic function by using a certain fitting technique of linguistic functions, then find the corresponding function value as an approximate value found.

References

1. Lian S (2009) Principles of imprecise-information processing. Science Press, Beijing
2. Lee H (1998) The interpolation mechanism of fuzzy control. Sci China (Series E) 28(3):259–267
3. Ross TJ (2010) Fuzzy logic with engineering applications, 3rd ed. Wiley, London
4. Jang JSR, Sun CT, Mizutani E (1997) Neuro-Fuzzy and soft computing. Prentice Hall, Upper Saddle River, NJ, pp 342–245, 382–385

References

1. ...
2. ...
3. ...

Chapter 17
Approximate Reasoning and Computation Based on the Approximate Evaluation Principle of Flexible Linguistic Functions

Abstract Starting from the relation between flexible linguistic rules and flexible linguistic functions, this chapter reveals the mathematical essence of approximate reasoning, and then presents an approximate reasoning method and two approximate evaluation methods of numerical functions. Besides, it in principle makes a comparison between and commentary on the methods of approximate reasoning and computation given in this book and the traditional fuzzy methods.

Keywords Flexible linguistic functions · Approximate reasoning · Approximate evaluation

Approximate reasoning is originally a kind of logical inference with flexible linguistic rules, but from mathematical essence of flexible linguistic rules (see Sect. 13.5) and the relation between flexible linguistic rules and flexible linguistic functions (see Sect. 13.7), a flexible linguistic rule is a pair of corresponding values of corresponding flexible linguistic function. Thus, the approximate reasoning with flexible linguistic rule(s) can be induced into the approximate evaluation of corresponding flexible linguistic function, or to say, we can utilize the principle of the approximate evaluation of flexible linguistic functions to realize the approximate reasoning with flexible linguistic rule(s). This opens up a new approach for approximate reasoning. Besides, utilizing flexible linguistic function, the approximate evaluation of corresponding background numerical function can also be realized.

17.1 Mathematical Essence of Approximate Reasoning

Approximate reasoning can be divided as approximate reasoning with single rule and approximate reasoning with multiple rules. Approximate reasoning with single rule is to deduce corresponding flexible linguistic value as conclusion, B', from known flexible rule $A \rightarrow B$ and the approximate value A' of its antecedent flexible linguistic value A. Approximate reasoning with multiple rules is to deduce

© Springer Science+Business Media Singapore 2016

S. Lian, *Principles of Imprecise-Information Processing*,

DOI 10.1007/978-981-10-1549-6_17

corresponding flexible linguistic value as conclusion, B, from known flexible rules $A_1 \to B_1$, $A_2 \to B_2$, ..., $A_m \to B_m$ and linguistic value as evidentiary fact, A. Viewed from the relationship between the flexible linguistic rules and the flexible linguistic function, approximate reasoning with single rule just is the approximate evaluation of a flexible linguistic function with single pair of corresponding values; and approximate reasoning with multiple rules just is the approximate evaluation of a flexible linguistic function with multiple pairs of corresponding values. If we view these known valued pairs as a local flexible linguistic function on the universe, then approximate reasoning with multiple rules is tantamount to that in the situation of only knowing a local flexible linguistic function on the universe to do the evaluation computation of global flexible linguistic function.

Thus, it can be seen that the mathematical essence of approximate reasoning is the approximate evaluation of flexible linguistic functions. Thus, approximate reasoning with flexible linguistic rules can be realized by the approximate evaluation of corresponding flexible linguistic function [1].

17.2 Approximate Reasoning with Single Rule Based on the Approximate Evaluation Principle of Flexible Linguistic Functions

1. AT method of the approximate reasoning with single rule with single condition

From the mathematical essence of approximate reasoning and the AT method of the approximate evaluation of monovariate flexible linguistic function with single pair of corresponding values in Sect. 16.2, we have the approximate reasoning method with single rule with single condition.

Suppose there is flexible rule $A \to B$ and A' approximating to A, then we can derive the corresponding B' by the following steps:

(1) Find out the peak-value points ξ_A and $\xi_{A'}$ of flexible linguistic values A and A', use ξ_A and $\xi_{A'}$ to obtain the distance $d_{A'A}$ between extended cores $X_{A'}$ and X_A;
(2) Compute distance $d_{BB'}$ by using formula (16.1) or (16.2);
(3) Determine the orientation of $Y_{B'}$ relatively to Y_B according to the orientation correspondence relation decided by rule $A \to B$, then use the suitable formulas in Eqs. (16.3)–(16.8) to obtain critical points $s_{B'}^-$ and $s_{B'}^+$ and core–boundary points $c_{B'}^-$ and $c_{B'}^+$ of linguistic value B', and obtaining the extended core of B', and then write out consistency function $c_{B'}(y)$ of B' (if requiring); or use Eqs. (16.9)–(16.11) do directly translation transformation of consistency function $c_B(y)$ of linguistic value B to obtain $c_{B'}(y)$ of B'.

Thus, we obtain a new method of approximate reasoning, we may as well call the method to be AT method of approximate reasoning with single rule with single condition, or simply, AT **reasoning**. It can be seen that AT reasoning is actually a pure numerical computation process.

Example 17.1 Suppose there are flexible rule "if the furnace temperature is low, then open the air door big" and fact "furnace temperature is rather low." Try to do approximate reasoning by using AT method and make a decision on controlling the air door.

Solution Let the range of furnace temperature be $U = [100, 1000]$, the range of opening degree of the air door be $V = [0, 100]$, "low" and "big" be separately linguistic values on U and V, and the both be semi-peak values, of them "low" be a positive semi-peak and "big" be a negative semi-peak. Thus, the orientation correspondence relation between the antecedent and consequent linguistic values of the rule is positive-negative. We denote linguistic values "low" and "rather low" on U as A and A', respectively, and denote linguistic value "big" on V as B. We define the positive critical point of A to be $s_A^+ = 500$, positive core–boundary point $c_A^+ = 300$, and peak-value point $\xi_A = 100$; define the positive critical point of A' to be $s_{A'}^+ = 600$, positive core–boundary point $c_{A'}^+ = 400$, and peak-value point $\xi_{A'} = 200$; and define the negative critical point of B to be $s_B^- = 50$, negative core–boundary point $c_B^- = 80$, and peak-value point $\xi_B = 100$. The corresponding consistency functions are as follows:

$$c_A(x) = \frac{500 - x}{200}, x \in U$$

$$c_{A'}(x) = \frac{600 - x}{200}, x \in U$$

$$c_B(y) = \frac{y - 50}{30}, y \in V$$

The graphs of these functions are shown in Fig. 17.1.

It can be seen that flexible linguistic value A' is located at the positive side of A. Thus, B' to be obtained should be located at the negative side of B. And it is easy to obtain that median points $m_A^+ = 400$ and $m_B^- = 65$, approximate radius $r_A^+ = m_A^+ - \xi_A = 300$, approximate radius $r_B^- = \xi_B - m_B^- = 35$, and $d_{A'A} = \xi_{A'} - \xi_A = 100$, so from Eq. (16.2), the distance

Fig. 17.1 The consistency function of flexible linguistic values "low" and "rather low"

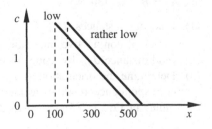

Fig. 17.2 The consistency
function of flexible linguistic
values "big" and "B'"

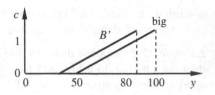

$$d_{B'B} = d_{A'A} \frac{r_R^-}{r_A^+} - 100 \times 35/300 \approx 11.7$$

Further, from Eq. (16.9), the consistency function of B' is

$$c_{B'}(y) = c_{B'}(y + d_{B'B}) = \frac{y - 38.3}{30}, \quad y \in V$$

The graph is shown in Fig. 17.2.

Viewed from the distance and function curve, linguistic value B' can be named "rather big." That is to say, the decision obtained by approximate reasoning is open the air door rather big.

It needs to be noted that although AT reasoning is realized by using the approximation evaluation method of a flexible linguistic function, we do not need to know corresponding flexible linguistic function $Y = f(X)$ in implementation, only need viewing the known rule $A \rightarrow B$ as a pair of corresponding values, (A, B), of corresponding linguistic function $Y = f(X)$. Besides, in AT reasoning, the function $c_{A'}(x)$ is required must be displacement functions of $c_A(x)$. However, in the actual reasoning, we do not have to write out the expressions of $c_{A'}(x)$, and only needing to know the peak-value point of flexible linguistic value A'.

2. AT method of approximate reasoning with a conjunction-type rule with multiple conditions

Based on the mathematical essence of approximate reasoning and the AT method of approximate evaluation of a multivariate flexible linguistic function with single pair of values in Sect. 16.3, we have the AT method of approximate reasoning with a conjunction-type rule with multiple conditions.

Let $A = A_1 \wedge A_2 \wedge \cdots \wedge A_n$, $A' = A_1' \wedge A_2' \wedge \cdots \wedge A_n'$, $A' \subset U = U_1 \times U_2 \times \ldots \times U_n$, and $B \subset V$, where $A_i, A_i' (i = 1, 2, \cdots, n)$ and B are all one-dimensional semi-peak values. Then, the AT method of approximate reasoning with the conjunction-type rule $A \rightarrow B$ is as follows:

(1) Find out distances $d_{A_1 A_1'}, d_{A_2 A_2'}, \ldots, d_{A_n A_n'}$ between corresponding component values of conjunctive linguistic values A and A';
(2) Find distance $d_{BB'}$ by using formula (16.23);
(3) Determine the orientation of $Y_{B'}$ relatively to Y_B according to the orientation correspondence relation decided by rule $A \rightarrow B$, then choose suitable formulas in Eqs. (16.3)–(16.8) to do corresponding translation transforming of

critical points s_B^- and s_B^+ and core–boundary points c_B^- and c_B^+ of linguistic value B to obtain critical points $s_{B'}^-$ and $s_{B'}^+$ and core–boundary points $c_{B'}^-$ and $c_{B'}^+$ of linguistic value B', and obtaining the extended core of B', and then write out the consistency function $c_{B'}(y)$ of linguistic value B' (if requiring); or do directly translation transformation of consistency function $c_B(y)$ of linguistic value B with Eqs. (16.9)–(16.11) to obtain consistency function $c_{B'}(y)$ of linguistic value B'.

3. AT method of approximate reasoning with a non-conjunction-type rule with multiple conditions

In the above, we have developed an approximate reasoning method with a conjunction-type flexible rule with multiple conditions on the basis of the approximate evaluation principle of a multivariate flexible linguistic function. However, the rule with multiple conditions also have of disjunction-type and synthesis-type, even more general rule with a composite linguistic value formed by multiple kinds of operations. Then, for these non-conjunction-type rules, can the approximate evaluation principle of flexible linguistic functions also be used to realize the corresponding approximate reasoning?

Actually, for disjunction-type rule $A_1 \vee A_2 \vee \cdots \vee A_n \to B$, if composite linguistic value $A_1 \vee A_2 \vee \cdots \vee A_n$ is viewed as an atomic linguistic value on product space $U_1 \times U_2 \times \cdots \times U_n$, that is, set $A_1 \vee A_2 \vee \cdots \vee A_n = A$, then this rule becomes a rule with single condition, $A \to B$. And if set also $A_1' \vee A_2' \vee \cdots \vee A_n' = A'$, then from the relationship between the flexible rules and the flexible linguistic function, in principle, AT method of approximate reasoning with a rule with single condition can be used to realize the approximate reasoning with disjunction-type rules. In implementation, when computing distance $d_{B'B}$, since A and A' are actually composite linguistic values, and it is known from Sect. 6.8 that $s_{A'A} = \max\{s_{A_1'A_1}, s_{A_2'A_2}, \ldots, s_{A_n'A_n}\}$ and $D_{A'A} = \min\{D_{A_1'A_1}, D_{A_2'A_2}, \ldots, D_{A_n'A_n}\}$, while $D_{A_i'A_i} = \dfrac{d_{A_i'A_i}}{r_{A_i}}$ ($i = 1, 2, \ldots, n$); thus, here the computation formula should be

$$d_{B'B} = \min\left\{ \frac{d_{A_1'A_1}}{r_{A_1}}, \frac{d_{A_2'A_2}}{r_{A_2}}, \ldots \frac{d_{A_n'A_n}}{r_{A_n}} \right\} \cdot r_B \qquad (17.1)$$

Similarly, for synthesis-type flexible rule $A_1 \oplus A_2 \oplus \cdots \oplus A_n \to B$, we can also use AT method to realize approximate reasoning. Of course, the computation formula of distance $d_{B'B}$ here is

$$d_{B'B} = \sum_{i=1}^{n} w_i \frac{d_{A_i'}}{r_{A_i}} \cdot r_B, \quad \sum_{i=1}^{n} w_i = 1 \qquad (17.2)$$

Further, for more general rule with a composite linguistic value, $E(A_1, A_2, \ldots, A_n) \to B$, its antecedent $E(A_1, A_2, \ldots, A_n)$ can also be viewed as an atomic linguistic value, that is, setting $E(A_1, A_2, …, A_n) = A$, and setting fact $E(A_1', A_2', …, A_n') = A'$, then treat which as approximate reasoning with a rule with single condition. Of course, when determining approximate degree $s_{A'A}$, needing to compute the distance $d_{A'A}$ by using corresponding formulas of operations $\wedge$, $\vee$ and $\oplus$ from top to down layer by layer according to the structure of $E(A_1, A_2, …, A_n)$.

Example 17.2 Suppose there is a flexible rule with a composite linguistic value

$$(A \wedge B \wedge C) \vee (D \oplus E \oplus F) \wedge (G \vee H \vee I) \to J$$

and fact

$$(A' \wedge B' \wedge C') \vee (D' \oplus E' \oplus F') \wedge (G' \vee H' \vee I')$$

To be asked to do approximate reasoning using AT method.

Solution Set $(A \wedge B \wedge C) \vee (D \oplus E \oplus F) \wedge (G \vee H \vee I) = A_1$, then $(A' \wedge B' \wedge C') \vee (D' \oplus E' \oplus F') \wedge (G' \vee H' \vee I') = A_1'$. Thus, the original rule and fact become

$$A_1 \to J \text{ and } A_1'$$

Thus, by AT method, we have

$$d_{J'J} = d_{A_1'A_1} \frac{r_J}{r_{A_1}}$$

However, here A_1 is a composite flexible linguistic value, from the logical operation order and brackets, the structure of the top layer of A_1 is

$$A_2 \vee A_3$$

where $A_2 = A \wedge B \wedge C$, $A_3 = (D \oplus E \oplus F) \wedge (G \vee H \vee I)$. Thus,

$$d_{A_1'A_1} = \min\left\{\frac{d_{A_2'A_2}}{r_{A_2}}, \frac{d_{A_3'A_3}}{r_{A_3}}\right\} \cdot r_{A_1}$$

while

$$d_{A_2'A_2} = \max\left\{\frac{d_{A'A}}{r_A}, \frac{d_{B'B}}{r_B}, \frac{d_{C'C}}{r_C}\right\} \cdot r_{A_2}$$

Set $A_4 = D \oplus E \oplus F$, $A_5 = G \vee H \vee I$, then $A_3 = A_4 \wedge A_5$. Thus,

$$d_{A_3'A_3} = \max\left\{\frac{d_{A_4'A_4}}{r_{A_4}}, \frac{d_{A_5'A_5}}{r_{A_5}}\right\} \cdot r_{A_3}$$

where

$$d_{A_4'A_4} = w_1 \frac{d_{D'D}}{r_D} \cdot r_D + w_2 \frac{d_{E'E}}{r_E} \cdot r_E + w_3 \frac{d_{F'F}}{r_F} \cdot r_F, \quad \sum_{i=1}^{3} w_i = 1$$

$$d_{A_5'A_5} = \min\left\{\frac{d_{G'G}}{r_G}, \frac{d_{H'H}}{r_H}, \frac{d_{I'I}}{r_I}\right\} \cdot r_{A_5}$$

Now substitute the above all distances successively backward into the corresponding expressions, then $d_{J'J}$ is lastly obtained. Further corresponding flexible linguistic value J' can be obtained. As for the approximate radii in them, they can all be computed by the expression of definition of approximate radii of atomic linguistic values in Sect. 6.8.

4. AT method of approximate reasoning with a non-typical flexible rule

AT method is for the typical flexible rules. For a non-typical flexible rule, we can firstly transform it into a typical rule, and then using AT method to do approximate reasoning, then, transforming backward the resulting flexible linguistic value obtained by using original transformation, the eventual resulting flexible linguistic value can be obtained.

For example, let $A_1 \wedge A_2 \wedge \cdots \wedge A_n \rightarrow B$ be a conjunction-type non-typical flexible rule, where $A_i \subset U_i$ ($i = 1, 2, \cdots, n$) and $B \subset V$ are multidimensional property-type or relation-type flexible linguistic values. Then, we can take transformations $\varphi_1, \varphi_2,\ldots, \varphi_n,$ and ψ to transform multidimensional flexible linguistic values $A_1, A_2, \ldots, A_n$ and B separately into one-dimensional flexible linguistic values $A_1, A_2, \ldots, A_n$ and B, thus the original non-typical flexible rule $A_1 \wedge A_2 \wedge \cdots \wedge A_n \rightarrow B$ is transformed into typical flexible rule $A_1 \wedge A_2 \wedge \cdots \wedge A_n \rightarrow B$. For the latter, obviously, AT method can be used to do the approximate reasoning. Let one-dimensional flexible linguistic value obtained be B'. Then, using backwardly the original transformation ψ to B', the eventual result—multidimensional flexible linguistic value B' follows.

17.2.1 Approximate Reasoning with an N–L or L–N Rule

Since its antecedent is numerical value, so for an N–L rule, the usual distance $d_{A'A}$ with respect to L–L rule becomes the distance $d_{x'x}$ between the numerical value x' as evidence fact and numerical value x in the rule's antecedent, and the approximate radius r_A also becomes r_x. Thus, the computation formula of corresponding distance $d_{B'B}$ becomes

$$d_{B'B} = d_{x'x} \frac{r_B}{r_x} \tag{17.3}$$

(Actually, from the definitions of the distance and approximate radius about flexible linguistic values, it can be seen that distance $d_{x'x}$ is also tantamount to distance $d_{A'A}$ between flexible linguistic values A' and A with separately peak-value points x' and x, and approximate radius r_x is also tantamount to approximate radius r_A). As to the orientation of B', it can also be determined by the rule itself. Thus, the approximate reasoning with an N–L rule can also use AT method. Of course, when reasoning, the approximate radius r_x of numerical value x need to be determined according to practical problem.

Similarly, since its consequent is a numerical value, so for an L–N rule, the usual distance $d_{B'B}$ with respect to L–L rule is also the distance $d_{y'y}$ between y' as the resulting numerical value and numerical value y in the rule's consequent. Its computation formula is

$$d_{y'y} = d_{A'A} \frac{r_y}{r_A} \tag{17.4}$$

When approximate reasoning, it only is needed to treat conclusion y of rule as the peak-value point ξ_B of conclusion linguistic value B of the usual L–L rule, then y' to be found in approximate reasoning is also tantamount to peak-value point $\xi_{B'}$ of corresponding B'. Thus, from the computation formula of $\xi_{B'}$, we have

$$y' = y - d_{y'y} \quad (\text{when } y' < y) \tag{17.5}$$

$$y' = y + d_{y'y} \quad (\text{when } y' > y) \tag{17.6}$$

$$y' = y \quad (\text{when } y' = y) \tag{17.7}$$

Thus, the approximate reasoning with an L–N rule can also use AT method. Of course, when reasoning, the approximate radius r_y of numerical value y and the magnitude relation between y' and y need to be determined according to the practical problem.

17.3 Approximate Reasoning with Multiple Rules Based on the Approximate Evaluation Principle of Flexible Linguistic Functions

From the approximate evaluation method of flexible linguistic functions and the mathematical essence of approximate reasoning with multiple rules, we can have three approaches of approximate reasoning with multiple rules:

(1) **By using AT method**. That is, firstly choose the A_k from antecedent linguistic values $A_1, A_2, \ldots, A_n$ of all rules, which is closest to linguistic value A as fact (actually, if $A_1, A_2, \ldots, A_n$ are all basic flexible linguistic values or composite

linguistic values composed by basic flexible linguistic values, then A can only be approximate to one flexible linguistic value among them), then do approximate reasoning with corresponding rule $A_k \rightarrow B_k$ and fact A, and by using AT method to obtain corresponding B_k'. Then, this B_k' is just the B to be found.

(2) **By using a global linguistic function**. That is, firstly treat rules $A_1 \rightarrow B_1$, $A_2 \rightarrow B_2$, ..., $A_n \rightarrow B_n$ as pairs of corresponding values, (A_1, B_1), (A_2, B_2), ..., (A_n, B_n), of a unknown global linguistic function, that is, viewing $\{(A_1, B_1), (A_2, B_2), ..., (A_n, B_n)\}$ as a local linguistic function, then with this, by using knowledge discovery techniques to obtain this global linguistic function $Y = f(X)$, then use directly the $Y = f(X)$ to find $f(A) = B$. Thus, the corresponding approximate reasoning is also realized.

(3) **By using linguistic interpolation**. That is, firstly treat rules $A_1 \rightarrow B_1$, $A_2 \rightarrow B_2$, ..., $A_n \rightarrow B_n$ as pairs corresponding values, (A_1, B_1), (A_2, B_2), ..., (A_n, B_n), of a global linguistic function, then with these valued pairs and known linguistic value A, using the approximate computation method of certain kinds of "linguistic interpolation" to directly obtain the approximate value B' of corresponding linguistic function. Thus, this approximate evaluation of linguistic function just realizes the corresponding approximate reasoning.

It can be seen that approach (1) reduces actually the approximate reasoning with multiple rules to the approximate reasoning with single rule. However, if the linguistic function formed by corresponding rule set is discontinuous, then the method cannot be normally used. Besides, AT method also requires all antecedent linguistic values of flexible rules to be semi-peak linguistic values.

17.4   Approximate Evaluation of Numerical Functions Based on the Approximate Evaluation Principles of Flexible Linguistic Functions

The evidence fact A' in AT method is a linguistic value (an atomic linguistic value or a composite linguistic value). However, in practical problems, some evidence facts may be numerical value x (including scalar or vector). Then, in this case, must numerical value x be converted into linguistic value A'? The answer is negative. The reason is that since in AT method, actually as long as knowing peak-value point $\xi_{A'}$ of linguistic value A', distance $d_{B'B}$ can be obtained at once. So for numerical evidence fact x, it can be directly treated as peak-value point $\xi_{A'}$ of corresponding linguistic value A'. That is to say, a numerical fact met in AT method can directly participate in reasoning and does not need to be converted into a linguistic value.

On the other hand, from Eqs. (16.12), (16.13), and (16.14) in Chap.16, it can be seen that in AT method, peak-value point $\xi_{B'}$ of conclusion linguistic value B' can be directly obtained, then, when a practical problem eventually needs is a numerical value, this $\xi_{B'}$ can be treated as one choice for the eventual numerical value, or it can even be the first choice. Therefore, speaking in this sense, linguistic value B' neither need to be converted into a numerical value in the approximate reasoning with AT method.

Approximate reasoning with AT method does not need to convert numerical evidence facts into linguistic values and also does not need to convert a linguistic value conclusion into a numerical value. This means AT method can be directly used in the approximate evaluation computation of corresponding background functions. Besides, from Sect. 16.4, it is known that utilizing linguistic interpolation, the approximate evaluation computation of corresponding background numerical functions can also be realized.

Thus, the approximate computation method based on the approximate evaluation principles of flexible linguistic functions can be used for automatic control. As a matter of fact, in actual automatic control systems, the input quantity and the output quantity are generally always exact numerical values. For this, the traditional fuzzy control generally uses the methods of fuzzification and defuzzification to treat. Then, now if the corresponding control system uses AT method or linguistic interpolation method to realize approximate computing, then these two procedures of fuzzification and defuzzification can be omitted (the examples on the approximate computing and control using AT method can see a demonstration in Sect. 18.6).

17.5   Utilizing Exact Evaluation of a Flexible Linguistic Function to Realize Approximate Evaluation of a Numerical Function

Actually, by the relation between linguistic function and numerical function, as well as N–L and L–N conversions, we can utilize directly the exact evaluation of a flexible linguistic function to realize the approximate evaluation of the corresponding background numerical function (including the multivalued background function). Specifically speaking, first convert the known number (vector) $x_0 = (x_1, x_2, ..., x_n)$ $\in U = U_1 \times U_2 \times \cdots \times U_n$ into flexible linguistic value (vector) $A = (A_1, A_2, ..., A_n) = A_1 \wedge A_2 \wedge \cdots \wedge A_n$ on the domain $Ls(U)$ of the corresponding flexible linguistic function $Y = f(X)$ through N–L conversion; then find out the corresponding value of linguistic function $B = f(A)$ from A and $Y = f(X)$; finally convert flexible linguistic value B into number $y_0' \in V$ through L–N conversion. Then, the y_0' is just an approximate value of y_0 which is the value of background function of flexible linguistic function $y = f(x)$ at $x = x_0$, namely $y_0' \approx y_0 = f(x_0)$. The process and principle of this method of approximate evaluation computing is shown in Fig. 17.3.

It can be seen that here function value B is an exact value of flexible linguistic function $Y = f(X)$ at independent variable $X = A$. So the "approximation" of the

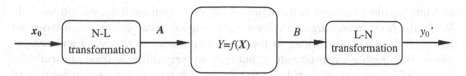

Fig. 17.3 Diagram of principle of approximate evaluation of a numerical function realized by utilizing exact evaluation of a flexible linguistic function

approximate evaluation actually is mainly showed on the final L–N conversion. From the conversion relations between linguistic values and numerical values and the semantics of flexible linguistic function, if the flexible linguistic values A and B and flexible linguistic function $Y = f(X)$ are proper for the corresponding practical problem, then the y_0' obtained from x_0 given is reasonable, effective and sufficient for requirement.

The kind of approximate evaluation computation method based on exact evaluation or exact inference has another characteristic, that is, only needing to know the extended cores of corresponding flexible linguistic values, then all reasoning and computation can be done by "geometric method" but do not need the consistency functions of the flexible linguistic values, do also not need the adjoint degreed functions and measured functions of the corresponding rules. So the method of approximate evaluation computation is actually simpler and more convenient.

Actually, from the relation between flexible rules and flexible linguistic functions, the exact evaluation of a flexible linguistic function is also the exact reasoning with corresponding flexible rules, so this exact evaluation of numerical function utilizing the exact evaluation of flexible linguistic function is also the exact evaluation of numerical function realized by utilizing natural inference with data conversions in Sect. 15.1. Therefore, this exact evaluation method of numerical function, in theory, is completely feasible for those problems (such as the action control of robots and certain process control) without exact mathematical models but can be described and dealt with by flexible linguistic functions or flexible rules. Even more general, for any numerical function (including multiple valued function and vector valued function) relation this method can be used to find its function value as long as the corresponding flexible linguistic function can be written out.

17.6   A Comparison of Approximate Reasoning and Computation in This Book and Fuzzy Reasoning and Computation

We used five chapters from Chaps. 13–17 to reveal the principles of the approximate reasoning and computation with flexible linguistic values and gave multiple approaches of approximate reasoning and computation such as natural inference (with data conversion(s)), reasoning with degrees in the frame of logic, AT method and interpolation methods in the frame of flexible linguistic function and so forth. While the fuzzy reasoning and corresponding approximate computation based on fuzzy set theory is also belong to the kind of approximate reasoning and computation with flexible linguistic values, but they are very different from our methods. In the following, we will make a simple comparison of the two in aspects of principles, methods, effects, etc.

1. Rationales

It is well known that the basic principle of traditional fuzzy reasoning is CRI (compositional rule of inference) proposed by Zadeh, that is, using the composition of fuzzy relations called to realize approximate reasoning. Specifically speaking, it is that first represent a fuzzy rule (i.e., flexible rule we call) $A \rightarrow B$ into a fuzzy relation (i.e., flexible relation we call) R (as defining $R = (A^c \times V) \cup (U \times B)$), and viewing fuzzy set A' approximating to A also as a fuzzy relation and to do composition of A' and R, then treating the obtained result B' (also a fuzzy set) as the result deduced by rule $A \rightarrow B$ and fact A'. This principle of inference is represented by an equation just is

$$B' = A' \circ R \tag{17.8}$$

The membership functional representation form of the equation is

$$\mu_{B'}(y) = f(\mu_R(x,y), \mu_{A'}(x)) \tag{17.8'}$$

where $\mu_R(x, y)$ is implicational operator so-called.

However, research of us shows that rule $A \rightarrow B$ can merely be represented as a binary rigid relation, but a binary flexible relation, or, a binary fuzzy relation (see Sects. 13.5 and 13.6). In fact, in theory, rule $A \rightarrow B$ is equivalent to a binary rigid relation, but the binary rigid relation cannot be definitely written in general, so can only be represented by universal relation $core(A)^+ \times core(B)^+$ or supp $(A) \times supp(B)$.

Besides, we see that to obtain approximate conclusion B', the rule $A \rightarrow B$ and fact A' would be used of course, but the conduct of directly making the two intersect is improper perhaps. Because $A \rightarrow B$ is only a rule with linguistic values but a linguistic function, and the fact A' does not match the antecedent A of $A \rightarrow B$, then, in this situation, doing directly composition of A' and $A \rightarrow B$ is unreasonable no matter from the logic or mathematics. No wonder CRI is not compatible with traditional *modus ponens*, despite CRI is called generalized modus ponens in fuzzy set theory.

Actually, the reasoning with one rule $A \rightarrow B$ is an inference on properties, while the reasoning with two rules $A \rightarrow B$ and $B \rightarrow C$ just is an inference on relations. In CRI, the minor premise, namely fact A', is extended as a binary relation, but the binary relation is only a pseudo binary relation, which is still a monadic relation really. In addition, the composition of relations is conditional, not any two relations can be composed; as for composing two rules of binary relation, they must satisfy transitivity.

Also, viewed from over ten kinds of implicational operators and over ten kinds of formulas of relation composition, it cannot be helped the objectivity of fuzzy reasoning arousing suspicion. In fact, from the implicational operator (presented by Zadeh)

$$\mu_R(x, y) = \max\{1 - \mu_A(x), \min\{\mu_A(x), \mu_B(y)\}\} \tag{17.9}$$

and the formula of relation composition

$$\mu_{B'}(y) = \min\{\mu_{A'}(x), \max\{1 - \mu_A(x), \min\{\mu_A(x), \mu_B(y)\}\}\} \tag{17.10}$$

it is not hard to see that $\min\{\mu_{A'}(x), \max\{1 - \mu_A(x), \min\{\mu_A(x), \mu_B(y)\}\}\}$ is actually the truth-degree of premise $(A \to B) \wedge A'$ of reasoning, while also treating it as the truth-degree of conclusion B' is then entirely artificially set.

Besides, fuzzy reasoning does not consider the orientation of linguistic value A' relatively to antecedent linguistic value A of rule $A \to B$, but which is closely related to the accuracy of conclusion linguistic value B' and the exactness of corresponding number y'. Because viewed from the perspective of flexible linguistic functions, the approximate reasoning is also the approximate evaluation of a flexible linguistic function with a single pair of corresponding values.

Correspondingly, our approximate reasoning and computation approaches have logical and mathematical rationales.

In fact, our natural inference is just usual *modus ponens*; and our reasoning with degrees is following inference rule "truth-degree-level(degree-level)-UMP", and in which the numerical computation models and methods are founded naturally on the bases of the mathematical essence of flexible rules and approximate reasoning, and under the constraints of the logical semantics of rules and the inference rules of "near-true" and "rough-true". So it is logical. Just because this, it is completely compatible with *modus ponens* in traditional logical, which is really the thinning and generalization of latter.

Our AT method is then presented based on the numerical models and the approximate evaluation principle of linguistic functions in viewpoint of mathematics completely, so its rationality is obvious (for specific demonstration see Sect. 16.2.3). As to other reasoning and computation methods (as interpolation method and approximate global function method) based on linguistic functions, since they are presented at the level of linguistic functions, so they have solid mathematical basis, and these methods are all closely related to mathematical backgrounds of practical problems, and our approximate computation with the adjoint measured functions of rules is then directly applying the approximate background functions of corresponding rules to be realized. Therefore, the rationality of these methods is also beyond doubt.

Actually, from the perspective of logic, our natural inference, reasoning with degrees, and AT method are all the reasoning in the sense of near-true (truth-degree > 0.5), that is, near-true inference we said (see Sect. 11.6); while fuzzy reasoning can roughly be counted as the reasoning in the sense of degree-true (truth-degree > 0), that is, degree-true inference we said (see Sect. 11.5), and, the fuzzy reasoning with multiple rules (such as Mamdani-style inference and Sugeno-style inference) can roughly be included in the parallel reasoning with degrees we said (see Sect. 15.6). From previous Chaps. 11–15, we see that near-true inference is consistent with the natural logical semantics of propositions, so it tallies with inference mechanism of human brain; while degree-true inference is not consistent with the natural logical semantics of propositions (it corresponds to extended logical semantics of propositions), hence it does not tally with inference mechanism of human brain, particularly, the parallel reasoning with degrees based on the degree-true inference is more different from the inference mechanism of human brain; in addition, after parallel reasoning with degrees, for the synthesizing of multiple conclusions, there is not a unified method or model having theoretical basis.

2. Efficiencies and effects

Comparing near-true inference and degree-true inference, we see that using near-true inference, one and the same problem can be solved by reasoning once, but if using degree-true inference, then it needs to be done by (parallel) reasoning generally twice or multiply. That is to say, the efficiency of our natural inference, reasoning with degrees and AT method in the sense of near-true, generally speaking, will be higher than that of fuzzy reasoning in the sense of degree-true.

Also, let us observe ranges of truth-degrees, the range of truth-degrees of near-true inference is $(0.5, \beta]$ $(\beta \geq 1)$, and the range of truth-degrees of degree-true inference is $(0, \beta]$; while the range of truth-degrees of fuzzy reasoning is $[0, 1]$. Note that the 1 in $[0, 1]$ here is equivalent to $[1, \beta]$ in $(0.5, \beta]$. From the three ranges of truth-degrees, it can roughly be observed that the effect (i.e., accuracy) of our reasoning methods, generally speaking, will be better than that of fuzzy reasoning.

In the following, we will further analyze the efficiencies and effects of the two kinds of reasoning.

First, fuzzy reasoning treats rule $A \rightarrow B$ as couple or implication relation (namely, "if A then B, else $B \vee \neg B$"). As couple relation, the rule $A \rightarrow B$ is represented as the Cartesian product of corresponding fuzzy sets, $A \times B$, the geometric space this fuzzy set occupied in real is the region supp(A) $\times$ supp(B) in corresponding universe of discourse (as shown in Fig. 17.4a). While in our inference methods, the rule $A \rightarrow B$ is then further represented as smaller region core $(A)^+ \times$ core$(B)^+$ (as shown in Fig. 17.4b).

As implication relation, rule $A \rightarrow B$ is also the following two correspondences:

$$\begin{cases} A \mapsto B \\ \neg A \mapsto B \vee \neg B \end{cases}$$

Which is represented as region $(A^c \times V) \cup (U \times B)$ (as shown in Fig. 17.4c) in fuzzy reasoning. Since the evidence fact A' of the reasoning is an approximate value

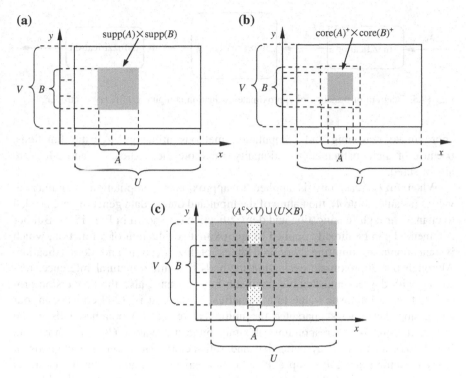

Fig. 17.4 The spaces rule $A \to B$ occupies in different methods

of A, so the above second correspondence "$\neg A \mapsto B \vee \neg B$" is really superfluous. That is to say, when rule $A \to B$ is treated as implication relation, there is redundancy in fuzzy reasoning. Yet our reasoning methods only use correspondence $A \mapsto B$ in implication relation like the usual logical inference. Therefore, the reasoning can be implemented, but also there is no any redundancy.

Secondly, seeing from processes, fuzzy reasoning is based on region supp $(A) \times$ supp(B) or $(A^c \times V) \cup (U \times B)$ to realize approximate reasoning and computation. From the analysis above, it can be known that region $(A^c \times V) \cup (U \times B) - \text{core}(A)^+ \times \text{core}(B)^+$ is practically useless in reasoning. Then, the effect is conceivable finding the approximate value B' or corresponding number y' based on such a space much bigger than the actual requirement.

Yet our natural inference and reasoning with degrees realize approximate reasoning and computation in the subregion $\text{core}(A)^+ \times \text{core}(B)^+$ of$(A \times V) \cup (U \times B)$; our AT method realizes approximate reasoning and computation round the peak-value point (ξ_A, ξ_B) in region $\text{core}(A) \times \text{core}(B)$. From Sect. 13.5 we known that region $\text{core}(A)^+ \times \text{core}(B)^+$ is the smallest space that includes the background function or background correlation of rule $A \to B$. Thus, viewed only from problem-solving space, the error caused by our approximate reasoning methods is certainly not more than that by fuzzy reasoning, on the whole. As to our other

Fig. 17.5 Diagram of principle of approximate evaluation computing of fuzzy controller

approximate reasoning and computation methods utilizing linguistic functions, because of their mathematical rationality, therefore their efficiency and effect are also assured.

When fuzzy reasoning is applied to approximate computation of numerical values (as auto-control), the nature of the input and output data generally are needed to change through fuzzification and defuzzification (as shown in Fig. 17.5). But our AT method can be directly applied to approximate evaluation of a function, which is tantamount to omitting two procedures of fuzzification and defuzzification. Although our approximate evaluation methods that utilize natural inference, reasoning with degrees and flexible linguistic function have also the conversion procedure from a linguistic value to a numerical value, that is, L–N conversion, our conversion can always guarantee the numerical result (y_0') obtained falls in the extended core of corresponding flexible linguistic value (B) (as shown in Fig. 17.6a), so it is always effective and satisfactory. However, the methods of fuzzy reasoning and follow-up defuzzification cannot guarantee that the obtained numerical result (y_0') falls within the extended core of corresponding flexible linguistic value (B) (as shown in Fig. 17.6b), as a result, there would occur the phenomenon of sometimes valid, but some other times invalid as well as valid for some problems but invalid for some other problems. We think, it just is an important cause of the fuzzy control being not reliable and stable enough.

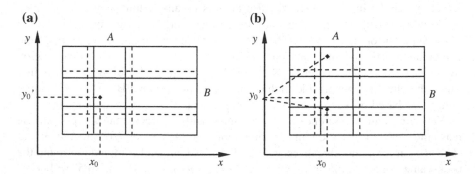

Fig. 17.6 Examples of the positions of numerical result (y_0') obtained by two different approximate computing models. **a** The region between two *vertical dotted lines* is support set of flexible linguistic value A, and between two *solid lines* is the extended core of A. **b** The region between two *horizontal dotted lines* is support set of flexible linguistic value B, and between two *solid lines* is the extended core of B

Actually, fuzzy controller is a kind of interpolator essentially (this is just the cause of fuzzy reasoning can be used to approximate computation to solve the practical problems such as automatic control). But viewed from approximate computation, this interpolate method is inefficient (the cause is as stated above).

The approximate computing model of a fuzzy controller is considered to be able to further develop into a fuzzy logic system which can realize "universal approximator," but this kind of approximate computing systems has a severe problem—its principle cannot be clearly explained (see Sect. 16.4.3). But the kinds of approximate computing models we gave in Chaps. 15–17 only need to add, respectively, a function of reducing dynamically granule sizes of flexible linguistic values, then they can all realizing the "universal approximator," and the functions approximated are more widespread, for example, which can also being a multiple valued function and vector function.

3. Research methods

In research of fuzzy reasoning, the mathematical backgrounds of practical problems are consider rarely, the orientation of an approximate linguistic value is still not considered, but it is tried in the same way that at the level of logic and language to research a fuzzy logic method that can treat the practical problems in different poses and with different expressions.

However, since $(A \rightarrow B) \wedge A' \Rightarrow B'$ is not a valid argument form, so it is impossible to deduce B' directly from $A \rightarrow B$ and A'. On the other hand, logical relation cannot reflect, or in other words, covers the orientation relation when one linguistic value approaches to another linguistic value, but the orientation relation is of great importance to the B' to be obtained.

We know that a linguistic valued rule actually summarizes infinite number of practical binary relations—correlations or functions. And relatively to the practical binary relation summarized by rule $A \rightarrow B$, the binary relation $core(A)^+ \times core(B)^+$ is an universal relation, while binary relation $(A^c \times V) \cup (U \times B)$ is a bigger universal relation. Therefore, the approximate reasoning at the level of linguistic value should have a guide of the lower-level numerical values, i.e., the mathematical background. Otherwise, with only logic or mathematical methods, it is difficult to obtain a desired effect. On the other hand, the approximation of flexible linguistic value A' to A also involves the orientation problem; different orientations may bring different inference results. However, in the methods of pure logical inference, the information about orientation cannot be reflected, so A' approximates to A no matter from which direction, the result B' is all the same. Obviously, the efficiency of the result of such approximate reasoning can only go by luck.

But our approximate reasoning methods are just guided by the relevant mathematical background information of practical problems, where linguistic values only have the effect of macroscopic positioning, while the specific computations are also different with different problems.

4. Scopes and abilities

The rules with multiple conditions involved in fuzzy reasoning are only two types of conjunction and disjunction, but our flexible rules have a type of synthesis-type rule except for the two types of rules. That is to say, compared with fuzzy reasoning, our methods have more wide scope of application and more powerful processing ability.

17.7 Summary

In this chapter, starting from the relation between flexible linguistic rules and flexible linguistic functions, we revealed the mathematical essence of approximate reasoning and then presented an approximate reasoning method and two approximate evaluation methods of numerical functions. Besides, we also made a comparison between and commentary on, in principle, the methods of approximate reasoning and computation given in this book and the traditional fuzzy methods.

The main points and results of the chapter are as follows:

- The mathematical essence of approximate reasoning with flexible linguistic rules is the approximate evaluation of a flexible linguistic function, so the approximate evaluation principles of flexible linguistic functions can be used to realize the approximate reasoning. Thus, for approximate reasoning with single rule, there is a series of AT reasoning method; and for approximate reasoning with multiple rules, we then can translate it into approximate reasoning with single rule; besides, we can also use the methods of approximate global linguistic function or linguistic interpolation. These approximate reasoning methods can also be used for the approximate evaluation of corresponding numerical functions.

- Approximate reasoning at the level of linguistic value is really the summarization of approximate computation at the level of numerical value, which should be guided by the related mathematical background; A pure logic or pure mathematical approximate reasoning method without the consideration of the mathematical background of rules is significant only when the sizes of the corresponding flexible linguistic values are sufficiently small, while in usual situation, the effect of the approximate reasoning only depends on luck.

- Utilizing the exact evaluation of flexible linguistic functions and the corresponding N–L and L–N conversions, the approximate evaluation of a corresponding background function can be realized, its effectiveness is decided by the properness of corresponding flexible linguistic values and flexible linguistic function.

- In theory, a system of approximate evaluation computation of a numerical function with a flexible linguistic function or flexible linguistic rule set which can reduce dynamically granule sizes of linguistic values is just a "universal approximator" that can approximate any (non-chaotic) continuous function on a measurement space.

Reference

1. Shiyou L (2009) Principles of imprecise-information processing. Science Press, Beijing

Part VI
Imprecise-Problem Solving and Imprecise-Knowledge Discovery: Application of Imprecise-Information Processing

Chapter 18
Imprecise-Problem Solving, and Anthropomorphic Computer Application Systems

Abstract This chapter expounds the basic techniques of imprecise-problem solving—the engineering application of the imprecise-information processing and then discusses the anthropomorphic computer application systems with imprecise-information processing ability.

Keywords Imprecise-problem solving · Flexible classifying · Flexible judging · Flexible decision making · Flexible control · Flexible pattern recognition · Human–computer interface · Natural language processing · Anthropomorphic computer application systems

In the preceding chapters, the basic principles and methods of the imprecise-information processing were expatiated. In this chapter, we discuss how to use these principles and methods to solve practical problems, and then to build anthropomorphic computer application systems, which have the ability to deal with imprecise information.

18.1 Imprecise Problems and Their Solving

Automatic solution to difficult problem, that is, the so-called problem solving, is an important research issue of artificial intelligence. In addition to the intellectual problems, broadly speaking, the problems here include the practical problems or engineering problems such as classifying, recognition, judging, decision making, forecasting, diagnosis, control, dispatching, planning and programming, designing, and explaining. Of numerous practical problems, some are imprecise, that is, in which the imprecise information is contained. For example, if the classes in a problem of classifying or recognition are flexible classes, then the problem is imprecise. For another example, when the condition, goal, or conclusion in a problem of judging, decision, control, programming, forecasting, or diagnosis are flexible linguistic values or flexible sets, then the problem is imprecise. There are some problems (as process control), which are themselves precise, but the knowledge and data used in the process to solve them are flexible linguistic values or flexible linguistic rules or functions [1].

The processes and methods of solving imprecise problems are similar to that of the usual precise-problem solving, but the imprecise-problem solving will involve imprecise-information processing. That is to say, in the imprecise-problem solving, flexible linguistic values, flexible sets, membership functions, consistency functions, flexible linguistic rules, or flexible linguistic functions, etc., would be used, and the interconversion between numerical values and flexible linguistic values may be involved. In fact, according to the nature and characteristics of a problem and the representation form of relevant knowledge, the imprecise-problem solving can be reduced to two basic ways of computing with membership or consistency functions (or flexible sets or their extended cores) and reasoning and computing with flexible linguistic rules or flexible linguistic functions.

18.2   Flexible Classifying with Membership Functions (or Extended Cores)

We refer to the classifying in which the classifications are flexible classes as **flexible classifying**.

Pattern recognition is a kind of classifying, which can be generally classified into two types of statistical pattern recognition and structural pattern recognition. Statistical pattern recognition respects the numerical feature of objects to describe a pattern or a pattern class awaiting recognition as an n-dimensional vector, called feature vector. Structural pattern recognition respects the structural feature of the objects to describe a pattern or a pattern class awaiting recognition as a character string having certain kind of tree structure. The pattern classes in usual pattern recognition are rigid classes. Then, if the pattern classes in a pattern recognition are flexible classes, then we say the pattern recognition is a **flexible pattern recognition**. Flexible pattern recognition is a kind of flexible classifying. The flexible classifying, alike usual rigid classifying, firstly there have to exist the knowledge about known classifications, then the corresponding classifying can be done with these knowledge. In the usual statistical pattern recognition, the knowledge about known classifications is shown as the discrimination functions of the classifications; in the usual structural pattern recognition, the knowledge about known classifications is shown as the descriptive rules of the classifications (a kind of grammar rules of formal language).

Next, we take flexible pattern recognition as an example to examine flexible classifying. In this section, we consider firstly flexible pattern recognition with feature vectors, that is, flexible statistical pattern recognition.

18.2.1 Usual Flexible Classifying

In traditional statistical pattern recognition, the discrimination functions are real functions of n variables defined on the feature space (that is, measurement space) of objects awaiting classifying. For instance, let Ω be a feature space, and let $x = (x_1, x_2, \ldots, x_n) \in \Omega$, and $g_i(x)$ be the discrimination function of class $\omega_i \subset \Omega$ ($i = 1, 2, \ldots, c$), then we have classifying rules as follows:

$$\text{If for } \forall \quad j \neq I, g_i(x) > g_j(x), \quad \text{then } x \in \omega_i \ (i = 1, 2, \ldots, c) \qquad (18.1)$$

This is the usual statistical pattern recognition method. Then, when here class ω_i is a basic flexible class on Ω, this recognition is just a flexible recognition or flexible classifying.

So flexible classifying also needs to define the discrimination functions of the corresponding flexible classes. It can be seen from the relation between membership functions and flexible classes that the membership functions are also the discrimination functions of flexible classes.

Let $\pi = \{A_1, A_2, \ldots, A_n\}$ be a flexible partition of feature space U, and $A_1, A_2, \ldots, A_n$ be basic flexible classes of U. We know that for any flexible class $A_i \subset \{A_1, A_2, \ldots, A_n\}$ and any object $x \in U$, we can always say x belongs to A_i with a certain degree ($m_A(x)$), which can be symbolically represented as $x \in_{m_{A_i}(x)} A$. However, usual classifying requires the class that x belongs to be only one choice rather than ambiguous. Therefore, at this time we need to make a selection from basic flexible classes $A_1, A_2, \ldots, A_n$. Naturally, the flexible class with the biggest membership-degree should be chosen as the class that x belongs to (this is just the principle of maximum membership in fuzzy set theory), that is, x should be classified as the flexible class A_k with membership-degree $m_{A_k}(x) = \max\{m_{A_1}(x), m_{A_2}(x), \ldots, m_{A_n}(x)\}$, that is, $x \in A_k$. Thus, we have the following flexible classifying decision rules:

$$\text{If for } \forall \quad j \neq i, m_{A_i}(x) > m_{A_j}(x), \quad \text{then } x \in A_i \ (i = 1, 2, \ldots, c) \qquad (18.2)$$

Certainly, these rules can also be formulated as:

$$\text{If } m_{A_i}(x) = \max\{m_{A_1}(x), m_{A_2}(x), \ldots, m_{A_n}(x)\}, \quad \text{then } x \in A_i \ (i = 1, 2, \ldots, c) \qquad (18.3)$$

Thus, flexible classifying is also the classifying with membership functions.

Observing membership-degree $m_A(x) = \max\{m_{A_1}(x), m_{A_2}(x), \ldots, m_{A_n}(x)\}$, it can be seen that necessarily $m_{A_i}(x) \geq 0.5$. Because if $m_{A_i}(x) < 0.5$, then from the complementary law of degrees, should $m_{A_i}(x) = m_{\neg A_k}(x) > 0.5$, which is obviously contradictory to $m_{A_k}(x) = \max\{m_{A_1}(x), m_{A_2}(x), \ldots, m_{A_n}(x)\}$ (from the relation between basic linguistic values, $\neg A_k$ is a semi-peak value of basic flexible linguistic value A_l adjacent to A_k, so $\neg A_k$ is also in π). Thus, with respect to the basic flexible classes, $m_{A_k}(x) > 0.5$ also equals to say $m_{A_k}(x) = \max\{m_{A_1}(x), m_{A_2}(x), \ldots, m_{A_n}(x)\}$.

Based on the analysis above, we give the following definition.

Definition 18.1 Let $\{A_1, A_2, \ldots, A_n\}$ be a flexible partition of feature space U. For any $x \in U$, if $m_{A_k}(x) > 0.5$, then we call object x belongs to flexible set $A_k \subset \{A_1, A_2, \ldots, A_n\}$, denote as $x \in A_k$.

Thus, we have another kind of decision rules of flexible classifying:

$$\text{If } m_{A_k}(x) > 0.5, \quad \text{then } x \in A_i \ (i = 1, 2, \ldots, c) \tag{18.4}$$

$$\text{if } m_{A_k}(x) < 0.5, \quad \text{then } x \in \neg A_i \ (i = 1, 2, \ldots, c) \tag{18.5}$$

As thus, $m_A(x) = 0.5$ is just the plane of demarcation between flexible classes A_k and $\neg A_k$.

Actually, the plane of demarcation $m_{A_k}(x) = 0.5$ is just the plane of demarcation $g_i(x) - g_j(x) = 0$ between adjacent flexible classes constructed by discrimination functions $g_i(x)$ and $g_j(x)$ $(j \neq i)$ in usual pattern recognition. We show this fact by one-dimensional flexible classes A_k and $\neg A_k$ below. From the general expression of membership function of one-dimensional flexible classes (see Eq. (2.5) in Sect. 2.2.2), it can be seen that the expressions of membership functions of flexible classes A_k and $\neg A_k$ at part of intersection are separately $\frac{s_{A_k}^+ - x}{s_{A_k}^+ - c_{A_k}^+}$ and $\frac{x - s_{\neg A_k}^-}{c_{\neg A_k}^- - s_{\neg A_k}^-}$. Whereas the two expressions are just the discrimination functions of flexible classes A_k and $\neg A_k$ separately (at part of intersection), which are separately tantamount to usual discrimination functions $g_i(x)$ and $g_j(x)$. Thus, from the plane of demarcation $g_i(x) - g_j(x) = 0$, we can have the plane of demarcation between A_k and $\neg A_k$ as

$$\frac{s_{A_k}^+ - x}{s_{A_k}^+ - c_{A_k}^+} - \frac{x - s_{\neg A_k}^-}{c_{\neg A_k}^- - s_{\neg A_k}^-} = 0$$

In consideration of $c_{A_k}^+ = s_{\neg A_K}^-$ and $s_{A_k}^+ = c_{\neg A_k}^-$, thus from the equation above, we can deduce

$$2\frac{x - s_{\neg A_k}^-}{c_{\neg A_k}^- - s_{\neg A_k}^-} - 1 = 0,$$

namely

$$\frac{x - s_{\neg A_k}^-}{c_{\neg A_k}^- - s_{\neg A_k}^-} = 0.5$$

that is

$$m_{\neg A_k}(x) = 0.5 \quad \left(s_{\neg A_K}^- < x < c_{\neg A_k}^- \right)$$

thus,

$$\frac{s_{A_k}^+ - x}{s_{A_k}^+ - c_{A_k}^+} = 0.5$$

that is

$$m_{A_k}(x) = 0.5 \quad \left(c_{A_k}^+ (< x < s_{A_k}^+ \right)$$

That is to say, $m_{A_k}(x) - m_{\neg A_k}(x) = 0$ is equivalent to $m_{A_k}(x) = 0.5$ or $m_{\neg A_k}(x) = 0.5$.

Conversely, from $m_{A_k}(x) = 0.5$ and $m_{\neg A_k}(x) = 0.5$, it follows obviously that $m_{A_k}(x) - m_{\neg A_k}(x) = 0$.

Yet on the other hand, we know, $m_{A_k}(x) = 0.5$ is originally the median plane of flexible class A_k (also flexible class $\neg A_k$); hence, it is just the plane of demarcation between extended core core(A_k) and core($\neg A_k$). Thus, the above analysis shows that we in actual can take directly the plane of demarcation between extended cores of adjacent flexible classes as the plane of demarcation between corresponding flexible classes. This means that the flexible classifying can also be realized by extended cores of flexible classes but without membership functions.

Note that the above flexible classifying method is actually for the relatively complement flexible classes. For relatively opposite flexible classes (see Sect. 8.6), then the decision condition $m_{A_k}(x) > 0.5$ of classifying needs to be changed to $m_{A_k}(x) > 0$, and $m_{A_k}(x) < 0.5$ be changed to $m_{A_k}(x) < 0$.

In the above, we introduced the flexible classifying method with membership functions (or extended cores). It can be seen that as long as having the membership functions (or extended cores) of flexible pattern classes, the main problem of flexible classifying is then solved. Thus, the main work of flexible classifying is to obtain the membership functions of the corresponding flexible classes (which is similar to the main work—try to obtain the discrimination functions—of classifications of the usual statistical pattern recognition). Then, for the flexible classes with direct membership functions, the problem becomes very simple; but for the flexible classes without direct membership functions, then needing by certain method to indirectly obtain their membership functions. Next, we introduce a kind of method to obtain the membership functions (or extended cores) of the related flexible classes by decomposing linguistic values and further realize flexible pattern recognition.

18.2.2   An Example of Flexible Pattern Recognition

As we know, the basic strokes of Chinese characters are horizontal, vertical, left-falling, right-falling, dot, hook, etc., which are basic components of Chinese characters; any Chinese character is all assembled by these basic components. Therefore, these basic strokes are also basic linguistic values on the character pattern space of Chinese characters. Since these strokes do not have strict and rigid writing standards (for instance, "horizontal" does not have to be absolutely level, "vertical" does not have to be absolutely upright, and as for "left-falling" and "right-falling", they are more less strict), therefore, these basic strokes are actually also the basic flexible linguistic values on the character pattern space of Chinese characters. Then, any character (character pattern) made up of these basic flexible linguistic values is naturally a composite flexible linguistic value on the character pattern space of Chinese characters. Similarly, the 26 letters in English are also the basic flexible linguistic values on the character pattern space of English language, viewed from the character pattern, and any English word is also a composite flexible linguistic value made up of these basic flexible linguistic values. Besides, 10 Arabic numbers are also the basic flexible linguistic values on the character pattern space of numbers, viewed from the character pattern, and any number is also just a composite flexible linguistic value made up of these basic flexible linguistic values.

Examining these the formation of these composite flexible linguistic values, it can be seen that the writing order of the basic linguistic values should be logical relation of "and" one by one, so such composite linguistic values are the combined linguistic values made up of its basic values, logically. Thus, viewed from the character pattern, every Chinese character is a combined value made up of basic strokes, every English word is a combined value made up of basic letters, and every number character is a combined value made up of basic numbers. Since different combined values have different make-up manners, therefore, different structure forms are formed.

For instance, for numeric character 6, we write it as the "standard form" as shown in Fig. 18.1a, from its structural characteristics, we then cut it up into two parts (as shown in Fig. 18.1b). It can be seen that these two parts can be made up of directed line segments a, b, c, and d (as shown in Fig. 18.1c). Therefore, directed line segments a, b, c, and d can be treated as basic components, while the whole numeric character 6 can be viewed to be formed by line segments a, b, c, and

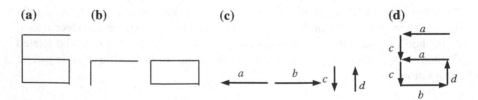

Fig. 18.1 Illustration of the formation of numeric character 6

Fig. 18.2 Structure tree of standard 6

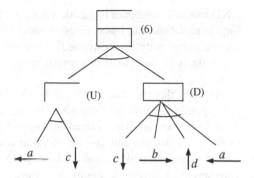

d according to the writing order of numeric character 6, linking from the beginning to the end and forming first the upper part then the lower part (as shown in Fig. 18.1d). That is to say, this standard numeric character 6 is formed by standard directed line segments *a*, *b*, *c*, and *d* according to certain order and structure. Thus, here *a*, *b*, *c*, and *d* are just basic linguistic values describing numeric character 6, and numeric character 6 is described as *accbda* by the 4 basic linguistic values, whose structure is shown in Fig. 18.2.

The basic linguistic values *a*, *b*, *c*, and *d* given above are standard vertical and horizontal directed line segments. However, the numeric character 6 usually written by people is actually not of that standard form, in which the segments are only approximate *a*, *b*, *c*, and *d*, or in other words, flexible *a*, *b*, *c*, and *d*. For distinguishing, we denote flexible *a*, *b*, *c*, and *d* in boldface letter as **a**, **b**, **c**, and **d**. Thus, the numeric character 6 we usually write is "flexible shaped 6", while **a**, **b**, **c**, and **d** are its basic flexible linguistic values, "flexible shaped 6" is a combined value of these basic flexible linguistic values. Thus, we only need to change the above *a*, *b*, *c*, and *d* that describe "standard 6" into **a**, **b**, **c**, and **d**, and the structure tree of "flexible shaped 6" (as shown in Fig. 18.3) is obtained.

From the figure, it can be seen that the structure tree of this combined flexible linguistic value "flexible shaped 6" is an "AND tree" in form. The root node of this

Fig. 18.3 Structure tree of "flexible shaped 6"

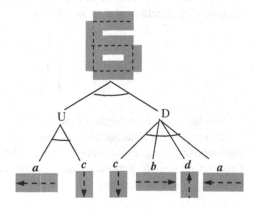

AND tree is a combined linguistic value, the leaf nodes are the corresponding basic linguistic values, and intermediate nodes have double identities: component values for the upper node and combined values for the lower nodes. The whole structure tree reflects the structural characteristics of the corresponding combined linguistic value.

As this "flexible shaped 6" can be decomposed into a logical combination of corresponding component values, then, as long as the membership functions of flexible classes that basic values correspond to are given, the membership function of flexible class that this combined value corresponds to can be obtained. We use still "flexible shaped 6" and a, b, c, d to denote the flexible classes that they correspond to as flexible linguistic values.

Let the membership functions of basic flexible classes a, b, c, and d be $m_a(x)$, $m_b(y)$, $m_c(z)$, and $m_d(u)$. Then, from the structure tree of "flexible shaped 6", we have

$$m_U(x,z) = \min\{m_a(x), m_c(z)\},$$
$$m_D(x,y,z,u) = \min\{m_c(z), m_b(y), m_d(u), m_a(x)\},$$
$$m_6(x,y,z,u) = \min\{m_U(x,z), m_D(x,y,z,u)\}$$
$$= \min\{m_a(x), m_c(z), m_c(z), m_b(y), m_d(u), m_a(x)\}$$
$$= \min\{m_a(x), m_b(y), m_c(z), m_d(u)\}$$

That is, the membership function of "flexible shaped 6" is

$$m_6(x,y,z,u) = \min\{m_a(x), m_b(y), m_c(z), m_d(u)\}$$

We now give the expressions of membership functions of basic flexible classes a, b, c, and d. We define a, b, c, and d all as single-point-core flexible sets. Taking a as an instance, let absolute horizontal directed line segment a_1 be the cluster center of a, then, the degree x of the angle between a stroke s with a_1 is the measurement of stroke s, and the a is a flexible class in measurement space $[-180, 180]$. Let the two directed line segments that have included angle of $\pm 20°$ with a_1 are separately the positive and negative critical elements of a, then the expression of membership function of a is as follows

$$m_a(x) = \begin{cases} \frac{x+20}{20}, & -20 \leq x < 0 \\ 1, & x = 0 \\ \frac{20-x}{20}, & 0 < x \leq 20 \end{cases} \qquad (18.6)$$

Similarly, the membership functions of b, c, and d can also be given.

Thus, we have obtained the membership function of "flexible shaped 6" through decomposing linguistic values. Then, take this function as the discrimination function of flexible pattern class "flexible shaped 6", the recognition of numeric character 6 can be realized.

It can be seen from Fig. 18.3 that we can also actually use geometric method to judge whether a line segment is in the extended core of a flexible line segment, then to judge whether a whole combination of awaiting recognition line segments belongs to "flexible shaped 6".

Of course, the above is only an example. But from that it can be seen that for numeric characters, letters, written words, figures, pictures, and even sceneries, etc., this kind of method can all be used to realize the corresponding flexible pattern recognition. The point of this method is two: firstly, the considered pattern classes can be reduced to certain kinds of flexible linguistic values; secondly, try to use the method of decomposing linguistic value to obtain the membership functions (or extended cores) of related flexible classes.

Actually, in usual, this type of pattern recognition of written characters and figures is more dealt with by using the method of structural pattern recognition. Structural pattern recognition generally uses rules and inferences to realize recognition. But here we then used function computation to realize the recognition.

The usual classifying actually generally does not know or does not need to construct the classes themselves, only the discrimination function, even the line or plane of demarcation are needed, while the membership function of a flexible class in flexible classifying is also tantamount to the flexible class itself. So the difference between the two lies in that such a flexible class is a subspace of a continuous measurement space, while a usual class is then a set of some non-continuously distributed points, and the classes are also disjoint. Usual classifying problems are all that known some non-continuously-distributed points with class-label, then use certain method to derive the lines or planes of demarcation or the discrimination function of corresponding classes. Then, can such methods be applied to flexible classifying? That is, known some non-continuously-distributed points with class-labels and membership-degrees, then use certain method to derive the membership functions of corresponding flexible classes. This problem is also tantamount to finding the membership function of a flexible concept from a subset of instances of a known flexible concept, or in other words, known a part of corresponding values of a membership function to find the whole corresponding values. If this problem can be solved, then this is using the method of finding membership functions to obtain line and plane of demarcation. Then, can flexible rope, flexible line, flexible plate, and flexible plane be treated as flexible classes to obtain the corresponding lines and planes of demarcation? That is a problem that needs our study.

18.2.3 Flexible Classifying of Multiple Conclusions

The above flexible classifying are all the usual flexible classifying of which the classifying conclusion only contains one flexible classification. But if the conclusion is set as or required to be multiple flexible classifications, then the conclusion of corresponding flexible classifying are the multiple flexible classes with membership-degree:

$$(A_1, d_{A_1}), (A_2, d_{A_2}), \ldots, (A_m, d_{A_m})$$

here A_1, A_2, …, A_m are flexible classifications as conclusion set in advance, $d_{A_i} = m_{A_i}(x)$ ($i = 1, 2, \ldots, m$). And the classifying process is

$$x \rightarrow \begin{cases} m_{A_1}(x) \\ m_{A_2}(x) \\ \ldots \\ m_{A_m}(x) \end{cases} \rightarrow \begin{cases} x \in (A_1, d_{A_1}) \\ x \in (A_2, d_{A_2}) \\ \ldots \\ x \in (A_m, d_{A_m}) \end{cases} \tag{18.7}$$

We call the flexible classifying of which the conclusion contains multiple flexible classifications to be the flexible classifying of multiple conclusions, and call usual the flexible classifying of which the conclusion contains only one flexible classification to be the flexible classifying of one conclusion.

By the way, the classifying problem is actually a typical problem. It actually represents a class of problems. In fact, in addition to recognition problems, other problems such as diagnosis, forecasting, and judging can be summed up as or in essence are also classifying problem.

18.3 Flexible Judging with Consistency Functions

We refer to the judging whose result is flexible linguistic value as **flexible judging**.

18.3.1 Usual Flexible Judging

1. Some concepts on flexible judging Like the usual judging, flexible judging also involves the objects, indexes, bases, models, ways, and methods of judging.

Objects of judging are the objects being judged, which can be all kinds of things, such as a paper, a class, a performance, a product, a design, or a software system, an engineering project, an enterprise, a school, even the economic system of a country or the whole world.

Indexes of judging are the items of the content of judging, also the features of objects judged. In order to judge an object as accurately as possible, usually the feature of the object is decomposed into multiple or multilevel more specific and detailed features, thus forming a feature tree, that is, a judging index system. For instance, the feature tree shown by Fig. 18.4 can be viewed as an index system. From the relationship between features and measurement spaces, an index system also has a corresponding measurement space tree.

Fig. 18.4 An example of
feature tree

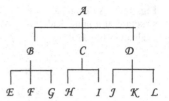

Bases of judging are the formulation of the feature values really possessed by objects judged, which can be numerical values or linguistic values. Numerical values are also a kind of score, and linguistic values are a kind of grade. For example, the hundred-mark system and the five-grade marking system to evaluate students' academic performance in schools nowadays are, respectively, numerical values and linguistic values we speak here. The numerical values need a formula or a function to synthesize, and we refer to this kind of function as **numerical evaluation function** of objects judged. The linguistic values also need a formula or function to synthesize, and we refer to this kind of function as the **linguistic-valued evaluation function** of objects judged. Flexible judging can also be realized by adopting (the set of) flexible rules called evaluation rules and through reasoning (some linguistic-valued evaluation functions can also be represented as the form of a set of production rules). Numerical evaluation functions, linguistic-valued evaluation functions, and evaluation rules are the models of flexible judging.

With the evaluation functions, the numerical feature values or linguistic feature values of objects judged can be computed or calculated, according to this then a decision can be made for objects. So flexible judging with evaluation functions also needs to have the corresponding decision functions and decision rules. The flexible judging with evaluation rules needs to have the corresponding inference mechanism, which we will introduce in Sect. 18.5.

2. Numerical Evaluation Functions and Linguistic-Valued Evaluation Functions

A judging system can have only one numerical evaluation function. This numerical evaluation function is just the numerical function in the measurement space that the evaluation index system corresponds to, which is just the function summarized and described by corresponding flexible linguistic function. Therefore, the numerical evaluation function is actually a compound function formed by multiple or multi-layer subfunctions, whose hiberarchy is completely the same as that of the judging index system. For instance, the construction of the numerical evaluation function to which the evaluation index system shown in Fig. 18.4 corresponds is shown in Fig. 18.5. Where the 8 characters at bottom are separately the variables on 8 bottom-layer measurement ranges, they are 8 independent variables of evaluation function; u_1, u_2, and u_3 at middle layers are the variables on 3 middle-layer measurement ranges, and they are separately the functions of corresponding bottom-layer independent variables, namely

Fig. 18.5 An example of the structure of a numerical evaluation function

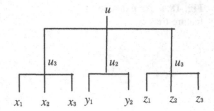

$$u_1 = f_1(x_1, x_2, x_3), u_2 = f_2(y_1, y_2), u_3 = f_3(z_1, z_2, z_3)$$

The u at the top is a variable on the top-layer measurement range A, which is a function of variables u_1, u_2, and u_3 on middle measurement range, namely

$$u = f_0(u_1, u_2, u_3)$$

To compound these two layers of functions, we have

$$u = f(x_1, x_2, x_3, y_1, y_2, z_1, z_2, z_3)$$

This is the evaluation function of the example system, whose range is $\mathcal{A}$, and the domain is $\mathcal{E} \times \mathcal{F} \times \mathcal{G} \times \mathcal{H} \times \mathcal{I} \times \mathcal{J} \times \mathcal{K} \times \mathcal{L}$. If every layer function is all weighted sum form, then the expression of the evaluation function is

$$u = w_1x_1 + w_2x_2 + w_3x_3 + w_4y_1 + w_5y_2 + w_6z_1 + w_7z_2 + w_8z_3$$

where $w_1 + w_2 + w_3 + w_4 + w_5 + w_6 + w_7 + w_8 = 1$.

A judging system also can have only a linguistic-valued evaluation function. This linguistic-valued evaluation function is just the flexible linguistic function on the measurement space to which the evaluation index system corresponds. Generally speaking, the linguistic-valued evaluation function to which an evaluation index system corresponds is represented as a linguistic-valued net.

3. **Discrimination Functions and Decision Rules**

(1) **Discrimination function based on a numerical evaluation function**
Judging with a numerical evaluation function is actually to reclassify the values of evaluation function of objects judged. For this purpose, we need to design a discrimination function for each of the preset judging results. Here, this discrimination function is just the consistency function of every final linguistic value as result. Since the linguistic values on the top measurement space are the final result linguistic values, so the consistency functions of all linguistic values on top measurement space are just the discrimination functions of the corresponding judging results. Thus, for the judging way with a numerical evaluation function, the work we need to do is to design corresponding numerical evaluation function according to the

judging index system, then to design all the linguistic values and their consistency functions, that is, discrimination functions, on the range of the numerical evaluation function, that is, the top measurement space. The specific steps are, for a given numerical evaluation function, firstly find its range U, then define corresponding basic flexible linguistic value A_i ($i = 1$, 2, ..., n) on U, thus the consistency functions:

$$c_{A_i}(u) \quad (i = 1, 2, \ldots, n)$$

follows, then, substitute evaluation function $u = f(x_1, x_2, \ldots, x_m)$ into them, we have

$$c_{A_i}(x_1, x_2, \ldots, x_m) \quad (i = 1, 2, \ldots, n) \tag{18.8}$$

These are the discrimination functions based on a numerical evaluation function. It can be seen that these discrimination functions are originally monovariate functions on the top one-dimensional measurement space U, but through evaluation function $u = f(x_1, x_2, \ldots, x_m)$, which are transformed into multivariate functions on the product of all measurement spaces at bottom layer.

(2) **Discrimination functions based on a linguistic-valued evaluation function**

The flexible judging with a linguistic-valued evaluation function is to calculate the concerned data of various preset conclusion linguistic values through linguistic-valued evaluation function according to the relevant data of objects judged, then to decide final conclusion linguistic values. It can be seen that for every preset conclusion linguistic value, its extended consistency function on the measurement spaces at bottom layer in index system can be obtained through corresponding linguistic-valued evaluation function, and on the basis of these consistency functions the final conclusion linguistic values can be chosen, so these consistency functions are just the discrimination functions based on a linguistic-valued evaluation function.

(3) **Decision rules**

With the discrimination functions, flexible judging problem is the same as flexible classifying. Let $U = U_1 \times U_2 \times \cdots \times U_n$ be the measurement space at bottom layer in a judging index system, and let $A_1, A_2, \ldots, A_m$ be the preset basic flexible linguistic values as the judging conclusion linguistic values on the top measurement space. With discrimination functions $c_{A_i}(x)$ ($i = 1, 2, \ldots, m$), we have flexible decision rules:

$$\text{If } c_{A_i}(x) > 0.5, \quad \text{then } A_i(x) \ (i = 1, 2, \ldots, m) \tag{18.9}$$

It can be seen that these flexible decision rules are actually also tantamount to the conversion rules from numerical values (vectors included) to flexible linguistic values.

From the above stated, it can be seen that the flexible judging with evaluation functions (including numerical evaluation function and linguistic-valued evaluation function) is all really based on the evaluation computation of consistency functions of conclusion flexible linguistic values.

Note that the above flexible judging methods are actually for the flexible linguistic values with negation. For the flexible linguistic values with opposite, the judging conditions $c_{A_i}(x) > 0.5$ should be changed into $c_{A_i}(x) > 0$ ($i = 1, 2, \ldots, m$).

18.3.2 Flexible Judging of Multiple Conclusions

The above flexible judging is usual flexible judging of which the judging conclusion only contains one flexible linguistic value. But if the judging conclusion is set as or required to be multiple linguistic values, then the conclusion of corresponding flexible judging is multiple flexible linguistic values with degree:

$$(A_1, d_{A_1}), (A_2, d_{A_2}), \ldots, (A_m, d_{A_m})$$

here A_1, A_2, ..., A_m are preset conclusion flexible linguistic values, $d_{A_i} = c_{A_i}(x)$ ($i = 1, 2, \ldots, m$). And the judging process is

$$x \rightarrow \begin{cases} c_{A_1}(x) \\ c_{A_2}(x) \\ \cdots \\ c_{A_m}(x) \end{cases} \rightarrow \begin{cases} (A_1, d_{A_1})(x) \\ (A_2, d_{A_2})(x) \\ \cdots \\ (A_m, d_{A_m})(x) \end{cases} \tag{18.10}$$

Thus, the decision rules of this kind of flexible judging are tantamount to the one-to-many conversion from numerical values (vectors included) to flexible linguistic values with degree.

We call the flexible judging of which the conclusion contains multiple flexible linguistic values to be the flexible judging of multiple conclusions, and call the flexible judging of which the conclusion contains only one linguistic value to be the flexible judging of one conclusion.

18.4   Flexible Programming Problem Solving

1. What is flexible programming?

The usual programming is to find the extreme value of a function under constraint conditions. The general mode is

$$\begin{cases} \min \quad y = f(x) \\ \text{s.t.} \end{cases}$$

where $y = f(x)$ is called objective function, and s.t. (abbreviation of subject to) followed by an equation or inequality is the constraint condition.

Obviously, the objective functions and constraint conditions of traditional programming problems are precise and rigid. However, the objective functions or constraint conditions of some practical problems can also be flexible: either the objectives are flexible, or the constraints are flexible, or the two are both flexible. Flexible objective means that the objective is not an exact number but may be a flexible number. Flexible constraint means that the constraint is not a strict equation or inequality, but a flexible equation or flexible inequality, that is, a flexible set. For example, the objective "profit being high" and the constraint "cost being low" are separately a flexible objective and a flexible constraint.

In order to distinguish, we refer to the programming whose objective is flexible objective or whose constraint is flexible constraint as **flexible programming**. Thus, viewed from the nature of the objective and constraint, flexible programming problems can be summarized into the following 4 types:

① rigid objective–rigid constraint type
② rigid objective–flexible constraint type
③ flexible objective–rigid constraint type
④ flexible objective–flexible constraint type

It can be seen that where type ① is the usual programming, that is, rigid programming, while the latter three are all flexible programming.

Next, we discuss the latter 3 kinds of flexible programming problems, which involve two aspects of description and solution.

2. Description of flexible programming problems

Actually, the flexible constraint is just that the corresponding "equal" is not strictly equal but approximately equal; the corresponding "not equal" is not strictly greater than or less than, but approximately greater than or approximately less than. Thus, the range of values of decision variable x is not an ordinary set, but a flexible set. Therefore, for a flexible constraint, we can use a flexible set to describe it. For example, let A be a flexible subset in the universe of discourse which represents "approximately equal to 0", then, we can use

$$\text{s.t.} \quad x \in A$$

to describe the constraint condition of "x is approximately equal to 0." For another example, if flexible set B stands for "low (cost)", then "(cost) x is low" and the constraint condition can be described as

$$\text{s.t.} \quad x \in B$$

Similarly, a flexible objective means the objective is not an exact number, but a flexible number. Then, the corresponding objective function is an N-L flexible linguistic function of "$y =$ about $f(x)$". This flexible linguistic function is a certain kind of flexible line or flexible plane. That is to say, for a flexible objective function, we can use a flexible linguistic function to describe.

3. Solving of flexible programming problems

We know that in the sense of flexible classifying (of complementary classes), a flexible set is completely stood for by its extended core (see Proposition 7.1 in Sect. 7.3.1). Thus, $x \in X \Leftrightarrow x \in \text{core}(X)^+$. Hence, $\text{core}(X)^+$ can be used to replace $x \in X$. As thus, the original flexible constraint becomes a rigid constraint. For instance, the above

$$\text{s.t.} \quad x \in A$$
$$\text{s.t.} \quad x \in B$$

can be rewritten as

$$\text{s.t.} \quad x \in \text{core}(A)^+$$
$$\text{s.t.} \quad x \in \text{core}(B)^+$$

That is to say, for a programming problem with flexible constraint we can use the method of "first flexible then rigid" to solve.

We also know that an N-L flexible linguistic function of "$y =$ about $f(x)$" always has a center line or center plane, while this center line or center plane is the usual function (in fact, this kind of flexible linguistic function is just constructed from such center line or center plane). Thus, we can first solve the programming problem with the center line or center plane as objective function, then make the result obtained into a flexible number, while this flexible number is just the objective value to be found.

That is to say, for a programming problem with flexible objective we can use the method of "first rigid then flexible" to solve.

To sum up the above analysis, we have the following conclusions:

① For a flexible programming problem of rigid objective and flexible constraint, we can use the method of "first flexible then rigid" to solve.

② For a flexible programming problem of flexible objective and rigid constraint, we can use the method of "first rigid then flexible" to solve.

③ For a flexible programming problem of flexible objective and flexible constraint, we can use both the methods of "first flexible then rigid" and "first rigid then flexible" to solve.

Finally, we point out that the flexible programming problems with flexible sets with opposite and their solving methods are similar to the above stated.

18.5   Flexible Linguistic Rule/Function-Based Systems

In this section, we consider the problem solving with flexible linguistic rules or flexible linguistic functions

From the previous Chaps. 15–17, we know that the methods of approximate reasoning with flexible linguistic rules have the reasoning with degrees, AT method, interpolation method and approximate global function method, and the methods of approximate evaluation of flexible linguistic functions have AT method, interpolation method and approximate global function method, and utilizing these methods of approximate reasoning and computing about linguistic values the approximate evaluation of corresponding background numerical functions can be realized. Besides, utilizing the exact reasoning with flexible linguistic rules or the exact evaluation of flexible linguistic functions, the approximate evaluation of corresponding background numerical functions can also be realized. Further examining and comparing these methods, it is not hard to see that although these methods of reasoning and computing have respective characteristics, they have all the order and process of "data conversing → processing (reasoning and computing) → data conversing"; therefore, the corresponding computer program systems can adopt uniform structure form—**flexible linguistic rule/function-based system**.

1. Architecture of a flexible linguistic rule/function-based system

A flexible linguistic rule/function-based system includes four basic components of "reasoning-computing" engine, flexible linguistic rule/function base, and input and output interfaces. Besides, there is a temporary component—dynamic database. Its architecture is shown in Fig. 18.6.

- The "reasoning-computing" engine is a program module, whose function is to execute (approximate) reasoning, computing, and system control. In general, for natural inference and reasoning with degrees, corresponding "reasoning-computing" engine can be design as a common program module similar to the inference engine in usual production systems; but for the exact evaluation of AT method and interpolation method, corresponding "reasoning-computing" engine can be design as the program module executing corresponding algorithms.

Fig. 18.6 Architecture of a
flexible linguistic
rule/function-based system

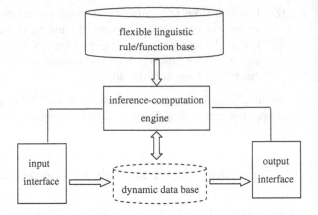

- The flexible linguistic rule/function base is a kind of data file that organizes and stores flexible linguistic rules (and adjoint functions) or flexible linguistic functions.
- The dynamic database is also called global database, synthetic database, etc., which is a kind of temporary dynamic data structure, used for storing source data, intermediate results, and final results, etc.
- Input and output interfaces are in charge of the conversion of the natures and formats of data. For example, for reasoning with degrees, input interface converts different form inputs of numerical data, pure linguistic values and so on into linguistic values with degree, then puts them into dynamic database for use in reasoning, while output interface converts the final result of reasoning into required linguistic value with degree, pure linguistic value, or number as the output of system. And for the approximate evaluation computing of background numerical functions, input interface then converts numbers into flexible linguistic values, output interface, then conversely, converts flexible linguistic values into numbers.

2. **Models and algorithms of reasoning and computation**

Models and algorithms of reasoning and computation in different flexible linguistic rule/function-based systems will be different from specific methods of reasoning and computation used. In the following, we introduce the corresponding models and algorithms in the case of reasoning with degrees.

Reasoning with degrees with flexible linguistic rules is a kind of imprecise reasoning. In reasoning with degrees, besides symbol matching and deduction, computation and transmission of degree are also involved. In addition, if the same result is deduced many times, then related synthesis method (e.g., mean, weighted sum, maximum, minimum) of final degree of this result also needs to be given. Similar to uncertain reasoning, in reasoning with degrees the related adjoint degreed functions, threshold, and the synthesis methods of degree computation and transmission together can also be called a kind of imprecise-reasoning model.

Similar to the usual production system, a flexible linguistic rule/function-based system can have two basic run ways of forward reasoning and backward reasoning. In the following, we give the two basic algorithms of the engine of reasoning with degrees for reference.

Forward reasoning algorithm

① Put initial linguistic values with degree into dynamic database.

② Take linguistic values with degree in the dynamic database to match objective condition, if the objective condition is satisfied, then the reasoning succeeds, end.

③ Take the antecedents of all rules in the rule base to match separately linguistic values with degree in dynamic database, organize matched succeed rules into a set of await-using rules.

④ If the set of await-using rules is empty, then reasoning fails, quit.

⑤ Select a rule from set of await-using rules by using a certain strategy to do the corresponding computation of degrees, then construct a corresponding linguistic value with degree from the obtained results and add it into dynamic database.

⑥ Cancel the set of await-using rules, turn to ②.

Backward reasoning algorithm

① Put the initial linguistic values with degree into dynamic database, put the objective condition into the OPEN table and CLOSED table in the dynamic database.

② If the OPEN table is empty, then the reasoning succeeds; then starting from the leaf nodes of the search tree recoded by CLOSED table, compute layer-by-layer degrees of all nodes, treat the root node and its degree as the resulting linguistic value with degree, and insert it into the dynamic database, end.

③ Move out the first node in OPEN table to match initial linguistic values with degree in the dynamic database; if the match succeeds, then turn to ②.

④ Take the consequents of all rules in the rule set to match with objective condition, take the antecedent of the rule which is first matched successfully and is unused as a new node, and put it with a pointer directing its father node into OPEN table and CLOSED table separately, then turn to ③.

⑤ If this node is the objective condition, then reasoning fails, quit.

⑥ Move the father node of this node back to OPEN table to replace this node and its brother nodes, turn to ③.

3. Building of systems

The flexible linguistic rule/function-based systems are a kind of software system of imprecise-problem solving with flexible linguistic rules or flexible linguistic functions. The work of building this type of system includes the acquisition of flexible linguistic rules or flexible linguistic functions, the establishment of rule/function bases, the design of "reasoning-computing" engine, and the design of input–output interfaces. It can be seen that building of this type of system is very similar to that of traditional expert/knowledge systems. Hence, the proper tools and environments of expert system or knowledge engineering can be employed to simplify and accelerate the development of these systems. If there is a common shell system, then for a specific problem solving, only the corresponding rule/function base need to be set up or the corresponding "reasoning-computing" engine need to be selected. Thus, when the rules in rule base are the classifying rules of a field or engineering problem, or when the "reasoning-computing" engine is classifying-oriented, this system is just a flexible classifying system; and when the rules in rule base are the evaluation rules of a field or object, or when the "reasoning-computing" engine is judging-oriented, this system is just a flexible judging system; and when the rules in rule base are the decision rules of a field or problem, or when the "reasoning-computing" engine is decision-oriented, this system is just a flexible decision system. With the support of the shell system, the main work of building flexible classifying, flexible judging, and flexible decision systems are to acquire related classifying rules or functions, judging rules or functions, and decision rules or functions of flexible linguistic values.

About the acquisition of flexible linguistic rules and flexible linguistic functions, we will have a special discussion in Chap. 19. But since flexible-linguistic-valued net is another representation form of a flexible linguistic rule set or a flexible linguistic function, so the former can be converted into the latter when the flexible linguistic valued net is known. For example, from the flexible-linguistic-valued net (see Fig. 18.3) describing the character pattern of "flexible shaped 6" in number recognition problem in Sect. 18.2.2 the following rule set follows:

$$6 ::= UD$$
$$U ::= ab$$
$$D ::= cbda$$

Here, the production rules are similar to the grammatical rules in formal language, which are the combination rules of graph objects. Actually, this method using production rules to describe the construction of objects is just the method usually used in structural pattern recognition.

Building a flexible linguistic rule/function-based system, the relevant practical problem can be solved through reasoning or computation by using this software system.

18.6   Principle of Flexible Control with Examples

We refer to the decision expressing in flexible linguistic value as **flexible decision**. We refer to the control whose decision is a flexible decision as **flexible control**. The general structure of a flexible control system with flexible rules or flexible linguistic function is shown in Fig. 18.7.

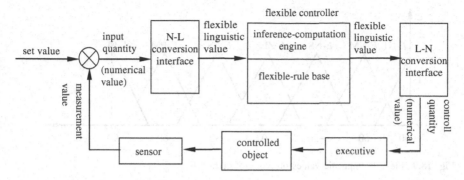

Fig. 18.7 Structure 1 of a flexible control system

From the figure, it can be seen that flexible control system is different from other control systems is that it use flexible controller; the flexible controller together with two data converting interfaces is a flexible linguistic rule/function-based system, of which the working process is as follows: firstly, convert input quantities (numerical values) into flexible linguistic values, then do reasoning and computing and obtaining result flexible linguistic value, and then convert the flexible linguistic value into a control quantity (a numerical value).

But if the "reasoning-computing" use the methods of AT, interpolation, or approximate computation with measured function, then the N-L and L-N conversions are not needed. Thus, the corresponding flexible control system can be simplified as what is shown in Fig. 18.8.

We give several examples of flexible control below.

Let the input quantities of a certain control system be error e and change rate Δe of error, let the controlling quantity be u, and let their ranges of values be $E = [-90, 90]$, $\Delta E = [-45, 45]$ and $U = [-18, 18]$ respectively. As shown in Figs. 18.9, 18.10 and 18.11, suppose the 7 basic linguistic values of negative large, negative medium, negative small, zero, positive small, positive medium, and

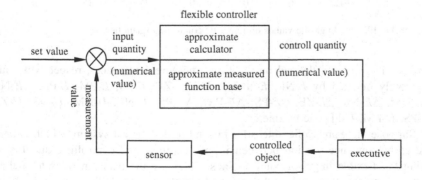

Fig. 18.8 Structure 2 of a flexible control system

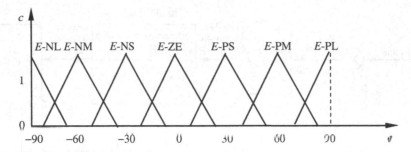

Fig. 18.9 Flexible linguistic values on range of errors, E

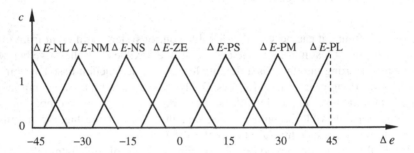

Fig. 18.10 Flexible linguistic values on range of rates of error change, ΔE

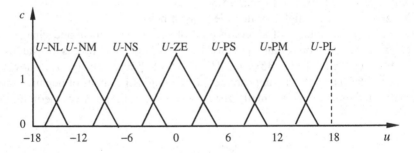

Fig. 18.11 Flexible linguistic values on range of controlling quantities, U

positive large are all defined on the three measurement spaces, respectively, and separately denoted by E-NL, E-NM, E-NS, E-ZE, E-PS, E-PM, E-PL; ΔE-NL, ΔE-NM, ΔE-NS, ΔE-ZE, ΔE-PS, ΔE-PM, ΔE-PL; U-NL, U-NM, U-NS, U-ZE, U-PS, U-PM, U-PL one by one.

Suppose there are 15 flexible rules in total in the control system, which express the correspondence relation between input quantities and controlling quantity by using the flexible linguistic values. These rules are the decision rules to make a certain control with respect to the states of object controlled. The rule set is shown

Table 18.1 A set of flexible control rules

U	ΔE						
E	ΔE-NL	ΔE-NM	ΔE-NS	ΔE-ZE	ΔE-PS	ΔE-PM	ΔE-PL
E-NL				U-PL			
E-NM				U-PM			
E-NS				U-PS	U-PS		
E-ZE	U-PL	U-PM	U-PS	U-ZE	U-NS	U-NM	U-NL
E-PS			U-NS	U-NS			
E-PM				U-NM			
E-PL				U-NL			

in Table 18.1. The linguistic values located at the first row and the first column in the table are two antecedent linguistic values of a rule, while the other linguistic values in the table are the consequent linguistic values of the rules which correspond to corresponding antecedent linguistic values.

Now split each of the rules whose antecedent linguistic values include full-peak value into the rule whose antecedent linguistic values are all semi-peak value. For instance, the rule located at row 5 and column 3

$$\text{If } e \text{ is } E\text{-PS and } \Delta e \text{ is } \Delta E\text{-NS,} \quad \text{then } u \text{ is } U\text{-NS}$$

after being split, the following 2 rules can be resulted (may not be all)

① If e is E-PS⁻ and Δe is ΔE-NS⁺, then u is U-NS⁺
② If e is E-PS⁺ and Δe is ΔE-NS⁻, then u is U-NS⁻

Thus, the critical points, median points, core–boundary points, and peak-value points of E-PS⁻ and E-PS⁺ on measurement space E are one by one:

$$s_{E\text{-PS}^-}^- = 10, m_{E\text{-PS}^-}^- = 15, c_{E\text{-PS}^-}^- = 20, \xi_{E\text{-PS}^-} = 30$$
$$s_{E\text{-PS}^+}^+ = 50, m_{E\text{-PS}^+}^+ = 45, c_{E\text{-PS}^+}^+ = 40, \xi_{E\text{-PS}^+} = 30$$

Then the consistency functions are

$$c_{E\text{-PS}^-}(e) = \frac{1}{10}e - 1, \quad 10 < e \le 30$$
$$c_{E\text{-PS}^+}(e) = -\frac{1}{10}e + 5, \quad 30 \le e < 50$$

And the critical points, median points, core–boundary points, and peak-value points of ΔE-NS⁻ and ΔE-NS⁺ on measurement space ΔE are one by one:

$$s^-_{\Delta E\text{-NS}^-} = -25, m^-_{\Delta E\text{-NS}^-} = -21.5, c^-_{\Delta E\text{-NS}^-} = -20, \xi_{\Delta E\text{-NS}^-} = -15$$
$$s^+_{\Delta E\text{-NS}^+} = -5, m^+_{\Delta E\text{-NS}^+} = -7.5, c^+_{\Delta E\text{-NS}^+} = -10, \xi_{\Delta E\text{-NS}^+} = -15$$

Then, the consistency functions are

$$c_{\Delta E\text{-NS}^-}(\Delta e) = \frac{1}{5}\Delta e + 5, \quad -25 < \Delta e \leq -15$$

$$c_{\Delta E\text{-NS}^+}(\Delta e) = -\frac{1}{5}\Delta e - 1, \quad -15 \leq \Delta e < -5$$

Also, the critical points, median points, core–boundary points, and peak-value points of U-NS$^-$ and U-NS$^+$ on measurement space U are one by one:

$$s^-_{U\text{-NS}^-} = -10, m^-_{U\text{-NS}^-} = -9, c^-_{U\text{-NS}^-} = -8, \xi_{U\text{-NS}^-} = -6$$
$$s^+_{U\text{-NS}^+} = -2, m^+_{U\text{-NS}^+} = -3, c^+_{U\text{-NS}^+} = -4, \xi_{U\text{-NS}^+} = -6$$

Then, the consistency functions are

$$c_{U\text{-NS}^-}(u) = \frac{1}{2}u + 5, \quad -10 < u \leq -6$$
$$c_{U\text{-NS}^+}(u) = -\frac{1}{2}u - 1, \quad -6 \leq u < -2$$

Now, suppose input quantities $e = 20$ and $\Delta e = -8$, it is asked to give the corresponding control quantity u.

In the following, we will use separately 4 kinds of different methods of approximate reasoning and computation to obtain the corresponding control quantity u.

(1) By method of natural inference with data conversion(s)

Substituting input quantities $e = 20$ and $\Delta e = -8$ into the consistency functions of the antecedent linguistic values of two rules above, it follows that

$$c_{E\text{-PS}^-}(20) = \frac{1}{10} \times 20 - 1 = 1 > 0.5$$

$$c_{\Delta E\text{-NS}^+}(-8) = -\frac{1}{5} \times (-8) - 1 = 0.6 > 0.5$$

$$c_{E\text{-PS}^+}(20) = 0 < 0.5$$

$$c_{\Delta E\text{-NS}^-}(-8) = 0 < 0.5$$

Consequently, we have facts E-PS$^-$ and ΔE-NS$^+$, and know that only premises "e is E-PS$^-$" and "Δe is ΔE-NS$^+$" of rule ① are satisfied. Thus, by rule (①)

$$\text{If } E\text{-PS}^- \text{ and } \Delta E\text{-NS}^+ \quad \text{then } U\text{-NS}^+$$

and facts

$$E\text{-PS}^- \quad \text{and} \quad \Delta E\text{-NS}^+$$

the conclusion U-NS$^+$ is deduced directly.

This is a result of a linguistic value, but process control requires numerical control quantity, so it needs to convert the flexible linguistic value U-NS$^+$ into a numerical value as control quantity u. It is known from L-N converting rule that arbitrary number $u_0 \in \text{core}(U\text{-NS}^+)^+ = [-6, -3)$ can all be taken as the value of control quantity u such as -4, -5 or -6. Here, we take $u = -3.2$.

(2) By method of reasoning with degrees

Similarly, from $e = 20$ and $\Delta e = -8$ as well as $c_{E\text{-PS}^-}(20) = 1$ and $c_{\Delta E\text{-NS}^+}(-8) = 0.6$, we know that only rule ① can be used, and facts (linguistic values with degree) (E-PS$^-$, 1.0) and (ΔE-NS$^+$, 0.6) follow. Thus, we use the following reasoning with degrees

$$(E\text{-PS}^- \wedge \Delta E\text{-NS}^+ \;\rightarrow\; U\text{-NS}^+\,; f_d(d_{E\text{-PS}^- \wedge \Delta E\text{-NS}^+}))$$

$$\frac{(E\text{-PS}^- \wedge \Delta E\text{-NS}^+\,, d_{E\text{-PS}^- \wedge \Delta E\text{-NS}^+})}{(U\text{-NS}^+\,, d_{U\text{-NS}^+})}$$

We use Eq. (14.12) in the reference models of the adjoint degreed functions of a rule with multiple conditions in Sect. 14.3.1, that is,

$$d_B = \frac{\beta_B - 0.5}{\beta_\wedge - 0.5}(d_A - 0.5) + 0.5$$

where $\beta_\wedge = \min\{\beta_{A_1}, \beta_{A_2}, \beta_{A_3}, \beta_{A_4}\}$, $d_A = \min\{d_{A_1}, d_{A_2}, d_{A_3}, d_{A_4}\}$, $0.5 < d_{A_j} \leq \beta_{A_j}$ ($j = 1, 2, 3, 4$).

Suppose the maximum of consistency-degrees of E-PS$^-$ is $\beta_{E\text{-PS}^-} = 1.5$, and the maximum of consistency-degrees of ΔE-NS$^+$ is $\beta_{\Delta E\text{-NS}^+} = 1.5$; And Suppose the maximum of consistency-degrees of U-NS$^+$ is $\beta_{U\text{-NS}^+} = 1.5$. Then, the actual adjoint degreed function of the rule ① above is

$$d_u = \frac{2 - 0.5}{2 - 0.5}(d_{e \wedge \Delta e} - 0.5) + 0.5$$

here $d_{e \wedge \Delta e} = \min\{d_e, d_{\Delta e}\}$, $d_e = c_{E\text{-PS}^-}(e)$ and $d_{\Delta e} = c_{\Delta E\text{-NS}^+}(\Delta e)$, $(d_e, d_{\Delta e}) \in (0.5, 2] \times (0.5, 2])$.

Also, from (E-PS$^-$, 1.0) and (ΔE-NS$^+$, 0.6), it follows that

$$(E\text{-PS}^-, 1.0) \wedge (\Delta E\text{-NS}^+, 0.6) = (E\text{-PS}^- \wedge \Delta E\text{-NS}^+, \min\{1, 0.6\})$$
$$= (E\text{-PS}^- \wedge \Delta E\text{-NS}^+, 0.6)$$

Consequently, by reasoning with degrees, we have the conclusion: $(U\text{-NS}^+, 0.6)$.

This is a result of a linguistic value with degree, so it needs to convert $(U\text{-NS}^+, 0.6)$ into a numerical value u. Thus, substituting $d_u = 0.6$ into the inverse function of $c_{U\text{-NS}^+}(u)$

$$c_{U\text{-NS}^+}^{-1}(d_u) = -2d_u - 2$$

we have

$$u = c_{U\text{-NS}^+}^{-1}(d_u) = -2d_u - 2 = -2 \times 0.6 - 2 = -3.2$$

namely

$$u = -3.2$$

This is just the final controlling quantity.

(3) **By AT method**

It can be seen that the input quantities $e = 20 \in \text{core}(E\text{-PS}^-)^+ = (15, 30]$ and $\Delta e = -8 \in \text{core}(\Delta E\text{-NS}^+)^+ = [-15, -7.5)$. So set 20 be $\xi_{E\text{-PS}^{-\prime}}$ and -8 be $\xi_{\Delta E\text{-NS}^{+\prime}}$, that is, take 20 and -8, respectively, as peak-value points of the flexible linguistic values $E\text{-PS}^{-\prime}$ and $\Delta E\text{-NS}^{+\prime}$ that are, respectively, approximate to $E\text{-PS}^-$ and $\Delta E\text{-NS}^+$ on the measurement spaces E and ΔE (as shown in Figs. 18.12 and 18.13). And on the other hand, known $\xi_{E\text{-PS}^-} = 30$ and $\xi_{\Delta E\text{-NS}^+} = -15$. Thus, distances $d_{E\text{-PS}^{-\prime}E\text{-PS}^-} = 30 - 20 = 10$ and $d_{\Delta E\text{-NS}^{+\prime}\Delta E\text{-NS}^+} = -8 - (-15) = 7$. Since $E\text{-PS}^{-\prime}$ and $\Delta E\text{-NS}^{+\prime}$ are separately approximate to the premise conditions $E\text{-PS}^-$ and $\Delta E\text{-NS}^+$ of rule ①, so the above rule ① would be used for approximate reasoning.

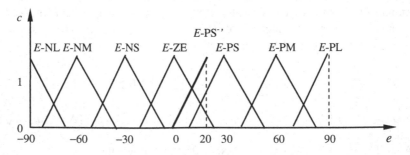

Fig. 18.12 The $E\text{-PS}^{-\prime}$ approximating to $E\text{-PS}^-$

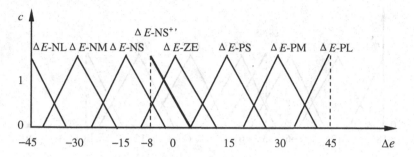

Fig. 18.13 The $\Delta E\text{-NS}^{+\prime}$ approximating to $\Delta E\text{-NS}^+$

Thus, according to AT method, the distance $d_{U\text{-NS}^{+\prime}U\text{-NS}^+}$ is required. From the peak-value point and median point gave in the problem, we have the approximate radii

$$r_{E\text{-PS}^-} = \xi_{E\text{-PS}^-} - m_{E\text{-PS}^-}^- = 30 - 15 = 15$$
$$r_{\Delta E\text{-NS}^+} = m_{\Delta E\text{-NS}^+}^+ - \xi_{\Delta E\text{-NS}^+} = -7.5 - (-15) = 7.5$$
$$r_{U\text{-NS}^+} = m_{U\text{-NS}^+}^+ - \xi_{U\text{-NS}^+} = -3 - (-6) = 3$$

Thus, by Formula (16.23), we have the distance

$$d_{U\text{-NS}^{+\prime}U\text{-NS}^+} = \max\left\{\frac{d_{E\text{-PS}^{-\prime}E\text{-PS}^-}}{r_{E\text{-PS}^-}}, \frac{d_{\Delta E\text{-NS}^{+\prime}\Delta E\text{-NS}^+}}{r_{\Delta E\text{-NS}^+}}\right\} \cdot r_{U\text{-NS}^+}$$
$$= \max\left\{\frac{10}{15}, \frac{7}{7.5}\right\}$$
$$= \frac{7}{7.5} \times 3$$
$$= 2.8$$

Since $U\text{-NS}^+$ is a positive semi-peak value, so by rule ①, $U\text{-NS}^{+\prime}$ ought to be located at the positive side of $U\text{-NS}^+$ (as shown in Fig. 18.14). Thus, by Formula (16.13) we have

$$\xi_{U\text{-NS}^{+\prime}} = \xi_{U\text{-NS}^+} + d_{U\text{-NS}^{+\prime}U\text{-NS}^+} = -6 + 2.8 = -3.2$$

If this $\xi_{U\text{-NS}^{+\prime}}$ is just taken as control quantity, then

$$u = -3.2$$

Of course, other numbers in extended core $\text{core}(U\text{-NS}^{+\prime})^+$ are also be taken as control quantity.

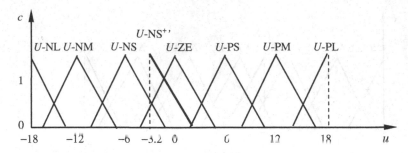

Fig. 18.14 The U-NS$^{+\prime}$ approximating to U-NS$^+$

(4) By method of computation of approximate measure function

We use Eq. (14.20) in the reference models of approximate measure functions of a rule with multiple conditions in Sect. 14.3.2, then the adjoint measured function of the above rule ① is

$$u = \frac{\xi_{U\text{-NS}^+} - s^+_{U\text{-NS}^+}}{\xi_{E\text{-PS}^-} - s^-_{E\text{-PS}^-}}(e - m^-_{E\text{-PS}^-}) + m^+_{U\text{-NS}^+}, \quad m^-_{E\text{-PS}^-} < e \le \xi_{E\text{-PS}^-}, \xi_{\Delta E\text{-NS}^-} \le \Delta e < m^+_{\Delta E\text{-NS}^+}$$

$$u = \frac{\xi_{U\text{-NS}^+} - s^+_{U\text{-NS}^+}}{\xi_{\Delta E\text{-NS}^+} - s^+_{\Delta E\text{-NS}^+}}(\Delta e - m^+_{\Delta E\text{-NS}^+}) + m^+_{U\text{-NS}^+}, \quad m^-_{E\text{-PS}^-} < e \le \xi_{E\text{-PS}^-}, \xi_{\Delta E\text{-NS}^-} \le \Delta e < m^+_{\Delta E\text{-NS}^+}$$

Substituting $s^-_{EPS^-} = 10$, $m^-_{EPS^-} = 15$, $\xi_{EPS^-} = 30$, $s^+_{\Delta ENS^+} = -5$, $m^+_{\Delta ENS^+} = -7.5$, $\xi_{\Delta ENS^+} = -15$, $s^+_{UNS^+} = -2$, $m^+_{UNS^+} - 3$ and $\xi_{UNS^+} = -6$ separately into the expressions of right side of the above two equations, then we have

$$u = -\frac{1}{5}e, \quad 15 \le e \le 30, \quad -15 \le \Delta e \le -5$$

$$u = \frac{2}{5}\Delta e, \quad 15 \le e \le 30, \quad -15 \le \Delta e \le -5$$

Similarly, we can have the adjoint measured function of rule ② to be

$$u = \frac{1}{5}e - 12, \quad 30 \le e \le 45, \quad -21.5 \le \Delta e \le -15$$

$$u = -\frac{1}{5}\Delta e - 13.5, \quad 30 \le e \le 45, \quad -21.5 \le \Delta e \le -15$$

From

$$c_{E\text{-PS}^-}(20) = \frac{1}{10} \times 20 - 1 = 1 > 0.5$$

$$c_{\Delta E\text{-NS}^+}(-8) = -\frac{1}{5} \times (-8) - 1 = 0.6 > 0.5$$

$$c_{E\text{-PS}^+}(20) = 0 < 0.5$$

$$c_{\Delta E\text{-NS}^-}(-8) = 0 < 0.5$$

it can be seen that input quantities $e = 20$ and $\Delta e = -8$ can satisfy and also only satisfy the premise conditions of rule ①, so the adjoint measured function of rule ① can be used for computing. Also since

$$\min\{c_{E\text{-PS}^-}(20), c_{\Delta E\text{-NS}^+}(-8)\} = \min\{1, 0.6\} = c_{\Delta E\text{-NS}^+}(-8)$$

So we use the second measure function of the above rule ① to compute control quantity. Substituting input quantities directly into the function expression, we have

$$u = \frac{2}{5}\Delta e = \frac{2}{5} \times (-8) = -3.2$$

namely

$$u = -3.2$$

Substituting $u = -3.2$ into the consistency function of flexible linguistic value $U\text{-NS}^+$ on space U, we have consistency-degree

$$c_{U\text{-NS}^+}(-3.2) = -\frac{1}{2} \times (-3.2) - 1 = 0.6 > 0.5$$

That shows that control quantity -3.2 is just $U\text{-NS}^+$ in terms of linguistic values. This is consistent with the reasoning result of the above (2) and (3).

In the above, the working process of a flexible controller was introduced through examples. It can be seen that here the methods of reasoning and computation all have mathematical and logical basis, so the accurateness of final result—control quantity will depends mainly on the accurateness of the field rules. If the field rules themselves were inaccurate or even wrong, then the reasoning result would naturally be inaccurate.

It would need to be noted that as the traditional fuzzy controller, for some control problems, the flexible controller here can also be further simplified into a control table, so as to be realized directly in hardware. In fact, if the accuracy requirement of control quantity of a practical problem can be met, the measurement spaces which input and output quantities belong to can become two discrete point sets by discretization, then the control function realized by a controller can be simplified into a digital list—control table. Thus, control can be directly performed through consulting the table. And this flexible controller that only has one control table can completely realized in hardware.

18.7   Anthropomorphic Computer Application Systems with Imprecise-Information Processing Ability

Examining the principles and methods of imprecise-problem solving in the previous sections, it can be found that the flexible classifying with membership functions and the flexible judging with consistency functions, abstractly, are the conversion from numerical values to flexible linguistic values, and the reasoning with flexible linguistic rules and the evaluation of flexible linguistic functions are the transformation between flexible linguistic values. That is to say, the interconversion between numerical values and flexible linguistic values, and the transformation between flexible linguistic values are the basic techniques of imprecise-information processing used in imprecise-problem solving. Associating our human information processing from these techniques and methods of solving imprecise problems on computer systems, it is not hard to find that the method of dealing with imprecise information of human is similar to that in engineering problems (actually, the latter just simulates the former).

In fact, in the process of daily information processing (including perceiving, thinking and expressing) in human brain, the interconversion between numerical values and flexible linguistic values, and the transformation between flexible linguistic values are also performed from time to time. For instance, in bitterly cold winter, when you go out of door and feel cold, you might say "it's so cold today." Then, this "cold" is just the flexible linguistic value converted from a numerical value of the outdoor temperature (as $-10\,^\circ\mathrm{C}$) through your nervous system (including sensory nerve and cerebral cortex) which reflects your macro psychological feeling (this conversion from numerical values to flexible linguistic values, i.e., from external stimuli to the psychological feeling, is a process of flexible classifying). For another example, when you hear the imperative sentence "walk faster!", you probably would stride forward with an appropriate speed according to the situation on the spot. Here, "fast" is just a flexible linguistic value, while your walking speed is a measurable value, that is, a numerical value. Thus, in the process from you hearing the flexible linguistic value "faster" to you changing your pace, there is a conversion from a flexible linguistic value to a numerical value in your nervous system (this conversion from numerical values to flexible linguistic values, i.e., from "faster" heard to the specific walking speed, is a process of flexible control). For additional example, "if it is cold, you wear more clothes" is one of our common sense of life, which is also a flexible linguistic rule. Then, when performing the rule, human brain does a transformation from flexible linguistic value "cold" to flexible linguistic value "more" through reasoning.

Now that there occurs the interconversion between numerical values and flexible linguistic values, and the transformation between flexible linguistic values in the process of human's perceiving, thinking, and expressing, then we can use these imprecise-information processing techniques and the machine understanding and generation techniques of imprecise information to simulate functionally the process of imprecise-information processing of human (on the computers), and to develop and build the computer application systems, or more general intelligent systems (especially, intelligent robots), which can perceive, think, and express in flexible linguistic values as human, so as to realize the anthropomorphic computer application systems and intelligent systems with imprecise-information processing ability.

18.8   Summary

In this chapter, we expounded the basic techniques of imprecise-problem solving—the engineering application of the imprecise-information processing, and then discussed the anthropomorphic computer application systems with imprecise-information processing ability.

The main points and results of the chapter are as follows:

- Imprecise problems are the practical problems or engineering problems containing or involving imprecise information or knowledge, of them, flexible classifying, flexible judging, flexible decision, and flexible programming are typical problems. Imprecise-problem solving can be reduced to two basic ways of computing with membership or consistency functions (or flexible sets or their extended cores) and reasoning and computing with flexible linguistic rules or flexible linguistic functions.
- The classifying whose classifications are flexible classes is flexible classifying, which can be classified as the usual flexible classifying of single conclusion and the special flexible classifying of multiple conclusions. Flexible classifying can be realized by taking the membership functions of the corresponding flexible classes as the discriminant functions or taking the plane of demarcation between extended cores of corresponding flexible classes as the plane of demarcation or taking the flexible linguistic rules or flexible linguistic functions as the decision rules or decision functions.
- The judging whose conclusions are flexible linguistic values is flexible judging, which can be classified as the usual flexible judging of single conclusion and the special flexible judging of multiple conclusions. Flexible judging can be realized by taking the consistency functions of the corresponding flexible linguistic values as the decision functions, can also be realized by taking the related flexible linguistic rules as the evaluation rules and through reasoning.

- The decision expressed by flexible linguistic values is flexible decision. Flexible decision can generally be realized through reasoning or computation by using flexible linguistic rules or flexible linguistic function.
- The programming whose objective functions or constraints are flexible objectives or flexible constraints is flexible programming. Flexible programming can be solved by methods of "first flexible then rigid," "first rigid then flexible," and the combination of the two.
- Flexible linguistic rule/function-based system is a type of common problem solving system, which solves corresponding problem through (approximate) reasoning and computation with relevant flexible linguistic rules or flexible linguistic functions. The architecture and building process and method of the problem solving system are similar to usual expert systems or knowledge-based systems.
- The interconversion between numerical values and flexible linguistic values, and the transformation between flexible linguistic values are the basic techniques of imprecise-information processing used in imprecise-problem solving, and using these techniques, we can also build the anthropomorphic computer application systems and intelligent systems with imprecise-information processing ability.

Reference

1. Lian S (2009) Principles of imprecise-information processing. Science Press, Beijing

Chapter 19
Imprecise-Knowledge Discovery

Abstract This chapter further discusses firstly the formation mechanism and essential characteristic of flexible concepts and discriminates between the flexible concept and the vague (fuzzy) concept; then, it explores imprecise-knowledge discovery, proposes some ideas and approaches, and presents some specific algorithms and methods.

Keywords Imprecise-knowledge discovery · Data mining · Machine learning

Knowledge discovery, or data mining, is an important research area of the current artificial intelligence technology. In this chapter, we further discuss firstly the formation mechanism and essential characteristic of flexible concepts as well as the relation and distinction between flexible concepts and vague (fuzzy) concepts on the basis of previous chapters and then discuss techniques and methods of imprecise-knowledge discovery.

19.1 Instinct of Clustering and Summarizing and Natural Classifying Mechanism of Human Brain and Flexible Concepts

The author infers after studies that clustering and summarizing is an intrinsic instinct of human brain and that the brain also has a natural classifying mechanism [1]. For example, we can see different colors when visible lights of different wavelengths come to our eyes, e.g., the light of wavelengths of 622–760 nm is red and that of 492–577 nm is green. We know that light waves originally have no color, but only are of different lengths, yet we see corresponding colors after they come to our eyes. These different colors are actually the result of human brain's clustering and summarizing of light waves according to their wavelengths. Therefore, the process from lights coming to our eyes to see corresponding colors is actually a process of human brain classifying the lights.

© Springer Science+Business Media Singapore 2016   473
S. Lian, *Principles of Imprecise-Information Processing*,
DOI 10.1007/978-981-10-1549-6_19

Further, physiology tells us that optic nerve cells (called cone cells) sensitive to the three colors of red, green, and blue, respectively, are distributed on the retina of human eyes. When lights come to our eyes, those of different wavelengths are felt by different cone cells, and then, the corresponding stimuli are transmitted to the visual cortex of brain, thus forming visual senses—colors. This is to say, lights which come to eyes are allocated to corresponding color classes by the three kinds of cone cells. This shows that the mechanism of automatically classifying lights is what human brain has naturally, which is embodied in the structure of human's nervous system.

Similarly, acoustics and physiology tell us that the scale and tone of sound are determined by the frequencies of sound waves and different frequencies result in different scales. This indicates that the sound we hear is actually the result of brain's classifying of sound wave frequency signals received by ears. Obviously, this kind of classifying mechanism of human brain is also natural-born.

Actually, bitter and sweet of taste, fragrant and smelly of smell, soft and hard of sense of touch, the hot and cold feelings of body, and so on are all results of brain classifying corresponding sensory stimulus signals. And the corresponding classifying mechanisms are also innate. For example, specialized neurons which can separately feel hundreds of different flavors are just distributed on human's tongue, and there are also specialized neurons in the nose which can feel hundreds of thousands of different odors.

Now, we consider whether human brain also has a corresponding natural classifying mechanism for such feelings as "big," "much," "tall," "fast," "beautiful," and "good". It is not hard to see that these feelings are also the results of human brain clustering and summarizing corresponding numerical feature values, that is, some flexible classifications partitioned. When we face an object, the corresponding feeling we get is the result of our brain classifying the object according to the corresponding observations or appraisal values. We believe these classifications are not acquired by people individually clustering and summarizing later in life, and they are also embodied in the existing classifying mechanism of human brain. That is to say, these flexible concepts in the human brain are not acquired by humans consciously clustering and summarizing the magnitudes of features of things in life; rather, they are preset innately, while only names are given afterward to those corresponding flexible classes, respectively. In fact, it is also impossible that each individual personally clusters and summarizes from scratch after birth to formulate corresponding flexible concepts. This is because we cannot watch all people of all heights to cluster "tall" and "short," nor can we examine all people of all ages to cluster "old" and "young." Actually, infants and children learn these flexible concepts not by clustering and summarizing but just by remembering the names of corresponding concepts through a few examples. For example, the author of the book once heard a two-and-a-half-year-old child say "wind so strong" and "I'm tired." It can be seen that the two flexible concepts in the child's brain, "strong" of wind force and "tired" of body, are inherent, while the two words, "strong" and "tired," representing those concepts are what the child heard from adults around.

Color, sound, odor, flavor, and so on are psychological feelings formed from the direct stimulation of sensory organs by certain kinds of physical signals, which are the clustering and summarizing of the body's physiological quantities formed by the outside objective physical quantities. We may as well call these feelings the first-level feelings or primary feelings and on the basis of which there are also higher-level feelings, such as second-level, third-level, and even n-level. For example, "big," "much," "tall," "fast," "beautiful," "noble," "diligent," and "excellent" are some high-level feelings. A higher-level feeling is a feeling formed on the basis of lower-level feelings. For example, a beautiful painting is formed by various colors and a piece of pleasing music is formed by various sounds. This is to say, the aesthetics of works of paintings and music are formed on the basis of colors and sounds, respectively, while the quality of the works of an artist is the reflections of his/her artistic level; thus, the feeling of the artistic level of an artist is based on the quality of his/her works. It can be seen that high-level feelings result from the psychological quantities in the human brain formed by objective things (such as numbers, shapes, words, behaviors, events), which represent the information of certain semantics, logic, or image. That is, high-level feelings are the clustering and summarization of psychological quantities. Unlike primary feelings, high-level feelings seem to be formed through a process of judging and evaluation, and there exist no specialized sensory neurons for these feelings. But like primary feelings, when we face an object and get a certain high-level feeling (and then describe the feeling in corresponding words), this is actually as classifying the object according to the observation or appraisal values.

Actually, the psychological quantities stated in the above are just a kind of marks, i.e., a kind of subjective measurement values, given after people estimating things; and some psychological quantities and physiological quantities can also be represented by the corresponding objective measurement values. Therefore, our high-level feelings and primary feelings are flexible concepts on these measurement values, while the linguistic expressing of these feelings is flexible linguistic values on the corresponding measurement ranges.

To sum up, we believe that the mechanisms of human brain flexibly clustering and flexibly classifying continuous magnitudes of features of things are innate and that these mechanisms are gradually formed and developed in human being's or even living being's evolution with the increase of the quantity of information that the nervous system processes. This is to say, the natural flexible clustering and flexible classifying functions of human brain are adapted to its information environment, which is just as the body structure and flying ability of birds are fitted with the physical environment of air.

19.2   Further Understanding Flexible Concepts and Flexible Treating Mechanism of Human Brain

1. From flexible clustering to flexible classifying

By the decision rule "if $m_{A_k}(x) > 0.5$, then $x \in A_i$" of flexible classifying in Sect. 18.2, any x whose membership-degree $m_A(x) > 0.5$ is all regarded as belonging to flexible class A. As thus, those objects with membership degrees between 1 and 0.5 which are originally not core members of flexible class A are actually also treated as core members of A in flexible classifying. By the decision rule "if $c_{A_i}(x) > 0.5$, then $A_i(x)$" of flexible judging in Sect. 18.3, all x whose consistency-degrees $c_A(x) > 0.5$ are regarded as having flexible linguistic value A. As thus, those objects with consistency-degrees between 1 and 0.5 which originally did not completely have flexible linguistic value A are actually treated as completely having value A in flexible judging. Actually, the judging is essentially a kind of classifying. Thus, viewed from the angle of classifying, flexible concepts are also flexible.

On the other hand, from Sect. 7.3.1, we know that the conversion from a flexible linguistic value to a numerical value is not unique, but has a range of values—the extended core. That is to say, not only the numbers in core (i.e., the core members) can be treated as the representative of corresponding flexible linguistic value, but those numbers that are outside the core in the extended core, originally peripheral members, can also be treated as the representatives of corresponding flexible linguistic values. Thus, this kind of L-N conversion has a certain elasticity; that is, it is also flexible. Consideration of a flexible decision (such as flexible control) is always expressed as a certain numerical value, that is, to do an L-N conversion, in the specific implementation. Therefore, viewed from the decision-making process, flexible concepts are still flexible.

Now, we see that the "flexible" of the flexible concepts not only is shown in the formation (clustering) of flexible concepts, but is also shown in people's applying of them (like classifying, judging, and decision). Therefore, only from the two aspects, the formation and application, of flexible concepts, can we understand flexible concepts fully and entirely.

2. Rigid-ening of flexible linguistic values

However, the flexible classifying, flexible judging, and flexible decision making are having their bottom lines. In fact, the median point, median line, or median plane in the boundary region of two adjacent flexible classes is just the bottom line of corresponding flexible treating. That is to say, in the case that the bottom line is not touched, classifying, judging, and decision making are flexible, but once the bottom line is touched, they then become rigid.

Fig. 19.1 An illustration of rigid-ened basic flexible classes in two-dimensional space

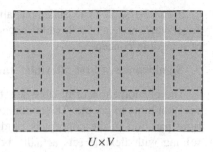

$$U \times V$$

In fact, from

$$x \in A \Leftrightarrow x \in \text{core}(A)^{+}$$

it can be seen that if viewed from classifying or membership relation, a flexible set A is actually equivalent to its extended $\text{core}(A)^{+}$ or is stood for completely by its extended $\text{core}(A)^{+}$. While the latter is a rigid set, this is tantamount to saying that flexible set A is rigid-ened. Thus, the basic flexible classes of the original measurement space are also rigid-ened in the sense of classifying. For instance, basic flexible classes rigid-ened in two-dimensional measurement space are shown in Fig. 19.1. Here, the white straight lines were the original median lines of the boundary of two adjacent flexible classes, and now, they are the boundary lines of two adjacent rigid-ened flexible classes (the blocks encircled by white lines in the figure are the extended cores of corresponding flexible sets, and black broken line blocks are the cores of corresponding flexible sets).

On the other hand, from $A(x) \Leftrightarrow x \in \text{core}(A)^{+}$, it can be seen that flexible linguistic value A is actually equivalent to rigid linguistic value $\text{core}(A)^{+}$, or in other words, it is completely replaced by rigid linguistic value $\text{core}(A)^{+}$. That is tantamount to saying that the flexible linguistic value A is rigid-ened. Thus, original basic flexible linguistic values on measurement space are also rigid-ened in the sense of possessive relation. For instance, the basic flexible linguistic values rigid-ened in one-dimensional space are shown in Fig. 19.2. It can be seen that the median points of two adjacent flexible linguistic boundaries now become the demarcation points of two adjacent rigid-ened flexible linguistic values.

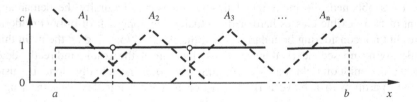

Fig. 19.2 An illustration of rigid-ened basic flexible linguistic values in one-dimensional space

It can be seen from statement above that the core and support set are only conceptual models of a flexible concept, while extended core is the practical model of a flexible concept.

3. Significance of first "flexible" then "rigid"

The phenomenon of flexible sets being rigid-ened in classifying, judging, and decision hints to people: The continuous measurement space considered difficult to rigid partition in the concept is rigidly partitioned in practice; thus, the problem solving with flexible sets actually becomes the problem solving with rigid sets. Thus, there occur questions:

(1) Now that the rigid partition has been realized finally, why did we take a circuitous route—first flexible parting and then rigid parting?
(2) Now that the problem is finally solved on the rigid set, why did we take a circuitous route—first professing it as being solved on the flexible set and then it changes to be solved on the rigid set?
(3) Since it is so, what's the meaning of using the flexible sets? Why not directly determine the demarcation point (line or plane) between two sets from the beginning?

We say that flexible partition is not redundant.

Since the points in a measurement space are continuous, the points between two adjacent flexible classes are all transitional points that "belong to this class to a certain degree and also belong to that class to a certain degree," and also, the closer to the central position between two classes a point is, the more difficult it is to determine the membership of that point. Therefore, in this situation, it is very difficult to directly find the demarcation point (line or plane) between two sets. However, we cannot directly look for the demarcation point (line or plane). Instead, we determine the cores of two classes firstly, that is, determine those points that belong completely to corresponding classes. For one-dimensional flexible classes, this is to determine their core-boundary points. Obviously, to judge whether an object has a certain property completely is much easier than to judge whether an object has just two properties with half-and-half, that is, whether it is located at the middle between two classes. That is to say, to determine the core-boundary points of two classes is relatively easier than to determine the midpoint between the two classes. Additionally, as long as the core-boundary points of two adjacent classes are determined, then the region between the two core-boundary points of the two classes is their public boundary. And once the public boundary region of two classes is obtained, the midpoint of this region is just naturally the demarcation point of two flexible classes hardened. Actually, the approach of only considering cores but not considering boundaries is just the skill of "focusing on the main thing while giving up secondary things." Thus, looking at the result, indirectly determining the midpoint through determining the cores is actually done by using flexible partition to achieve a rigid partition and by using flexible clustering to achieve rigid clustering; because if we do not use the flexible partition, then it is very difficult to rigidly partition continuous objects.

On the other hand, when classifying, judging, and decision making, the flexible classes obtained from flexible clustering are used as rigid classes. This embodies a kind of flexibility and elasticity to deal with the problem, and it is a show of "having a hard core despite one's mild exterior." Although objects located between two classes are both in the two classes, but in problem solving, generally it is always required that the corresponding result is not ambiguous—it must be in one class or the other, not both. Therefore, we have to use the approach of "first flexible then rigid."

Actually, in the final analysis, the strategy of "first flexible then rigid" is originally possessed by the human brain. In fact, flexible clustering is a strategy that human brain deals with the continuity (or uniform chain similarity); on the other hand, the logic of human brain thinking is of "middle" excluded, so its decision can only be that there occurs rigidness in flexibleness.

In summary, the approach of "first flexible then rigid" is actually a correct, scientific, and clever strategy and technique to solve the problems with continuity (or uniform chain similarity), and it is not an unnecessary move, nor is it deliberately mystifying.

4. **Flexible treating and complexity**

Let us also make an analysis of the connection between the treating way of flexible clustering and flexible partition used by the human brain and the complexity of corresponding things.

From the above statement, we see that flexible clustering and flexible partitioning are a kind of expedient measures adopted by the human brain when faced with continuous magnitudes of features of things that are hard to be rigidly clustered and partitioned. That is to say, if there were no continuous distribution or change of magnitudes of features of things, the human brain would not do flexible clustering and flexible partitioning. In other words, the flexible treating mechanism of the human brain is only related to continuity of magnitudes of features of things (or the uniform chain similarity of things).

Of course, for complex things, the human brain would still use clustering and partitioning, because by doing so, we can describe things in bigger granules. Thus, not only the amount of information would be reduced, but also the complexity of problems would be lowered, and we can also analyze problems, understand, and grasp things at a higher level.

Actually, from the methodology point of view, for those complex or large-scale problems, the strategies to be adopted first should be classifying, graduating, blocking, grouping, and so on, known as "divide and conquer." Actually, people follow and use this principle consciously and unconsciously in many areas. For instance, the zoning and ranking in administration and management, the modularizing and data abstracting in software development, and the categorizing in scientific research are all effective means and methods used by people in solving complex or large-scale problems.

Therefore, a complex problem should be analyzed and researched with appropriate big granule or large scale. Otherwise, we would "see the trees but not the forest," thereby making the problems difficult to solve. However, such big granule and large scale are not necessarily flexible granule and flexible scale, or fuzzy granule and fuzzy scale. When and only when there also exists "continuous" in the complexity, big and flexible granules and scales should be used.

To sum up, there is no necessary connection between flexible treating and complexity, and the real cause of the human brain flexibly treating corresponding information is the continuity of magnitudes of features of things (or the uniform chain similarity of things), rather than complexity or largeness of things.

19.3   Discrimination Between Flexible Concept and Vague (Fuzzy) Concept

1. Why is the "flexible concepts" taken as appellation?

As stated in Sect. 2.1, the denotation of a flexible concept is a flexible class, that is, the boundary of which is a flexible region with characteristic of "belonging to this and also belonging to that," "degree changing gradually," and "transiting smoothly" rather than a rigid line with characteristic of "belonging to either this or that" and "degree changing sharply." This characteristic of the denotations of flexible concepts can be abstractly illustrated as follows (see Fig. 19.3). Besides, the center region and boundary region of the denotation of one and the same flexible concept also have a slight difference in different people. That is to say, the center and boundary regions of the denotation of a flexible concept are elastic to a certain extent. In a word, viewed from the static characteristic of denotations, flexible concepts have a kind of "flexibleness" or "elasticity."

Let us also examine the situations of flexible concepts in practical applications. From the last section, flexible concepts also show a certain flexibleness in such processes as classifying, judging, decision making, etc. For example, the objects close to the center region of a flexible class within the boundary region are

Fig. 19.3 The sketch map of denotation characteristics of flexible concepts. These three figures represent respectively a denotation of a flexible concept. The blackness of each point represents the degree of the point belonging to the corresponding flexible class

originally the peripheral members of the flexible class, but they usually are also included in the corresponding class as core members. Similarly, the number objects close to the center region of a flexible class are usually also treated as the representatives of the corresponding linguistic value for use in decision making. That is to say, such classifying and decision making are really "flexible classifying" and "flexible decision making." For another example, in process control with flexible linguistic values, the system first needs to convert the input measurements into corresponding flexible linguistic values, which is really a flexible classifying procedure. Then, for a decision of flexible linguistic values, the system flexibly selects a number in the center region of the corresponding flexible class or among the peripheral members close to the center region as the control quantity to output.

Further, we find that the width or position of the center region and boundary region of the denotation of a flexible concept would change dynamically for certain reasons. That is, the center and boundary regions have certain expanding or contracting. For example, when having a cold, we feel "cold" in the temperature that is not cold to us normally. This is really because the center region or boundary region of "cold" extends toward higher temperatures. While once recovered, the center region or boundary region of "cold" would contract back, and thus, our feeling of cold returns to normal.

To sum up, whether viewed from the static characteristics in formation or from the dynamic characteristics in application, the flexible concepts all have flexibleness. Therefore, we call this kind of concepts to be the "flexible concepts."

2. What is vague concept?

As the term suggests, a vague concept should be such a concept whose connotation is not yet or completely clear to people, which is the abstraction of things that people have not yet a clear understanding of at the moment, or only have a superficial knowledge. For example, "life" and "intelligence" are just such concepts. Since which essential attributes they have are not yet or completely clear, their denotation is not yet or completely definite. Besides, whether a concept is vague or not may vary from person to person. In fact, a concept can be clear to some people but vague to some other people. For instance, "genes" and "black hole" may be clear concepts to professionals or experts but vague concepts to laymen.

However, we see that although the denotations of vague concepts are not definite, in general, their denotations have no such characteristics of "degree changing gradually" and "transiting smoothly" as shown in Fig. 19.3. Besides, vague concepts have no what dynamic characteristics in application.

3. Flexible concepts or vague (fuzzy) concepts?

However, for a long time, flexible concepts have been viewed as vague concepts (which also be called fuzzy concepts after fuzzy sets). The reason is that their denotative boundaries are unsharp.

It can be seen from the analyses of characteristics of flexible concepts and vague concepts that viewed from "denotative boundaries are unsharp," flexible concepts seem to be a kind of vague concepts. However, the outstanding characteristics of "degree changing gradually," "transiting smoothly," and "expanding and contracting dynamically" of the denotative boundaries make also them become a kind of special vague concept—flexible concepts. That is to say, although flexible concepts have vagueness (fuzziness), this is not their outstanding and key characteristic. Therefore, it is inappropriate to call flexible concepts as vague (fuzzy) concepts only by reason of "denotative boundaries are unsharp." It is a similar distinction as saying that although mankind has also some characteristics of animals, we cannot broadly call mankind to be just an animal; after all, the characteristics human beings have are far beyond the capabilities of general animals.

Indeed, the denotation of the flexible concept has no definite definition in general. However, viewed from the formation of a flexible concept, the denotative boundary being indefinite is not because that people do not know clearly the things covered by which as well as their characteristics, but only an expediency in treatment. Just because of this, unsharp denotative boundaries do not affect people's understanding and application of this kind of concepts. In fact, in daily information exchange, a flexible linguistic value represents mainly its core members and those peripheral members close to the center region. But for other objects in the boundary region, people then, in general, use another flexible concept such as "slightly ××," "comparatively ××," and "very ××" to describe and sometimes also use "medium" to represent the objects in the common boundary region of two relatively negative flexible concepts. For example, if 1.75 m is a non-core member of "tall," then we can say that it is "relative tall," and 1.70 m, which is between "tall" and "short," can then be said to be "average." Thus, although denotative boundaries are unsharp, for every flexible concept in a universe, people know what they refer to. Conversely, every object in a universe also can be represented by a certain flexible linguistic value on the universe. Thus, it can be seen that people actually know fairly well concepts with "unsharp denotative boundaries." This is also why people do not consciously try to find where the accurate boundaries of these concepts are, but can still understand and grasp them accurately and can handle them unmistakably. Obviously, in these places, flexible concepts and general vague (fuzzy) concepts are not the same. Thus, viewed from people's grasp and use of them, the flexible concepts seem to differ from real vague (fuzzy) concepts. Just think: a two-year-old child understands what is mean to "run quickly," so can we say that "quickly" is vague? Of course, people also use some real vague concepts, but cannot grasp them accurately (otherwise, they are not vague).

To sum up, we have the following conclusions:

① Viewed from denotative boundaries being unsharp, flexible concepts can be regarded as a kind of vague (fuzzy) concept, but the outstanding characteristics of "degree changing gradually," "transiting smoothly," and "expanding and contracting dynamically" of which make also them become actually flexible concepts.

② Viewed from formation and application, flexible concepts are not real vague (fuzzy) concepts.

We then examine the meaning of the words. It is obvious that "vague" or "fuzzy" does not reflect the objective facts of the "gradual change" and "continuous" of the properties of things, nor does it expresses the outstanding characteristics of "degree changing gradually," "transiting smoothly," and "expanding and contracting dynamically" of the flexible boundary of the corresponding flexible classes, even less to embody the non-rigid characteristic of flexible concepts when they are used; however, "flexible" just reflects, expresses, and embodies these facts and characteristics.

The analysis above shows that concepts based on numerical values without definite definitions in the human brain but can still be grasped accurately and be used flexibly by people should be called flexible concepts, but calling or viewing them as vague (fuzzy) concepts is not appropriate, which is absolutely a misunderstanding. This misunderstanding not only confuses flexible concepts with real vague concepts, but also loses essential characteristics of flexible concepts; hence, it is disadvantageous to, even will incorrect guidance for, people's research on flexible concepts. The root of bringing about the misunderstanding is that the objective basis, formation principle, and characteristic in the practical application of flexible concepts are not considered, but only "denotative boundaries being unsharp," a non-essential characteristic, are seen one sidedly, in isolation and statically.

19.4   Flexible Cluster Analysis

Cluster analysis is using mathematical method to divide a given set of objects (data points in a measurement space) into several classes. Conventional clustering methods can be classified as two types: dividing by a threshold and clustering with centers.

Dividing by a threshold is usually called graded clustering or pedigree clustering, whose basic conduct is according to a set threshold to divide a set awaiting partition with certain similarity measurements between every two points to realize a partition of the set. Visually speaking, dividing by a threshold is to define a similarity function firstly to compute the similarity degree of adjacent two points and

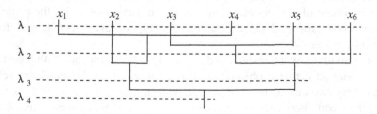

Fig. 19.4 Example 1 of dividing by a threshold

Fig. 19.5 Example 2 of method of dividing by a threshold

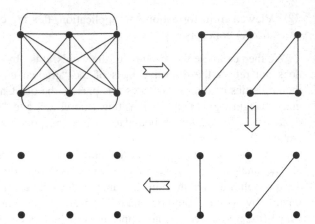

form a similarity relation net and then rigidly divide the relation net according to a set threshold for similarity degrees, thus forming several rigid classes (see examples in Figs. 19.4 and 19.5). The number of classes obtained by using dividing by a threshold is determined by the magnitude of the threshold, the extreme case being a point is a class, or all points are a class. Of course, in specific operations, people also propose many methods such as "biggest tree method." Besides, people also transform the similarity problem into fuzzy equivalence problem through the transitive closure of similar matrix, but the validity of results obtained by this method has no guarantee in theory.

Clustering with centers is also called dynamic clustering method, whose basic conduct is to search the optimum partition of a set awaiting partition with the similarity measurements of every class center to all other points by employing a certain algorithm in the constraint of clustering criterion. Here, the center of a class can be a representative point, mean point, or point set as the core of the class, and these class centers are dynamically changeable; the clustering criterion requires the similarity degree between members within a class should be biggest, but similarity degree between members from distinct classes should be smallest, generally which are given in the form of an objective function or a criterion function. C-mean algorithm just is a classic algorithm in clustering with centers. Visually speaking, clustering with centers is to define a similarity measure and take some points as clustering centers, then compute the similarity degrees between the rest points and all center points, thus forming a system of clusters on similarity, and then rigidly divide the system of clusters according to the similarity degree by the principle of similarity maximum, and decide the membership of every point, and forming several rigid classes.

Though dividing by a threshold and clustering with centers are both based on the similarity relation between objects, the former is based on the similarity relations between every two objects in an object set, while the latter is based on the similarity relations between every center object and other objects. Since the center of a class actually represents a property, the practice of clustering with centers is taking the property represented by the class center as the standard to gather objects having similar property into a class. Therefore, clustering with centers is really a kind of clustering on property, while dividing by a threshold is a kind of clustering on relation. The name of an object class formed by clustering on property is just the name of the corresponding property. But the name of an object class formed by clustering on relation is not the name of the corresponding relation ("similar" or "approximate"), but which need to be renamed, and the name is decided according to the common property of objects in the class. As thus, all classes obtained from relation clustering are actually also property classes. Property is explicit and visual, and the clustering based on property can be helped by intuition. Relation is usually implicit, and clustering based on relation is hard to be helped by intuition.

The approximation (similarity) relation is originally a kind of flexible relation, but the given data sets awaiting partition usually are not continuous, so the traditional cluster analysis, that is, the above stated two types of methods, is actually a kind of rigid clustering method realized by cutting the weak link on similarity chain or similarity net. If a data set awaiting partition is continuous, then from the Sect. 2.1, we know that only the method of flexible clustering can be used [now we see that the flexible clustering is also a kind of clustering on property based on the approximation (similarity) relation]. Only in some special situations, rigid clustering is then used to partition a data set.

Then, for a non-continuous data set, can the method of flexible clustering still be used to do flexible partition?

After the appearance of fuzzy sets, people proposed fuzzy cluster analysis. The idea is to cluster a data set into one and another fuzzy subset, then if required, rigid-en the obtained fuzzy subsets into rigid subsets. Since data sets are usually non-continuous, this kind of fuzzy subset is actually an instance subset of flexible set we said. An important algorithm of fuzzy clustering is fuzzy C-mean algorithm. There the membership function $\mu_j(x)$ is used in place of the feature function $t_j(x)$ in the original C-mean algorithm, but we cannot see how this membership function is related to the corresponding fuzzy set.

In the following, we give a method of utilizing flexible clustering to partition a data set. The basic idea of this method is to construct c classes to collect or carve up the points in the data set. The requirement is that there is no intersection among cores of c classes and it should guarantee that every point in the data set given at least falls into the support set of a certain flexible class. The method can be separated into the following 4 specific algorithms.

1. Point clustering algorithm

This algorithm takes points as clustering centers to construct flexible classes by using point-based flexible clustering, and the basic steps are as follows:

① For data set D considered, give c points as clustering centers and corresponding core radii and support set radii, to construct the membership functions of corresponding flexible points (flexible circles or flexible spheres).

② Compute the membership degrees of every point in D to all flexible points (flexible circles or flexible spheres) and relegate all points into corresponding flexible classes according to the principle of maximum membership.

③ Compute criterion function, if requirement is satisfied, then end.

④ Adjust clustering center points, or core radii and support set radii, turn to ②.

2. Line clustering algorithm

This algorithm takes lines as clustering centers to construct flexible classes by using line-based flexible clustering, and the basic steps are as follows:

① For data set D considered, give c lines as clustering centers and corresponding core radii and support set radii, to construct the membership functions of corresponding flexible lines (flexible bands or flexible ropes).

② Compute the membership degrees of every point in D to all flexible lines (flexible bands or flexible ropes) and relegate all points into corresponding flexible classes according to the principle of maximum membership.

③ Compute criterion function, if requirement is satisfied, then end.

④ Adjust clustering center lines or core radii and support set radii, turn to ②.

3. Plane clustering algorithm

This algorithm takes planes as clustering centers to construct flexible classes by using plane-based flexible clustering, and the basic steps are as follows:

① For data set D considered, give c planes as clustering centers and the corresponding core radii and support set radii, to construct the membership functions of corresponding flexible planes (flexible plates).

② Compute the membership degrees of every point in D to all flexible planes (flexible plates) and then relegate all points into corresponding flexible classes according to the principle of maximum membership.

③ Compute criterion function, if requirement is satisfied, then end.

④ Adjust clustering center planes or core radii and support set radii, turn to ②.

4. Block clustering algorithm

This algorithm takes points as clustering centers to construct flexible classes by using point-based square flexible clustering, and the basic steps are as follows:

① For data set D considered, give c points as clustering centers and corresponding core radii and support set radii, to construct the membership functions of corresponding flexible squares.

② Compute the membership degrees of every point in D to all flexible squares and relegate all points into corresponding flexible classes according to the principle of maximum membership.

③ Compute criterion function, if requirement is satisfied, then end.

④ Adjust clustering center points or core radii and support set radii, turn to ②.

The membership functions of the corresponding flexible classes in these clustering algorithms can be constructed and obtained by methods given in Chap. 3.

Besides the above partition based on relation and partition based on property (that is, dividing by a threshold and clustering with centers), there is another method of partitioning a data set, that is, to use the sample data whose classifications are known to induce the discrimination function of each classification, the geometric interpretation of which is to induce the curve, and surface or hypersurface which can partition the product space of the space that data belong to and the space of classification labels. We may as well refer to this method of partitioning a data set through samples as the **induction with samples**. The characteristics of sample induction method are that the number and labels of classes of a data set are known (while the classes and the number in dividing by a threshold and clustering with centers methods are not known beforehand but need to be formed and determined in partition). The common characteristic of the three methods partitioning a data set is that they all need to adjust associative parameters over and over again, which is actually also doing searches in corresponding spaces. Viewed with the eye of learning, these searches are also machine learning. Specifically speaking, methods of dividing by a threshold and clustering with centers are learning without a teacher or unsupervised learning, while sample induction method is learning with a teacher or supervised learning.

19.5   Flexible Linguistic Rule Discovery

Flexible linguistic rules used in problem solving can generally be given by domain experts from experience, which can also be discovered and acquired by machines directly from relevant sample data through self-learning. The relevant sample data may be pure numerical values, may also be pure linguistic values, or may also be the hybrids of numerical values and linguistic values. For pure numerical sample data, we can adopt statistical method to induce and summarize corresponding flexible linguistic rules (set). The specific ideas and methods will be introduced in the discovery of flexible linguistic function in the next section. For non-pure numerical samples, we can use a machine learning method called "decision tree learning" to obtain the corresponding flexible linguistic rules (set).

Decision tree learning is an important rule discovery method. For a known instance set, through decision tree learning, some rules with linguistic values can be induced usually. An instance in an instance set is a tuple or record consisted of several attribute values of an object, in which one term is taken as the conclusion or decision, the rest is the condition or premise of this decision term. Decision tree learning is to induce a general association relation, that is, decision rules, from the association facts between those known conditions and conclusions in an instance set. Here, an attributive value is generally a certain kind of linguistic values. Therefore, every data term in the instance set has a corresponding measurement space, and this linguistic attributive value corresponds to a subset in the corresponding measurement space. Then, if this linguistic attributive value is a flexible linguistic value, then the corresponding subset is also a flexible subset; conversely, if the subset in the measurement space is a flexible subset, then the corresponding linguistic attributive value is a flexible linguistic value. Thus, for such an instance set consisted of flexible linguistic attributive values, the rules obtained through decision tree learning are also flexible linguistic rules.

The above analysis shows that the machine learning method of decision tree learning can be used to discover and acquire flexible linguistic rules. But the precondition is that the attributive values in the instance set are flexible linguistic values. Thus, if the known attributive values in an instance set are already linguistic values or flexible linguistic values, then decision tree learning can be performed directly. From the decision rule of possessive relation (see Sect. 18.3), it is known that in this situation, a flexible linguistic attributive value is in essence tantamount to a rigid linguistic attributive value. Therefore, this kind of decision tree learning with known flexible linguistic attributive values is not different from the usual decision tree learning. But if there are pure numerical data terms in known instances, then such numerical attributive values need to be converted into flexible linguistic values and then to conduct decision tree learning. The converting method is also that of converting a numerical value into a flexible linguistic value stated previously in Sect. 7.3.1.

Since numerical attributive values can be converted into flexible linguistic attributive values, then, for a pure numerical data set, we can convert it into a data set of flexible linguistic values by N-L conversion, taking the latter as the instance set of decision tree learning, further to induce and abstract corresponding flexible linguistic rules from it. This can be viewed as another approach to discover flexible linguistic rules with pure numerical samples.

Certainly, prior to data conversion, it is necessary to firstly flexibly partition the measurement space $[a_i, b_i]$ that each numerical datum x_i belongs to and define the flexible linguistic values on it. The method of flexibly partitioning space $[a_i, b_i]$ can see Sect. 6.2 for reference.

Besides, in the process of discovering flexible linguistic rules by decision tree learning stated above, the acquiring of adjoint functions of rules can also be done at the same time. In fact, in the converting process from numerical values to linguistic values, the instances of the correspondence between the antecedent consistency-degree and the consequent consistency-degree of a rule have already been produced; in the mean time, there exist also already the instances of the correspondence between the measures of antecedent and consequent of the rule. The former can be used to construct or adjust the adjoint degreed functions of rules, and the latter can be used to construct or adjust the adjoint measured functions of rules.

Specifically speaking, in the process of acquiring an instance set, also, note down the correspondence relations between linguistic values in every instance

$$(A_1, A_2, \ldots, A_{n-1}) \mapsto A_n$$

and the correspondence relation between corresponding original numerical values

$$(x_1, x_2, \ldots, x_{n-1}) \mapsto x_n$$

and correspondence relation between degrees

$$(d_1, d_2, \ldots, d_{n-1}) \mapsto d_n$$

Thus, taking these numerical correspondence relations and degree correspondence relations as samples, the corresponding adjoint measure functions and adjoint degree functions can be constructed or optimized. Of course, these samples can also be used directly in machine learning to obtain corresponding approximate adjoint measure functions and approximate adjoint degree functions of rules.

19.6 Flexible Linguistic Function Discovery

The flexible linguistic function discovery is to induce and summarize a corresponding approximate flexible linguistic function from some known sample data (since flexible rules are also a representation form of flexible linguistic functions, the flexible rule discovery in last section also belongs to the flexible linguistic function discovery). The flexible linguistic function discovery is a new problem, which has no ready theories and methods yet. According to the characteristics of sample data, we will give some specific ideas and methods below for reference.

1. Inducing directly a L-L function by statistical method

For pure numerical sample data, according to their characteristics, we can induce and summarize the corresponding flexible linguistic function by statistical method directly. The specific steps and method are as follows:

① Do flexible partition of the measurement spaces which are taken as the domain and range of a flexible linguistic function, then convert known pure numerical samples into corresponding flexible linguistic valued samples;

② Induce the correspondence relation between corresponding flexible linguistic values with sample data by using statistical method (which can also be complemented by visual technique);

③ If the induction succeeds, then the result obtained can just be treated as a corresponding (approximate) L-L flexible linguistic function, end;

④ According to current case, refine the partition of domain and range, or adjust appropriately the original basic flexible linguistic values, then turn to ②.

Obviously, generally speaking, if the domain is partitioned sufficiently minute, then we can always obtain the correspondence relation between the corresponding flexible linguistic values. Therefore, the above stated method is feasible in theory.

Of course, to be a flexible linguistic function finally, it might be necessary to optimize the obtained result from induction (e.g., to appropriately merge or divide relevant flexible subsets). Since the result obtained by induction is actually a group of flexible rules, if there occurs redundancy, decision tree learning can be used for reduction.

Since linguistic functions are closely related to numerical functions, while the numerical function discovery has had many mature theories and methods, we can use for reference the method of numerical function discovery to develop the methods of flexible linguistic function discovery, and we can also use numerical function discovery to indirectly realize flexible linguistic function discovery. Next, we give two specific methods.

2. Discovering a N-L flexible linguistic function by flexible fitting

The so-called flexible fitting is using an N-L flexible linguistic function to fit related known data points.

Let (a_i, b_i) $(a_i \in U, b_i \in V, i = 1, 2, \ldots, n)$ be the known sample data. The basic steps of flexible fitting for which are as follows:

① According to the characteristics of the sample data, choose one numerical function $y = f(x)$ $(x \in U, y \in V)$ as the cluster center and define the corresponding flexible number $(r) = (f(x))$ (that is, "about $f(x)$") and its consistency function $c_{(r)}(y)$;

② Take a_i and b_i from data set $\{(a_1, b_1), (a_2, b_2), \ldots, (a_n, b_n)\}$, compute $f(a_i), (f(a_i)) = (r_i)$ and $c_{(r_i)}(b_i)$ $(i = 1, 2, \ldots, n)$;

③ If $\min_{1 \le i \le n} c_{(r_i)}(b_i) > 0.5$, then end, $Y = (f(x))$ is the N-L function to be found;

④ Appropriately adjust relevant parameters of $f(x)$, turn to ②.

3. Constructing an approximate flexible linguistic function by an approximate numerical function

This method is first to discover a corresponding approximate numerical function from the sample data and then use the approximate numerical function to obtain the corresponding flexible linguistic function. The specific steps are as follows:

Let (a_i, b_i) $(a_i \in U, b_i \in V, i = 1, 2, \ldots, n)$ be the known sample data.

Firstly, use certain method to do function fitting for data points (a_i, b_i) $(i = 1, 2, \ldots, n)$ to obtain an approximate numerical function $y = f(x)$ $(x \in U, y \in V)$;

Then, utilize the approximate numerical function $y = f(x)$ to construct corresponding linguistic function:

① Constructing L-L function

For the flexible linguistic value $X \subset U$, set $y_1 = \min\limits_{x \in core(X)^+} f(x)$ and $y_2 = \max\limits_{x \in core(X)^+} f(x)$ and construct flexible interval $((y_1], [y_2))$, thus obtaining flexible linguistic value $Y = ((y_1], [y_2)) \subset V$.

In this way, to each flexible linguistic value $X \in L(U)$, there corresponds a unique flexible linguistic value $Y \subset V$. Thus, we have an L-L flexible linguistic function $Y = F(X)$, $X \in L(U)$.

② Constructing N-L function

For $x \in U$, set $Y = (f(x))$ (i.e., "about $f(x)$").

Thus, to each $x \in U$, there corresponds a unique flexible linguistic value $Y \subset V$. Thus, we have an N-L flexible linguistic function $Y = F(x)$, $x \in U$.

Notes: The above function fitting is in a broad sense, and the specific methods include the mathematical methods of interpolation method, regression method, fitting method, and so on as well as the machine learning methods of neural network and SVM (short for support vector machine). As to what method to choose should be decided by the characteristics of the sample data. Strategically speaking, the fitting can be global or local such as sectional or piecewise. Generally speaking, the function of the latter is simpler and more precise, so it is more preferred.

The above methods of acquiring linguistic functions are for pure numerical samples. For non-pure numerical samples, the method of decision tree learning introduced in the previous section can be used. We can also convert them into pure numerical data and then use the above stated methods. The method of converting is as follows:

Let the known sample data be (A_i, B_i) or (x_i, B_i) $(A_i \subset U, x_i \in U, B_i \in V, i = 1, 2, \ldots, n)$. Then, separately replacing A_i and B_i by standard instances a_i and b_i of flexible linguistic values A_i and B_i, thus obtains numerical sample (a_i, b_i) $(a_i \in U, b_i \in V, i = 1, 2, \ldots, n)$.

Here, the standard instance is the numerical value in the support set of corresponding flexible linguistic value which can stand for the flexible linguistic value. Generally, taking the peak-value point of a flexible linguistic value as its standard instance, can also take the corresponding core-boundary point or a certain point in the extended core. Of course, it is best to take the point with the highest probability according to the background knowledge of the practical problem.

As to the samples of type (A_i, b_i) $(A_i \subset U, b_i \in V)$, the converting can be omitted, because after conversion, it is still difficult to obtain the corresponding numerical function and linguistic function. But, from which the local L-N linguistic function $y = F(X)$, $X \in Ls(U) \subset L(U)$ can be directly obtained.

19.7   Flexible Function Discovery

As a kind of special flexible linguistic functions, the flexible functions, i.e., flexible-numbered functions, of course can be discovered by using the methods of flexible linguistic function discovery given in last section. But since the particularity of flexible numbers—the kind of flexible linguistic values—that which and rigid numbers, that is, real numbers, can be converted to each other, and the operations of which can be reduced to the operations of real numbers, therefore, we can utilize the discovery methods of usual functions plus data conversions to realize the discovery of flexible functions. In the following, we give a method of constructing approximate flexible function by combining data conversions with neural network.

We know that a flexible number (x_0) can rigid-en as rigid number x_0, and conversely, a rigid number y_0 can also flexible-en as flexible number (y_0) (see Sect. 10.5). On the other hand, for a functional relation indicated by sample data of real numbers, generally speaking, it can be approximated with a neural network. Thus, for the sample data consisted of pairs of flexible numbers, we can combine data conversions with neural network to construct corresponding approximate flexible function, and the specific method is as follows:

Suppose there is a set of sample data of flexible numbers: $(((a_{1_i}), (a_{2_i}), \ldots, (a_{n_i})), (b_i))$ $(a_{j_i} \in U_j, b_i \in V, i = 1, 2, \ldots, m, j = 1, 2, \ldots, n)$. Firstly, convert the set of sample data of flexible number, $(((a_{1_i}), (a_{2_i}), \ldots, (a_{n_i})), (b_i))$ $(i = 1, 2, \ldots, m)$, into set of sample data of rigid number, $((a_{1_i}, a_{2_i}, \ldots, a_{n_i}), b_i)$ $(i = 1, 2, \ldots, m)$, through L-N conversion; then according to the characteristics of these data of real number, construct a proper neural network and train it with these data (the neural network trained successfully is tantamount to a corresponding approximate numerical function) and then add interfaces of L-N and N-L conversions to the neural network trained successfully, and thus, we obtain a approximate flexible function indicated by original sample data of flexible numbers, $(y) = f((x_{1_i}), (x_{2_i}), \ldots, (x_{n_i}))$ (as shown in Fig. 19.6).

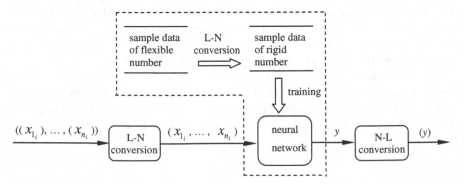

Fig. 19.6 Diagram of an approximate flexible function combining data conversions and neural network

19.8 Summary

In this chapter, we further discussed firstly the formation mechanism and essential characteristic of flexible concepts and discriminated between the flexible concept and the vague (fuzzy) concept, and then, we explored imprecise-knowledge discovery, proposed some ideas and approaches, and presented some specific algorithms and methods.

The main points and results of the chapter are as follows:

- The "flexible" of the flexible concepts not only is shown on the forming (clustering) of flexible concepts, but also is reflected on the employing (classifying, judging, and decision) by people of it. Therefore, only when from the two aspects of formation and application to understand the flexible concepts, can we understand the flexible concepts fully and entirely.
- The concepts in human brain, which are based on numerical values and are not defined definitely but can be grasped accurately and be used flexibly by people, are really a kind of flexible concepts, so calling or regarding them as vague (fuzzy) concepts is not appropriate.
- There is no necessary connection between the flexible treating of related information by human brain and the complexity of things. The real cause of human brain doing flexible treating is the continuity of the quantities of a feature of things (less strictly speaking, it is the uniform chain similarity of things), rather than complicatedness or largeness of things.
- For a discrete set of points in a measurement space, we can realize its partition by flexible clustering in the corresponding measurement space, which is called flexible cluster analysis. For different dimensional measurement spaces, the 4 flexible clustering algorithms of point clustering, line clustering, plane clustering, and block clustering are presented here.
- We can use the method of decision tree learning to discover flexible rules.
- We can use the statistical method to induce and summarize directly corresponding flexible linguistic functions from sample data.
- We can use the method of flexible fitting to discover N-L flexible linguistic functions and also can construct an approximate linguistic function from an approximate numerical function.
- We can use the method of combining data conversions and neural network to construct an approximate flexible function.

Reference

1. Lian S (2009) Principles of imprecise-information processing. Science Press, Beijing

Part VII
Quantifiable Rigid Linguistic Values and Information Processing with Degrees: Extension of Imprecise Information

Chapter 20
Several Measures of Sets and Flexible Sets

Abstract This chapter introduces multiple measures such as size and share for ordinary sets and flexible sets and analyzes and expounds their properties, relations, and operations. And it presents the concepts of partial inclusion, partial equality, and partial correspondence and introduces the measures of inclusion-degree, equality-degree, and correspondence-rate. In particular, it founds and presents the sufficient and necessary conditions for these relations to satisfy transitivity and the corresponding transitive formulas of inclusion-degrees (equality-degrees) and correspondence-rates.

Keywords Size of a flexible set · Share of a flexible set · Partial inclusion · Partial equality · Partial correspondence · Inclusion-degree · Equality-degree · Correspondence-rate

This chapter gives several measures for ordinary sets and flexible sets, which include size, share, inclusion-degree, equality-degree, and correspondence-rate.

20.1 Size of a Set

1. Definition and properties of the size of a set

Definition 20.1 We call the amount of elements of a set X (when X is a discrete set), or length, area, volume, or hyper volume of a set X (when X is a continuous set) to be the size of the set X, denotes as $|X|$ [1].

Example 20.1

① Let $X = \{\text{all English letters}\}$, then $|X| = 26$.
② Let $X = \{\text{all integers}\}$, then $|X| = \infty$.

© Springer Science+Business Media Singapore 2016   497
S. Lian, *Principles of Imprecise-Information Processing*,
DOI 10.1007/978-981-10-1549-6_20

③   Let $X = \{$all real numbers$\}$, then $|X| = \infty$.
④   Let $X = [a, b]$, then $|X| = b - a$.
⑤   Let $X = [a, b] \times [c, d]$, then $|X| = (b - a) \times (d - c)$.

It can be seen that "size" is also a kind of numerical feature of a set, which is a kind of absolute measure of a set. For the discrete sets, the size here is completely the same as the concept of cardinal number in the traditional set theory, but for the continuous sets, the size is not the same thing as the cardinal number, and the sizes of continuous sets of different dimension are not comparable.

From the definition, the size of a set obviously has the following basic properties.

Theorem 20.1 *Let U be a universal set, and $A, B {\subseteq} U$.*

(i) $|A| \leq |U|$ *(boundedness)*
(ii) if $A \subset B$ then $|A| \leq |B|$ *(monotony)*
(iii) $|A| + |A'| = |U|$ *(complementarity)*
(iv) $|\varnothing| = 0$

It can be seen that for universal set U, the size of subsets actually also defines a function on its power set; that is, set function $\mu: 2^U \rightarrow [0, \beta], (\beta = |U|).$

2. Size of a compound set

Let U be a universal set, and $A, B \in 2^U$, we consider the sizes of compound sets $A \cap B$, $A \cup B$ and A^c. Of course, the most basic method is to compute specifically according to the operation definitions of sets and the definitions of the size of a set. We now consider whether we can utilize the sizes of component sets to indirectly compute the size of a compound set.

From the definitions of operations of intersection ($\cap$), union ($\cup$), and complement (^{c}) of sets, it follows easily that

$$|A \cup B| = |A| + |B| - |A \cap B| \tag{20.1}$$

$$|A^c| = |U| - |A| \tag{20.2}$$

From the two equalities, it can be seen that the size of the union can be reduced as the size of the intersection. But since the elements in intersection $A \cap B$ are related to the relation between component sets A and B, so for the size $|A \cap B|$, we are unable to give a general computation formula based on the size of component sets; that is, there does not exist a common operation $\triangle$ such that

$$|A \cap B| = |A| \triangle |B|$$

Thus, there does not exist a common operation ∇ such that

$$|A \cup B| = |A| \nabla |B|$$

Only in some special situations there are some computation formulas of sizes. In fact, it is not hard to see:

(1) When $A \cap B = \phi$,

$$|A \cup B| = |A| + |B| \qquad (20.3)$$

(2) If $A \subset B$, then

$$|A \cap B| = |A|, \quad |A \cup B| = |B| \qquad (20.4)$$

(3) If there exists linearly ordered set $\langle L, \subset \rangle \subset 2^U$, then for arbitrary A, $B \in \langle L, \subset \rangle$,

$$|A \cap B| = \min\{|A|, |B|\} \qquad (20.5)$$

$$|A \cup B| = \max\{|A|, |B|\} \qquad (20.6)$$

3. Size of a Cartesian product

Firstly, from the definition of the size of a set, it is not hard to see that for any two sets X and Y, always

$$|X \times Y| = |X| \cdot |Y| \qquad (20.7)$$

That is, the size of a product is equal to the product of sizes of factors.

Since the Cartesian product of n sets satisfies

$$\underset{i=1}{\overset{n}{\times}} X_i = (X_1, X_2 \times \cdots \times X_{n-1}) \times X_n$$

Therefore, the above formula can also be generalized. That is, for any $n(n > 1)$ sets $X_1, X_2, \ldots, X_n$,

$$|X_1 X_2 \times \cdots \times X_n| = |X_1| \cdot |X_2| \cdot \cdots \cdot |X_n| \qquad (20.8)$$

4. Size of an orthogonal compound subset

In the following, we consider the size of a compound set formed by the orthogonal subsets in Cartesian product $U \times V$.

Let $A \subset U$ and $B \subset V$, then in Cartesian product $U \times V$, A is extended as $A \times V$, and B is extended as $B \times U$, and the two are orthogonal. It is not hard to see that $A \times V \cap B \times U = A \times B$, and obviously

$$|A \times B| = |A| \cdot |B| \tag{20.9}$$

Thus,

$$|A \times V \cap U \times B| = |A| \cdot |B| \tag{20.10}$$

Further,

$$
\begin{aligned}
|A \times V \cup U \times B| &= |A \times V| + |U \times B| - |A \times V \cap U \times B| \\
&= |A| \cdot |V| + |U| \cdot |B| - |A| \cdot |B| \tag{20.11}
\end{aligned}
$$

Generally, let set $A_i \subset U_i (i = 1, 2, \ldots, n)$, denote $U_1 \times U_2 \times \cdots U_{i-1} \times A_i \times U_{i+1} \times \cdots \times U_n$ as A_i, then

$$A_1 \cap A_2 \cap \cdots \cap A_n = A_1 \times A_2 \times \cdots \times A_n$$

Further,

$$|A_1 \cap A_2 \cap \cdots \cap A_n| = |A_1| \cdot |A_2| \cdots \cdot |A_n| \tag{20.12}$$

$$
\begin{aligned}
&|A_1 \cup A_2 \cup \cdots \cup A_n| \\
&= \sum_{1 \le i \le n} |A_i| - \sum_{1 \le i \le j \le n} \left(|A_i| \cdot |A_j|\right) + \sum_{1 \le i \le j \le k \le n} \left(|A_i| \cdot |A_j| \cdot |A_k|\right) \\
&\quad - \cdots + (1)^{n+1} |A_1| \cdot |A_2| \cdots \cdot |A_n| \tag{20.13}
\end{aligned}
$$

20.2 Sizes of a Flexible Set

1. Definition and properties of the size of a flexible set

Let U be a universal set and U be a discrete set, and A be a flexible subset of U. Since the membership-degree $m_A(x) = 1$ of element x in core $\text{core}(A)$, so every element x in core $\text{core}(A)$ all can be regarded as a full member of A. Yet the element x in boundary $\text{boun}(A)$ only belongs to A with degree $m_A(x)$, so these elements cannot be regarded as the full members of A. Then, how do we calculate the size of flexible set A? Considering the membership-degree of element x in $\text{core}(A)$ is 1, which is just regarded as one element of A, so for any $x \in \text{boun}(A)$, it is appropriate to take numerical value $m_A(x)$ as the element amount of x relative to A. For instance, when $m_A(x) = 0.8$, then x is regarded as 0.8 member of A, when $m_A(x) = 0.5$, then x is regarded as 0.5 or a half member of A. Thus, for flexible set A in discrete universe of discourse U, we use membership-degree $m_A(x)$ to convert and count its size. Thus, we have

$$|A| = \sum_{x \in \text{supp}(A)} m_A(x)$$

We then consider the size of continuous flexible subset A of U when universal set U is continuous set $[a, b]$. Similarly, every element in core(A) of flexible set A is a full member of A, while element x in boun(A) is only a partial member of A. Therefore, we still use the membership degree to convert and count the size of flexible set A. But since here flexible set A is a flexible subset of one-dimensional continuous set $[a, b]$, its size is described by length but not by the number of elements. Based on this characteristics of A, we first equally divide interval $[a, b]$ into n small intervals: $[a_1, b_1], [a_2, b_2], \ldots, [a_n, b_n]$ ($a_1 = a$, $b_n = b$), for every small interval $[a_i, b_i]$ ($i = 1, 2, \ldots, n$), take the membership-degree of its center point; that is, $m_A(\frac{a_i+b_i}{2})$, as the uniform membership-degree of all points in $[a_i, b_i]$; thus, it can be seen that

$$m_A(\frac{a_i + b_i}{2}) \cdot \left\| [a_i, b_i] \right\| = m_A(\frac{a_i + b_i}{2}) \cdot (b_i - a_i)$$

is just the approximate value of size $\| [a_i, b_i] \|$ of interval $[a_i, b_i]$. Further, $\sum_{i=1}^{n} m_A(\frac{a_i+b_i}{2}) \cdot (b_i - a_i)$ is the approximate value of |A|. Now set $\| [a_i, b_i] \| = d \rightarrow 0$, then we have

$$\lim_{d \to 0} \sum_{i=1}^{n} m_A\left(\frac{a_i + b_i}{2}\right) \cdot (b_i - a_i) = \int_a^b m_A(x) dx$$

this is just the length of flexible set A; thus,

$$|A| = \int_a^b m_A(x) dx = \int_U m_A(x) dx$$

Thus, the computation problem of the size of flexible subset of continuous set $[a, b]$ is solved. It can be seen that generalizing this formula into general n-dimensional measurement space U, it would be

$$|A| = \iint \cdots \int_U m_A(x_1, x_2, \ldots, x_n) dx_1 dx_2 \ldots dx_n$$

To sum up, we give the definition below.

Definition 20.2 Let U be an n-dimensional measurement space, A be a flexible subset of U. We denote the size of A by $|A|$.

(i)   When U is a discrete set,

$$|A| = \sum_{x \in U} m_A(x) \tag{20.14}$$

(ii) When U is a continuous set,

$$|A| = \underset{U}{\iint \cdots \int} m_A(x_1, x_2, \ldots, x_n) dx_1 dx_2 \ldots dx_n \tag{20.15}$$

From the definition, it can be seen that for any flexible set $A \subset U$, always

$$|\text{core}(A)| < |A| < |\text{supp}(A)| \tag{20.16}$$

For the size of flexible sets, there are the following properties.

Theorem 20.2 *Let U be an n-dimensional measurement space, and A and B be the flexible subsets of U, then*

(i) $|A| \leq |U|$ *(boundedness)*
(ii) *if $A \subset B$ then $|A| \leq |B|$ (monotony)*
(iii) $|A| + |A^c| = |U|$ *(complementarity)*

Proof From the definition of the size of a flexible set, (i) and (ii) hold viously, we only prove (iii). By Definition 20.2, when U is a discrete set,

$$
\begin{aligned}
|A| + |A^c| &= \sum_{x \in U} m_A(x) + \sum_{x \in U} (1 - m_A(x)) \\
&= \sum_{x \in U} [m_A(x) + (1 - m_A(x))] \\
&= \sum_{x \in U} 1 \\
&= |U|
\end{aligned}
$$

When U is a continuous set,

$$
\begin{aligned}
|A| + |A^c| &= \int_U m_A(x)\mathrm{d}x + \int_U (1 - m_A(x))\mathrm{d}x \\
&= \int_U [m_A(x) + (1 - m_A(x))]\mathrm{d}x \\
&= \int_U 1\mathrm{d}x \\
&= |U|
\end{aligned}
$$

(2) Relation between the sizes of a flexible set and its extended core

We find that the sizes of a flexible set and its extended core have the following relation.

Theorem 20.3 *Let A be a flexible subset of n-dimensional measurement space U, if A is not a Cartesian product, then*

$$
\left|\mathrm{core}(A)^+\right| = |A| \tag{20.17}
$$

That is, the size of the extended core of a flexible set is equal to the size of the flexible set.

Proof From Corollary 5.1 in Sect. 5.6, it is known that the extended cores of flexible Cartesian products cannot be expressed in a uniform expression of operations, so Eq. (20.17) is not for flexible Cartesian products. Next, we prove that for single one-dimensional flexible set, two-dimensional flexible set, and two-dimensional flexible intersection and flexible union, Eq. (20.17) is tenable.

(1) Let A be a flexible subset of one-dimensional space $U = [a, b]$, whose membership function is

$$
m_A(x) = \begin{cases}
0, & a \leq x \leq s_A^- \\
\frac{x - s_A^-}{c_A^- - s_A^-}, & s_A^- < x < c_A^- \\
1, & c_A^- \leq x \leq c_A^+ \\
\frac{s_A^+ - x}{s_A^+ - c_A^+}, & c_A^+ < x < s_A^+ \\
0, & s_A^+ \leq x \leq b
\end{cases}
$$

Then, by the definition of the size of a flexible set, should be

$$|A| = \int_a^b m_A(x)dx$$

$$= \int_{s_A^-}^{c_A^-} \frac{x - s_A^-}{c_A^- - s_A^-}dx + \int_{c_A^-}^{c_A^+} 1dx + \int_{c_A^+}^{s_A^+} \frac{s_A^+ - x}{s_A^+ - c_A^+}dx$$

$$= \frac{c_A^- - s_A^-}{2} + (c_A^+ - c_A^-) + \frac{s_A^+ - c_A^+}{2}$$

$$= \frac{s_A^+ + c_A^+}{2} - \frac{c_A^- + s_A^-}{2}$$

$$= m_A^+ - m_A^-$$

$$= |\mathrm{core}(A)^+|$$

Actually, this result can also be obtained by geometrical approach. We know that the value of definite integral $\int_a^b m_A(x)dx$ is the area of the curved side trapezoid determined by function $m_A(x)$ (as shown in Fig. 20.1). But from the figure, it can be seen that the two opposite vertex triangles over interval $\left[s_A^-, c_A^-\right]$ should be congruent, the two opposite vertex triangles over interval $\left[c_A^+, s_A^+\right]$ also be congruent. Therefore, the area of the rectangle S encircled by the three broken straight lines over interval $\left[m_A^-, m_A^+\right]$ just equals to the area of the curved side trapezoid determined by integrand function $m_A(x)$ on the coordinate axes, whereas the area of this rectangle S is obviously equal to

$$(m_A^+ - m_A^-) \cdot 1 = m_A^+ - m_A^-$$

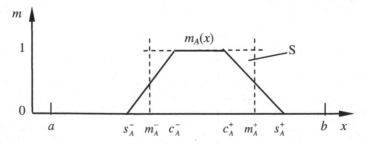

Fig. 20.1 Illustration of the relation between the sizes of flexible set A and its extended core

Thus, we have

$$\int_a^b m_A(x)\mathrm{d}x = m_A^+ - m_A^-$$

(2) Let $A \times V$ be a bar flexible subset in two-dimensional space $U \times V$ ($U = [a, b]$, $V = [c, d]$) (as shown in Fig. 5.5 of Sect. 5.5.2), whose membership function is $m_{A \times V}(x, y) = m_A(x)$ ($x \in U, y \in V$). From the definition of the size of a flexible set, $|A \times V| \int_{\mathrm{supp}(A \times V)} m_{A \times V}(x, y)\mathrm{d}x\mathrm{d}y = \int_{\mathrm{supp}(A)} m_A(x)\mathrm{d}x \int_V \mathrm{d}y$, whereas from (1), $\int_{\mathrm{supp}(A)} m_A(x)\mathrm{d}x = m_A^+ - m_A^-$, yet $\int_V \mathrm{d}y = (d - c)$; thus, $|A \times V| = (m_A^+ - m_A^-)(d - c)$, whereas just $(m_A^+ - m_A^-)(d - c) = |\mathrm{core}(A \times V)^+|$.

(3) Let A be a circular flexible subset in two-dimensional space U (as shown in Fig. 3.9 in Sect. 3.3), and let the membership function of A be $m_A(x, y)$, whose graph is shown in Fig. 3.11a, then $|A| = \int_{\mathrm{supp}(A)} m_A(x, y)\mathrm{d}x\mathrm{d}y$. However, from the geometric interpretation of double integral, it can be seen that the size $|A|$ of circular flexible set is numerically equal to the volume of the round platform shown in Fig. 3.11a. Any longitudinal section of this round platform that passes the center axis is a trapezoidal plane. Similar to the trapezoid in Fig. 20.1, by segmenting and patching of this trapezoidal plane, a rectangle plane can also be obtained, of which the area equal to that of the trapezoidal plane. Thus, a cylinder can be obtained whose volume equal to that of the round platform. And the volume of this cylinder is numerically $|\mathrm{core}(A)|$.

(4) Let $A \times V \cap U \times B$ be an orthogonal compound flexible set in two-dimensional space $U \times V$ (as shown in Fig. 5.5 of Sect. 5.5.2), whose membership function is $m_{A \times V}(x, y) = m_A(x)$ ($x \in U, y \in V$), whose graph is shown in Fig. 5.9. Similarly, size $|A \times V \cap U \times B|$ is numerically equal to the volume of the prismoid shown in Fig. 5.9. To find the volume of this prismoid, we firstly patch up a wedge separately on the two inclined planes on the direction of y-axis, making it a semi-prismoid with two right planes on the direction of y-axis. It can be imagined that this semi-prismoid is also a graph similar to the bar flexible subset in (2), that is, which is a truncated ridge. We cut this truncated ridge into three sections, that is, three shorter truncated ridges. From (2), we can know their volumes are one by one $(m_A^+ - m_A^-)$ $(c_B^- - s_B^-)$, $(m_A^+ - m_A^-)(c_B^+ - c_B^-)$, $(m_A^+ - m_A^-)(s_B^+ - c_B^+)$. Now, as long as subtracting the volume of the wedges in the first and rear two truncated ridges from the sum of the three volumes, we can obtain the volume the original prismoid we try to find. It can be seen that the volume of the wedge is just half of the volume of the truncated ridge it belongs to; that is, $(m_A^+ - m_A^-)$ $(c_B^- - s_B^-)/2$ and $(m_A^+ - m_A^-)(s_B^+ - c_B^+)/2$, whereas it is just that

$$(c_B^- - s_B^-)/2 = c_B^- - m_B^-, \quad (s_B^+ - c_B^+)/2 = m_B^+ - c_B^+$$

Therefore, the volume of the original prismoid is

$$\left(m_A^+ - m_A^-\right)\left(c_B^- - m_B^-\right) + \left(m_A^+ - m_A^-\right)\left(c_B^+ - c_B^-\right)$$
$$+ \left(m_A^+ - m_A^-\right)\left(m_B^+ - c_B^+\right) = \left(m_A^+ - m_A^-\right)\left(m_B^+ - m_B^-\right)$$

It can be seen that the right-hand side of this equation just is the size of the extended core of $A \times V \cap U \times B$, that is, $\left|\text{core}(A \times V \cap U \times B)^+\right|$. Thus,

$$|A \times V \cap U \times B| = \left|\text{core}(A \times V \cap U \times B)^+\right|$$

The fact that the size of a flexible set equals to that of its extended core has great significance. It can not only make size's computation simpler, but also provide a basis for the follow-up relevant theories.

3. **Size of a compound flexible set**

It is not hard to see that similar to the rigid sets, the sizes of compound flexible sets also have the following two basic relational expressions:

$$|A^c| = |U| - |A| \tag{20.18}$$

$$|A \cup B| = |A| + |B| - |A \cap B| \tag{20.19}$$

As to the size of intersection $A \cap B$, it needs to be analyzed specifically; there is no a general compound computation rule and only in some special situations can the compound computation be performed.

Theorem 20.4 *Let U be an n-dimensional measurement space, and A and B be flexible subsets of U.*

(i) *If $A \cap B = \emptyset$, then*

$$|A \cup B| = |A| + |B| \tag{20.20}$$

(ii) *If $A \subset B$, then*

$$|A \cap B| = |A|, \quad |A \cup B| = |B| \tag{20.21}$$

(iii) *If there exists linearly ordered set $\langle L, \subset \rangle \subset 2^U$, then for arbitrary A, $B \in \langle L, \subset \rangle$,*

$$|A \cap B| = \min\{|A|, |B|\} \tag{20.22}$$

$$|A \cup B| = \max\{|A|, |B|\} \tag{20.23}$$

4. The size of a flexible Cartesian product and the size of an orthogonal compound flexible set

Let flexible set $A_i \subset U_i$ ($i = 1, 2, \ldots, n$). From the definition of the size of a flexible set,

$$|A_1 \times A_2 \times \cdots \times A_n| = \underset{\text{supp}(A_1 \times A_2 \times \cdots \times A_n)}{\int\!\!\int \cdots \int} m_{A_1 \times A_2 \times \cdots \times A_n}(x_1, x_2, \ldots, x_n) dx_1 dx_2 \ldots dx_n$$

$$\tag{20.24}$$

Then, for the Cartesian product of flexible sets, can its size be indirectly computed by using the sizes of its factor flexible sets?

Let flexible sets $A \subset U$ and $B \subset V$, then

$$
\begin{aligned}
|A \times B| &= \iint_{\text{supp}(A \times B)} m_{A \times B}(x, y) dx dy = \iint_{\text{supp}(A \times B)} [w_1 m_A(x) + w_2 m_B(y)] dx dy \\
&= \iint_{\text{supp}(A \times B)} [w_1 m_{A \times V}(x, y) + w_2 m_{U \times B}(x, y)] dx dy \\
&= \iint_{\text{supp}(A \times B)} w_1 m_{A \times V}(x, y) dx dy + \iint_{\text{supp}(A \times B)} w_2 m_{U \times B}(x, y) dx dy \\
&= w_1 \iint_{\text{supp}(A \times B)} m_{A \times V}(x, y) dx dy + w_2 \iint_{\text{supp}(A \times B)} m_{U \times B}(x, y) dx dy \\
&= w_1 \int_{\text{supp}(A)} m_A(x) dx \int_{\text{supp}(B)} dy + w_2 \int_{\text{supp}(A)} dx \int_{\text{supp}(B)} m_B(y) dy \\
&= w_1 \cdot |A| \cdot |\text{supp}(B)| + w_2 \cdot |\text{supp}(A)| \cdot |B|
\end{aligned}
$$

$$\tag{20.25}$$

where $w_1, w_2 \in (0, 1)$, $w_1 + w_2 = 1$.

20 Several Measures of Sets and Flexible Sets

Generally, for flexible set $A_i \subset U_i$ $(i = 1, 2, \ldots, n)$, we have

$$|A_1 \times A_2 \times \cdots \times A_n| = w_1 \cdot |A_1| \cdot |\text{supp}(A_2 \times A_3 \times \cdots \times A_n)|$$
$$+ w_2 \cdot |A_2| \cdot |\text{supp}(A_1 \times A_3 \times \cdots \times A_n)|$$
$$\cdots$$
$$+ w_n \cdot |A_n| \cdot |\text{supp}(A_1 \times A_2 \times \cdots \times A_{n-1})| \qquad (20.26)$$

where $w_1, w_2, \ldots, w_n \in (0,1)$, $w_1 + w_2 + \cdots + w_n = 1$.
On the other hand, from $\text{core}(A)^+ \times \text{core}(B)^+ \subset A \times B$, then

$$|A \times B| > |\text{core}(A)^+ \times \text{core}(B)^+|$$

while

$$|\text{core}(A)^+ \times \text{core}(B)^+| = |\text{core}(A)^+| \cdot |\text{core}(B)^+| = |A| \cdot |B|$$

Therefore, we have

$$|A \times B| > |A| \cdot |B| \qquad (20.27)$$

Then generally, for flexible set $A_i \subset U_i$ $(i = 1, 2, \ldots, n)$, we have

$$|A_1 \times A_2 \times \cdots \times A_n| > |A_1| \cdot |A_2| \cdot \cdots \cdot |A_n| \qquad (20.28)$$

Next, we consider the computation problem of sizes of compound flexible sets formed by the orthogonal flexible subsets in a product space.

Let U and V both be one-dimensional measurement spaces, and A and B separately be the flexible subsets of U and V. In product space $U \times V$, A is extended into $A \times V$, and B is extended into $U \times B$. Then, it is not hard to see that $A \times V \cap U \times B \subset U \times V$ is also a flexible set, while $\text{supp}(A \times V \cap U \times B) = \text{supp}(A \times B)$, so by the definition of the size of a flexible set,

$$|A \times V \cap U \times B| = \iint_{\text{supp}(A \times B)} \min\{m_A(x), m_B(y)\} dx dy \qquad (20.29)$$

where $\min\{m_A(x), m_B(y)\}$ is the membership function of intersection $A \times V \cap U \times B$. Then, the size of union $A \times V \cup U \times B$ is

$$|A \times V \cup U \times B| = |A \times V| + |U \times B| - |A \times V \cap U \times B|$$
$$= \iint_{\text{supp}(A \times V)} m_A(x) dx dy + \iint_{\text{supp}(U \times B)} m_B(y) dx dy$$
$$- \iint_{\text{supp}(A \times B)} \min\{m_A(x), m_B(y)\} dx dy \qquad (20.30)$$

On the other hand, by Theorem 20.3, it is known that the size of a flexible set is equal to that of its extended core. Thus,

$$|A \times V \cap U \times B| = |\text{core}(A \times V \cap U \times B)^+|$$

while

$$\text{core}(A \times V \cap U \times B)^+ = \text{core}(A)^+ \times \text{core}(B)^+$$

$$|\text{core}(A)^+ \times \text{core}(B)^+| = |\text{core}(A)^+| \cdot |\text{core}(B)^+|$$

Therefore,

$$|A \times V \cap U \times B| = |\text{core}(A \times V \cap U \times B)^+| = |\text{core}(A)^+ \times \text{core}(B)^+|$$
$$= |\text{core}(A)^+| \cdot |\text{core}(B)^+|$$

And then, by

$$\text{core}(A \times V \cup U \times B)^+ = \text{core}(A \times V)^+ + \text{core}(U \times B)^+ - \text{core}(A \times V \cap U \times B)^+$$

$$|\text{core}(A \times V)^+| = |\text{core}(A)^+| \cdot |V|$$
$$|\text{core}(U \times B)^+| = |\text{core}(B)^+| \cdot |U|$$

and it follows that

$$|A \times V \cup U \times B| = |\text{core}(A \times V \cup U \times B)^+|$$
$$= |\text{core}(A \times V)^+| + |\text{core}(U \times B)^+| - |\text{core}(A \times V \cap U \times B)^+|$$
$$= |\text{core}(A)^+| \cdot |V| + |\text{core}(B)^+| \cdot |U| - |\text{core}(A)^+| \cdot |\text{core}(B)^+|$$

Thus, we obtain a group of indirect computation formulas of the sizes of orthogonal compound flexible subsets, namely

$$|A \times V \cap U \times B| = |\text{core}(A)^+| \cdot |\text{core}(B)^+| \tag{20.31}$$

$$|A \times V \cup U \times B| = |\text{core}(A)^+| \cdot |V| + |\text{core}(B)^+| \cdot |U| - |\text{core}(A)^+| \cdot |\text{core}(B)^+| \tag{20.32}$$

Generally, let flexible set $A_i \subset U_i$ ($i = 1, 2, \ldots, n$), and denote $\times U_1 \times U_2 \times \cdots \times U_{i-1} \times A_i \times U_{i+1} \times \cdots \times U_n$ as $\boldsymbol{A}_i$, then

$$|\boldsymbol{A}_1 \cap \boldsymbol{A}_2 \cap \cdots \cap \boldsymbol{A}_n| = |\text{core}(A_1)^+| \cdot |\text{core}(A_2)^+| \cdot \cdots \cdot |\text{core}(A_n)^+| \quad (20.33)$$

$$|\boldsymbol{A}_1 \cup \boldsymbol{A}_2 \cup \cdots \cup \boldsymbol{A}_n|$$
$$= \sum_{1 \leqslant i \leqslant n} |\text{core}(A_i)^+| - \sum_{1 \leq i \leq j \leq n} (|\text{core}(A_i)^+| \cdot |\text{core}(A_j)^+|)$$
$$+ \sum_{1 \leq i \leq j \leq k \leq n} (|\text{core}(A_i)^+| \cdot |\text{core}(A_j)^+| \cdot |\text{core}(A_k)^+|)$$
$$- \cdots + (1)^{n+1} |\text{core}(A_1)^+| \cdot |\text{core}(A_2)^+| \cdot \cdots \cdot |\text{core}(A_n)^+| \quad (20.34)$$

20.3 Share of a Set

1. Definition and properties of the share of a set

Definition 20.22 Let U be a bounded set as a universal set, and $A \subseteq U$, set

$$\text{shar}(A) = \frac{|A|}{|U|} \quad (20.35)$$

to be called the share of subset A in universal set U, or simply the share of set A.

Example 20.2 Let $U = \{a, b, c, d, e\}$ and $A = \{a, b, c\}$, then

$$\text{shar}(A) = 3/5 = 0.6$$

It can be seen that the share $\text{shar}(\cdot)$ can also be viewed as a function on power set 2^U. Then, when U is a bounded discrete set, the range of $\text{shar}(x)$ is $\{0, \frac{1}{n}, \frac{2}{n}, \ldots, \frac{n-1}{n}, 1\}$; when U is a bounded continuous set, the range of $\text{shar}(x)$ is $[0,1]$.

Obviously, share is also a numerical feature of a set, and it is a relative measure of a set. It is not hard to prove that the share satisfies the following basic properties.

Theorem 20.5 *Let U be a bounded set as a universal set.*

(i) *Boundedness*: $\text{shar}(\varnothing) = 0$ *and* $\text{shar}(U) = 1$;
(ii) *Monotony*: $\forall A, B \in 2^U$, *if* $A \subset B$, *then* $\text{shar}(A) \leq \text{shar}(B)$.

2. **Shares of a compound set**

From the definition of the share, the shares of compound subsets $A \cap B$, $A \cup B$, and $A^c \in 2^U$ should be

$$\text{shar}(A \cap B) = \frac{|A \cap B|}{|U|} \tag{20.36}$$

$$\text{shar}(A \cup B) = \frac{|A \cup B|}{|U|} \tag{20.37}$$

$$\text{shar}(A^c) = \frac{|A^c|}{|U|} \tag{20.38}$$

This is the method to compute the share of a compound set from the definition of the share. Next, we discuss the method to compute the share of a compound set with the share of its component sets. It can be seen that the share computation of a compound set in fact is completely decided by its size, so from the previous compound computation of sizes, it is known that for arbitrary $A, B \in 2^U$, there are also the following two basic relational expressions:

$$\text{shar}(A^c) = 1 - \text{shar}(A) \tag{20.39}$$

$$\text{shar}(A \cup B) = \text{shar}(A) + \text{shar}(B) - \text{shar}(A \cap B) \tag{20.40}$$

Since there is no a general computation formula for the size $|A \cap B|$, so for share $\text{shar}(A \cap B)$, there is no a general computation formula either. However, in special situations, there are compound computation formulas for shares.

Theorem 20.6 *Let U be a universal set, and A and B are subsets of U, then*

(i) *If $A \cap B = \emptyset$, then*

$$\text{shar}(A \cup B) = \text{shar}(A) + \text{shar}(B) \tag{20.41}$$

(ii) *If $A \subset B$, then*

$$\text{shar}(A \cap B) = \text{shar}(A), \quad \text{shar}(A \cup B) = \text{shar}(B) \tag{20.42}$$

(iii) *If there exists linearly ordered set $\langle L, \subset \rangle \subset 2^U$, then for arbitrary A, $B \in \langle L, \subset \rangle$,*

$$\text{shar}(A \cap B) = \min\{\text{shar}(A), \text{shar}(B)\} \tag{20.43}$$

$$\text{shar}(A \cup B) = \max\{\text{shar}(A), \text{shar}(B)\} \tag{20.44}$$

Proof We prove the first equality of (iii):

$$\text{shar}(A \cap B) = \frac{|A \cap B|}{|U|}$$

$$= \frac{\min\{|A|, |B|\}}{|U|}$$

$$= \min\left\{\frac{|A|}{|U|}, \frac{|B|}{|U|}\right\}$$

$$= \min\{\text{shar}(A), \text{shar}(B)\}$$

Similarly, the second equality of (iii) can also be proved.

Besides, from Eq. (20.39), we have

$$\text{shar}(A) + \text{shar}(A^c) = 1 \tag{20.45}$$

That is, the sum of shares of complementary sets is 1. More generally, we have the following theorem.

Theorem 20.7 *Let U be a universal set, and $A_i \subset U$, $\bigcup_i A_i = U$, $\bigcap_i A_i = \varnothing$, $i = 1$, 2, ..., n, then*

$$\sum_{i=1}^{n} \text{shar}(A_i) = 1 \tag{20.46}$$

We call equalities (20.45) *and* (20.46) *to be the* **complement law of shares**.

From the definition of the share and the share computation formulas of compound sets, we can also have the following two theorems.

Theorem 20.8 *Let A and B be two subsets of one and the same universal set. Then,*

$$\text{shar}(A \cap B) = 1, \quad \text{if and only if} \quad \text{shar}(A) = 1 \text{ and } \text{shar}(B) = 1.$$

Proof

(1) Let $A, B \subseteq U$. Suppose $\text{shar}(A \cap B) = 1$. By the definition of the share, $\frac{|A \cap B|}{|U|} = 1$, thus $|A \cap B| = |U|$, since $A, B \subseteq U$, then must $A = U$ and $B = U$. Thus, it follows that $\text{shar}(A) = 1$ and $\text{shar}(B) = 1$.

Conversely, suppose $\text{shar}(A) = 1$ and $\text{shar}(B) = 1$. Then, it follows that $A = U$ and $B = U$. Thus, $A \cap B = U$. Therefore, it follows that $\text{shar}(A \cap B) = 1$.

(2) Let $A \subseteq U$ and $B \subseteq V$. Suppose $\text{shar}(A \cap B) = 1$. By the definition of the share, $\frac{|A \cap B|}{|U \times V|} = 1$, while $|A \cap B| = |A \times B| = |A| \cdot |B|$. Thus, $\frac{|A \cap B|}{|U \times V|} = \frac{|A| \cdot |B|}{|U \times V|} = \frac{|A| \cdot |B|}{|U| \cdot |V|} = 1$, and then $\frac{|A|}{|U|} \cdot \frac{|B|}{|V|} = 1$; thus, $\frac{|A|}{|U|} = 1$ and $\frac{|B|}{|V|} = 1$, so $|A| = |U|$ and $|B| = |V|$, since

$A \subseteq U$ and $B \subseteq V$, so must $A = U$ and $B = V$. Thus, it follows that shar$(A) = 1$ and shar$(B) = 1$.

Conversely, Suppose shar$(A) = 1$ and shar$(B) = 1$. Then, $A = U$ and $B = V$. Thus, $A \cap B = A \times B = U \times V$. Therefore, it follows that $(A \cap B) = $ shar$(U \times V) = 1$.

Theorem 20.9 *Let A and B be two subsets of one and the same universal set. If* shar$(A) = 1$ *or* shar$(B) = 1$, *then* shar$(A \cup B) = 1$.

Proof

(1) Let $A, B \subseteq U$, and shar$(A) = 1$ or shar$(B) = 1$. By the definition of the share, it must follow that $A = U$ or $B = U$; therefore, $A \cup B = U$; thus, shar$(A \cup B) = 1$.
(2) Let $A \subseteq U$, $B \subseteq V$, and shar$(A) = 1$ or shar$(B) = 1$. By the definition of the share, it must follow that $A = U$ or $B = V$; therefore, $A \cup B = (A \times V) \cup (B \times U) = U \times V$, thus, shar$(A \cup B) = 1$.

3. **Share of a Cartesian product and share of an orthogonal compound set**

Let $A \subset U$ and $B \subset V$, then $A \times B \subset U \times V$, and from $|A \times B| = |A| \cdot |B|$, we have

$$\text{shar}(A \times B) = \frac{|A \times B|}{|U \times V|} = \frac{|A| \cdot |B|}{|U| \cdot |V|} = \frac{|A|}{|U|} \cdot \frac{|B|}{|V|} = \text{shar}(A) \cdot \text{shar}(B)$$

That is,

$$\text{shar}(A \times B) = \text{shar}(A)\text{shar}(B) \tag{20.47}$$

That is to say, the share of a product is equal to the product of shares of factors. Let $A \subset U$ and $B \subset V$. From $A \times V \cap B \times U = A \times B$, it follows that

$$\text{shar}(A \times V \cap B \times U) = \text{shar}(A \times B) = \text{shar}(A)\text{shar}(B)$$

$$\begin{aligned} \text{shar}(A \times V \cup U \times B) &= \text{shar}(A \times V) + \text{shar}(U \times B) - \text{shar}(A \times V \cap B \times U) \\ &= \text{shar}(A)\text{shar}(V) + \text{shar}(U)\text{shar}(B) - \text{shar}(A)\text{shar}(B) \\ &= \text{shar}(A) \cdot 1 + 1 \cdot \text{shar}(B) - \text{shar}(A)\text{shar}(B) \\ &= \text{shar}(A) + \text{shar}(B) - \text{shar}(A)\text{shar}(B) \end{aligned}$$

Denote $A \times V$ as $\boldsymbol{A}$ and $U \times B$ as $\boldsymbol{B}$, then

$$\text{shar}(\boldsymbol{A} \cap \boldsymbol{B}) = \text{shar}(\boldsymbol{A})\text{shar}(\boldsymbol{B}) \tag{20.48}$$

$$\text{shar}(\boldsymbol{A} \cup \boldsymbol{B}) = \text{shar}(\boldsymbol{A}) + \text{shar}(\boldsymbol{B}) - \text{shar}(\boldsymbol{A})\text{shar}(\boldsymbol{B}) \tag{20.49}$$

That means the share of the intersection of extended sets is equal to the product of the shares of the original sets, and that the share of the union of extended sets is equal to the difference between sum and product of the shares of the original sets.

Generally, let set $A_i \subset U_i (i = 1, 2, \ldots, n)$, denote $U_1 \times U_2 \times \cdots \times U_{i-1} \times A_i \times U_{i+1} \times \cdots \times U_n$ as A_i. Then

$$\mathrm{shar}(A_1 \cap A_2 \cap \cdots \cap A_n) = \mathrm{shar}(A_1)\mathrm{shar}(A_2)\ldots\mathrm{shar}(A_n) \tag{20.50}$$

$$\mathrm{shar}(A_1 \cup A_2 \cup \cdots \cup A_n) = \sum_{1 \le i \le n} \mathrm{shar}(A_i) - \sum_{1 \le i \le j \le n} \mathrm{shar}(A_i)\mathrm{shar}(A_j)$$
$$+ \sum_{1 \le i \le j \le k \le n} \mathrm{shar}(A_i)\mathrm{shar}(A_j)\mathrm{shar}(A_k)$$
$$- \cdots + (1)^{n+1}\mathrm{shar}(A_1)\mathrm{shar}(A_2)\ldots\mathrm{shar}(A_n)$$
$$\tag{20.51}$$

20.4   Share of a Flexible Set

1. Definitions and properties of the Shares of a flexible set

By the definition of the size of a flexible set above, we directly give the definition of the share of a flexible set.

Definition 20.4 Let U be an n-dimensional measurement space, and A is a flexible subset of U, set

$$\mathrm{shar}(A) = \frac{|A|}{|U|} \tag{20.52}$$

to be called the share of flexible subset A in universal set U, or simply the share of flexible set A.

From this definition and the definition of the size of a flexible set, we have the following formulas:

(i) When U is a discrete set,

$$\mathrm{shar}(A) = \frac{\sum\limits_{x \in \mathrm{supp}(A)} m_A(x)}{|U|} \tag{20.53}$$

(ii) When U is a continuous set,

$$\text{shar}(A) = \frac{\int \cdots \int_{\text{supp}(A)} m_A(x_1, x_2, \ldots, x_n) dx_1 dx_2 \ldots dx_n}{|U|} \tag{20.54}$$

About the shares of flexible sets, there are the following basic properties.

Theorem 20.10 *Let A and B be flexible subsets of n-dimensional measurement space U. Then,*

(i)   shar$(A) < 1$ *(boundedness)*
(ii) *If* $A \subset B$, *then* shar$(A) <$ shar(B) *(monotony)*
(iii) shar$(\text{core}(A)^+) =$ shar(A) *(here A is not a flexible product set)*

Properties (i) and (ii) are obvious; property (iii) can be obtained from the definition of the share of a flexible set and Theorem 20.3.

2. Share of a compound flexible set

(1) Directly computing based on the definition of share

Based on the definition of the share of a flexible set,

$$\text{shar}(A^c) = \frac{|A^c|}{|U|} \tag{20.55}$$

$$\text{shar}|A \cap B| = \frac{|A \cap B|}{|U|} \tag{20.56}$$

$$\text{shar}|A \cup B| = \frac{|A \cup B|}{|U|} \tag{20.57}$$

(2) Indirectly computing based on the shares of component sets

The share computation of the compound flexible sets has similar results to the ordinary compound sets. That is, for arbitrary flexible subsets A, B in space U, always

$$\text{shar}(A^c) = 1 - \text{shar}(A) \tag{20.58}$$

$$\text{shar}(A \cup B) = \text{shar}(A) + \text{shar}(B) - \text{shar}(A \cap B) \tag{20.59}$$

In particular, the following theorem follows.

Theorem 20.11 *Let A and B be flexible subsets of n-dimensional measurement space U.*

(i)   *If $A \cap B = \emptyset$, then*

$$\text{shar}(A \cup B) = \text{shar}(A) + \text{shar}(B) \tag{20.60}$$

(ii) *If $A \subset B$, then*

$$\text{shar}(A \cap B) = \text{shar}(A) \quad and \quad \text{shar}(A \cup B) = \text{shar}(B) \tag{20.61}$$

(iii) *If there exists linearly ordered set $\langle L, \subset \rangle \subset 2^U$, then for arbitrary A, $B \in \langle L, \subset \rangle$, then*

$$\text{shar}(A \cap B) = \min\{\text{shar}(A), \text{shar}(B)\} \tag{20.62}$$

$$\text{shar}(A \cup B) = \max\{\text{shar}(A), \text{shar}(B)\} \tag{20.63}$$

Besides, from Eq. (20.54), we have

$$\text{shar}(A) + \text{shar}(A^c) = 1 \tag{20.64}$$

That is, the sum of the shares of complementary flexible sets is 1. More generally, we have the following theorem.

Theorem 20.12 *Let $A_1, A_2, \ldots, A_n$ be a group of flexible subsets of n-dimensional measurement space U, they just form a flexible partition of space U, then*

$$\sum_{i=1}^{n} \text{shar}(A_i) = 1 \tag{20.65}$$

*So, the shares of flexible sets still obey the **complement law of shares**.*

3. Shares of flexible Cartesian product and orthogonal compound flexible sets

Let flexible set $A_i \subset U_i (i = 1, 2, \ldots, n)$. By the definition of the share of a flexible set, it follows that

$$
\begin{aligned}
\text{shar}(A_1 \times A_2 \times \cdots \times A_n) &= |A_1 \times A_2 \times \ldots \times A_n| / |U_1 \times U_2 \times \cdots \times U_n| \\
&= \underset{\text{supp}(A_1 \times A_2 \times \cdots \times A_n)}{\int\!\!\int \cdots \int} m_{A_1 \times A_2 \times \cdots \times A_n}(x_1, x_2, \ldots, x_n) dx_1 dx_2 \ldots dx_n / |U_1 \times U_2 \times \cdots \times U_n|
\end{aligned}
\tag{20.66}
$$

But from Eq. (20.25), for flexible sets $A \subset U$ and $B \subset V$, then

$$|A \times B| = w_1 \cdot |A| \cdot |\text{supp}(B)| + w_2 \cdot |\text{supp}(A)| \cdot |B|$$

Thus, we have

$$
\begin{aligned}
\text{shar}(A \times B) &= |A \times B|/|U \times V| \\
&= (w_1 \cdot |A| \cdot |\text{supp}(B)| + w_2 \cdot |\text{supp}(A)| \cdot |B|)/|U \times V| \\
&= (w_1 \cdot |A| \cdot |\text{supp}(B)|)/|U \times V| + (w_2 \cdot |\text{supp}(A)| \cdot |B|)/|U \times V| \\
&= w_1 \text{shar}(A)\, \text{shar}(\text{supp}(B)) + w_2 \text{shar}(\text{supp}(A)) \text{shar}(B)
\end{aligned}
$$

$$(20.67)$$

where $w_1, w_2 \in (0,1)$, $w_1 + w_2 = 1$.

Generally, from Eq. (20.26), for flexible sets $A_i \subset U_i (i = 1, 2, \ldots, n)$,

$$
\begin{aligned}
\text{shar}(A_1 \times A_2 \times \cdots \times A_n) &= w_1 \text{shar}(A_1) \text{shar}(\text{supp}(A_2 \times A_3 \times \cdots \times A_n)) \\
&\quad + w_2 \text{shar}(A_2) \text{shar}(\text{supp}(A_1 \times A_3 \times \cdots \times A_n)) \\
&\qquad \cdots \\
&\quad + w_n \text{shar}(A_n) \text{shar}(\text{supp}(A_1 \times A_2 \times \cdots \times A_{n-1}))
\end{aligned}
$$

$$(20.68)$$

where $w_1, w_2, \ldots, w_n \in (0,1)$, $w_1 + w_2 + \cdots + w_n = 1$.

On the other hand, from Eq. (20.28), that is

$$
|A_1 \times A_2 \times \cdots \times A_n| > |A_1| \cdot |A_2| \cdot \cdots \cdot |A_n|
$$

we have

$$
\text{shar}(A_1 \times A_2 \times \cdots \times A_n) > \text{shar}(A_1) \text{shar}(A_2) \ldots \text{shar}(A_n) \qquad (20.69)
$$

Next, we consider the computation problem of the share of compound flexible sets formed by orthogonal flexible subsets in a product space.

Let U and V both be one-dimensional measurement spaces, and A and B be separately the flexible subsets of U and V. From the definition of the share of a flexible set,

$$
\begin{aligned}
\text{shar}(A \times V \cap U \times B) &= |A \times V \cap U \times B|/|U \times V| \\
&= \iint_{\text{supp}(A \times B)} \min\{m_A(x), m_B(y)\} dx dy/|U \times V|
\end{aligned}
$$

$$(20.70)$$

$$\text{shar}(A \times V \cup U \times B) = \text{shar}(A \times V) + \text{shar}(U \times B) - \text{shar}(A \times V \cap U \times B)$$

$$= \iint\limits_{\text{supp}(A \times V)} m_A(x)\mathrm{d}x\mathrm{d}y / |U \times V|$$

$$+ \iint\limits_{\text{supp}(U \times B)} m_B(y)\mathrm{d}x\mathrm{d}y / |U \times V|$$

$$- \iint\limits_{\text{supp}(A \cap B)} \min\{m_A(x), m_B(y)\}\mathrm{d}x\mathrm{d}y / |U \times V|$$

$$(20.71)$$

But on the other hand, from the above section, we have already known that

$$|A \times V \cap U \times B| = |\text{core}(A)^+| \cdot |\text{core}(B)^+|$$

Consequently,

$$\begin{aligned}
\text{shar}|A \times V \cap U \times B| &= (|\text{core}(A)^+| \cdot |\text{core}(B)^+|)/|U \times V| \\
&= (|\text{core}(A)^+| \cdot |\text{core}(B)^+|)/(|U| \cdot |V|) \\
&= (|\text{core}(A)^+|/|U|) \cdot (|\text{core}(B)^+|/|V|) \\
&= (|A|/|U|) \cdot (|B|/|V|) \\
&= \text{shar}(A)\text{shar}(B)
\end{aligned}$$

And from

$$|A \times V| = |\text{core}(A \times V)^+|, \quad |U \times B| = |\text{core}(U \times B)^+|$$

it follows that

$$\text{shar}(A \times V) = |A \times V|/|U \times V| = |A|/|U| = \text{shar}(A)$$

$$\text{shar}(U \times B) = |U \times B|/|U \times V| = |B|/|V| = \text{shar}(B)$$

Thus,

$$\begin{aligned}
\text{shar}(A \times V \cup U \times B) &= \text{shar}(A \times V) + \text{shar}(U \times B) - \text{shar}(A \times V \cap U \times B) \\
&= \text{shar}(A) + \text{shar}(B) - \text{shar}(A)\text{shar}(B)
\end{aligned}$$

As thus, we obtain a group of indirect computation formulas of the shares of orthogonal compound flexible sets, namely

$$\text{shar}(A \times V \cap U \times B) = \text{shar}(A)\text{shar}(B) \tag{20.72}$$

$$\text{shar}(A \times V \cup U \times B) = \text{shar}(A) + \text{shar}(B) - \text{shar}(A)\text{shar}(B) \tag{20.73}$$

Generally, let flexible sets $A_i \subset U_i (i = 1, 2, \ldots, n)$, denote $U_1 \times U_2 \times \cdots \times U_{i-1} \times A_i \times U_{i+1} \times \cdots \times U_n$ as A_i, then

$$\text{shar}(A_1 \cap A_2 \cap \cdots \cap A_n) = \text{shar}(A_1)\text{shar}(A_2)\ldots\text{shar}(A_n) \tag{20.74}$$

$$\begin{aligned}
\text{shar}(A_1 \cup A_2 \cup \cdots \cup A_n) = &\sum_{1 \le i \le n} \text{shar}(A_i) - \sum_{1 \le i < j \le n} \text{shar}(A_i)\text{shar}(A_j) \\
&+ \sum_{1 \le i < j < k \le n} \text{shar}(A_i)\text{shar}(A_j)\text{shar}(A_k) \\
&- \cdots + (1)^{n+1}\text{shar}(A_1)\text{shar}(A_2)\ldots\text{shar}(A_n)
\end{aligned} \tag{20.75}$$

20.5   Inclusion-Degree and Equality-Degree

1. Partial inclusion and partial equality

"Inclusion" and "equality" are two relations between sets, but "intersection" can also be viewed as inclusion to some degree or equality to some degree, or in other words "partial inclusion" and "partial equality."

Definition 20.5 Let A and B be two sets.

(i) If $A \cap B \ne \varnothing$, then we say set A partially contains set B and set B partially contains set A, or to say that set A and set B are partially equal.

(ii) If $A \cap B \ne \varnothing$ and $A \not\subset B$, $B \not\subset A$, then we say that set A properly partially contains set B and set B properly partially contains set A, or to say that set A and set B are properly partially equal.

Definition 20.6 Let A and B be two flexible sets.

(i) If $\text{coer}(A)^+ \cap \text{coer}(B)^+ \ne \varnothing$, then we say that flexible set A partially contains flexible set B and flexible set B partially contains flexible set A, or to say that flexible set A and flexible set B are partially equal.

(ii) If $\text{coer}(A)^+ \cap \text{coer}(B)^+ \ne \varnothing$ and $\text{coer}(A)^+ \not\subset \text{coer}(B)^+$, $\text{coer}(B)^+ \not\subset \text{coer}(A)^+$, then we say that flexible set A properly partially contains flexible set B and flexible set B properly partially contains flexible set A, or to say that flexible set A and flexible set B are properly partially equal.

Relatively to the partial inclusion, usual inclusion "⊂" is a complete inclusion. Conversely speaking, the complete inclusion is a special case of the partial inclusion. Likewise, relatively to the partial equality, usual equality "=" is a complete equality. Conversely speaking, the complete equality is a special case of the partial equality.

Obviously, the relations of partial inclusion and partial equality between sets (including flexible sets) all satisfy symmetry. But it is not hard to see that they are not necessarily to satisfy transitivity.

Theorem 20.13

(1) *Let A, B, and C be three different non-empty sets. Then, a sufficient and necessary condition for that sets A, B, and C are of partial inclusion in order and the partial inclusion relation satisfies transitivity is*

$$(A \cap B) \cap (B \cap C) \neq \varnothing \tag{20.76}$$

(2) *Let A, B, and C be three different non-empty flexible sets, then a sufficient and necessary condition for that flexible sets A, B, and C are of partial inclusion in order and the partial inclusion relation satisfies transitivity is*

$$(\mathrm{core}(A)^{+} \cap \mathrm{core}(B)^{+}) \cap (\mathrm{core}(B)^{+} \mathrm{core}(C)^{+}) \neq \varnothing \tag{20.77}$$

Proof We prove (1) first.

Sufficiency: Suppose $(A \cap B) \cap (B \cap C) \neq \varnothing$. since $(A \cap B) \cap (B \cap C) = A \cap B \cap C = A \cap C \cap B$, so $A \cap C \cap B \neq \varnothing$, and then $A \cap C \neq \varnothing$. While $(A \cap B) \cap (B \cap C) \neq \varnothing$ is equivalent to $A \cap B \neq \varnothing$ and $B \cap C \neq \varnothing$, while $A \cap B \neq \varnothing$ and $B \cap C \neq \varnothing$ are equivalent to A being partially contained in B and B being partially contained in C; additionally, $A \cap C \neq \varnothing$ is equivalent to A being partially contained in C. Thus, as long as A is partially contained in B and B is partially contained in C, then A is partially contained in C.

Necessity (by contraposition):

Suppose $(A \cap B) \cap (B \cap C) \neq \varnothing$. Then $(A \cap B) \neq \varnothing$ or $(B \cap C) \neq \varnothing$; thus, A is not partially contained in B or B is not partially contained in C. Also since $(A \cap B) \cap (B \cap C) = A \cap C \cap B$, so also $A \cap C \neq \varnothing$ or $B \neq \varnothing$; but known $B \neq \varnothing$, so only $A \cap C \neq \varnothing$; thus, A is not partially contained in C.

Thus, (1) is proved.

Similarly, (2) can also be proved.

Example 20.3 Let U, V, and W be three different measurement spaces, and A, B, and C are three subsets or flexible subsets that are pairwise orthogonal in product space $U \times V \times W$. It is not hard to verify that A, B, and C are of partial inclusion in order, and the partial inclusion relation satisfies transitivity.

It is clear that partial equality relation and its transitivity have also the sufficient and necessary condition similar to Theorem 20.13.

2. Definitions of the inclusion-degree and equality-degree

Since there are different cases of "parts," we can use certain measures to quantify "partial inclusion" and "partial equality."

Definition 20.25 Let U be a bounded set as a universal set, and $A, B \subseteq U$. Set

$$\text{cont}(A, B) = \frac{|A \cap B|}{|A|} \tag{20.78}$$

to be called the degree of set A being contained in set B, or the degree of set B containing set A, short for the degree of inclusion, written inclusion-degree, of B to A.

From Eq. (20.78), it can be seen that when $A \cap B = \varnothing$, $\text{cont}(A, B) = 0$; when $A \cap B = A$, $\text{cont}(A, B) = 1$; when $A \cap B \neq \varnothing$ and $A \cap B \neq B$, $0 < \text{cont}(A, B) < 1$, which indicates that set A is contained in set B with degree $\text{cont}(A, B)$. It can be seen that the measure of inclusion-degree not only can quantify the "partial inclusion" between sets, but also can describe "non-inclusion" and "complete inclusion" between sets; that is, it uniforms "non-inclusion," "complete inclusion," and "partial inclusion" between sets.

Note that inclusion-degree portrays the approximate degree for the inclusion relation, but not the membership-degree or consistency-degree for inclusion relation. In fact, inclusion relation is a rigid relation, while "basically containing," "rather containing," and "nearly containing" are flexible relations. Inclusion-degree is a kind of measure for inclusion; on the range of this measure, the flexible linguistic values of "basically containing," "rather containing," and "nearly containing" can be defined.

Similarly, we can also use a certain measure to quantify the "partial equality" between sets.

Definition 20.8 Let U be a bounded sct as a universal set, and A and $B \subseteq U$. Set

$$\text{equa}(A, B) = \frac{|A \cap B|}{|A \cup B|} \tag{20.79}$$

to be called the degree of equality, written equality-degree, between set A and set B.

From Eq. (20.79), it can be seen that when $A \cap B \neq \varnothing$, $\text{equa}(A, B) = 0$; when $A \cap B \neq \varnothing$ and $A = B$, $\text{equa}(A, B) = 1$; when $A \cap B \neq \varnothing$ and $A \neq B$, then $A \cap B \subset A \cup B$, thus, $0 < \text{equa}(A, B) < 1$, which indicates that set A and set B are equal with degree $\text{equa}(A, B)$. This shows that the measure of equality-degree not only can quantify "partial equality" between sets, but also can describe "non equality" and "complete equality" between scts; that is, it uniforms the "non equality," "complete equality," and "partial equality" between sets. And we can define the flexible linguistic values of "(being) basically equal," "(being) rather equal," and "(being) nearly equal."

Definition 20.9 Let U be a bounded measurement space, and A and B be flexible subsets of U, set

$$\text{cont}(A, B) = \frac{\left|\text{core}(A)^+ \cap \text{core}(B)^+\right|}{\left|\text{core}(A)^+\right|} \tag{20.80}$$

to be called the degree of flexible set A being contained in flexible set B, or the degree of flexible set B containing flexible set A, simply the degree of inclusion, written inclusion-degree, of flexible set B to flexible set A.

Of course, conceptually speaking, it should be

$$\text{cont}(A, B) = \frac{|A \cap B|}{|A|} \tag{20.81}$$

and

(i) When U is a discrete set,

$$\text{cont}(A, B) = \frac{\sum_{x \in \text{supp}(A \cap B)} m_{A \cap B}(x_1, x_2, \ldots, x_n)}{\sum_{x \in \text{supp}(A)} m_A(x_1, x_2, \ldots, x_n)} \tag{20.82}$$

(ii) When U is a continuous set,

$$\text{cont}(A, B) = \frac{\int \cdots \int_{\text{supp}(A \cap B)} m_{A \cap B}(x_1, x_2, \ldots, x_n) dx_1 dx_2 \ldots dx_n}{\int \cdots \int_{\text{supp}(A)} m_A(x_1, x_2, \ldots, x_n) dx_1 dx_2 \ldots dx_n} \tag{20.83}$$

Definition 20.10 Let U be a bounded measurement space, and A and B be the flexible subsets of U, set

$$\text{equa}(A, B) = \frac{\left|\text{core}(A)^+ \cap \text{core}(B)^+\right|}{\left|\text{core}(A)^+ \cup \text{core}(B)^+\right|} \tag{20.84}$$

to be called the degree of equality, written equality-degree, between flexible set A and flexible set B.

Similarly, conceptually speaking, it should be

$$\text{equa}(A, B) = \frac{|A \cap B|}{|A \cup B|} \qquad (20.85)$$

and

(i) When U is a discrete set,

$$\text{equa}(A, B) = \frac{\sum_{x \in \text{supp}(A \cap B)} m_{A \cap B}(x_1, x_2, \ldots, x_n)}{\sum_{x \in \text{supp}(A \cup B)} m_{A \cup B}(x_1, x_2, \ldots, x_n)} \qquad (20.86)$$

(ii) When U is a continuous set,

$$\text{equa}(A, B) = \frac{\int \cdots \int_{\text{supp}(A \cap B)} m_{A \cap B}(x_1, x_2, \ldots, x_n) dx_1 dx_2 \ldots dx_n}{\int \cdots \int_{\text{supp}(A \cup B)} m_{A \cup B}(x_1, x_2, \ldots, x_n) dx_1 dx_2 \ldots dx_n} \qquad (20.87)$$

It can be seen that the inclusion-degree, equality-degree, and share of sets are all defined on the basis of the sizes of sets. Then, are there connections between them? The answer is affirmative. In fact, for arbitrary A, $B \in U$, since

$$\text{cont}(A, B) = \frac{|A \cap B|}{|A|} = \frac{|A \cap B|/|U|}{|A|/|U|} = \frac{\text{shar}(A \cap B)}{\text{shar}A}$$

Therefore,

$$\text{cont}(A, B) = \frac{\text{shar}(A \cap B)}{\text{shar}A} \qquad (20.88)$$

Similarly, it can follow that

$$\text{equa}(A, B) = \frac{\text{shar}(A \cap B)}{\text{shar}(A \cup B)} \qquad (20.89)$$

The two equalities reveal the relation between the share of sets and the inclusion-degree and equality-degree between sets. They can also be viewed as another definition of the inclusion-degree and equality-degree between sets. Since the inclusion-degree and equality-degree between flexible sets are defined based on the their extended cores, the latter are rigid sets, so there is also such relations between the share of flexible sets and the inclusion-degree and equality-degree between flexible sets.

3. Basic properties of inclusion-degree and equality-degree

Inclusion-degree and equality-degree are two kinds of measures of the relations between sets. From the definition, it is not hard to verify that the inclusion-degree and equality-degree between sets (including flexible sets) have the following basic properties.

Theorem 20.14 *Let U be a bounded measurement space, and A and B be subsets or flexible subsets of U, then*

(i) $0 \le \text{cont}(A, B) \le 1$.
(ii) $\text{cont}(A, B) = 0$, *if and only if* $A \cap B = \varnothing$.
(iii) $\text{cont}(A, B) = 1$, *if and only if* $A \cap B = A$, *that is,* $A \subset B$.

Theorem 20.15 *Let U be a bounded measurement space, and A and B are the subsets or flexible subsets of U, then*

(i) $0 \le \text{equa}(A, B) \le 1$.
(ii) $\text{equa}(A, B) = 0$, *if and only if* $A \cap B = \varnothing$.
(iii) $\text{equa}(A, B) = 1$, *if and only if* $A = B$.

Theorem 20.16 *Let U, V, and W are three different measurement spaces, and A, B, and C are three subsets or flexible subsets that are pairwise orthogonal in product space* $U \times V \times W$, *then*

$$\text{cont}(A, C) = \text{cont}(A, B) \cdot \text{cont}(B, C) \tag{20.90}$$

Proof Let A, B, and C be rigid subsets. Thus,

$$\text{cont}(A, B) \cdot \text{cont}(B, C) = \frac{|A \cap B|}{|A|} \times \frac{|B \cap C|}{|B|}$$

$$= \frac{|A \cap B| \times |B \cap C|}{|A| \times |B|} = \frac{|(A \cap B) \cap (B \cap C)|}{|A| \times |B|} \quad \text{(because } A \text{ and } B \text{ are orthogonal,}$$
$$\text{and } B \text{ and } C \text{ are orthogonal, so}$$
$$A \cap B \text{ and } B \cap C \text{ are also}$$
$$\text{orthogonal)}$$

$$= \frac{|A \cap B \cap C|}{|A| \times |B|} = \frac{|(A \cap C) \cap B|}{|A| \times |B|} = \frac{|A \cap C| \times |B|}{|A| \times |B|} \quad (A \cap C \text{ and } B \text{ are orthogonal)}$$

$$= \frac{|A \cap C|}{|A|} = \text{cont}(A, C)$$

that is,

$$\text{cont}(A, C) = \text{cont}(A, B) \cdot \text{cont}(B, C)$$

Similarly, we can prove when A, B, and C are flexible subsets, also $\text{cont}(A, C) = \text{cont}(A, B) \cdot \text{cont}(B, C)$.

It is clear that partial equality relation has also the property similar to Theorem 20.16.

20.6 Partial Correspondence and Correspondence-Rate

In Sect. 9.1, we discussed the correspondence between two sets; in this section, we further discuss partial correspondence between two sets and its measure.

Definition 20.11 Let A and B be two sets. If to partial $x \in A$, or in other wards, to each $x \in A_1 \subset A \cdot (A_1 \neq \varnothing)$, there corresponds a $y \in B$, then we say that set A is partially corresponded by set B, or set B corresponds to a part of set A. This relationship between sets A and B is called the partial correspondence.

In the sense of membership, a flexible set is completely represented by its extended core, so the partial correspondence between flexible sets is the following definition.

Definition 20.12 Let A and B be two flexible sets. if to partial $x \in \text{core}(A)^+$, or in other wards, to each $x \in A_1 \subset \text{core}(A_1)^+ (A_1 \neq \varnothing)$, there corresponds a $y \in \text{core}(B)^+$, then we say that flexible set A is partially corresponded by flexible set B, or flexible set B corresponds to a part of flexible set A. This relationship between flexible sets A and B is called the partial correspondence.

Relatively to the partial correspondence, the correspondence "$\mapsto$" in Sect. 9.1 is an complete correspondence. Conversely speaking, the complete correspondence $\mapsto$ is a special case of the partial correspondence.

It is not hard to see that partial correspondence (relation) between sets is not necessarily to satisfy transitivity.

Theorem 20.17 *Let A, B, and C are three non-empty sets, there is partial correspondence relation f_{A-B} from A to B, and there is partial correspondence relation f_{B-C} from B to C, then A is partially corresponded by C, or C corresponds to a part of set A, if and only if,*

$$f_{A-B}(A) \cap f_{B-C}^{-1}(C) \neq \varnothing \tag{20.91}$$

Proof This theorem is to say that the intersection of the image of partial correspondence relation f_{A-B} and the inverse image of partial correspondence relation f_{B-C} is not empty is a sufficient and necessary condition for that the partial correspondence relation between sets A, B, and C has transitivity. Because $f_{A-B}(A) \cap f_{B-C}^{-1}(C) \subset B$, $f_{A-B}(A) \cap f_{B-C}^{-1}(C) \neq \varnothing$ means that there is at least an $y \in B$, which is the image of a certain $x \in A$ as well as the inverse image of a certain $z \in C$. Then, through this y, the z in C just corresponds to the x in A. Thus, A is partially corresponded by C.

Otherwise, if $f_{A-B}(A) \cap f_{B-C}^{-1}(C) \neq \varnothing$, then it means that there is no such $y \in B$, which is the image of a certain $x \in A$ as well as the inverse image of a certain $z \in C$. Thus, any $x \in A$ could not be corresponded by elements in C. Therefore, A is not (partially) corresponded by C.   ∎

Next, we analyze the relationship between the "inclusion" and "correspondence" between two sets.

Let set A be partially contained in set B, then there is $x \in A$, and this $x \in B$ too; thus $x \longmapsto x$ is just a correspondence between the elements of A and B. This shows that set A is partially corresponded by set B. Thus, partial inclusion can be viewed as a special case of partial correspondence. Similarly, the inclusion between sets can also be viewed as a special case of correspondence between sets.

Definition 20.13 Let A and B be two sets, $C \subseteq A \times B$ be a correspondence relation from A to B, and $A_1 = \{x \mid (x, y) \in C\}$. Set

$$\text{corr}(A, B) = \frac{|A_1|}{|A|} \tag{20.92}$$

to be called the rate of correspondence, simply written correspondence-rate, of set B to set A; or, the correspondence-rate of correspondence (relation) from A to B.

Definition 20.13' Let A and B be two flexible sets, $C \subseteq \text{core}(A)^+ \times \text{core}(B)^+$ be a correspondence relation from A to B, and $A_1 = \{x \mid (x, y) \in C\}$. Set

$$\text{corr}(A, B) = \frac{|A_1|}{|\text{core}(A)^+|} \tag{20.93}$$

to be called the rate of correspondence, simply written correspondence-rate, of flexible set B to flexible set A; or, the correspondence-rate of correspondence (relation) from A to B.

Actually, for a flexible set, its size is equal to the size of its extended core, so $\frac{|A_1|}{|\text{core}(A)^+|}$ is also equal to $\frac{|A_1|}{|A|}$.

Since $\frac{|A_1|}{|A|} = \text{shar}(A_1)$, the correspondence-rate of set B to set A is just the share of subset A_1 in A.

Obviously, when $A_1 \neq \varnothing$, $\text{corr}(A, B) = 0$; when $A_1 \neq \varnothing$ and $A_1 \subset A$, $0 < \text{corr}(A, B) < 1$; when $A_1 = A$, $\text{corr}(A, B) = 1$. And vice versa.

As thus, the "correspondence-rate" unifies "correspondence"(i.e., "complete correspondence"), "partial correspondence," and "non correspondence" between sets, and which can be used to define these three rigid linguistic values. In particular, "partial correspondence" is quantified into number set $(0, 1)$, which can be called the range of correspondence-rate. Further, we can also define the flexible linguistic values: "rather corresponding," "basically corresponding," and "roughly corresponding" on the range $(0, 1)$ of correspondence-rates. These flexible linguistic values as relationship are some flexible relations.

Actually, a correspondence is also a transformation, and the correspondence-rate is also transformation rate. Therefore, for the partial correspondence relation with transitivity, the partial correspondence from A to C is also the product of the partial correspondence from A to B and the partial correspondence from B to C. Thus, the correspondence-rate of C to A is also the product of the correspondence-rate of B to A and the correspondence-rate of C to B.

Theorem 20.18 *Let A, B, and C be three non-empty sets. If A, B, and C are partial correspondence in order, and the partial correspondence relation satisfies transitivity, then*

$$\text{corr}(A, C) = \text{corr}(A, B) \cdot \text{corr}(B, C) \tag{20.94}$$

We call Eq. (20.94) to be the transitive formula of correspondence-rates.

In the following, we analyze the relation between the inclusion-degree and correspondence-rate between sets.

Let sets A and B be the two subsets of universal set U, then,

$$\text{cont}(A, B) = \frac{|A \cap B|}{|A|}$$

$$\text{corr}(A, B) = \frac{|A_1|}{|A|}, \quad A_1 = \{x | (x, x) \in R \subset A \times B\}$$

Now we consider if there is $A_1 = A \cap B$?

(1) From the expression of definition of A_1, it can be seen that when $A \cap B = \varnothing$, necessarily $A_1 = \varnothing$; and vice versa.
(2) Suppose $A \cap B \neq \varnothing$, take $\forall x \in A \cap B$, then $x \in A$ and $x \in B$; thus, $x \in A_1$; take $\forall x \in A_1$, from the expression of definition of A_1, then $x \in A$ and $x \in B$, thus, $x \in A \cap B$.

From (1) and (2), we have $A \cap B = A_1$. Consequently,

$$\text{cont}(A, B) = \text{corr}(A, B) \tag{20.95}$$

That is to say, inclusion-degree $\text{cont}(A, B)$ is a special case of correspondence-rate $\text{corr}(A, B)$. Thus, by Theorem 20.18, we have immediately the following theorem.

Theorem 20.19 *Let A, B, and C be three non-empty sets. If A, B, and C are partial inclusion in order, and the partial inclusion relation satisfies transitivity, then*

$$\text{cont}(A, C) = \text{cont}(A, B) \cdot \text{cont}(B, C) \tag{20.96}$$

We call Eq. (20.96) to be the transitive formula of inclusion-degrees.
Actually, Theorem 20.16 just verifies the correctness of the theorem.

20.7 Summary

In this chapter, we introduced multiple measures such as size and share for ordinary sets and flexible sets, and analyzed and expounded their properties, relations, and operations. We presented also the concepts of partial inclusion, partial equality, and partial correspondence and introduced the measures of inclusion-degree, equality-degree, and correspondence-rate. In particular, we found and presented the sufficient and necessary conditions for these relations to satisfy transitivity and the corresponding transitive formulas of inclusion-degrees (equality-degrees) and correspondence-rates.

The main points and results of the chapter are as follows:

- Size is a kind of measure of the element amount of a set; share is the ratio of the size of a set to the size of the universal set it is in.
- Size and share of the sets have boundedness, monotony, and complementarity.
- Size and share of a flexible set that is not a Cartesian product equal to the size and share of its extended core.
- Size and share of the compound sets (intersection and union) of subsets have no general compound computation formulas, but the size and share of the compound sets of subsets that forms a linearly ordered set have compound computation formulas.
- Size and share of compound sets of orthogonal subsets in a product spaces have general compound computation formulas.
- Size and share of the Cartesian product of rigid sets separately equal to the products of sizes and product of shares of its factor sets; size and share of the Cartesian product of flexible sets have no general indirect computation formulas, but the range of values of them can be estimated.
- Inclusion-degree and equality-degree are, respectively, a kind of quantification for the partial inclusion relation and partial equality relation between two sets, and correspondence-rate is a kind of quantification for the partial correspondence relation between two sets. For a relation satisfying transitivity, we can use the transitive formula to obtain indirectly corresponding measurement.

Reference

1. Lian S (2009) Principles of imprecise-information processing. Science Press, Beijing

Chapter 21
Quantifiable Rigid Linguistic Values and Related Theories

Abstract This chapter expounds briefly the basic theory of information processing with quantifiable rigid linguistic values, which is parallel or similar to the corresponding theory of information processing with flexible linguistic values. Besides, it introduces the quantifiable rigid linguistic values and their measures, partial possession and possessing rate, partial implication and implication-degree, and partial equivalence and equivalence-degree, in particular, finds and presents the judging condition of partial implication (equivalence) relation satisfying transitivity and the corresponding transitive formulas of implication-degrees (equivalence-degrees).

Keywords Quantifiable rigid linguistic values · Imprecise information · Partial implication · Implication-degree

The so-called quantifiable rigid linguistic values are words that can be replaced by numerical values stated in Sect. 1.1. Information processing with quantifiable rigid linguistic values is also a part of imprecise-information processing.

21.1 Quantifiable Rigid Linguistic Values and Their Mathematical Models

1. **Quantifiable features, quantifiable rigid linguistic values, and ordered rigid sets**

   We find that some features of objects can be described by numerical values as well as by linguistic values, but some other features can only be described by linguistic values or symbols. For instance, a person's height and weight can be described by numerical values as well as by linguistic values, while whose sex and hobby can only be described by linguistic values or symbols.

   We call the features of objects that can be described by numerical values to be quantifiable features and call the features that can be described only by linguistic values or symbols to be unquantifiable features.

© Springer Science+Business Media Singapore 2016
529
S. Lian, *Principles of Imprecise-Information Processing*,
DOI 10.1007/978-981-10-1549-6_21

Then, a person's height and weight are quantifiable features, while a person's sex and hobby are unquantifiable features.

Similarly, atmospheric pressure and temperature of the earth's surface, and the length, area, and volume of objects are all quantifiable features. People already gave corresponding measures for these features. There are also some features theoretically speaking should be quantifiable but have generally no corresponding measures. For instance, a person's character, appearance, will and morality, etc., are such quantifiable features.

For a quantifiable feature, we can design or define certain kind of measure as corresponding numerical value (the usual numerical values are actually all to be produced in this way), further obtaining the corresponding range of numerical values. If the range of numerical values (one-dimensional measurement space) of a feature is continuous, then we can do flexible clustering or flexible partitioning of it, and obtaining corresponding flexible subsets and flexible linguistic values. But if requiring, we can also do rigid clustering or rigid partitioning of a range of numerical values and obtaining corresponding rigid subsets and rigid linguistic values. Since the ranges of numerical values are ordered sets, so its rigid subsets are also ordered sets. The elements in an ordered set are similar but not the same, though the membership-degree of every element in an ordered rigid set to the ordered rigid set are all the same—all is 1 without exception, the degree of consistency of them with corresponding rigid linguistic value are not the same. In other words, elements in an ordered rigid set are equivalent relation for membership-degree, but a chain approximate or even uniform chain approximate relation for consistency-degree. Therefore, a rigid linguistic value that summarizes an ordered rigid set can also have consistency function. Based on this characteristic, we call the linguistic value that summarizes an ordered rigid set to be a **quantifiable rigid linguistic value**.

Examples of quantifiable rigid linguistic values are very many. For example, "positive" and "negative" of numbers; "surplus" and "deficit" in business; "rise" and "fall" of stock price are all quantifiable rigid linguistic values obtained by doing rigid partitioning of the corresponding range of numerical values. For another example, "fail," "pass," "medium," "good," and "excellence" representing one by one score sections [0, 59], [60, 69], [70, 79], [80, 89], and [90, 100] in hundred mark system are all quantifiable rigid linguistic values.

The above given quantifiable rigid linguistic values are all resulted from the rigid clustering or rigid partitioning of the existing ranges of numerical values. There is also another type of quantifiable rigid linguistic values that are prior to the corresponding measures and ranges of numerical values, that is, the linguistic values occur firstly and then the corresponding measures and ranges of numerical values are defined definitely. For instance, "possibly/probably/likely" is a quantifiable rigid linguistic value of this type, which obviously occurred before the "probability" in mathematics. Besides, "partial inclusion" of sets also belongs to this type of quantifiable rigid linguistic values. In fact, "partial inclusion" is "non-entire inclusion," so firstly it is a rigid linguistic value, whereas in previous Sect. 20.5, we defined the measure of "inclusion-degree" for "partial inclusion"; further, there is also range (0, 1) of inclusion-degrees. Actually, many rigid linguistic values in our daily language all belong to this type of quantifiable rigid linguistic values, which are the summarization of corresponding numerical feature values of objects, but have no definitely defined measures and ranges of numerical values.

Actually, the relation between a quantifiable rigid linguistic value and a range of numerical values is the relation of "summarized" and "quantified." Viewed from the range of numerical values to the quantifiable rigid linguistic value, a range of numerical values is summarized as a corresponding quantifiable rigid linguistic value; conversely, viewed from the quantifiable rigid linguistic value to the range of numerical values, a quantifiable rigid linguistic value is then quantified as a corresponding range of numerical values. For instance, range (0, 1) of probabilitics is summarized as "possibly"; viewed conversely, then "possibly" is quantified as the range (0, 1) of probabilities. Also, the range (0, 1) of inclusion-degrees is summarized as "partial inclusion," viewed conversely, then "partial inclusion" is quantified as the range (0, 1) of inclusion-degrees.

2. **Consistency functions of quantifiable rigid linguistic values**

   To establish the consistency function of a quantifiable rigid linguistic value, firstly the property such as monotone increasing, monotone deceasing, or convex of the consistency function needs to be determined according to the semantics of the linguistic value. That is tantamount to determining whether the linguistic value is left semi-peak value, right semi-peak value, or full-peak value.

Then, consider the number, that is, standard number, whose consistency-degree with the quantifiable rigid linguistic value is 1 in the corresponding ordered rigid set. Since the membership-degrees of all elements in an ordered rigid set are all 1, so the starting value of the consistency-degrees of all elements in an ordered rigid set with corresponding quantifiable rigid linguistic values is all 1. Thus, for a left semi-peak quantifiable rigid linguistic value, its left boundary point is just its standard number; for a right semi-peak quantifiable rigid linguistic value, its right boundary point is just its standard number; for a full-peak quantifiable rigid linguistic value, its left and right boundary points are both its standard numbers.

Next, determine the reference distance as a unit quantity. With the unit quantity, the expression of the consistency function can be written out.

Example 21.1 The graphs of consistency functions of 5 quantifiable rigid linguistic values that represent the academic achievements are shown in Fig. 21.1. Of them,

Fig. 21.1 Examples of the consistency functions of quantifiable rigid linguistic values

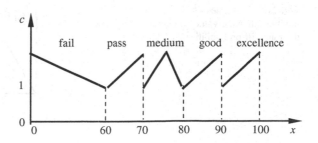

the expressions of consistency functions of "excellence," "medium," and "fail" are one by one:

$$c_{\text{excellence}}(x) = 1 + \frac{x - 90}{10}, \quad 90 \leq x \leq 100$$

$$c_{\text{medium}}(x) = \begin{cases} 1 + \dfrac{x - 70}{5}, & 70 \leq x \leq 75 \\ 1 + \dfrac{80 - x}{5}, & 75 \leq x < 80 \end{cases}$$

$$c_{\text{fail}}(x) = 1 + \frac{60 - x}{60}, \quad 0 \leq x < 60$$

Here, we take 10 as the unit quantity of "pass," "good," and "excellence", take 60 as the unit quantity of "fail," and take 5 as the unit quantity of "medium." Of course, other numbers can also be taken.

We see that although a quantifiable rigid linguistic value is the summarization of a rigid set, since its rigid set is an ordered set of numbers, so its consistency function may be monotone or convex, or even be triangular and semi-triangular like the consistency function of the flexible linguistic values.

3. **Relation between quantifiable rigid linguistic values and flexible linguistic values**

If a range (or sub range) of numerical values that a quantifiable rigid linguistic value corresponds to is also continuous, then we can define flexible linguistic values on it. For instance, on range (0, 1) of numerical values that "possibly" corresponds to, that is, range of probabilities, the flexible linguistic values "fairly probably," "very likely" can be defined (as shown in Fig. 21.2), on ranges [a, 0) and (0, b] of numerical values that "surplus" and "deficit" belong to, flexible linguistic values "small deficit" and "large surplus" and so on can be defined, respectively.

Actually, a range of numerical values of a quantifiable feature usually also corresponds to one quantifiable rigid linguistic value, only that for some features the linguistic value is not so obvious. For instance, temperature in fact indicates the degree of "hot," age indicates the degree of "old," height indicates the degree of "high," and speed indicates the degree of "fast." Therefore, speaking in this sense,

Fig. 21.2 Examples of flexible linguistic values on the range of probabilities

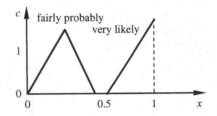

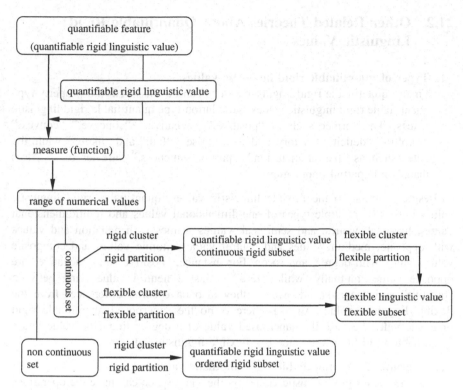

Fig. 21.3 Diagram of the origins and relation of quantifiable rigid linguistic values and flexible linguistic values

the range of numerical values of air temperature this quantifiable feature, as range [−50, 50] of temperature, just corresponds to quantifiable rigid linguistic value "hot to a certain degree." Conversely speaking, the "hot to a certain degree" summarizes range [−50, 50] of temperature. Similarly, range [1, 200] of age corresponds to "old to a certain degree," range (0, b] of height corresponds to "high to a certain degree," range (0, c] of speed corresponds to "fast to a certain degree". Thus, a flexible linguistic value defined on the range of numerical values of a certain feature can also be said as defined on the range of numerical values of the corresponding quantifiable rigid linguistic value.

Thus, all flexible linguistic values can be said as defined on the range of numerical values of corresponding quantifiable rigid linguistic values. Therefore, we can say that all flexible linguistic values originate from quantifiable rigid linguistic values. Such is just the relation between the quantifiable rigid linguistic values and the flexible linguistic values. Thus, the origins and relation of the quantifiable rigid linguistic values and the flexible linguistic values are shown in Fig. 21.3.

534   21 Quantifiable Rigid Linguistic Values and Related Theories

21.2 Other Related Theories About Quantifiable Rigid Linguistic Values

1. **Types of quantifiable rigid linguistic values**

   Firstly, quantifiable rigid linguistic values can be classified into property-type quantifiable rigid linguistic values and relation-type quantifiable rigid linguistic values. The former such as "positive," "negative," "concave," "convex," "surplus," "deficit," "victory," "defeat," "rise," "fall," and "likely," while the latter such as "partial inclusion," "partial sameness," "greater than," "less than," and "partial implication."

   Besides, similar to the flexible linguistic values, quantifiable rigid linguistic values have also multiple types of one-dimensional values and multidimensional values, full-peak values, and semi-peak values, values with negation and values with opposite, medium values and neutral values, atomic values and composite values, value with degree, and so on. For instance, "positive" and "negative" are opposite values mutually, while "zero" is just a neutral value lying between "positive" and "negative." However, they also have some differences from the flexible linguistic values, such as there is no the superposed quantifiable rigid linguistic value because the superposed value of a degree linguistic value and a quantifiable rigid linguistic value is a flexible linguistic value.

2. **Operations on quantifiable rigid linguistic values**

   Quantifiable rigid linguistic values on the same space can have the operations of conjunction, disjunction, and negation. These operations are reduced to operations of corresponding ordered rigid sets, that is, the usual set operations. But here we need to view the linguistic value to which empty set ϕ corresponds as a special quantifiable rigid linguistic value. Besides, for the quantifiable linguistic value that itself represents whole range of numerical values, the corresponding range of numerical values of its negation needs to do analysis specifically. Generally speaking, this kind of range of numerical values is the number set of a single point. For example, "possibly" represents range $(0, 1)$ of probabilities, its negative value "not possibly" represents then probability 0. While 0 can also be represented as set $\{0\}$, then, this single pointed number set is just the range of numerical values that "not possibly" corresponds to. From this example, we see that the negative value of a quantifiable rigid linguistic value is still a quantifiable rigid linguistic value. Thus, the conjunctions, disjunctions, and negations of the quantifiable rigid linguistic values on the same space are still quantifiable rigid linguistic values.

   Quantifiable rigid linguistic values on distinct spaces can also have the operations of conjunction, disjunction, and synthesis. These operations can be represented and implemented by using the corresponding operations of consistency functions and can also be reduced to operations of corresponding ordered rigid sets, that is, conjunction ($\land$), disjunction ($\lor$), and negation ($\neg$) operations are reduced separately to the intersection (∩), union (∪), and complement (^{c}) operations of corresponding ordered rigid sets, while synthesis operation (⊕) is then reduced to the product operation (×) of corresponding ordered rigid sets.

3. **Quantifiable rigid relations**

   The relation denoted by a relation-type quantifiable rigid linguistic value is a quantifiable rigid relation. For instance, the previously mentioned "partial inclusion," "partial sameness," "greater than," "less than," and "partial implication" are all quantifiable rigid relations.

From Fig. 21.3, it can be seen that from a relation-type quantifiable rigid linguistic value, the corresponding measure and range of numerical values can be obtained. Therefore, a quantifiable rigid relation is actually also a rigid relation that can be quantified by using certain kind of measure. For example, inclusion-degree can be used to quantify the "partial inclusion" between sets, so the "partial inclusion" relation between the sets is a quantifiable rigid relation. For another example, "sameness-degree" can be used to quantify the "partial sameness" between objects, so "partial sameness" relation is also a quantifiable rigid relation. Similarly, "greater than," "less than" between numbers, "partial sameness" between sets, and so on are all quantifiable rigid relations.

From Sect. 3.6, it is known that flexible relations (concepts) all can be reduced to flexible concepts on the one-dimensional measurement space. Actually, this one-dimensional measurement space is also the measurement space of corresponding quantifiable rigid relation. Therefore, a relation-type flexible linguistic value is also a flexible linguistic value on the measurement space of corresponding quantifiable rigid relation. For example, "similar" is a flexible linguistic value on the measurement space—range [0, 1] of sameness-degrees—of "partial sameness," and "far greater than" is a flexible linguistic value on the measurement space—range [1, +∞) of greater-than-degrees—of "greater than," and "close to inclusion" between sets is a flexible linguistic value on the measurement space—range (0, 1) of inclusion-degrees—of "partial inclusion." This shows that the flexible relations actually come on the basis of the quantifiable rigid relations.

4. **Other related topics**

- Since the upper half part of the graph of consistency function is also a triangular or semi-triangular, so a quantifiable rigid linguistic value also has a peak-value point.
- There is also relatively negative relation between quantifiable rigid linguistic values (which means a quantifiable rigid linguistic value also has a negative value), which corresponds to the relatively complement relation between corresponding rigid subsets. More generally, there is also complementary relation between quantifiable rigid linguistic values, which corresponds to the complementary relation between corresponding rigid subsets. And the complementary relation between rigid subsets corresponds to the rigid partition of a range of numerical values.

- Between quantifiable rigid linguistic values, there are also relations of inclusion, same-level, similarity, approximation, and so on.
- Quantifiable rigid linguistic values can also be quantified, that is, there are also rigid linguistic values with degree.
- Quantifiable rigid linguistic values can also mutually convert with the numerical values as well as the rigid linguistic values with degree. The converting method is similar to that of the flexible linguistic values.
- There can also be quantifiable rigid linguistic functions and quantifiable rigid linguistic valued vectors.

21.3 Relatively Opposite Quantifiable Rigid Linguistic Values and Relatively Opposite Rigid Sets

Definition 21.1 Let A, B, and C be three basic quantifiable rigid linguistic values adjacent one by one on one-dimensional measurement space U, which form a partition of U. If the semantics of A and C is opposite, B is non-A and non-C, then we say that A and C are opposite mutually, and call B a neutral value.

Likewise, we denote the opposite of a quantifiable rigid linguistic value A as $-A$, and denote a neutral value as *Neu*.

There are many examples of relatively opposite quantifiable rigid linguistic values. Examples, "positive" and "negative" are relatively opposite, while "0" is the neutral value; "concave" and "convex" are relatively opposite, while "flat" is the neutral value; "up" and "down" are relatively opposite, while "middle" is the neutral value; "left" and "right" are relatively opposite, while "middle" is the neutral value; "front" and "back" are relatively opposite, while "middle" is the neutral value; "support" and "opposite" are relatively opposite, while "neutrality" is the neutral value; "surplus" and "deficit" are relatively opposite, while "balance" is the neutral value; "victory" and "defeat" are relatively opposite, while "draw" is the neutral value; and "rise" and "fall" are relatively opposite, while "unchanged" is the neutral value.

The ranges of numerical values of features to which different relatively opposite quantifiable rigid linguistic values belong are also different. However, viewed abstractly, this type of range of numerical values is generally a symmetrical real interval $[-b, b]$ with center 0.

Let A and $-A$ be a pair of relatively opposite quantifiable rigid linguistic values on space $U = [-b, b]$, and *Neu* is the neutral value. Then, A, $-A$, and *Neu* separately represent sets $[-b, 0)$, $(0, b]$, and $\{0\}$. Hereafter we still use symbols A, $-A$, and *Neu* to denote sets represented by them.

Definition 21.2 Let A and $-A$ be a pair of relatively opposite quantifiable rigid linguistic values on space $U = [-b, b]$, and *Neu* be the neutral value. Then, we say corresponding sets A and $-A \subset U$ are opposite mutually, $\{0\}$ is the neutral set. In particular, we call $0 \in U$ the neutral point.

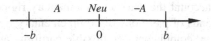

Fig. 21.4 Examples of relatively opposite quantifiable rigid linguistic values and their sets

As shown in Fig. 21.4, let A, Neu and $-A$ separately denote subsets $[-b, 0)$, $\{0\}$ and $(0, b]$. Thus, we have the following set "operations":

$$-(-A) = A \tag{21.1}$$

$$A^c = -A \cup Neu \tag{21.2}$$

$$(-A)^c = A \cup Neu \tag{21.3}$$

$$Neu^c = A \cup -A \tag{21.4}$$

$$A \cap -A = \emptyset \tag{21.5}$$

$$A \cup -A \cup Neu = U \tag{21.6}$$

$$A^c \cap (-A)^c = Neu \tag{21.7}$$

$$A^c \cup (-A)^c = U \tag{21.8}$$

From the above set operations, we have the following relational expressions of linguistic values:

$$-(-A) = A \tag{21.9}$$

$$\neg A = -A \vee Neu \tag{21.10}$$

$$\neg(-A) = A \vee Neu \tag{21.11}$$

$$\neg Neu = A \vee -A \tag{21.12}$$

$$\neg A \wedge \neg(-A) = Neu \tag{21.13}$$

Besides, the relatively opposite between quantifiable rigid linguistic values is all objective relatively opposite.

The relatively opposite is a relation between two quantifiable rigid linguistic values. More generally, among multiple quantifiable rigid linguistic values is just a mutually exclusive relation, which corresponds to the mutually exclusive relation among corresponding rigid subsets, and the latter corresponds to the mutually exclusive rigid partition of a measurement space.

The so-called mutually exclusive rigid partition, simply speaking, is that there is a neutral point (line or planc) that is neither this nor that between all adjacent rigid subsets in space.

Thus, to take into account the usual complementary rigid partition, a measurement space has four kinds of partitions of complementary rigid partition, mutually exclusive rigid partition, complementary flexible partition, and mutually exclusive flexible partition.

21.4 Quantifiable Rigid Propositions and Corresponding Logic and Inference

1. **Quantifiable rigid propositions with negation and their logic and inference**
   We call a proposition that contain quantifiable rigid linguistic values to be a quantifiable rigid proposition. For example,

   ```
   A company had a financial deficit the current year.
      5 is greater than 2.
   ```

   are two quantifiable rigid propositions.

   Generally, Let p: x_0 is A, be a quantifiable rigid proposition. Obviously, it can also be represented by possessive relational form $A(x_0)$ and membership relational form $x_0 \in A$.

   Since the linguistic value A therein is a rigid linguistic value, so the truth values of quantifiable rigid proposition $A(x_0)$ can be rigid linguistic truth values "true" or "false." Since A is also quantifiable, so its truth values can also be numerical value—truth-degree. This truth-degree (which refers to connotation truth-degree, same below) is also just consistency-degree $c_A(x_0)$, that is, $t(p) = c_A(x_0)$. Since consistency-degree $c_A(x_0) \in [1, \beta]$, the "true" of the quantifiable rigid proposition is the summarization of range $[1, \beta]$ of truth-degrees; viewed conversely, it is also that the "true" of a quantifiable rigid proposition is quantified as range $[1, \beta]$ of truth-degrees. Thus, **the "true" of the quantifiable rigid proposition is a quantifiable rigid linguistic truth value**.

   Let us conceive that if the quantifiable rigid proposition $A(x_0)$ is false, then its negation $\neg A(x_0)$ is true. Since the negation of a quantifiable rigid linguistic value is still a quantifiable rigid linguistic value, $\neg A$ is still a quantifiable rigid linguistic value, and $\neg A(x_0)$ is also a quantifiable rigid proposition; thus, the "true" of $\neg A(x_0)$ is also a quantifiable linguistic truth value. We denote the corresponding range of truth-degrees that it represents as $[1, \beta_0]$. The "true" of negative proposition $\neg A(x_0)$ is equivalent to the "false" of the original proposition $A(x_0)$. As thus, the "false" of original proposition $A(x_0)$ is also a quantifiable rigid linguistic truth value, and the "range of falseness-degrees" that it represents is the range $[1, \beta_0]$ of truth-degrees of $\neg A(x_0)$. To distinguish, we denote quantifiable rigid linguistic truth values "true" and "false" separately as $\mathbb{T}$ and $\mathbb{F}$.

   Since the conjunction, disjunction, and negation of the quantifiable rigid linguistic values with negation are still quantifiable rigid linguistic values with negation, the conjunctive, disjunctive, and negative propositions of the quantifiable rigid propositions with negation are still quantifiable rigid propositions with negation. Meanwhile, the "true" and "false" of quantifiable rigid propositions are quantifiable rigid linguistic truth values, so the logic based on the quantifiable rigid propositions with negation is a kind of quantifiable rigid linguistic truth valued relatively negative logic.

The computation methods of truth-degrees of the conjunctive and disjunctive compound quantifiable rigid propositions are the same as those of corresponding compound flexible propositions, that is,

$$t(p \wedge q) = \min\{c_A(x), c_B(y)\} = \min\{t(p), t(q)\} \tag{21.14}$$

$$t(p \vee q) = \max\{c_A(x), c_B(y)\} = \max\{t(p), t(q)\} \tag{21.15}$$

While the computation of truth-degree of negative quantifiable rigid proposition can only be

$$t(\neg p) = c_{\neg A}(x_0) \tag{21.16}$$

Thus, the truth-degree computation formulas of the implicational and equivalent quantifiable rigid propositions are

$$t(p \rightarrow q) = \max\{t(\neg p), t(q)\} \tag{21.17}$$

$$t(p \longleftrightarrow q) = \min\{\max\{t(\neg p), t(q)\}, \max\{t(p), t(\neg q)\}\} \tag{21.18}$$

Here, $t(p), t(q), t(\neg p)$ and $t(\neg q) \in [1, \beta]$; thus, $t(p \wedge q), t(p \vee q), t(p \rightarrow q)$ and $t(p \longleftrightarrow q) \in [1, \beta]$. And if $\neg A$ is the symbolic rigid linguistic value, then $c_{\neg A}(x_0)$ is meaningless. In such a case, $t(\neg p) = $ false.

It can be seen that the truth-degree computation formulas of conjunctive and disjunctive quantifiable rigid propositions actually define two kinds of operations on range $[1, \beta]$ of truth-degrees:

$$x \cdot y = \min\{x, y\} \tag{21.19}$$

$$x + y = \max\{x, y\} \tag{21.20}$$

Thus, based on range $[1, \beta]$ of truth-degrees, we can also construct corresponding truth-degree logic algebra, such as $\langle [1, \beta], \cdot, + \rangle$.

It can be seen that the quantifiable rigid linguistic truth values $\mathbb{T}$ and $\mathbb{F}$ are really the particular cases of the rigid linguistic truth values "true" and "false," that is, T and F, in the usual two-valued logic. Therefore, viewed from the level of linguistic truth values, the two-valued logic based on quantifiable rigid linguistic truth values $\mathbb{T}$ and $\mathbb{F}$ is "isomorphic" with the traditional two-valued logic. Thus, in this two-valued logic, there also are the inference rules and inference schemes that are similar to those in traditional two-valued logic. And because $\mathbb{T}$ and $\mathbb{F}$ are the detailing of non-symbolic rigid linguistic truth values T and F, so the rules of inference and inference in this two-valued logic can also be formulated by scope of truth-degrees like the rules of rough-true inference and the rough-true inference in Sect. 12.6.2. For example, its modus ponens

$$p \to q \ (\mathbb{T})$$
$$\frac{p \ (\mathbb{T})}{\therefore q \ (\mathbb{T})} \tag{21.21}$$

can be expressed by writing

$$p \to q, \ t(p \to q) > 1$$
$$\frac{p, \ t(p) > 1}{\therefore q, \ t(q) > 1} \tag{21.22}$$

Yet its universal modus ponens

$$A(x) \to B(y) \ \mathbb{T})$$
$$\frac{A(x_0) \ (\mathbb{T})}{\therefore B(y_0) \ (\mathbb{T})} \tag{21.23}$$

can be expressed by writing

$$A(x) \to B(y), t(A(x) \to B(y)) > 1$$
$$\frac{A(x_0), \ t(A(x_0)) > 1}{\therefore B(y_0), \ t(B(y_0)) > 1} \tag{21.24}$$

2. **Quantifiable rigid propositions with opposite and the corresponding logic**
For quantifiable rigid propositions with opposite, their conjunction and disjunction operations are completely the same as those of the above quantifiable rigid propositions with negation, while the opposite proposition is a proposition with the opposite value of the linguistic value of the original proposition as the linguistic value. Since the conjunction, disjunction, and opposite of quantifiable rigid linguistic values with opposite are still quantifiable rigid linguistic values with opposite, so the conjunction, disjunction, and opposite of quantifiable rigid propositions with opposite are still quantifiable rigid propositions with opposite. In the meantime, it is easy to see that the "true" and "false" of a quantifiable rigid proposition with opposite are also quantifiable rigid linguistic truth values (which can be separately denoted by T and F. But notice that here T and F are of a relatively opposite relation), so the logic based on the quantifiable rigid propositions with opposite is a kind of opposite-type logic of quantifiable rigid linguistic truth values.

21.5 Quantifiable Rigid Linguistic Rules and Quantifiable Rigid Linguistic Functions, and Corresponding Reasoning and Computation

We call the rule whose antecedent and consequent contain quantifiable rigid linguistic values to be a quantifiable rigid linguistic rule. For example,

```
If a listed company had a financial deficit, then its stock price
would fall
```

is a quantifiable rigid linguistic rule.

Since quantifiable rigid linguistic values have many similar properties, operations, and relations with flexible linguistic values, quantifiable rigid linguistic rules also have similar classifications, logic semantics, mathematical background, and mathematical essence with flexible linguistic rules. Further, these rules have also similar triple adjoint functions and the reasoning and computation with them.

A function whose independent variables or dependent variable take on quantifiable rigid linguistic values is called a quantifiable rigid linguistic function. The quantifiable rigid linguistic functions have also classifications, representations, mathematical background, and mathematical essence that are similar to those of the flexible linguistic functions. Thus, there are also similar approximate evaluation methods.

Speaking from connotation, information represented by quantifiable rigid linguistic value is also a kind of imprecise information. Therefore, there are also corresponding imprecise problems. And there are also the problems solving based on the consistency functions of the quantifiable rigid linguistic values or based on the quantifiable rigid linguistic rule-/function-based systems. Besides, there are the acquisition and discovery methods of quantifiable rigid rules and quantifiable rigid linguistic functions similar to those of flexible values and flexible linguistic functions.

21.6 Several Important Quantifiable Rigid Linguistic Values and Their Measures

This section introduces some important rigid linguistic values about sets, linguistic values, and propositions and introduces corresponding measures for them.

1. Partial possession and possessing rate

Definition 21.3 Let U be a set, and A be a subset of U. If part of objects in U possesses property A, i.e., property-type linguistic value A, that subset A stands for, then we say that A is of partial possession. If all of the objects in U possess property-type linguistic value A, then we say that A is of total possession. Set

$$\text{poss}(A) = \frac{|A|}{|U|} = \text{shar}(A) \tag{21.25}$$

to be called the rate of objects in U possessing corresponding property-type linguistic value A, simply, possessing rate of property-type linguistic value A.

This definition means that the possessed rate of a property-type linguistic value is numerically equal to the share of the corresponding set.

Obviously, generally speaking, $0 \leq \text{poss}(A) \leq 1$. If $A \neq \emptyset$ and $A \subset U$, then $0 < \text{poss}(A) < 1$. Whereas $A = U$, $\text{poss}(A) = 1$, and vice versa.

It is clear that the measure "possessing rate" unifies "partial possession," "total possession," and "non-possession," and it can also be used to define these three rigid linguistic values. In particular, "partial possession" is quantified as set (0, 1) of real numbers, which can be called the range of possessing rates. Further, on the range (0, 1) of possessing rates we can define flexible linguistic values such as "most possession," "few possession," and "very few possession."

2. **Partial implication and implication-degree**

We know that there are two most basic relations of implication and equivalence between propositions. The usual implication and equivalence are defined from the angle of logic by the relation between the truth values of propositions, which have nothing to do with the content of the propositions, so they are logical implication and logical equivalence. However, we find that considering from the content, there also exist implication and equivalence relations between propositions. We call the implication and equivalence relations based on the content of the propositions to be the semantic implication and equivalence relations.

(1) **Partial (semantic) implication**

Definition 21.4 Let p: x_0 is A and q: y_0 is B, be two propositions. If the corresponding set A is contained in set B, that is, $A \subseteq B$, then we say that proposition p semantically implies proposition q, denote $p = \triangleright q$.

Example 21.2 Let p: Zhang is a junior, and q: Zhang is a university student. Then, $p = \triangleright q$.

Definition 21.5 Let p: x_0 is A and q: y_0 is B, be two propositions. If the corresponding set A is corresponded by set B, that is, $A \mapsto B$, then we also say that proposition p semantically implies proposition q, denote $p = \triangleright q$.

Definition 21.6 Let p: x_0 is A and q: y_0 is B, be two propositions. If $p = \triangleright q$ and $q = \triangleright p$, then we say that proposition p and proposition q are semantically equivalent, denote $p \triangleleft = \triangleright q$.

Semantic implication and semantic equivalence are defined on the basis of the relation between corresponding sets of propositions, but it can be seen that:

① If proposition p semantically implies proposition q, then p is also certainly logically implies proposition q, that is, from $p = \rhd q$ necessarily $p \Rightarrow q$ follows, but conversely, not necessarily;

② If propositions p and q are semantically equivalent, then p and q also certainly be logically equivalent, that is, from $p\lhd = \rhd q$ necessarily $p \Leftrightarrow q$ follows, but conversely, not necessarily.

Definition 21.7 Let p: x_0 is A and q: y_0 is B, be two propositions. If the corresponding set A is partially contained in set B, then we say proposition p partially (semantically) implies proposition q, denote $p_{\ddot{-}}\rhd q$.

Example 21.3 Let p: Zhang is a youth, and q: Zhang is a university student. It is not hard to see that set {youth} is partially contained in set {university students}, so $p_{\ddot{-}}\rhd q$.

Example 21.4 Let p: Zhang is short, and q: Zhang is very fat. Since there is a part and only a part people are very fat in short people, set {short} is partially contained in set {very fat}. Thus, $p_{\ddot{-}}\rhd q$.

Definition 21.8 Let p: x_0 is A and q: y_0 is B, be two propositions. If the corresponding set A is partially corresponded by set B, then we say proposition p partially (semantically) implies proposition q, denote $p_{\ddot{-}}\rhd q$.

Relatively to the partial (semantic) implication $_{\ddot{-}}\rhd$, the (semantic) implication "$= \rhd$" above is a complete implication. Conversely speaking, the complete implication $= \rhd$ is a special case of the partial implication $_{\ddot{-}}\rhd$.

(2) **Implication-degree [1] and transitive formula of implication-degrees**

From Definitions 21.7 and 21.8, it can be seen that "partial implication" is a quantifiable rigid linguistic value, which represents a quantifiable rigid relation.

Definition 21.9 Let propositions p: x_0 is A, and q: y_0 is B. We use impl(p, q) to denote the degree of proposition p (semantically) implying proposition q, that is, the degree of implication, to write as implication-degree, of p to q. Then,

(1) If the corresponding set A is partially contained in set B, then

$$\mathrm{impl}(p, q) = \mathrm{cont}(A, B) \tag{21.26}$$

(2) If the corresponding set A is corresponded partially by set B, then

$$\mathrm{impl}(p, q) = \mathrm{corr}(A, B) \tag{21.27}$$

Example 21.5 Let p: Jack is a junior college student, and q: Jack is a college student, and let $A = \{$junior college students$\}$ and $B = \{$college students$\}$. Then,

$$\text{impl}(p, q) = \text{cont}(A, B) = |A \cap B|/|A| = |A|/|A| = 1$$

That is, the implication-degree of proposition p implying proposition q is 1.

Example 21.6 For propositions p and q in Example 21.3, suppose that 60 % of the youth are university students. Then,

$$\begin{aligned}
\text{impl}(p, q) &= \text{cont}(\{\text{youth}\}, \{\text{college students}\}) \\
&- |\{\text{youth}\} \cap \{\text{college students}\}|/|\{\text{youth}\}| \\
&= |\{\text{young college students}\}|/|\{\text{youth}\}| \\
&= 0.6
\end{aligned}$$

From the Eq. (21.27), it can be seen that $\text{impl}(p, q) = 0$ means p does not imply q; $\text{impl}(p, q) = 1$ means p completely implies q; $0 < \text{impl}(p, q) < 1$ means p implies q to a degree.

As thus, implication-degree unifies "partial implication," "complete implication," and "non-implication," and which can also be used to define these three rigid linguistic values. In particular, "partial implication" is quantified as set $(0, 1)$ of real numbers, which can be called range of implication-degrees. Further, we can define the flexible linguistic values such as "rather implying," "basically implying," and "nearly implying," etc., on the range $(0, 1)$ of implication-degrees.

From the definition of partial implication, when proposition p partly implies proposition q, proposition q also partly implies proposition p. That is to say, the relation of partial implication between propositions (including flexible propositions) satisfies (semantics) symmetry. But the partial implication is not necessarily to satisfy transitivity. However, since the partial implication between propositions is defined by the partial inclusion or partial correspondence between corresponding sets, so the transitivity of the former should be consistent with that of the latter. Thus, we have the following theorems.

Theorem 21.1 *Let propositions p: x_0 is A, q: y_0 is B, and r: z_0 is C. If the corresponding sets A, B, and C are partially contained in order and the partial inclusion relation satisfies transitivity, or A, B, and C are partially corresponded in order and the partial correspondence relation also satisfies transitivity, then the propositions p, q, and r are partially implied in order and the partial implication relation satisfies transitivity, that is, if $p \underset{\cdot\cdot}{\triangleright} q$ and $q \underset{\cdot\cdot}{\triangleright} r$, then $p \underset{\cdot\cdot}{\triangleright} r$.*

Theorem 21.2 *Let propositions p: x_0 is A, q: y_0 is B, and r: z_0 is C. If $p \underset{\cdot\cdot}{\triangleright} q$ and $q \underset{\cdot\cdot}{\triangleright} r$, and the partial implication relation satisfies transitivity, then*

$$\text{impl}(p, r) = \text{impl}(p, q) \cdot \text{impl}(q, r) \tag{21.28}$$

Proof When implication-degree is defined as inclusion-degree, by the Eq. (20.90) in Sect. 20.5, we have immediately the Eq. (21.28), and when implication-degree is defined as correspondence rate, by the Eq. (20.94) in Sect. 20.6, we have also immediately the Eq. (21.28).

We call Eq. (21.28) to be the transitive formula of implication-degrees.

3. **Partial equivalence and equivalence-degree**

Definition 21.10 Let p: x_0 is A and q: y_0 is B, be two propositions. If $p_{\dot{=}} \triangleright q$ and $q_{\dot{=}} \triangleright r$, then we say propositions p and q are partially semantically equivalent, denote $p \triangleleft_{\dot{=}} \triangleright q$.

Example 21.7 Let p: Zhang is a youth and q: Zhang is a university student. It is not hard to see that set {youth} is partly contained in set {college students}, and set {college students} is partly contained in set {youth}, which shows that propositions p and q are partially implied each other; thus, $p \triangleleft_{\dot{=}} \triangleright q$.

Similarly, "partial equivalence" is also a quantifiable rigid linguistic value, which represents a quantifiable rigid relation.

Definition 21.11 Let proposition p: x_0 is A and q: y_0 is B. Set

$$\text{equi}(p, q) = \text{same}(A, B) \tag{21.29}$$

to be called the degree of equivalence, written equivalence-degree, between proposition p and proposition q.

From Eq. (21.29), it can be seen that equi$(p, q) = 0$ means p and q are not equivalent; equi$(p, q) = 1$ means p and q are completely equivalent; $0 < \text{equi}(p, q) < 1$ means p and q are equivalent to a degree.

As thus, equivalence-degree unifies "partial equivalence," "equivalence" (i.e., "completely equivalence") and "non-equivalence," and which can also be used to define these three rigid linguistic values. In particular, the "partial equivalence" is quantified as set (0, 1) of real numbers, which can be called the range of equivalence-degrees. Further, we can also define the flexible linguistic values such as "(being) rather equivalent," "(being) basically equivalent," and "(being) nearly equivalence," etc., on the range (0, 1) of equivalence-degrees.

From definitions, it can be seen that the partial equivalence relation satisfies symmetry, however, is not necessarily to satisfy transitivity. But the partial equivalence relation has also the judgment conditions and properties similar to Theorems 21.1 and 21.2.

21.7 Summary

In this chapter, we expounded briefly the basic theory of information processing with quantifiable rigid linguistic values, which is parallel or similar to the corresponding theory of information processing with flexible linguistic values. But we should note the relation and distinction between quantifiable rigid linguistic values and flexible linguistic values:

- The linguistic value which summarizes an ordered rigid set is a quantifiable rigid linguistic value, and the linguistic value which summarizes a flexible subset is a flexible linguistic value.
- All flexible linguistic values can be said as defined on the range of numerical values of the corresponding quantifiable rigid linguistic values.

Besides, we introduced also the quantifiable rigid linguistic values and their measures: partial possession and possessing rate, partial implication and implication-degree, and partial equivalence and equivalence-degree, in particular, found and presented the judging condition of partial implication (equivalence) relation satisfying transitivity and the corresponding transitive formulas of implication-degrees (equivalence-degrees).

Reference

1. Lian S (2006) A method to obtain the truth values of implicational compound propositions based on implication-degree and implication-rate. Comput Eng Appl 42(7):69–115

Chapter 22
Methodology of Imprecise-Information Processing and Some Other Application Problems

Abstract This chapter proposes two basic ideas and techniques of imprecise-information processing—degree-introducing and appropriate granularity; discusses the relational inference with quantifiable rigid relations and flexible relations; proposes the method of knowledge representation with degrees and gives some representation schemes; discusses relational database models with degrees; and proposes the principles and methods of machine understanding of flexible concepts with knowledge base and automatic generation of flexible concepts by using object-oriented programming.

Keywords Imprecise-information processing · Degree-introducing · Appropriate granularity · Relational inference with degrees · Knowledge representation with degrees · Relational database with degrees

In this chapter, we will talk briefly basic ideas and techniques of imprecise-information processing and discuss several other application problems besides imprecise-problem solving.

22.1 Degree-Introducing and Appropriate Granularity—Basic Approach and Technique of Imprecise-Information Processing

Degree-introducing and appropriate granularity are basic approach and technique of imprecise-information processing.

The so-called degree-introducing is to use certain kind of measure to quantify information. For example, we use "inclusion-degree" to quantify "partial inclusion" and use "sameness-degree" to quantify "partial sameness," which are typical degree-introducing. For another example, we use "greater-than-degree" to quantify the "greater than" of numbers, use difference-degree, sameness-degree, and equivalence-degree to quantify the similarities and differences of numbers and

© Springer Science+Business Media Singapore 2016   547
S. Lian, *Principles of Imprecise-Information Processing*,
DOI 10.1007/978-981-10-1549-6_22

vectors, and use similarity-degree and approximate-degree to quantify similarity and approximate relations, and so on, which are all degree-introducing. Furthermore, we use the sameness-degree to define membership-degree and use a unit quantity to define consistency-degree, which are also degree-introducing, and we use membership-degree and consistency-degree to formulate the membership and consistency relations between an object and a flexible concept; this is also a kind of degree-introducing. Further, that we use the linguistic value with degree to portray attributes, states, and relations as well as behaviors of objects is also degree-introducing. Viewed from representation form $\{(r, m_A(x))|x \in U\}$, the flexible sets (and fuzzy sets) can also be called "the set with degrees" in a broad sense. Also, in the establishment of adjoint degreed functions of flexible rules, the idea and method of degree-introducing are also used [1, 2].

Flexible linguistic values and quantifiable rigid linguistic values are the origins of imprecise information, and they are also the objects of imprecise-information processing. We diagram the origins, relation, and modeling principles of these two classes of linguistic values as follows (see Fig. 22.1), from which we can see more clearly and visually the position and function of degree-introducing in imprecise-information processing. It can be seen that degree-introducing is a basic technique of imprecise-information processing.

Obviously, flexible linguistic values are also a kind of quantifiable linguistic values. Thus, flexible linguistic values and quantifiable rigid linguistic values can be uniformly called the "quantifiable linguistic values." Thus, imprecise-information processing can also be said as the information processing about quantifiable linguistic values.

The so-called appropriate granularity is according to specific problem to design the linguistic values whose granule sizes are appropriate. Appropriate linguistic values can be used to describe more accurately things and relations between them, and the results of information processing also are more effective. In general, the smaller the granules of linguistic values are, the more precision the things are portrayed. The design and selection of linguistic values can be classified as two types: static and dynamic. The latter is especially important in approximate computation problems (as flexible control) based on flexible rules or flexible linguistic functions. Static design is for a relevant measurement space to set basic linguistic values in advance, while dynamic design is then according to the requirement to set temporarily the corresponding linguistic values, because in some information processing, corresponding linguistic values and their sizes will change dynamically with the expand and contract and change of measurement spaces. The sizes and number of linguistic values need to be determined according to the characteristics of the specific problem, the relevant domain knowledge, and the actual requirements and, for dynamic linguistic values, need to design the corresponding data structures and algorithms, so to conduct real-time definition and partition of measurement spaces,

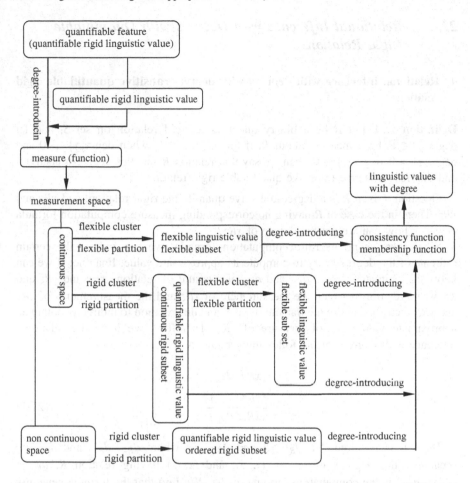

Fig. 22.1 Diagram of the principles of origins, relation, and modeling of quantifiable rigid linguistic values and flexible linguistic values

and to conduct real-time definition, addition, deletion, and adjustment of (cores and support sets, or extended cores) of relevant linguistic values, thus to make system can use flexibly appropriate linguistic value granules in the running process. It seems that there will be some work we can do here.

22.2 Relational Inference with Degrees

We know that relational inference is also a common inference form. Traditional relational inference is the inference with rigid relations. This section discusses a kind of relational inference with quantifiable rigid relation and flexible relation.

22.2.1 Relational Inference with Degrees with Quantifiable Rigid Relations

1. Relational inference with degrees with degree-transitive quantifiable rigid relations

Definition 22.1 Let R be a binary quantifiable rigid relation on set S, and let $\deg(x, y) \subset R$ be a measure about R. If for $\forall x, y, z \in S$, when $\deg(x, y) > 0$ and $\deg(y, z) > 0, \deg(x, z) > 0$, then we say that relation R satisfies degree-transitivity and that R is a degree-transitive quantifiable rigid relation.

Question: Given R is a degree-transitive quantifiable rigid relation, and xRy and yRz. Then, in the case of R having no corresponding measure computation formula $\deg(\cdot, \cdot)$, how can we get the degree of xRz?

Since there is no measure computation formula, we can only make certain estimate of the degree or try to compute its approximate value. Imagine: If we can know the estimated values $d_{R_{xy}}$ and $d_{R_{yz}}$ of measure of relations xRy and yRz and know the relation between the measurement of xRz and the measurements of xRy and yRz, then we can use the transitivity of R and this relation to indirectly obtain an approximate value $d_{R_{xz}}$ of measure of xRz. Therefore, we have the relational inference with degrees of the following scheme:

$$\frac{(xRy, d_{R_{xy}})}{(xRz, d_{R_{xz}})} \quad (yRz, d_{R_{yz}})$$

$$\text{(22.1)}$$

Here, it is required that $d_{R_{xy}} \geq 1$ and $d_{R_{yz}} \geq 1$, where $d_{R_{xy}}$, $d_{R_{yz}}$, and $d_{R_{xz}}$ are separately the degrees of (x, y), (y, z), and (x, z) having relation R; $d_{R_{xz}} = f(d_{R_{xy}}, d_{R_{yz}})$ is the computation formula of $d_{R_{xz}}$. We find that the formula generally can take

$$d_{R_{xz}} = d_{R_{xy}} + d_{R_{yz}} - d_{R_{xy}} d_{R_{yz}} \quad (22.2)$$

For example, the "greater than" of real numbers is a degree-transitive quantifiable rigid relation; then, with "greater than," we can do a relational inference with degrees.

For another example, if propositions p, q, and r are partially implied in order, and the partial implication relation satisfies transitivity, then from Definitions 21.7, 21.8, and 21.9, the partial implication satisfies degree-transitivity. Thus, by

Theorem 21.2 (see Sect. 21.6), the following relational inference with degrees would follow:

$$\frac{\begin{array}{c}(p \mathrel{\vdots\!\!\vartriangleright} q, \text{impl}(p, q))\\ (q \mathrel{\vdots\!\!\vartriangleright} r, \text{impl}(q, r))\end{array}}{\therefore (p \mathrel{\vdots\!\!\vartriangleright} r, \text{impl}(p, r))} \tag{22.3}$$

here, $\text{impl}(p, r) = \text{impl}(p, q)\text{impl}(q, r)$.

2. **Relational inference with degrees with degree-symmetric quantifiable rigid relation**

Definition 22.2 Let R be a binary quantifiable rigid relation on set S, and let $\deg(x, y) \in \mathbf{R}$ be a measure about R. For $\forall x, y \in S$, if $\deg(x, y) > 0$, then $\deg(y, x) > 0$, and $\deg(x, y) = \deg(y, x)$, then we say R satisfies the degree-symmetry and that R is a degree-symmetric quantifiable rigid relation.

Question: Given R is a degree-symmetric quantifiable rigid relation, and xRy; also, A is a quantifiable property, and object x has property A. Then, in the case of A having no corresponding consistency function $c_A(\cdot)$, how can we obtain the degree of object y having property A?

Imagine that if we can know the measure $\deg(x, y)$ of xRy or its estimated value $d_{R_{xy}}$ and the estimated value d_{A_x} of consistency-degree $c_A(x)$ and know the relation between $\deg(x, y)|d_{A_y}$ (symbol "|" denotes "or") and $d_{R_{xy}}$ and d_{A_x}, then we can use the symmetry of R and this relation to obtain the approximate value of corresponding $c_A(y)$. Therefore, we have the relational inference with degrees as the following scheme:

$$\frac{\begin{array}{c}(xRy, {}^{\circ}x, y)|d_{R_{xy}})\\ (A(x), d_{A_x})\end{array}}{(A(y), d_{A_y})} \tag{22.4}$$

Here, it is required that $d_{R_{xy}} \geq 1$ and $d_{A_x} > 0.5$ (when A is a flexible linguistic value) or $d_{A_x} \geq 1$ (when A is a quantifiable rigid linguistic value), $\deg(x, y)$ and $d_{R_{xy}}$ are the degrees of x and y having relation R; d_{A_x} and d_{A_y} are separately the degree of x and y having property A; $d_{A_y} = f(d_R, d_A)$ is the computation formula of d_{A_y}. The formula generally can be

$$d_{A_y} = (\deg(x, y)|d_{R_{xy}})d_{A_x} \tag{22.5}$$

Example 22.1 Let propositions p: x_0 is A and q: y_0 is B have partial equivalence relation; then, from Definitions 21.10 and 21.11, the partial equivalence satisfies the degree-symmetry. Thus, we can have the relational inference with degrees as the following scheme:

$$\frac{(p \vartriangleleft \vdots \vartriangleright q, \text{equi}(p, q))}{(p, \text{poss}(A))}$$

$$\therefore (q, \text{poss}(B))$$

(22.6)

where $\text{poss}(B) = \text{equi}(p, q)\text{poss}(A)$.

22.2.2   Relational Inference with Degrees with Flexible Relations

1. Relational inference with degrees with flexible transitive flexible relations

Definition 22.3 Let U be an n-dimensional measurement space, and let R be a binary flexible relation on product space $U \times U$, and let $c_R(x, y)$ be the consistency function of R. If for $\forall x, y, z \in U$, when $c_R(x, y) > 0$ and $c_R(y, z) > 0$, then $c_R(x, z) > 0$, then we say relation R satisfies flexible transitivity and that R is a flexible transitive flexible relation.

Question: Given R is a flexible transitive flexible relation, and xRy and yRz. Then, in the case of R having no corresponding consistency function $c_R(\cdot, \cdot)$, how do we obtain the degree of xRz?

Since there is no consistency function, we can only make certain estimate of the degree or try to compute its approximate value. Imagine that if we can know the estimated values $d_{R_{xy}}$ and $d_{R_{yz}}$ of the consistency-degrees of xRy and yRz, and the relation between the consistency-degree of relation xRz and the consistency-degrees of xRy and yRz, then we can use the transitivity of R and this relation to indirectly obtain the approximate value $d_{R_{xz}}$ of consistency-degree of relation xRz. Therefore, we have the relational inference with degrees as the following scheme:

$$\frac{(xRy, d_{R_{xy}})}{(yRz, d_{R_{yz}})}{(xRz, d_{R_{xz}})}$$

(22.7)

here, it is required that $d_{R_{xy}} > 0.5$ and $d_{R_{yz}} > 0.5$, where $d_{R_{xy}}$, $d_{R_{yz}}$, and $d_{R_{xz}}$ are separately the degrees of (x, y), (y, z), and (x, z) having relation R; $d_{R_{xz}} = f(d_{R_{xy}}, d_{R_{yz}})$ is the computation formula of $d_{R_{xz}}$.

For example, the "far greater than" relation of real numbers is a flexible transitive flexible relation; then, with "far greater than," we can do a relational inference with degrees.

2. Relational inference with degrees with flexible symmetric flexible relations

Definition 22.4 Let U be an n-dimensional measurement space, let R be a binary flexible relation on product space $U \times U$, and let $c_R(x, y)$ be the consistency function of R. If for $\forall x, y \in U$, when $c_R(x, y) > 0$ then $c_R(y, x) > 0$, and $c_R(x, y) = c_R(y, x)$, then we say R satisfies flexible symmetry and that R is a flexible symmetrical flexible relation.

Question: Given that R is a flexible symmetric flexible relation, and xRy; also, A is a quantifiable property, and object x has property A. Then, in the case of having no corresponding consistency function $c_A(\cdot)$ of A, how do we obtain the degree of object y having property A?

Imagine that if we can know the consistency-degree $c_R(x, y)$ of xRy or its estimated value $d_{R_{xy}}$ and the estimated value d_{A_x} of consistency-degree $c_A(x)$ and know the relation between $c_R(x,y)|d_{A_y}$ and $d_{R_{xy}}$ and d_{A_x}, we can then use the symmetry of R and this relation to obtain the approximate value of corresponding $c_A(y)$. Therefore, we have the relational inference with degrees as the following scheme:

$$\frac{(xRy, c_R(x,y)|d_{R_{xy}})}{(A(y), d_{A_y})} \quad (A(x), d_{A_x})$$

(22.8)

here, it is required that $c_R(x, y) > 0.5$, $d_{R_{xy}} > 0.5$, and $d_{A_x} > 0.5$ (when A is a flexible linguistic value) or $d_{A_x} \geq 1$ (when A is a quantifiable rigid linguistic value), $c_R(x, y)$, and $d_{R_{xy}}$ are the degrees of x and y having relation R; d_{A_x} and d_{A_y} are separately the degrees of x and y having property A; $d_{A_y} = f(c_R(x,y)|d_{R_{xy}}, d_{A_x})$ is the computation formula of d_{A_y}.

Example 22.2

$$\frac{(P_1 \text{ alike } P_2, \ 0.9) \quad (\text{pretty } (P_1), \ 1.2)}{(\text{pretty } (P_2), \ 1.08)}$$

here $1.08 = 0.9 \times 1.2$, that is, $d_{\text{pretty}_y} = d_{\text{alike}_{xy}} d_{\text{pretty}_x}$.

It can be seen that the key of relational inference with degrees is the degree computation formula. Then, how do we determine this computation formula? For the transitive relation and symmetric relation, is there a unified expression of functions for either? These are just the problems we need to study. But it is not hard to see that for many practical problems, the multiplication operation similar to the above examples is all applicable.

22.3 Knowledge Representation with Degrees

Using degrees, i.e., consistency-degrees, we can precisely depict usual knowledge representation schemes such as tuples, predicates, rules, frames, semantic, and networks to form the tuple with degrees, the predicate with degrees, the rule with degrees, the frame with degrees, the semantic net with degrees, etc. Next, we will exemplify them one by one.

1. **Tuple with degrees**
   The general form of the tuple with degrees is as follows

$$(\langle\text{object}\rangle, \langle\text{feature}\rangle, (\langle\text{linguistic value}\rangle, \langle\text{degree}\rangle)) \qquad (22.9)$$

For example

$$(\text{the apple, taste, (sweet, 0.95)})$$

This is just a tuple with degree, which can be interpreted as:

`This apple is comparatively sweet.`

2. **Predicate with degrees**
   Predicates are also linguistic values. Following the practice of linguistic values with degree, we can also attach a degree to a predicate, that is, refine it as a predicate with degree to precisely depict the feature of the corresponding object. According to the characteristics of the form of predicates, we can write a predicate with degree as

$$P_d(\langle\text{object}\rangle) \quad \text{or} \quad dP(\langle\text{object}\rangle) \qquad (22.10)$$

where P denotes predicate, d denotes degree, P_d is subscript denotation, and dP is multiplication denotation.
   For example,

$$\text{white}_{1.0}(\text{snow}) \quad \text{or} \quad 1.0\text{white (snow)}$$

is just a predicate with degree, which can be interpreted as follows: Snow is white. For another example,

$$\text{friends}_{1.15}(\text{Mike, Jack}) \quad \text{or} \quad 1.15 \text{ friends(Mike, Jack)}$$

is also a predicate with degree, which can be interpreted as:

`Mike and Jack are good friends.`

3. **Rule with degrees**

A production rule containing linguistic values with degrees is called a production rule with degree, simply, a rule with degree. Its general form is as follows:

$$(\langle object\rangle, \langle feature\rangle, (\langle linguistic\ value\rangle, \langle degree\rangle))$$
$$\rightarrow (\langle object\rangle, \langle feature\rangle, (\langle linguistic\ value\rangle, \langle degree\rangle)) \qquad (22.11)$$

For example,

$$(banana, color, (yellow, 0.7)) \rightarrow (banana, maturity, (ripe, 0.9))$$

is just a rule with degree, which can be interpreted as:

    If a banana is rather yellow, then it is comparatively ripe.

4. **Frame with degrees**

We call a frame containing linguistic values with degrees to be a frame with degrees.

For example, there is a frame with degrees describing date below.

frame name: $\langle date\rangle$

category: $(\langle dried\ fruit\rangle, 0.8)$
shape: (round, 0.7)
color: (red, 1.0)
taste: (sweet, 1.1)
usage: scope: (edible, officinal)

default: edible

5. **Semantic network with degrees**

We call a semantic network containing linguistic values with degrees a semantic network with degrees.

For example, there is a semantic network with degrees describing dog below (see Fig. 22.2).

Note that the "understand" in linguistic value with degree (can-understand, 0.9) in Fig. 22.2 is not that kind of adjectives or adverbs as attribute or adverbial representing the attributes, states, or relations of objects, but a verb as predicate representing the behavior of object, and the 0.9 is then to modify this verb, which play the role of adverb (adverbial) (actually, the degrees in linguistic values with

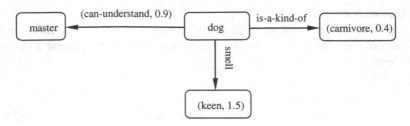

Fig. 22.2 An example of semantic network with degrees

degrees are all play the role of adverb), and which can be viewed as the conversion of corresponding adverb such as "well."

It can be seen that the thinking of depicting imprecision by degree is like depicting uncertainty by probability. This kind of method is used in the traditional knowledge representations, thus extending the expression ability of these knowledge representations.

Lastly, it should be pointed out that the representation with degrees above for imprecise knowledge is a kind of representation with respect to objects, that is, the degrees here are with respect to specific objects but they are not the mathematical models of corresponding flexible linguistic values; however, if an *object* in a tuple, predicate, rule, frame, semantic, or network is a variable, then corresponding *degree* is also a variable—it is the consistency function of corresponding flexible linguistic value.

22.4   Relational Database with Degrees

We know that as attribute values of objects, linguistic values are largely used in databases. But an object always has a certain linguistic value to a certain degree, so it would appear not fine or accurate enough that we only use pure linguistic values to describe things in some situations. For example, in a database of offender characteristics, naturally the data are hoped to be detailed and accurate. Therefore, we can attach a real number to a linguistic value in database to depict the degree of corresponding object having the linguistic value. Particularly, for the relational database, if the relation R is also a flexible relation (for example, "talented professionals" is a flexible relation), then every record should also have a consistency-degree with R. In the following, we introduce the degrees into the field of database and give several relational database models with degree for reference.

1. **Relational model of linguistic values with degrees**

   To depict objects in more detail and more accurately, we can attach a real number called the degree to a data item in usual relational databases to form a compound data item. Then, the relational database model based on this kind of compound data items is what we call the relational model of linguistic values with degrees.

(1) Data models

Definition 22.5 Let $D_1, D_2, \ldots, D_n$ be n one-dimensional measurement spaces as domains, and let $\mathscr{L}_i$ be a set of linguistic values on D_i ($i = 1, 2, \ldots, n$). Construct Cartesian product:

$$\mathscr{L}_1 \times \mathscr{L}_2 \times \cdots \times \mathscr{L}_n = \mathscr{L}_D$$

Take subset $R \subseteq \mathscr{L}_D$ such that it can form a relation on domain $\mathscr{L}_1 \times \mathscr{L}_2 \times \cdots \times \mathscr{L}_n$. If for every tuple in R

$$r = (A_1, A_2, \ldots, A_n)$$

a vector

$$d = (d_1, d_2, \ldots, d_n) \quad (d_i \in [\alpha_i, \beta_i], \alpha_i \leq 0, \beta_i \geq 1, i = 1, \ldots, n)$$

can be determined, then two-tuples

$$(R, d) \tag{22.12}$$

are called a relation of linguistic values with degrees on domain $\mathscr{L}_1 \times \mathscr{L}_2 \times \cdots \times \mathscr{L}_n$, denoted R_{Ld}. A tuple in R_{Ld} can be represented as an ordered pair of tuples.

$$(r, d) = ((A_1, A_2, \ldots, A_n), (d_1, d_2, \ldots, d_n)) \tag{22.13}$$

where $r = (A_1, A_2, \ldots, A_n)$ is called a tuple of linguistic values, and $d = (d_1, d_2, \ldots, d_n)$ is called a tuple of degrees.

It can be seen that the relation of linguistic values with degrees is actually a double relation. This kind of relation can be implemented by two associative "isomorphism" two-dimensional tables. For example, Table 22.1 is two two-dimensional tables of a relation of linguistic values with degrees.

Viewed from effect, the relation of linguistic values with degrees, R_{Ld}, is a detailing of the usual relation R.

(2) Relational operations

The relation of linguistic values with degrees, R_{Ld}, can also have all kinds of relational operations. But since the data items are compound ones, the parts of linguistic value and degree of a data item should be operated synchronically.

Table 22.1 An example of relation of linguistic values with degrees

Number	Age	Height	Face Shape	Hair Style	Hair Color	Skin Color	...
001	youth	tall	long	long	black	black	...
002	old	short	square	bald	white	yellow	...
003	middle	medium	circular	short	yellow	white	...
...	...	...	...	...	...	...	...

Number	Age	Height	Face Shape	Hair Style	Hair Color	Skin Color	...
001	0.9	1.2	1.0	0.8	0.9	0.7	...
002	0.8	0.9	0.9	0.8	0.7	1.0	...
003	0.9	1.0	0.9	1.0	0.7	0.9	...
...	...	...	...	...	...	...	...

2. Flexible-relational model with degrees

For a flexible relation, we can establish a kind of flexible-relational database model with degrees called.

(1) Data model

Definition 22.6 Let $D_1, D_2, \ldots, D_n$ be n one-dimensional measurement spaces as domains; construct Cartesian product:

$$D_1 \times D_2 \times \cdots \times D_n = D$$

Take subset $R \subseteq D$ such that it can form a flexible relation on domains $D_1 \times D_2 \times \cdots \times D_n$. If for every tuple in R

$$r = (x_1, x_2, \ldots, x_n)$$

a consistency-degree with flexible relation R

$$d(r) \in [\alpha, \beta] \quad (\alpha \le 0, \beta \ge 1)$$

can be determined, then two-tuples

$$(R, d(r)) \tag{22.14}$$

Table 22.2 An example of flexible relation with degrees

Number	Circulation	Number of Users	Using Period	Times Cited	...
001	20000	20	3	180	...
002	80000	105	4	350	...
003	100000	80	6	300	...
...	...	...	...	...	...

Number	Consistency-degree
001	0.9
002	1.2
003	1.0
...	...

are called a flexible relation with degrees on domain $D_1 \times D_2 \times \cdots \times D_n$, denoted R_d. A tuple of R_d can be represented as ordered pair

$$(r, d(r)) = ((x_1, x_2, \ldots, x_n), d(r)) \tag{22.15}$$

where $r = (x_1, x_2, \ldots, x_n) \in D$ is the tuple of flexible relation R, and $d(r) \in [\alpha, \beta]$ $(\alpha \leq 0,\ \beta \geq 1)$ is the consistency-degree of r with flexible relation R.

It can be seen that the flexible relation with degrees is actually a kind of two-layer relation, which can be implemented by two associative "homomorphism" two-dimensional tables. For example, Table 22.2 is an example of two-dimensional table of flexible relation with degrees about "excellent textbook."

It can be seen that flexible relation with degrees, R_d, is a kind of generalization of the usual relations (database).

(2) **Relational operations**

Let $\{R_1, d_1(r_1)\}$ and $\{R_2, d_2(r_2)\}$ be two flexible relations with degrees; then, the related operations are defined as follows:

(1) Intersection with degrees ($\cap$)

$$\{R_1, d_1(r_1)\} \cap \{R_2, d_2(r_2)\} = \{R_1 \cap R_2, \min(d_1(r_1), d_2(r_2))\}$$

(2) Union with degrees ($\cup$)

$$\{R_1, d_1(r_1)\} \cup \{R_2, d_2(r)\} = \{R_1 \cup R_2, \max(d_1(r_1), d_2(r_2))\}$$

(3) Other operations including selection, projection, connection, insertion, deletion, updating, etc., are basically the same as operations in usual relational
databases, and only the processing of degrees needs to be added. Obviously, in
these operations, degree $d(r)$ can also be treated as a condition.

3. **Flexible-relational model with degrees of linguistic values with degrees**
If a relation is a linguistic valued relation as well as a flexible relation, then the
two models above can be superposed into a kind of database model of flexible
relation with degrees of linguistic values with degrees.

(1) Data model

Definition 22.7 Let $D_1, D_2, \ldots, D_n$ be n one-dimensional measurement spaces as
domains, and $\mathcal{L}_i$ be a set of linguistic values on $D_i (i = 1, 2, \ldots, n)$. Construct
Cartesian product:

$$\mathcal{L}_1 \times \mathcal{L}_2 \times \cdots \times \mathcal{L}_n = \mathcal{L}_D$$

and take subset $R \subseteq \mathcal{L}_D$ such that it can form a flexible relation on domain
$\mathcal{L}_1 \times \mathcal{L}_2 \times \cdots \times \mathcal{L}_n$. If for every tuple in R

$$r = (A_1, A_2, \ldots, A_n)$$

a vector

$$d = (d_1, d_2, \ldots, d_n) \quad (d_i \in [\alpha_i, \beta_i], \alpha_i \leq 0, \beta_i \geq 1, i = 1, \ldots, n)$$

can be determined, and for (r, d) and R, a real number

$$d(r) \in [\alpha, \beta] \quad (\alpha \leq 0, \beta \geq 1)$$

can be determined, then compound two-tuples

$$((R, d), \ d(r)) \tag{22.16}$$

are called a flexible relation with degrees of linguistic values with degrees on
domain $\mathcal{L}_1 \times \mathcal{L}_2 \times \cdots \times \mathcal{L}_n$, denoted R_{d-Ld}. A tuple in R_{d-Ld} can be represented
as an ordered pair of compound tuples

$$((r, d), \ d(r)) = (((A_1, A_2, \ldots, A_n), (d_1, d_2, \ldots, d_n)), d(r)) \tag{22.17}$$

where $r = (A_1, A_2, \ldots, A_n)$ is a tuple of linguistic values, $d = (d_1, d_2, \ldots, d_n)$ is a
tuple of degrees, and $d(r) \in [\alpha, \beta]$ ($\alpha \leq 0$, $\beta \geq 1$) is the consistency-degree of
(r, d) with flexible relation R.

It can be seen that the flexible relation with degrees of linguistic values with
degrees is actually a kind of two-layer double relation, which can be implemented

Table 22.3 An example of flexible relation with degrees of linguistic values with degrees

Number	Circulation	Number of Users	Using Period	Times Cited	...
001	general	medium	short	more	...
002	larger	many	medium	many	...
003	large	more	long	many	...
...	...	...	...	...	...

Number	Circulation	Number of Users	Using Period	Times Cited	...
001	0.9	0.85	1.0	1.0	...
002	1.0	1.1	0.8	0.9	...
003	0.8	1.1	0.9	0.8	...
...	...	...	...	...	...

Number	Consistency-degree
001	0.9
002	1.2
003	1.0
...	...

by two associative "isomorphism" two-dimensional tables and another two-dimensional table that is "homomorphism" with them. For example, Table 22.3 is an example of the two-dimensional tables of the flexible relation with degrees of linguistic values with degrees about "excellent textbook."

It can be seen that flexible relation with degrees of linguistic values with degrees, R_{d-Ld}, is a kind of detail and generalization of the usual relation (database).

(2) **Relational operations**

Let $\{(R_1, r_1), d_1(r_1)\}$ and $\{(R_2, r_2), d_2(r_2)\}$ be two flexible relations with degrees of linguistic values with degrees; then, the related operations based on flexible relations are defined as follows:

(1) Intersection with degrees ($\cap$)

$$\{(R_1, r_1), d_1(r_1)\} \cap \{(R_2, r_2), d_2(r_2)\} = \{(R_1, r_1) \cap (R_2, r_2), \min(d_1(r_1), d_2(r_2))\}$$

(2) Union with degrees ($\cup$)

$$\{(R_1,\ r_1),\ d_1(r_1)\}\cup\{(R_2,\ r_2),\ d_2(r)\} = \{(R_1,\ r_1)\cup(R_2,\ r_2),\ \max(d_1(r_1),\ d_2(r_2))\}$$

(3) Other operations including selection, projection, connection, insertion, deletion, updating, etc., are basically the same as operations in usual relational
databases, and only the processing of degrees needs to be added.

In the above, we gave tentatively three kinds of relational database models with
degrees. It can be seen that this idea and method of degree-introducing are also
applicable for other types of databases (such as network and object-oriented).

It should be pointed out lastly that an important application of the database with
degrees is that it can be used to describe the sample data set in supervised machine
learning. In fact, the degrees here are the consistency-degrees between corresponding numerical objects and corresponding linguistic values, which can just be
treated as the supervisor's signal in sample data.

22.5 Some Ideas for Machine Understanding and Generation of Flexible Concepts and Flexible Propositions

22.5.1 Machine Understanding of Flexible Concepts with Knowledgebase

1. **Understanding of atomic concepts**
   Atomic flexible concepts are a kind of initial and basic flexible concepts directly
   founded on the basis of perceived information, which belongs to common sense
   or axiomatic knowledge. According to the analyses in Sect. 19.1, the formation
   mechanism of the atomic concepts should be innate of human brain. So we
   believe that people understand directly the atomic flexible concepts from perceptual knowledge. Thus, understanding of the atomic concepts can be implemented directly using their membership/consistency functions. For instance, for
   "tall," from its membership function, its core and support set can be known;
   thus, the scope of height of a certain person can be estimated. If one is also "very
   tall," then from the consistency function, one's height scope can be estimated.

   Viewed from the realization method, the understanding of the atomic perceptual
flexible concepts can be realized by conversion from numerical values to flexible
linguistic values; conversely, the execution of atomic flexible commands can also
be realized by conversion from flexible linguistic values to numerical values.

2. **Understanding of the compound concepts**

   Compound concepts are a kind of concepts compounded by its component concepts through logical connectives. Therefore, people's understanding of a compound concept should be achieved by combining the understanding of its component concepts and the understanding of logical connectives such as AND, OR, and NOT. It is clear that the meaning of a logical connective is its logical semantics. Therefore, the understanding of a compound concept is the understanding based on its component concepts and logical knowledge. The understanding of a compound concept can be realized by the computation of corresponding compound consistency function and also by the reasoning with the corresponding flexible rules.

3. **Understanding of the synthetic concepts**

   A synthetic concept is a concept compounded by its ingredient concepts through algebraic operations. The understanding of a synthetic concept can be realized by the computation of corresponding consistency function and also by reasoning with the corresponding synthesis-type flexible rules.

4. **Understanding of the derivative concepts**

   The so-called derivative concepts are a kind of concepts defined by known concepts. The understanding of a derivative concept is the understanding based on its definition. The derivative concepts is a kind of high-level concept based on the known concepts, which generally belong to specialized knowledge, defined by rules, and acquired by learning. Therefore, the understanding of the derivative concepts is the understanding based on specialized knowledge and linguistic knowledge, and the understanding according to the relationship between known concepts and defined concepts.

To sum up, the understanding of flexible concepts require to build two level of knowledge bases, that is, a knowledge base of primary concepts and a knowledge base of advanced concepts. The former stores the membership or consistency functions of the primary concepts, and the latter stores the membership or consistency functions of the derivative concepts and the flexible rules. With the two types of knowledge bases, we can build a machine understanding system of flexible concepts with knowledge bases. The structure of the system is shown in Fig. 22.3. Therein, the computing of function and the reasoning with rules are two main function components.

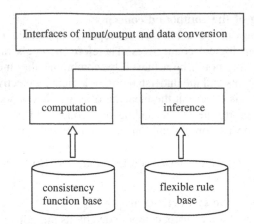

Fig. 22.3 Structure of an understanding system of flexible concept with knowledge base

22.5.2 Machine Generation and Application of Flexible Concepts with Object-Oriented Programs

Here, the so-called machine generation of flexible concepts is that machine can independently automatically generate the membership/consistency functions of relevant flexible concepts. Specifically speaking, the membership or consistency functions can be automatically formed from parameters given in advance. To achieve this, in the machine we can set a common parameters-undetermined computation program of the membership or consistency function. Once the parameter values are given, the program execute the corresponding computation of membership/consistency function, further obtaining corresponding membership-degree or consistency-degree. Thus, in effect, it is tantamount to the corresponding membership/consistency function being generated. It is not hard to see that this function can just be implemented by using object-oriented programs. As a matter of fact, designing a flexible concept class, then using a group of parameters, then an instance object, that is, a specific flexible concept, can be generated. Next, we give tentatively an example of a flexible concept class implemented by C++ programs.

```
      /*      definition of the flexible concept class    */
class Flexibleconcept
      {
      private:
         char* name; // name of a flexible concept
         float criticalpoint, coreboundarypoint; //critical points
and core-boundary point
      public:
         Flexibleconcept(char* name1, float criticalpoint1, float
coreboundarypoint1);
         ~ Flexibleconcept();
         float consistencyfunction(float x); //consistency function
      };
      /*      realization of the flexible concept class    */
      Flexibleconcept::Flexibleconcept(char*     name1,      float
criticalpoint1, float coreboundarypoint1)
      {
         name=new char[strlen(name)];
         strcpy(name, name1);
         criticalpoint=criticalpoint1;
         coreboundarypoint= coreboundarypoint1;
      }
      Flexibleconcept::~Flexibleconcept ()
      {
         delete name;
      }
      float Flexibleconcept:consistencyfunction(float x)
      {
         float consistencydegree=
      (x- criticalpoint)/(coreboundarypoint - criticalpoint);
         cout<<" consistencydegree: "<< consistencydegree;
      }
```

For simplicity, the above defined is only a flexible concept class of negative semi-peak value. Using this flexible concept class, then its instance objects, that is, specific flexible concepts, can be generated. For example, the statement below can generate an instance of flexible concept named "(weather) hot": hot.

```
Flexibleconcept hot("(weather) hot",20,28);
```

here, 20 and 28 are the critical point and core–boundary point of "(weather) hot." Next, we send a message to object hot:

```
hot. consistencyfunction(25);
```

That is, asking it to compute the consistency-degree of 25 °C with "(weather) hot." According to the critical point and core–boundary point for "(weather) hot" set above, the returned result of the system should be:

```
consistencydegree: 0.625
```

From this example, it can be seen that we can use object-oriented technology to build a flexible concept knowledgebase consisting of flexible concept classes with corresponding parameters, so as to simulate functionally generation and understanding of the flexible concepts. Obviously, the more flexible concepts there are in the knowledgebase, the richer is machine's relevant knowledge, and the stronger is the corresponding ability of imprecision information processing. And then, we can also build a flexible concept net according to logical relations between flexible concepts.

22.6 Summary

In this chapter, the work of us is as follows:

- From the perspective of methodology proposed, two basic ideas and techniques of imprecise-information processing—degree-introducing and appropriate granularity.
- Discuss the relational inference with quantifiable rigid relations and flexible relations and gave the corresponding schemes of relational inference with degrees for transitivity and symmetry relations.
- Propose the method of knowledge representation with degrees and gave some representation schemes such as tuple with degrees, predicate with degrees, frame with degrees, semantic network with degrees, and rule with degrees.
- Discussed relational database models with degrees and presented the relational model of linguistic value with degrees, flexible-relational model with degrees and flexible-relational model with degrees of linguistic values with degrees.
- Proposed a principle of machine understanding flexible concepts with knowledge base and presented a method of automatically generating flexible concepts by using object-oriented programming.

References

1. Lian S (2000) Degree theory—an information processing technique with degrees. Shanxi Science & Technology Publishing House, Shanxi
2. Lian S (2009) Principles of imprecise-information processing. Science Press, Beijing

References

1.
2.

Part VIII
Probabilities of Flexible Events, Believability-Degrees of Flexible Propositions and Reasoning with Believability-Degrees as Well as Modal Propositions: Overlap of and Correlation Between Imprecision and Uncertainty

Chapter 23
Probabilities of Random Flexible Events

Abstract This chapter introduces the concept of random flexible events, analyzes the corresponding probability computation principles, and then establishes corresponding probability computation models for various types of flexible events.

Keywords Random flexible events · Conceptual probability · Practical probability

What the previous chapters discussed are all pure imprecise-information processing, but there are also the phenomenon and problems of the crossing and overlapping of imprecision and uncertainty in the real world. In this chapter, we discuss random flexible events and their probabilities as well as flexible linguistic values on the probability range.

23.1 Random Flexible Events and Their Probabilities

23.1.1 Random Flexible Events

We know that the random event is a basic concept in probability theory. However, random events discussed in probability theory are usually some "rigid events" or "random rigid events." The so-called rigid events are ones that have rigid standards or demarcations. For instance, "raining" is a rigid event, since the demarcation of the event of "raining" is rigid, and raining or not raining is very clear cut. For another example, "draw an unqualified product from a batch of products" is also a rigid event since there must be a rigid standard on what is an unqualified product.

However, our information communication also involves a lot of random "flexible events." The so-called flexible events are ones that have no rigid standards or demarcations, or in other words that are described by flexible linguistic values [1]. For instance, "rain heavily" is a flexible event since how heavy should be regarded as heavy has not a rigid standard and demarcation. For another example, "draw a good product from a batch of product" is also a flexible event because "good" is only an adjective, and "good product" has not a rigid standard.

We know that for a rigid event, the probability of its occurrence can be calculated and the probability is a precise number. Then, for a flexible event, how its probability is represented and computed? In the following, we will discuss these problems.

23.1.2 Probability of a Flexible Event with Uniformly Distributed Random Variable

1. **Probability of a flexible event with uniformly distributed discrete random variable**

In the traditional probability theory, for uniformly distributed discrete random variable X, the probability of corresponding event e_A ($X \in A$) is generally computed by using the following formula:

$$P(e_A) = \frac{|A|}{|\Omega|} = \frac{k}{n}$$

where $\Omega = \{e_1, e_2, ..., e_n\}$ is a sample space, A is a set of basic events that event e_A contains, and $|\ |$ denotes the size of a set. What this equation formulates is the so-called classical probability.

Since the probabilities of all basic events are equivalent, or in other words, the probability distribution of random variable X on Ω is uniform, so the probability of event e_A should be equal to the ratio of the size $|A|$ of basic event set A to the size $|\Omega|$ of sample space, i.e., the share of A in Ω. Such is the basic principle of the classical probability.

When Ω is a set of consecutive integers and e_A is a flexible event, A would be a flexible set. And the elements in a flexible set all belong to this flexible set to some degrees. Therefore, the membership-degree of an element to the flexible set is the contribution of it in numerical to the size of the flexible set. Thus, the size of flexible set A should be

$$|A| = \sum_{x \in \sup p(A)} m_A(x)$$

That is, the element amount of flexible set A is the sum of element x in Ω by converting according to membership-degree $m_A(x)$. Thus, we have

$$P(e_A) = \frac{|A|}{|\Omega|} = \frac{\sum_{x \in \sup p(A)} m_A(x)}{|\Omega|} \tag{23.1}$$

This is the computation model of the probability of a flexible event with uniformly distributed discrete random variable.

Example 23.1 In the game of craps, sample space $\Omega = \{1, 2, 3, 4, 5, 6\}$. Here, integer $i \in (i = 1, 2, 3, 4, 5, 6)$ represents that result is i points.

We use random variable X to represent the number of points occurring in a throw of dice, then according to the classical probability,

$$P(X = i) = \frac{1}{6} \quad (i = 1, 2, 3, 4, 5, 6)$$

$$P(X = 2 \text{ or } X = 6) = \frac{1+1}{6} = \frac{2}{6}$$

$$P(X > 3) = \frac{1+1+1}{6} = \frac{3}{6}$$

The above three events are all rigid events. Examining sample space Ω, obviously, it is a continuous set in the sense of integers, so which can be regarded as a measurement space. Therefore, on it there can occur flexible events. For instance, "result is a big point" and "result is a small point" of the throw of dice are two flexible events on Ω. We define the membership functions of the corresponding flexible sets "big point" and "small point" of the two flexible events as follows:

$$m_{\text{big point}}(x) = \{(1,0), (2,0.2), (3,0.4), (4,0.6), (5,0.8), (6,1)\}$$
$$m_{\text{small point}}(x) = \{(1,1), (2,0.8), (3,0.6), (4,0.4), (5,0.2), (6,0)\}$$

Thus, from Eq. (23.1),

$$P(X \in \text{big point}) = \frac{0+0.2+0.4+0.6+0.8+1}{6} = \frac{3}{6} = \frac{1}{2}$$
$$P(X \in \text{small point}) = \frac{1+0.8+0.6+0.4+0.2+0}{6} = \frac{3}{6} = \frac{1}{2}$$

2. **Probabilities of flexible events with uniformly distributed continuous random variable**

In traditional probability theory, for uniformly distributed continuous random variable X, the probability of corresponding event e_A ($X \in A$) is generally computed by the following formula:

$$P(X \in A) = \frac{|A|}{|U|}$$

where U is a real number interval, and $A \subseteq U$ is measurable. What this equation formulates is the so-called geometric probability.

Since the distribution of the probability of random variable X on U is uniform, the probability of event e_A equals the ratio of the size $|A|$ of subset A to the size $|U|$ of U. Such is the basic principle of the geometric probability.

Just the same, when U is regarded as a measurement space and e_A is a flexible event, corresponding A is a continuous flexible set. And the size of a continuous flexible set equals the definite integral of its membership function on the support set. Thus, the size of A should be

$$|A| = \int\limits_{\sup p(A)} m_A(x)dx$$

Thus, we have

$$P(X \in A) = \frac{|A|}{|U|} = \frac{\int_{\sup p(A)} m_A(x)dx}{|U|} \tag{23.2}$$

This is the computation model of the probability of a flexible event with uniformly distributed continuous random variables.

Example 23.2 Suppose a certain random number generator can generate equiprobably any real number in interval [0, 10]. Now to compute: the probability of this random number generator producing a real number of "about 5" at one time.

Let X be a real number generated by the random number generator, then X is a random variable. Thus, "generating a real number of about 5" can be represented as: $X \in$ about 5. Clearly, this is a flexible event, suppose

$$m_{\text{about } 5}(x) = \begin{cases} 0, & x \le 4.6 \\ \dfrac{x - 4.6}{2}, & 4.6 < x < 4.8 \\ 1, & 4.8 \le x \le 5.2 \\ \dfrac{5.4 - x}{2}, & 5.2 < x < 5.4 \\ 0, & 5.4 \le x \end{cases}$$

It follows that

$$\int\limits_0^{10} m_{\text{about } 5}(x)dx = 0.6$$

$$|[0, 10]| = 10$$

Thus,

$$P(X \in \text{about } 5) = 0.6/10 = 0.06$$

23.1.3 Probability of a Flexible Event with Non-uniformly Distributed Random Variable

We know that in traditional probability theory, for non-uniformly distributed random variable X, the probability of the corresponding event e_A ($X \in A$) is generally computed by the following two formulas:

(1) For discrete random variable X

$$P(a_1 \leq X \leq b_1) = \sum_{x_k \in [a_1, b_1]} P_k$$

($\{P_k\}$ is the probability distribution sequence of X)

(2) For continuous random variable X

$$P(a_1 \leq X \leq b_1) = \int_{a_1}^{b_1} \rho(x) \mathrm{d}x$$

($\rho(x)$ is the distribution density function of X)

While when $e_A \colon X \in A$ is a flexible event, the corresponding set A is a flexible set, that is, flexible interval. Let $\mathrm{supp}(A) = [a_1, b_1]$. Imagine support set $[a_1, b_1]$ is divided into a number of small intervals $[a_{1_1}, b_{1_1}], [a_{1_2}, b_{1_2}], \ldots, [a_{1_n}, b_{1_n}]$, on every small interval, the probability density is close to uniform distribution; thus, we take $\rho(a_{1_i})$ as the uniform probability density on small interval $[a_{1_i}, b_{1_i}] (i = 1, 2, \ldots, n)$. Set $\Delta_i = \|[a_{1_i}, b_{1_i}]\|$, but because $[a_{1_i}, b_{1_i}]$ is the support set of a flexible linguistic value, the effective length of $[a_{1_i}, b_{1_i}]$ should be the integral of membership function $m_A(x)$ on $[a_{1_i}, b_{1_i}]$. Therefore, we then take $m_A(a_{1_i})$ as the uniform membership-degree on $[a_{1_i}, b_{1_i}]$, then

$$\|[a_{1_i}, b_{1_i}]\| \approx m_A(a_{1_i}) \Delta_i$$

Thus,

$$P(a_{1_i} \leq X \leq b_{1_i}) \approx \rho(a_{1_i}) m_A(a_{1_i}) \Delta_i$$

Therefore,

$$P(X \in A) \approx \sum_{i=1}^{n} \rho(a_{1_i}) m_A(a_{1_i}) \Delta_i$$

Set $\Delta_i \to 0$, then

$$\lim_{\Delta_i \to 0} \left(\sum_{i=1}^n \rho(a_{1_i}) m_A(a_{1_i}) \Delta_i \right) = \int_{a_1}^{b_1} \rho(x) m_A(x) \mathrm{d}x = \int_{\sup p(A)} \rho(x) m_A(x) \mathrm{d}x$$

Thus, we have

$$P(X \in A) \int_{\sup p(A)} \rho(x) m_A(x) \mathrm{d}x \qquad (23.3)$$

This is the probability of a flexible event with non-uniformly distributed continuous random variable.

However, unfortunately, for the flexible events with non-uniformly distributed discrete random variables, we have not found a reasonable method to compute their probabilities. But fortunately, we can compute their "practical probabilities" (see next section).

23.1.4   Probability on the Extended Core—The Practical Probability of a Flexible Event

In the above, we derived the computation formula of probability of a flexible event starting from the concept of flexible events. However, like flexible classifying, when we try to find the probability of a flexible event, actually the flexible event should be flexible classified firstly. And the probability to be found is just the probability of the flexible event that had been flexibly classified. For instance, to find the probability of a heavy rain, we do not do it in an isolated way, but find the probability of the flexible event that be classified as "heavy rain." Then, how do we compute the probability of this "flexible event" that is classified?

Since the flexible event after being classified actually already is a "rigid event" represented by the extended core of original flexible event, its probability is the probability of random variable on the extended core of the corresponding flexible linguistic value. Thus, for the flexible event e_A: $X \in A$, there would be following cases:

(1) If random variable X is uniform distribution, then

    ① When X is a discrete variable,

$$P(e_A) = P(X \in \mathrm{core}(A)^+) = \frac{|\mathrm{core}(A)^+|}{|\Omega|} \qquad (23.4)$$

where Ω is the sample space corresponding to random variable X

② When X is a continuous variable,

$$P(e_A) = P(X \in \text{core}(A)^+) = \frac{|\text{core}(A)^+|}{|U|} \qquad (23.5)$$

where U is the measurement space corresponding to random variable X

(2) If random variable X is non-uniform distribution, then

① When X is a discrete variable,

$$P(e_A) = P(X \in \text{core}(A)^+) = \sum_{x_k \in \text{core}(A)^+} P_i \qquad (23.6)$$

where $\{P_k\}$ is the probability distribution sequence of X.

② When X is a continuous variable,

$$P(e_A) = P(X \in \text{core}(A)^+) = \int_{\text{core}(A)^+} \rho(x)dx \qquad (23.7)$$

where $\rho(x)$ is the probability density function of X.

In order to distinguish, we call the probability of random variable X on the support set of flexible linguistic value A to be the **conceptual probability** of flexible event e_A, while call its probability on the extended core of flexible linguistic value A to be the **practical probability** of flexible event e_A. Thus, the previous Eqs. (23.1)–(23.3) are the computation formulas of conceptual probabilities, while the Eqs. (23.4)–(23.7) are the computation formulas of practical probabilities.

Example 23.3 Denote the random variable of daily rainfall as ξ. Let the range of values of ξ be interval $[0, 1000]$, and let probability density function be $\rho(x)$. Then, "rain lightly" is a flexible event on this interval. Suppose the support set and core of "light rain" be separately $(0, 30)$ and $(0, 20]$, then its extended core is $(0, 25]$. Thus, the conceptual probability of light rain on a certain day is

$$P(\xi \in \text{light rain}) = \int_0^{30} \rho(x)m_{\text{light rain}}(x)dx$$

The practical probability is

$$P(\xi \in \text{light rain}) = P(\xi \in (0, 25]) = \int_0^{25} \rho(x)dx$$

Later if not specifically specified, the probability of a flexible event always refers to its practical probability. But, from Theorem 20.3, the size of the extended core of a flexible set which is not a Cartesian product is equal to the size of the flexible set. So for a flexible event which is non-synthetic flexible event with uniformly distributed random variable, its conceptual probability is equal to the practical probability.

23.1.5 Linguistic Values on Probability Range

Obviously, "necessarily," "possibly," and "not possibly" are three basic rigid linguistic values on the range [0, 1] of probabilities (as shown in Fig. 23.1a). There are flexible linguistic values of "very probably/likely," "fairly probably/likely," and "not highly probably/likely," etc., on the range [0, 1] of probabilities, the graphs of whose membership functions are shown in Fig. 23.1b.

23.2 Probability of a Compound Flexible Event on the Same Space

Firstly, we call the flexible event e_A described by flexible linguistic value A on measurement space U to be a flexible event on measurement space U and call the compound flexible event made up of events on one and the same measurement space to be a compound flexible event on the same space.

In the following, we consider the probabilities of compound flexible events on the same space, $\neg e_A$, $e_A \wedge e_B$, $e_A \vee e_B$ and $e_A \rightarrow e_B$.

Let A and B be flexible linguistic values on one-dimensional measurement space U, X be a random variable on U, and $e_A: X \in A$ and $e_B: X \in B$ be flexible events on U. It can be seen that the negative flexible event $\neg e_A$ is actually also flexible event $e_{\neg A}: X \in A^c$, that is,

$$\neg e_A = e_{\neg A} \tag{23.8}$$

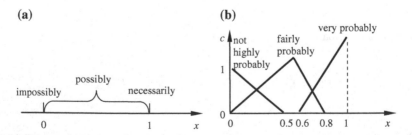

Fig. 23.1 a Basic linguistic values on the range of probabilities, **b** examples of flexible linguistic values on the range of probabilities

the conjunctive flexible event $e_A \wedge e_B$: $X \in A \wedge X \in B$ is also $e_{A \wedge B}$: $X \in A \cap B$; the disjunctive flexible event $e_A \vee e_B$: $X \in A \vee X \in B$ is also $e_{A \vee B}$: $X \in A \cup B$, namely

$$e_A \wedge e_B = e_{A \wedge B} \qquad (23.9)$$

$$e_A \vee e_B = e_{A \vee B} \qquad (23.10)$$

whereas implicational flexible event $e_A \rightarrow e_B$: $X \in A \rightarrow X \in B$ is the e_B in condition e_A, we adopt the notation in traditional probability theory to denote $e_A \rightarrow e_B$ as $e_B | e_A$ and call it the conditional flexible event.

Now, also by the definition of probability of the flexible event given in the section above, then

(1) When X is uniform distribution,

$$P(e_{\neg A}) = \frac{|core(A')^+|}{|U|} \qquad (23.11)$$

$$P(e_{A \wedge B}) = \frac{|core(A \cap B)^+|}{|U|} \qquad (23.12)$$

$$
\begin{aligned}
P(e_{A \vee B}) &= \frac{|core(A \cup B)^+|}{|U|} = \frac{|core(A)^+| + |core(B)^+| - |core(A \cap B)^+|}{|U|} \\
&= \frac{|core(A)^+|}{|U|} + \frac{|core(B)^+|}{|U|} - \frac{|core(A \cap B)^+|}{|U|} \\
&= P(e_A) + P(e_B) - P(e_{A \wedge B})
\end{aligned} \qquad (23.13)
$$

$$
\begin{aligned}
P(e_B | e_A) &= \frac{P(e_A \wedge e_B)}{P(e_A)} = \frac{P(e_{A \wedge B})}{P(e_A)} \\
&= \frac{|core(A \cap B)^+| / |U|}{|core(A)^+| / |U|} = \frac{|core(A \cap B)^+|}{|core(A)^+|}
\end{aligned} \qquad (23.14)
$$

Also, by the Theorem 20.3 and the definition of the inclusion-degree of flexible sets,

$$\frac{|core(A \cap B)^+|}{|core(A)^+|} = \frac{|A \cap B|}{|A|} = cont(A, B)$$

Thus, we have

$$P(e_B | e_A) = cont(A, B) \qquad (23.15)$$

That is, the probability of an implicational or conditional flexible event is numerically equal to the inclusion-degree between corresponding flexible sets.

Of course, Eq. (23.15) only is the definition of probability of conditional flexible event on the same space. Like the conditional probability in traditional probability theory, the probability of a conditional flexible event on the same space also has corresponding Bayes' formula. Besides, on the basis of probability of conditional flexible event on the same space, there is the corresponding total probability formula. These formulas are similar to the formulations in traditional probability theory, so it is unnecessary to go into details here.

(2) When X is non-uniform distribution and discrete,

$$P(e_{\neg A}) = P(x \in \text{core}(A')^+) = \sum_{x_k \in \text{core}(A')^+} P_i \qquad (23.16)$$

$$P(e_{A \wedge B}) = \sum_{x_k \in \text{core}(A \cap B)^+} P_i \qquad (23.17)$$

$$P(e_{A \vee B}) = \sum_{x_k \in \text{core}(A \cup B)^+} P_i = \sum_{x_k \in \text{core}(A)^+} P_i + \sum_{x_k \in \text{core}(B)^+} P_i - \sum_{x_k \in \text{core}(A \cap B)^+} P_i \qquad (23.18)$$

$$P(e_B | e_A) = \sum_{x_k \in \text{core}(A \cap B)^+} P_i / \sum_{x_k \in \text{core}(A)^+} P_i \qquad (23.19)$$

here $\{P_k\}$ is the distribution sequence of X.
(3) When X is non-uniform distribution and continuous,

$$P(e_{\neg A}) = \int_{\text{core}(A')^+} \rho_{A'}(x) dx \qquad (23.20)$$

$$P(e_{A \wedge B}) = \int_{\text{core}(A \cap B)^+} \rho_{A \cap B}(x) dx \qquad (23.21)$$

$$P(e_{A \vee B}) = \int_{\text{core}(A)^+} \rho_A(x) dx + \int_{\text{core}(B)^+} \rho_B(x) dx - \int_{\text{core}(A \cap B)^+} \rho_{A \cap B}(x) dx \qquad (23.22)$$

$$P(e_B | e_A) = \int_{\text{core}(A \cap B)^+} \rho_{A \cap B}(x) dx / \int_{\text{core}(A)^+} \rho(x) dx \qquad (23.23)$$

here $\rho(*)$ is the density function of X.

Since $\neg e_A = e_{\neg A}$, $e_A \wedge e_B = e_{A \wedge B}$, $e_A \vee e_B = e_{A \vee B}$, and $e_A \to e_B = e_B|e_A$; thus, we obtain a group of probability computation formulas of compound flexible events $\neg e_A$, $e_A \wedge e_B$, $e_A \vee e_B$ and $e_A \to e_B$.

Generally, let A_1, A_2, ..., A_n be flexible linguistic values on one-dimensional measurement space U, X be a random variable on U, and $e_{A_i}: X \in A_i$ ($i = 1, 2, ..., n$) be a flexible event on U, then

(1) When X is uniform distribution,

$$
\begin{aligned}
P(e_{A_1} \wedge e_{A_2} \wedge \cdots \wedge e_{A_n}) &= \frac{\left|\text{core}(A_1 \cap A_2 \cap \cdots \cap A_n)^+\right|}{|U|} \\
&= \frac{\left|\text{core}(A_1)^+\right|}{|U|} \frac{\left|\text{core}(A_1 \cap A_2)^+\right|}{\left|\text{core}(A_1)^+\right|} \cdots \frac{\left|\text{core}(A_1 \cap A_2 \cap \cdots \cap A_n)^+\right|}{\left|\text{core}(A_1 \cap A_2 \cap \cdots \cap A_{n-1})\right|} \\
&= P(e_{A_1})P(e_{A_2}|e_{A_1})\ldots P(e_{A_n}|e_{A_1} \wedge e_{A_2} \wedge \cdots \wedge e_{A_{n-1}})
\end{aligned}
$$

(23.24)

(2) When X is non-uniform distribution and discrete,

$$
\begin{aligned}
P(e_{A_1} \wedge e_{A_2} \wedge \cdots \wedge e_{A_n}) &= \sum_{x_k \in \text{core}(A_1 \cap A_2 \cap \cdots \cap A_n)^+} P_i \\
&= \sum_{x_k \in \text{core}(A_1)^+} P_i \frac{\sum_{x_i \in \text{core}(A_1 \cap A_2)^+} P_i}{\sum_{x_x \in \text{core}(A_1)^+} P_i} \cdots \frac{\sum_{x_i \in \text{core}(A_1 \cap A_2 \cap \cdots \cap A_n)} P_i}{\sum_{x_i \in \text{core}(A_1 \cap A_2 \cap \cdots \cap A_{n-1})} P_i} \\
&= P(e_{A_1})P(e_{A_2}|e_{A_1})\ldots P(e_{A_n}|e_{A_1} \wedge e_{A_2} \wedge \cdots \wedge e_{A_{n-1}})
\end{aligned}
$$

(23.25)

(3) When X is non-uniform distribution and continuous,

$$
\begin{aligned}
P(e_{A_1} \wedge e_{A_2} \wedge \cdots \wedge e_{A_n}) &= \int_{\text{core}(A_1 \cap A_2 \cap \cdots \cap A_n)^+} \rho(x)dx \\
&= \int_{\text{core}(A_1)^+} \rho(x)dx \frac{\int_{\text{core}(A_1 \cap A_2)^+} \rho(x)dx}{\int_{\text{core}(A_1)^+} \rho(x)dx} \cdots \frac{\int_{\text{core}(A_1 \cap A_2 \cap \cdots \cap A_n)^+} \rho(x)dx}{\int_{\text{core}(A_1 \cap A_2 \cdots \cap A_{n-1})^+} \rho(x)dx} \\
&= P(e_{A_1})P(e_{A_2}|e_{A_1})\ldots P(e_{A_n}|e_{A_1} \wedge e_{A_2}) \wedge \cdots \wedge e_{A_{n-1}})
\end{aligned}
$$

(23.26)

(4) For any X, always

$$
\begin{aligned}
P(e_{A_1} \vee e_{A_2} \vee \cdots \vee e_{A_n}) &= \sum_{i=1}^{n} P(e_{A_i}) - \sum_{1 \le i < j \le n} P(e_{A_i} \wedge e_{A_j}) \\
&+ \sum_{1 \le i < j < k \le n} P(e_{A_i} \wedge e_{A_j} \wedge e_{A_k}) - \cdots + (-1)^{n+1} P(e_{A_1}) \wedge e_{A_2} \wedge \cdots \wedge e_{A_n})
\end{aligned}
$$

(23.27)

$$P(e_B | e_{A_1} \wedge e_{A_2} \wedge \cdots \wedge e_{A_n})$$
$$= \frac{P(e_{A_1} \wedge e_{A_2} \wedge \cdots \wedge e_{A_n} \wedge e_B)}{P(e_{A_1} \wedge e_{A_2} \wedge \cdots \wedge e_{A_n})}, \quad P(e_{A_1} \wedge e_{A_2} \wedge \cdots \wedge e_{A_n}) \neq 0 \quad (23.28)$$

$$P(e_B | e_{A_1} \vee e_{A_2} \vee \cdots \vee e_{A_n})$$
$$= \frac{P((e_{A_1} \vee e_{A_2} \vee \cdots \vee e_{A_n}) \wedge e_B)}{P(e_{A_1} \vee e_{A_2} \vee \cdots \vee e_{A_n})}, \quad P(e_{A_1} \vee e_{A_2} \vee \cdots \vee e_{A_n}) \neq 0$$
$$(23.29)$$

In particular, when $A_1, A_2, \ldots, A_n$ are a group of complementary basic flexible linguistic values on U, flexible events $e_{A_1}, e_{A_2}, \ldots, e_{A_n}$ form **a group of complementary basic flexible events** on U.

Since

$$\mathrm{core}(A_1)^+ \cap \mathrm{core}(A_2)^+ \cap \cdots \cap \mathrm{core}(A_n)^+ = \varnothing$$

and in the situation that medium point $m_{A_i} (i = 1, 2, \ldots, n)$ is overlooked,

$$\mathrm{core}(A_1)^+ \cup \mathrm{core}(A_2)^+ \cup \cdots \cup \mathrm{core}(A_n)^+ = U$$

Thus,

$$P(e_{A_1} \wedge e_{A_2} \wedge \cdots \wedge e_{A_n}) = 0 \qquad (23.30)$$

$$P(e_{A_1} \vee e_{A_2} \vee \cdots \vee e_{A_n}) = P(e_{A_1}) + P(e_{A_2}) + \cdots + P(e_{A_n}) = 1 \qquad (23.31)$$

That is to say, flexible events $e_{A_1}, e_{A_2}, \ldots, e_{A_n}$ are really also a group of exclusive events.

From this, it follows that for any $A \subset U$,

$$P(e_A) \wedge P(e_{\neg A}) = 0 \qquad (23.32)$$

$$P(e_A) \vee P(e_{\neg A}) = P(e_A) + P(e_{\neg A}) = 1 \qquad (23.33)$$

Thus,

$$P(e_{\neg A}) = 1 - P(e_A) \qquad (23.34)$$

That is to say, flexible events e_A and $e_{\neg A}$ are conceptually complementary and are practically exclusive.

It is not hard to see that the above analysis and its result are also applicable to the flexible events on multidimensional space U.

Lastly, about the independence of flexible events on the same space, we have the following analysis and result.

Let A be a flexible linguistic value on measurement space U, X and Y be separately the random variables on U, and $e_A\colon X \in A$ and $e'_A\colon Y \in A$ be two flexible events on U. It can be seen that though the events e_A and e'_A are both with flexible linguistic value A, sine X and Y are two random variables; there is actually no dependent relation between the two events. Therefore,

$$P(e_A \wedge e'_A) = P(e_A)(e'_A)$$

Thus, flexible events e_A and e'_A are mutually independent.

Generally, **all flexible events that are with one and the same flexible linguistic value, but belong to different random variables, are all mutually independent**.

23.3 Probabilities of a Flexible Event with a Composite Linguistic Value and a Compound Flexible Event from Distinct Spaces

The above compound flexible events are all the events described by atomic linguistic values and the compound flexible events on the same space. In this section, we consider the flexible event described by a composite linguistic value and the compound flexible event made up of flexible events on distinct measurement spaces and their probabilities.

1. A flexible event with a combined linguistic value and its probability

Definition 23.1 Let A_1, A_2,..., A_n be flexible linguistic values on n-dimensional measurement space $U = U_1 \times U_2 \times \cdots \times U_n$, which denote separately mutually orthogonal flexible sets A_1, A_2, ..., A_n in U, and let $(X_1, X_2, ..., X_n)$ be a random vector (that is, n-dimensional random variable). We call the flexible event $e_{A_1 \wedge A_2 \wedge \cdots \wedge A_n}\colon (X_1, X_2, ..., X_n) \in A_1 \cap A_2 \cap \cdots \cap A_n$ and $e_{A_1 \vee A_2 \vee \cdots \vee A_n}\colon (X_1, X_2, ..., X_n) \in A_1 \cup A_2 \cup \cdots \cup A_n$ to be the flexible events with a combined linguistic value.

Example 23.4 Let $U = V = W = \{0, 1, 2, ..., 999, 1000\}$ be separately the ranges of scores of people's three attributions of facial features, facial shape, and skin, A be the flexible linguistic value: (facial features) "regular" on U, B be the flexible linguistic value: (facial shape) " proper" on V, and C be the flexible linguistic value: (skin) "bright and clean" on W, and let X, Y, and Z be separately random variables that represent the scores of facial features, facial shape, and skin. Then, the flexible events $e_{A \wedge B \wedge C}\colon (X, Y, Z) \in A \times V \times W \cap U \times B \times W \cap U \times V \times C$ (which means that the person has "regular (facial features)" and "proper (facial shape)" and "bright and clean (skin)") and $e_{A \vee B \vee C}\colon (X, Y, Z) \in A \times V \times W \cup U \times B \times W \cup U \times V \times C$ (which means that the person has "regular (facial features)" or "proper (facial shape)" or "bright and clean (skin)" are two flexible events with combined linguistic values, which separately take a photo of a person randomly from the photo database with index file $U \times V \times W$ who has "regular (facial features) and proper (facial shape) and bright and clean (skin)", and who has "regular (facial features) or proper (facial shape) or bright and clean (skin))". If we define "beautiful" = "regular (facial features)" and "proper (facial shape)" and "bright and clean (skin)," then $e_{\text{beautiful}} = e_{A \wedge B \wedge C}$; if we define "beautiful" = "regular (facial features)" or "proper (facial shape)" or "bright and clean (skin)," then $e_{\text{beautiful}} = e_{A \vee B \vee C}$.

Next, we analyze the probability computation of flexible events with a combined linguistic value.

Let A and B be two flexible linguistic values on measurement space $U \times V$, which denote separately flexible sets A and B on $U \times V$, and let flexible events $e_{A \wedge B}$: $(X, Y) \in A \cap B$ and $e_{A \vee B}$: $(X, Y) \in A \cup B$ be two flexible events with two-dimensional combined linguistic values. Then, by the definition of (practical) probability of a flexible event:

(1) When (X, Y) is uniform distribution,

$$
\begin{aligned}
P(e_{A \wedge B}) &= \frac{\left|\text{core}(A \cap B)^+\right|}{|U \times V|} \\
&= \frac{\left|\text{core}(A)^+ \times \text{core}(B)^+\right|}{|U \times V|} = \frac{\left|\text{core}(A)^+\right|\left|\text{core}(B)^+\right|}{|U||V|} \\
&= \frac{\left|\text{core}(A)^+\right|}{|U|} \frac{\left|\text{core}(B)^+\right|}{|V|} = P(e_A)P(e_B)
\end{aligned} \tag{23.35}
$$

$$
\begin{aligned}
P(e_{A \vee B}) &= \frac{\left|\text{core}(A \cup B)^+\right|}{|U \times V|} \\
&= \frac{\left|\text{core}(A)^+\right|}{|U \times V|} + \frac{\left|\text{core}(B)^+\right|}{|U \times V|} - \frac{\left|\text{core}(A \cap B)^+\right|}{|U \times V|} \\
&= \frac{\left|\text{core}(A)^+\right|}{|U|} + \frac{\left|\text{core}(B)^+\right|}{|V|} - \frac{\left|\text{core}(A)^+\right|}{|U|} \frac{\left|\text{core}(B)^+\right|}{|V|} \\
&= P(e_A) + (e_B) - P(e_A)(e_B)
\end{aligned} \tag{23.36}
$$

(2) When (X, Y) is non-uniform distribution,

① For discrete variable (X, Y),

$$
P(e_{A \vee B}) = \sum_{(x_k, y_j) \in \text{core}(A \times V \cap U \times B)^+} P_{ij} = \sum_{(x_k, y_j) \in \text{core}(A)^+ \times \text{core}(B)^+} P_{ij} \tag{23.37}
$$

$$P(e_{A\lor B}) = \sum_{(x_k,y_j)\in core(A\times V\cup U\times B)^+} P_{ij}$$

$$= \sum_{(x_k,y_j)\in core(A\times V)^+} P_{ij} + \sum_{(x_k,y_j)\in core(U\times B)^+} P_{ij} - \sum_{(x_k,y_j)\in core(A\times V\cap U\times B)^+} P_{ij}$$

$$= \sum_{x_k\in core(A)^+} P_i + \sum_{y_j\in core(B)^+} P_j - \sum_{(x_k,y_j)\in core(A\times V\cap U\times B)^+} P_{ij}$$

$$= P(e_A) + P(e_B) - P(e_{A\land B})$$

$$(23.38)$$

here $P_{ij} \in \{P_{kl}\}$, $\{P_{kl}\}$ is the joint distribution sequence of (X, Y); $P_i \in \{P_k\}$, $P_j \in \{P_l\}$, $\{P_k\}$, and $\{P_l\}$ are separately the distribution sequence of X and Y.

Then, as long as random variables X and Y are mutually independent, there would occur

$$\sum_{(x_k,y_j)\in core(A)^+\times core(B)^+} P_{ij} = \sum_{x_k\in core(A)^+} P_i \sum_{y_j\in core(B)^+} P_j$$

thus

$$P(e_{A\land B}) = \sum_{x_k\in core(A)^+} P_i \sum_{y_j\in core(B)^+} P_j = P(e_A)P(e_B) \qquad (23.39)$$

$$P(e_{A\lor B}) = \sum_{x_k\in core(A)^+} P_i + \sum_{y_j\in core(B)^+} P_j - \sum_{x_k\in core(A)^+} P_i \sum_{y_j\in core(B)^+} P_j$$

$$= P(e_A) + P(e_B) - P(e_A)P(e_B) \qquad (23.40)$$

② For continuous variable (X, Y),

$$P(e_{A\land B}) = \iint_{core(A\times V\cap U\times B)^+} \rho(x, y)\mathrm{d}x\mathrm{d}y = \iint_{core(A)^+\times core(B)^+} \rho(x, y)\mathrm{d}x\mathrm{d}y$$

$$(23.41)$$

$$P(e_{A\lor B}) = \iint_{core(A\times V\cup U\times B)^+} \rho(x, y)\mathrm{d}x\mathrm{d}y$$

$$= \iint_{core(A\times V)^+} \rho(x, y)\mathrm{d}x\mathrm{d}y + \iint_{core(U\times B)^+} \rho(x, y)\mathrm{d}x\mathrm{d}y - \iint_{core(A)^+\times core(B)^+} \rho(x, y)\mathrm{d}x\mathrm{d}y$$

$$(23.42)$$

where $\rho(x, y)$ is the joint density function of two-dimensional random variable (X, Y).

Then, as long as random variables X and Y are mutually independent, there would occur $\rho(x, y) = \rho_X(x) \cdot \rho_Y(y)$ ($\rho_X(x)$ and $\rho_Y(y)$ are separately the marginal density functions of random variables X and Y); thus,

$$
\begin{aligned}
P(e_{A \wedge B}) &= \iint_{\text{core}(A)^+ \times \text{core}(B)^+} \rho(x, y) \mathrm{d}x \mathrm{d}y = \iint_{\text{core}(A)^+ \times \text{core}(B)^+} \rho_X(x)\rho_Y(y)\mathrm{d}x\mathrm{d}y \\
&= \int_{\text{core}(A)^+} \rho_X(x)\mathrm{d}x \int_{\text{core}(B)^+} \rho_Y(y)\mathrm{d}y \\
&= P(e_A)P(e_B)
\end{aligned}
$$

$$(23.43)$$

$$
\begin{aligned}
P(e_{A \vee B}) &= \iint_{\text{core}(A \times V)^+} \rho(x, y)\mathrm{d}x\mathrm{d}y + \iint_{\text{core}(U \times B)^+} \rho(x, y)\mathrm{d}x\mathrm{d}y - \iint_{\text{core}(A)^+ \times \text{core}(B)^+} \rho(x, y)\mathrm{d}x\mathrm{d}y \\
&= \iint_{\text{core}(A \times V)^+} \rho_X(x)\rho_Y(y)\mathrm{d}x\mathrm{d}y + \iint_{\text{core}(U \times B)^+} \rho_X(x)\rho_Y(y)\mathrm{d}x\mathrm{d}y - \iint_{\text{core}(A)^+ \times \text{core}(B)^+} \rho_X(x)\rho_Y(y)\mathrm{d}x\mathrm{d}y \\
&= \int_{\text{core}(A)^+} \rho_X(x)\mathrm{d}x \int_V \rho_Y(y)\mathrm{d}y + \int_U \rho_X(x)\mathrm{d}x \int_{\text{core}(B)^+} \rho_Y(y)\mathrm{d}y - \int_{\text{core}(A)^+} \rho_X(x)\mathrm{d}x \int_{\text{core}(B)^+} \rho_Y(y)\mathrm{d}y \\
&= \int_{\text{core}(A)^+} \rho_X(x)\mathrm{d}x \cdot 1 + 1 \cdot \int_{\text{core}(B)^+} \rho_Y(y)\mathrm{d}y - \int_{\text{core}(A)^+} \rho_X(x)\mathrm{d}x \int_{\text{core}(B)^+} \rho_Y(y)\mathrm{d}y \\
&= \int_{\text{core}(A)^+} \rho_X(x)\mathrm{d}x + \int_{\text{core}(B)^+} \rho_Y(y)\mathrm{d}y - \int_{\text{core}(A)^+} \rho_X(x)\mathrm{d}x \int_{\text{core}(B)^+} \rho_Y(y)\mathrm{d}y \\
&= P(e_A) + P(e_B) - P(e_A)P(e_B)
\end{aligned}
$$

$$(23.44)$$

Thus, we obtain a group of the probability computation formulas of flexible events with combined linguistic values, $e_{A \wedge B}$ and $e_{A \vee B}$. Obviously, these formulas can also be generalized to flexible events with n-dimensional combined linguistic values, $e_{A_1 \wedge A_2 \wedge \cdots \wedge A_n}$ and $e_{A_1 \vee A_2 \vee \cdots \vee A_n}$. In particular, when random variables $X_1, X_2, \ldots, X_n$ are all uniform distribution and mutually independent,

$$P(e_{A_1 \wedge A_2 \wedge \cdots \wedge A_n}) = P(e_{A_1})P(e_{A_2}) \ldots P(e_{A_n}) \qquad (23.45)$$

$$
P(e_{A_1 \vee A_2 \vee \cdots \vee A_n}) = \sum_{i=1}^{n} P(e_{A_i}) - \sum_{1 \le i < j \le n} P(e_{A_i})P(e_{A_j}) + \sum_{1 \le i < j < k \le n} P(e_{A_i})P(e_{A_j})P(e_{A_k}) \\
- \cdots + (-1)^{n+1} P(e_{A_1})P(e_{A_2}) \ldots P(e_{A_n})
$$

$$(23.46)$$

2. Probability of a compound flexible event from distinct spaces

Here, the compound flexible event from distinct spaces refers to the flexible event made up of flexible events that are from distinct measurement spaces connected by logic relations $\wedge$ (and) or $\vee$ (or).

Example 23.5 Let $U = [a, b]$ and $V = [c, b]$ be separately the variation ranges of the temperature of the earth's surface and the wind power, A be a flexible linguistic value on U: (temperature) "high," and B be a flexible linguistic value on V: (wind) "strong," and let X and Y be separately random variables that represent temperature and wind power, and flexible events e_A: $X \in A$ and e_B: $Y \in B$ be separately "the temperature is high tomorrow" and "the wind is strong tomorrow." Then compound flexible events $e_A \wedge e_B$: $X \in A \wedge Y \in B$ and $e_A \vee e_B$: $X \in A \vee Y \in B$ are two compound flexible events from distinct spaces.

Next, we consider the probability of a compound flexible event from distinct spaces.

Let A and B be separately flexible linguistic values on measurement spaces U and V, X and Y be separately random variables on U and V, and e_A: $X \in A$ and e_B: $Y \in B$ be separately flexible events on the measurement space spaces U and V.

It can be seen that if extending A and B into the flexible linguistic values that denote separately sets $A \times V$ and $U \times B$, then the compound flexible event from distinct spaces, $e_A \wedge e_B$: $X \in A \wedge Y \in B$, is tantamount to the flexible event with combined linguistic value, $e_{A \wedge B}$: $(X, Y) \in A \times V \cap U \times B$, and $e_A \vee e_B$: $X \in A \vee Y \in B$ is tantamount to the flexible event with combined linguistic value, $e_{A \vee B}$: $(X, Y) \in A \times V \cup U \times B$. Namely, $e_A \wedge e_B = e_{A \wedge B}$, $e_A \vee e_B = e_{A \vee B}$. Thus, the probability computation formulas of flexible events with combined linguistic values, $e_{A \wedge B}$ and $e_{A \vee B}$, in the above are also the probability computation formulas of the compound flexible events from distinct spaces, $e_A \wedge e_B$ and $e_A \vee e_B$; that is,

$$P(e_A \wedge e_B) = P(e_{A \wedge B}) \tag{23.47}$$

$$P(e_A \vee e_B) = P(e_{A \vee B}) = P(e_A) + P(e_B) - P(e_{A \vee B}) \tag{23.48}$$

while when random variables X and Y are both uniform distribution and mutually independent,

$$P(e_A \wedge e_B) = P(e_A)P(e_B) \tag{23.49}$$

$$P(e_A \vee e_B) = P(e_A) + P(e_B) - P(e_A)P(e_B) \tag{23.50}$$

More generally,

$$P(e_{A_1} \wedge e_{A_2} \wedge \cdots \wedge e_{A_n}) = P(e_{A_1 \wedge A_2 \wedge \cdots \wedge A_n}) \tag{23.51}$$

$$P(e_{A_1} \vee e_{A_2} \vee \cdots \vee e_{A_n}) = P(e_{A_1 \vee A_2 \vee \cdots \vee A_n}) = \sum_{i=1}^{n} P(e_{A_i}) - P(e_{A_1 \wedge A_2 \wedge \cdots \wedge A_n}) \tag{23.52}$$

while when random variables X_1, X_2, ..., X_n are all uniform distribution and mutually independent,

$$P(e_{A_1} \wedge e_{A_2} \wedge \cdots \wedge e_{A_n}) = P(e_{A_1})P(e_{A_2})...P(e_{A_n}) \qquad (23.53)$$

$$P(e_{A_1} \vee e_{A_2} \vee)... \vee e_{A_n}$$

$$= \sum_{i=1}^{n} P(e_{A_i}) - \sum_{1 < i < j \leq n} P(e_{A_i})P(e_{A_j}) + \sum_{1 \leq i < j < k \leq n} P(e_{A_i})P(e_{A_j})P(e_{A_k})$$

$$- \cdots + (-1)^{n-1} P(e_{A_1})P(e_{A_2})...P(e_{A_n})$$

$$(23.54)$$

3. A flexible event with a synthetic linguistic value and its probability

Definition 23.2 Let A_1, A_2, ..., A_n be the flexible linguistic values on the n-dimensional measurement space $U = U_1 \times U_2 \times \cdots \times U_n$, which denote separately mutually orthogonal flexible sets $A_1, A_2, ..., A_n$ in U, and $(X_1, X_2, ..., X_n)$ be a random vector (that is, an n-dimensional random variable). We call the flexible event $e_{A_1 \oplus A_2 \oplus \cdots \oplus A_N}$: $(X_1, X_2, ..., X_n) \in A_1 \times A_2 \times \cdots \times A_n$ to be a flexible event with a synthetic linguistic value.

Example 23.6 Based on the flexible linguistic values A, B, and C in the Example 23.4, we define the flexible event with a synthetic linguistic value, $e_{A \oplus B \oplus C}$: $(X, Y, Z) \in A \times B \times C$, whose meaning is that this person has "regular (facial features)" plus "proper (facial shape)" plus "bright and clean (skin)," which takes randomly a photograph of a person from the photograph database with $U \times V \times W$ as the index file while this person has "regular (facial features)" plus "proper (facial shape)" plus "bright and clean (skin)." If we define "beautiful" = "regular (facial features)" $\oplus$ "proper (facial shape)" $\oplus$ "bright and clean (skin)," then $e_{\text{beautiful}} = e_{A \oplus B \oplus C}$.

Since the extended core $core(A_1 \times A_2 \times \cdots \times A_n)^+$ of flexible Cartesian product $A_1 \times A_2 \times \cdots \times A_n$ cannot be represented in a general expression of operations, the probability of flexible event with a synthetic linguistic value, $e_{A_1 \oplus A_2 \oplus \cdots \oplus A_N}$, also has no a common computation formula, but it needs specific analysis for specific problems. Therefore, in the following we consider the conceptual probability of a flexible event with a synthetic linguistic value, and for its practical probability we also make an analysis and estimate.

(1) Let random variable $(X_1, X_2, ..., X_n)$ is uniform distribution. By the definition of the conceptual probability of a flexible event (see Eqs. 23.1 and 23.2):

When $X_1, X_2, \ldots, X_n$ are discrete variables,

$$
\begin{aligned}
P_{\text{concept}}(e_{A_1 \oplus A_2 \oplus \cdots \oplus A_N}) &= \frac{|A_1 \times A_2 \times \cdots \times A_n|}{|\Omega_1 \times \Omega_2 \times \cdots \times \Omega_n|} \\
&= \frac{\displaystyle\sum_{(x_1, x_2, \ldots, x_n) \in \operatorname{supp}(A_1 \times A_2 \times \cdots \times A_n)} m_{A_1 \times A_2 \times \cdots \times A_n}(x_1, x_2, \ldots, x_n)}{|\Omega_1 \times \Omega_2 \times \cdots \times \Omega_n|}
\end{aligned}
\tag{23.55}
$$

When $X_1, X_2, \ldots, X_n$ are continuous variables,

$$
\begin{aligned}
P_{\text{concept}}(e_{A_1 \oplus A_2 \oplus \cdots \oplus A_N}) &= \frac{|A_1 \times A_2 \times \cdots \times A_n|}{|U_1 \times U_2 \times \cdots \times U_n|} \\
&= \frac{\int_{\operatorname{supp}(A_1 \times A_2 \times \cdots \times A_n)} m_{A_1 \times A_2 \times \cdots \times A_n}(x_1, x_2, \ldots, x_n) d(x_1, x_2, \ldots, x_n)}{|U_1 \times U_2 \times \cdots \times U_n|} \\
&= \frac{\int_{\operatorname{supp}(A_1 \times A_2 \times \cdots \times A_n)} \sum_{i=1}^{n} w_i m_{A_i}(x_i) dx_1 dx_2, \ldots, dx_n}{|U_1 \times U_2 \times \cdots \times U_n|} \\
&= \frac{\sum_{i=1}^{n} w_i \int_{\operatorname{supp}(A_i)} m_{A_i}(x_i) dx_i}{|U_1 \times U_2 \times \cdots \times U_n|}
\end{aligned}
\tag{23.56}
$$

(2) Let random variable $(X_1, X_2, \ldots, X_n)$ is non-uniform distribution. By the definition of the conceptual probability of the corresponding flexible event (see Eq. 23.3), when $X_1, X_2, \ldots, X_n$ are continuous variables,

$$
\begin{aligned}
P_{\text{concept}}(e_{A_1 \oplus A_2 \oplus \cdots \oplus A_N}) &= \int_{\operatorname{supp}(A_1 \times A_2 \times \cdots \times A_n)} \rho(x_1, x_2, \ldots, x_n) m_{A_1 \times A_2 \times \cdots \times A_n}(x_1, x_2, \ldots, x_n) d(x_1, x_2, \ldots, x_n) \\
&= \int_{\operatorname{supp}(A_1 \times A_2 \times \cdots \times A_n)} \rho(x_1, x_2, \ldots, x_n) \sum_{i=1}^{n} w_i m_{A_i}(x_i) d(x_1, x_2, \ldots, x_n) \\
&= \sum_{i=1}^{n} w_i \int_{\operatorname{supp}(A_i)} \rho(x_i) m_{A_i}(x_i) dx_i \\
&= \sum_{i=1}^{n} w_i P_{\text{concept}}(e_{A_i})
\end{aligned}
\tag{23.57}
$$

But it is a pity that when $X_1, X_2, \ldots, X_n$ are discrete variables, the corresponding probability $P_{\text{concept}}(e_{A_1 \oplus A_2 \oplus \cdots \oplus A_N})$ cannot be computed.

Next we analyze the practical probability of flexible event $e_{A_1 \oplus A_2 \oplus \cdots \oplus A_N}$. When random variables $X_1, X_2, \ldots, X_n$ are all uniform distribution,

$$
P(e_{A_1 \oplus A_2 \oplus \cdots \oplus A_N}) = \frac{|\text{core}(A_1 \times A_2 \times \cdots \times A_n)^+|}{|U_1 \times U_2 \times \cdots \times U_n|}
\tag{23.58}
$$

From the stated above, $\text{core}(A_1 \times A_2 \times \cdots \times A_n)^+$ cannot be expanded. But since the points in $\text{core}(A_1 \times A_2 \times \cdots \times A_n)^+$ are always more than points in core $(A_1)^+ \times \text{core}(A_1)^+ \times \cdots \times \text{core}(A_n)^+$, it follows that

$$\left|\text{core}(A_1 \times A_2 \times \cdots A_n)^+\right| > \left|\text{core}(A_1)^+ \times \text{core}(A_2)^+ \times \cdots \times \text{core}(A_n)^+\right|$$

$$\frac{\left|\text{core}(A_1 \times A_2 \times \cdots \times A_n)^+\right|}{\left|U_1 \times U_2 \times \cdots \times U_n\right|} > \frac{\left|\text{core}(A_1)^+ \times \text{core}(A_2)^+ \times \cdots \times \text{core}(A_n)^+\right|}{\left|U_1 \times U_2 \times \cdots \times U_n\right|}$$

while

$$\frac{\left|\text{core}(A_1)^+ \times \text{core}(A_2)^+ \times \cdots \times \text{core}(A_n)^+\right|}{\left|U_1 \times U_2 \times \cdots \times U_n\right|} = \frac{\left|\text{core}(A_1)^+\right|}{\left|U_1\right|} \frac{\left|\text{core}(A_2)^+\right|}{\left|U_2\right|} \cdots \frac{\left|\text{core}(A_n)^+\right|}{\left|U_n\right|}$$

$$= P(e_{A_1})P(e_{A_2})\dots P(e_{A_n})$$

$$= P(e_{A_1 \wedge A_2 \wedge \cdots \wedge A_n})$$

Therefore,

$$P(e_{A_1 \oplus A_2 \oplus \cdots \oplus A_N}) > P(e_{A_1 \wedge A_2 \wedge \cdots \wedge A_n}) \qquad (23.59)$$

When random variables $X_1, X_2, \ldots, X_n$ are all non-uniform distribution,

$$P(e_{A_1 \oplus A_2 \oplus \cdots \oplus A_N}) = \iint_{\text{core}(A_1 \times A_2 \times \cdots \times A_n)^+} \cdots \int \rho(x_1, x_2, \ldots, x_n) dx_1 dx_2 \ldots dx_n$$

$$(23.60)$$

Likewise, this expression also cannot be further expanded, but we can know

$$\iint_{\text{core}(A_1 \times A_2 \times \cdots \times A_n)^+} \cdots \int \rho(x_1, x_2, \ldots, x_n) dx_1 dx_2 \ldots dx_n > \int_{\text{core}(A_1)^+} \rho_{X_1}(x_1) \int_{\text{core}(A_2)^+} \rho_{X_2}(x_2) \cdots \int_{\text{core}(A_n)^+} \rho_{X_n}(x_n)$$

while

$$\int_{\text{core}(A_1)^+} \rho_{X_1}(x_1) \int_{\text{core}(A_2)^+} \rho_{X_2}(x_2) \cdots \int_{\text{core}(A_n)^+} \rho_{X_n}(x_n) = P(e_{A_1})P(e_{A_2})\dots P(e_{A_n})$$

$$= P(e_{A_1 \wedge A_2 \wedge \cdots \wedge A_n})$$

Thus, also

$$P(e_{A_1 \oplus A_2 \oplus \cdots \oplus A_N}) > P(e_{A_1 \wedge A_2 \wedge \cdots \wedge A_n})$$

This expression is tantamount to

$$P(e_{A_1 \wedge A_2 \wedge \cdots \wedge A_n}) < P(e_{A_1 \oplus A_2 \oplus \cdots \oplus A_N})$$

Further, it is not hard to see that

$$P(e_{A_1 \oplus A_2 \oplus \cdots \oplus A_N}) < P(e_{A_1 \vee A_2 \vee \cdots \vee A_n})$$

As thus, to sum up, we have

$$P(e_{A_1 \wedge A_2 \wedge \cdots \wedge A_n}) < P(e_{A_1 \oplus A_2 \oplus \cdots \oplus A_N}) < P(e_{A_1 \vee A_2 \vee \cdots \vee A_n}) \qquad (23.61)$$

Thus, generally speaking, we can also take

$$\frac{P(e_{A_1 \wedge A_2 \wedge \cdots \wedge A_n}) + P(e_{A_1 \vee A_2 \vee \cdots \vee A_n})}{2}$$

as an approximate value of $P(e_{A_1 \oplus A_2 \oplus \cdots \oplus A_N})$.

4. **Probability of a conditional flexible event from distinct spaces**

Let $e_A : X \in A$ and $e_B : Y \in B$ be separately a flexible event on measurement spaces U and V. We consider the probability of implicational flexible event $e_A \rightarrow e_B$: $X \in A \rightarrow Y \in B$, that is, the probability of conditional flexible event $e_B | e_A : X \in A |$ $Y \in B$.

(1) Random variables X and Y are uniform distribution

From Definition 20.13′, the correspondence-rate of flexible sets A to B: $\text{corr}(A, B) = \frac{|A_1|}{|\text{core}(A)^+|}$, where $A_1 = \{x | (x, y) \in R\}$ and $R \subseteq \text{core}(A)^+ \times \text{core}(B)^+$. From this, it can be seen that in the case of $A_1 \neq \varnothing$, that is, random variables X and Y are not mutually independent, when random variable $X \in A_1 \subset U$, the corresponding random variable $Y \in V$. That is, when event e_A occurs, event e_B also occurs at the same time. Let $e_{A_1} : X \in A_1$. Then $P(e_{A_1}) = \frac{|A_1|}{|U|} = P(e_A \wedge e_B)$. Thus,

$$\frac{|A_1|}{|\text{core}(A)^+|} = \frac{|A_1|/|U|}{|\text{core}(A)^+|/|U|} = \frac{P(e_A \wedge e_B)}{P(e_A)} = P(e_B | e_A)$$

Conversely, it is

$$P(e_B | e_A) = \frac{|A_1|}{|\text{core}(A)^+|} = \text{corr}(A, B) \qquad (23.62)$$

(2)   The random variables X and Y are non-uniform distribution

   ① For the discrete random variables X and Y, when X and Y are not mutually independent,

$$P(e_B|e_A) = \sum_{x_k \in A_1} P_i \bigg/ \sum_{x_k \in \text{core}(A)^+} P_i \qquad (23.63)$$

where $A_1 = \{x|(x, y) \in R\} \neq \varnothing, R \subseteq \text{core}(A)^+ \times \text{core}(B)^+, P_i \in \{P_k\}, \{P_k\}$ is the distribution sequence of X.

   ② For the continuous random variables X and Y, when X and Y are not mutually independent,

$$P(e_B|e_A) = \int_{A_1} \rho(x)dx \bigg/ \int_{\text{core}(A)^+} \rho(x)dx \qquad (23.64)$$

where $A_1 = \{x|(x, y) \in R\} \neq \varnothing, R \subseteq \text{core}(A)^+ \times \text{core}(B)^+$, $\rho(x)$ is the density function of X.

From the above stated, more generally, when random vector $(X_1, X_2, \ldots, X_n)$ and random variable Y are not mutually independent,

$$P(e_B|e_{A_1} \wedge e_{A_2} \wedge \cdots \wedge e_{A_n}) = \begin{cases} \dfrac{|A_b|}{\left|\text{core}(A_1 \cap A_2 \cap \cdots \cap A_n)^+\right|} \\[2mm] \dfrac{\sum_{x \in A_b} P_x}{\sum_{x \in \text{core}(A_1 \cap A_2 \cap \cdots \cap A_n)} P_x} \\[2mm] \dfrac{\iint_{A_b} \cdots \int \rho(x_1, x_2, \ldots, x_n)dx_1 dx_2 \ldots dx_n}{\iint_{\text{core}(A_1 \cap A_2 \cap \cdots \cap A_n)^+} \cdots \int \rho(x_1, x_2, \ldots, x_n)dx_1 dx_2 \ldots dx_n} \end{cases} \qquad (23.65)$$

where $A_b = \{x \mid (x, y) \in R\} \neq \varnothing, R \subseteq \text{core}(A_1 \cap A_2 \cap \cdots \cap A_n)^+ \times \text{core}(B)^+$; A_i is the flexible set of corresponding flexible linguistic value A_i $(i = 1, 2, \ldots, n)$ in product space $U_1 \times U_2 \times \cdots \times U_n$; the three operational expressions correspond separately to the probability computations of conditional flexible events $e_B|e_{A_1} \wedge e_{A_2} \wedge \cdots \wedge e_{A_n}$ when random variables are separately uniform distribution, non-uniform discrete distribution and non-uniform continuous distribution. Besides,

$$P(e_B|e_{A_1} \vee e_{A_2} \vee \cdots \vee e_{A_n}) = \frac{P((e_{A_1} \vee e_{A_2} \vee \cdots \vee e_{A_n}) \wedge e_B)}{P(e_{A_1} \vee e_{A_2} \vee \cdots \vee e_{A_n})}, P(e_{A_1} \vee e_{A_2} \vee \cdots \vee e_{A_n}) \neq 0$$

$$(23.66)$$

$$P\left(e_B \mid e_{A_1 \wedge A_2 \wedge \cdots \wedge A_n}\right) = \frac{P\left(e_{A_1 \wedge A_2 \wedge \cdots \wedge A_n} \wedge e_B\right)}{P\left(e_{A_1 \wedge A_2 \wedge \cdots \wedge A_n}\right)}, P\left(e_{A_1 \wedge A_2 \wedge \cdots \wedge A_n}\right) \neq 0 \qquad (23.67)$$

$$P\left(e_B \mid e_{A_1 \vee A_2 \vee \cdots \vee A_n}\right) = \frac{P\left(e_{A_1 \vee A_2 \vee \cdots \vee A_n} \wedge e_B\right)}{P\left(e_{A_1 \vee A_2 \vee \cdots \vee A_n}\right)}, P\left(e_{A_1 \vee A_2 \vee \cdots \vee A_n}\right) \neq 0 \qquad (23.68)$$

$$P\left(e_B \mid e_{A_1 \oplus A_2 \oplus \cdots \oplus A_N}\right) = \frac{P\left(e_{A_1 \oplus A_2 \oplus \cdots \oplus A_n} \wedge e_B\right)}{P\left(e_{A_1 \oplus A_2 \oplus \cdots \oplus A_n}\right)}, P\left(e_{A_1 \oplus A_2 \oplus \cdots \oplus A_n}\right) \neq 0 \qquad (23.69)$$

Of course, Eqs. (23.62)–(23.69) only are the definitions of probability of a conditional flexible event from distinct spaces. Like the conditional probability in traditional probability theory, the probability of a conditional flexible event from distinct spaces also has corresponding Bayes' formula. Besides, on the basis of probability of a conditional flexible event from distinct spaces, there are the corresponding total probability formula and independence of flexible events, which is similar to the formulations in traditional probability theory, so it is unnecessary to go into details here.

5. A flexible event with opposite and its probability

The flexible events discussed in previous sections are actually all the flexible events described by flexible linguistic values with negation—which we may as well call it as the flexible event with negation; thus, the flexible event described by flexible linguistic values with opposite is the flexible event with opposite.

Similar to flexible events with negation, flexible events with opposite can also have multiple types such as basic event, compound event on the same space, and compound event from distinct spaces and so forth, and they also have the corresponding properties. Since a linguistic value with opposite is completely stood for by its support set, the probability of a flexible event with opposite is determined by its support set. As thus the extended cores in the previous probability computation formulas of flexible events with negation are only needed to be changed into the support sets, we can obtain the probability computation formulas of the corresponding flexible events with opposite, and these formulas are the computation formulas of the practical probabilities of flexible events with opposite as well as the conceptual probabilities of them. In other words, such as usual "rigid" events, the flexible events with opposite do not distinguish conceptual probability and actual probability.

23.4 Summary

In this chapter, we introduced the concept of random flexible events, analyzed the corresponding probability computation principles, and then established corresponding probability computation models for various types of flexible events.

This chapter has also the following results:

- Probability of a flexible event can be separated as conceptual probability and practical probability, which are separately the probability on the support set and extended core of the corresponding flexible set.
- Compound flexible events have the separation of same space and distinct space, and the former has no general probability computation formulas, while the latter has.
- Flexible events with a synthetic linguistic value are a kind of special flexible events with composite linguistic value from distinct spaces; for their practical probabilities, we generally can only give the estimated values.

Reference

1. Lian S (2009) Principles of imprecise-information processing. Science Press, Beijing

Chapter 24
Degrees of Believability of Flexible Propositions and Reasoning with Believability-Degrees

Abstract This chapter introduces a measure called believability-degree for the believability of propositions, and on the basis of the theories of traditional probability and flexible event's probability presents the computation principles and methods of the believability-degrees of compound propositions on the same space and from distinct spaces, in particular, the computation principles and methods of propositions with a composite linguistic value and implicational propositions, thus founding the believability-degree theory of propositions. Then, it discusses the inference on the believability of propositions and presents an uncertain reasoning scheme based on believability-degrees—reasoning with believability-degrees and the corresponding reasoning model. And then, it discusses the dual reasoning about uncertainty and imprecision for the believability and truth of propositions simultaneously and presents a scheme and model of dual reasoning. Besides, it also discusses the correlation between partial implication, uncertain implicational proposition, relational inference with implication-degrees, and reasoning with believability-degrees.

Keywords Believability-degrees · Reasoning with believability-degrees · Partial implication · Uncertain reasoning

Truth is a so important attribute of the propositions that almost all logics are connected to the truth of the propositions. Besides truth, the propositions also have another attribute—believability. The believability of propositions involves uncertainty, it also involves imprecision, and it is related to the truth of the propositions. In this chapter, we examine the measure and the reasoning about the believability of propositions (referring mainly to flexible propositions).

© Springer Science+Business Media Singapore 2016
S. Lian, *Principles of Imprecise-Information Processing*,
DOI 10.1007/978-981-10-1549-6_24

24.1 Believability and Degree of Believability of a Proposition

1. Believability—another attribute of propositions

Examine the following statements:

...... × ×'s words are unbelievable......

......has half believing to some hearsay

......doubt about this academic view......

It can be seen that here "words," "hearsay," and "view" all refer to propositions from a logical point of view, while "unbelievable," "half-believing," and "doubt" are all for the truth of the corresponding propositions. "Unbelievable" is that one cannot believe a person's words (a proposition) to be true, "half-believing" and "doubt" are both suspicions about the truth or correctness of some speech (a proposition). So, the believability is actually also an attribute of propositions (including flexible propositions).

Obviously, if a proposition being true is certain or necessary, then which is completely believable, and vice versa; if a proposition being true is impossible, then which is completely unbelievable, and vice versa; and if a proposition being true is possible, then which is believable with some degree, and vice versa. Yet "necessary" and "impossible" can be viewed as two extreme cases of "possible." Thus, the believability of a proposition can be said as the possibility of the proposition being true [1]. Of course, here, true includes rigid true and flexible true (that is, rough-true).

2. Linguistic expressions of believability

We know that the linguistic expressions of possibility have necessarily, probably/likely, improbably/unlikely, highly probably/likely, fairly probably/likely, not very probably/likely, etc. Then, the linguistic description of a proposition being true has just: necessarily (true), probably/likely (true), improbably/unlikely (true), highly probably/likely (true), fairly probably/likely (true), not very probably/likely (true), etc. Correspondingly, the linguistic expressions of the believability of propositions have: completely believable, believable with some degree, completely unbelievable, highly believable, fairly believable, not very believable, etc. In these words stated above, the necessarily, probably/likely, and improbably/unlikely are three rigid linguistic values that express possibility, and they are three basic linguistic values, while highly probably/likely, fairly probably/likely, not very probably/likely, etc., are flexible linguistic values. These two kinds of linguistic values are all linguistic values on the range [0, 1] of probabilities. Correspondingly, completely believable, believable with some degree, and completely unbelievable are three rigid linguistic values that express the believability, and they are three basic linguistic values, while highly believable, fairly believable, and not very believable are flexible linguistic values. These two kinds of linguistic values should be the linguistic values on the measurement space of the believability of propositions.

3. **Degree of believability—numerical value of the believability of a proposition**

Since believability is the possibility of a proposition being true, and the number that describe the magnitude of possibility is probability, the numerical value of the believability of a proposition is numerically equal to the probability of the proposition being true. Therefore, this numerical value can be called the degree of believability of a proposition.

Definition 24.1 We call the practical probability of an event described by a proposition p to be the probability of the proposition being true, denote $P(p)$. Set

$$c(p) = P(p) \tag{24.1}$$

to be called the degree of believability of proposition p, simply written believability-degree.

Example 24.1 Let proposition p: The heads of a coin is upward. Since the probability of heads being upward when tossing a coin is always 0.5, the probability of proposition p being true is 0.5, that is, $P(p) = 0.5$. Thus, $c(p) = 0.5$.

By the definition, if A is a quantifiable linguistic value, that is, a flexible linguistic value or a quantifiable rigid linguistic value, on measurement space U, then for the random variable x that takes values from U, the believability-degree of proposition $A(x)$ is

$$c(A(x)) = \frac{|A|}{|U|} = \text{shar}(A) \tag{24.2}$$

That means for a proposition with quantifiable linguistic value, its believability-degree is numerically equal to the share of the corresponding set.

Example 24.2 Let proposition q: Zhang is tall. In the situation of the height of Zhang being unknown, the height h of Zhang is just a random variable, so the truth value of proposition q cannot be determined. Thus,

$$c(q) = \frac{|H|}{|[a, b]|}$$

here, H is the flexible set to which the flexible linguistic value "tall" corresponds, and interval $[a, b]$ is the range of height of humans.

It needs to be noted that though we define the probability of a proposition being true as the believability-degree of the proposition, how to compute this probability is complex. It's because the truthfulness of a proposition is not only related to the proposition itself, but also related to the situations such as the background and context of the proposition. For instance, for the believability-degree of proposition

q in the above Example 24.2, except the given computation methods, there are at least the following cases and methods of calculation.

① If for all people in a certain scope, then

$$c(q) = \frac{s_1}{s_2}$$

where s_1 denotes the total of tall persons, and s_2 denotes the total of relevant peoples.

② If for a certain classification (e.g., three classes of tall, middle, and short), then

$$c(q) = \frac{1}{s}$$

where s denotes the total of classes.

③ If for the maker Pe of proposition q, then

$$c(q) = \text{the probability of Pe telling the truth}$$

That means $c(q)$ may be a kind of frequency or estimated value.

In a word, just like the usual probability, the degree of believability of a proposition is also not always absolute and objective, which can also be relative and subjective. Therefore, the computation of the believability-degree of a proposition should be determined by specific problem.

Since believability-degree is defined by probability, the problem to find a believability-degree is also reduced to the problem to find the corresponding probability, and the range of believability-degrees is the same as the range of probabilities, which is also real interval [0, 1]. Besides, the types, properties, formulas, theorems, and so forth of probabilities are also true for believability-degrees.

Comparing Eqs. (24.2) and (21.25), it can be seen that for a quantifiable linguistic value A, the believability-degree of simple proposition $A(x)$ is numerically equal to possessing-rate of A, that is,

$$c(A(x)) = \text{poss}(A) \tag{24.3}$$

and vice versa.

The believability of propositions is relative to the uncertainty of information, and believability-degree is portraying and measuring the strength of uncertainty of information.

4. Linguistic values on believability-degree range

Now, examining the relation between the numerical values of the believability, that is, believability-degrees, of propositions and the linguistic values previously given, it can be seen that completely believable, believable with some degree, and completely unbelievable are also three basic linguistic values on the range [0, 1] of

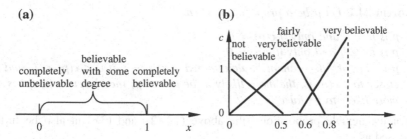

Fig. 24.1 a Rigid linguistic values on range of believability-degrees, **b** examples of flexible linguistic values on the range of believability-degrees

believability-degrees of propositions (as shown in Fig. 24.1a). And slightly believable, not very believable, and very believable are the flexible linguistic values on the range [0, 1] of believability-degrees (as shown by Fig. 24.1b).

If describing the believability of flexible propositions by using flexible linguistic values, then the uncertainty of imprecision is described by imprecision again. Here, imprecision and uncertainty cross two times: First is the uncertainty of imprecision, and second is the imprecision of uncertainty, which is just as that the probability of occurrence of a flexible event is described by a flexible linguistic value.

Examining the linguistic values on the range of believability-degrees and the linguistic values on the range of probabilities in Sect. 23.1.5, it can be seen that the two are just one to one correspondence, that is, completely believable ⟷ necessarily, believable with some degree ⟷ possibly, and completely unbelievable ⟷ not possibly. The correspondence relation between linguistic values is also the correspondence relation between the believability of a proposition and the possibility of it being true.

5. Relation between believability and truth of a proposition

A proposition has two attributes of truth and believability, and then, for one and the same proposition, what is the relation of its believability and truth?

From the definition of believability-degree, it can be seen that in ideal situation, if a proposition is true (also including rough-true and near-true, the same below), then its believability-degree is certainly 1; if it is false (also including rough-false and near-false, the same below), then its believability-degree is certainly 0; and if its truth value cannot be determined, then its believability-degree is certainly between 0 and 1. Conversely, if the believability-degree of a proposition is 1, then it is surely true; if its believability-degree is 0, then it is impossible to be true; if the believability-degree is between 0 and 1, then that it is true or false cannot be determined, but the closer to 1 the believability-degree is, the more likely true is the proposition, the closer to 0 the believability-degree is, the more likely false is the proposition.

To sum up the analysis above, the relation between the truth and believability of a proposition can be stated as the following theorem.

Theorem 24.1 *Let p be a proposition. Then,*

(1) *p is true if and only if c(p) = 1;*
(2) *p is false if and only if c(p) = 0;*
(3) *p is true or false cannot be determined if and only if 0 < c(p) < 1, and the closer to 1 c(p) is, the more likely to be true is p; the closer to 0 c(p) is, the more likely to be false is p.*

For the flexible propositions, the above (1), (2), and (3) can also be further expressed as

$$(1)\ t(p) > 0.5 \Leftrightarrow c(p) = 1 \tag{24.4}$$

$$(2)\ t(p) < 0.5 \Leftrightarrow c(p) = 0 \tag{24.5}$$

$$(3)\ t(p) > 0.5 \text{ or } t(p) < 0.5 \text{ can not be determined} \Leftrightarrow 0 < c(p) < 1 \tag{24.6}$$

The relation between the truth and believability of a flexible proposition described in linguistic values is as follows:

A proposition is true if and only if it is completely believable; a proposition is false if and only if it is completely unbelievable; and a proposition is true or false cannot be determined if and only if it is believable with some degree.

The theorem above reveals the relation between the truth and believability of a proposition, so we can use one of the two to obtain or estimate another. Since believability-degrees can always be obtained (when objective believability-degrees cannot be obtained, subjective believability-degrees can be given by experience), for a proposition whose truth value can not be determined, we can use its believability-degree to estimate its truth.

24.2 Believability-Degree of a Compound Flexible Proposition on the Same Space

We call the compound proposition in which the flexible linguistic values are on the same space to be a compound flexible proposition on the same space. A compound proposition on same space is also the proposition that describe a compound flexible event on the same space. Thus, the believability-degree of a compound flexible proposition on the same space is reduced to the probability of a flexible compound event on the same space.

24.2.1 Believability-Degrees of Basic Compound Flexible Propositions

Let propositions p: $A(x)$ and q: $B(y)$, and A and B be flexible linguistic values on measurement space U.

By the definition of the believability-degree of a proposition, the believability-degree of a compound proposition is also the probability of a compound proposition being true. Thus,

$$c(\neg p) = P(\neg p) = 1 - P(p) = 1 - c(p)$$
$$c(p \wedge q) = P(p \wedge q)$$
$$c(p \vee q) = P(p \vee q) = P(p) + P(q) - P(p \wedge q) = c(p) + c(q) - c(p \wedge q)$$

Thus, we have the computation formulas of the believability-degrees of the compound propositions as follows:

$$c(\neg p) = 1 - c(p) \tag{24.7}$$

$$c(p \wedge q) = P(p \wedge q) \tag{24.8}$$

$$c(p \vee q) = c(p) + c(q) - c(p \wedge q) \tag{24.9}$$

It can be seen that now the problem is reduced to how to find $P(p \wedge q)$?

From Sect. 23.2, we know that when p and q describe flexible events on one and the same measurement space, $P(p \wedge q)$ has no general computation formula about $P(p)$ and $P(q)$. Merely in the following special cases, $P(pq)$ can be definitely obtained. In fact,

If p and q are relatively negative, that is, $p = \neg q$, then by $P(p \wedge q) = P(\neg q \wedge q) = 0$, it follows that $c(p \wedge q) = 0$.

If $P(p) = 1$ and $P(q) = 1$, then by $P(p \wedge q) = 1$, it follows that $c(p \wedge q) = 1$.

If $P(p) = 0$ or $P(q) = 0$, then by $P(p \wedge q) = 0$, it follows that $c(p \wedge q) = 0$.

Consequently, we have the following theorem.

Theorem 24.2 *Let proposition p: A(x) and q: B(x), and A and B be linguistic values on the measurement space U.*

(1) *If p and q are relatively negative, then $c(p \wedge q) = 0$, $c(p \vee q) = c(p) + c(q)$;*
(2) *If $c(p) = 1$ and $c(q) = 1$, then $c(p \wedge q) = 1$;*
(3) *If $c(p) = 0$ or $c(q) = 0$, then $c(p \wedge q) = 0$.*

Besides, by $c(\neg p) = 1 - c(p)$, we have

$$c(p) + c(\neg p) = 1 \tag{24.10}$$

Equation (24.10) formulates the relation between the believability-degrees of relatively negative propositions, and we call them the complement law of believability-degrees.

From the complement law of believability-degrees, we have

$$\text{If } c(p) = 1, \text{ then } c(\neg p) = 0;$$
$$\text{If } c(p) = 0, \text{ then } c(\neg p) = 1;$$
$$\text{If } c(p) \in (0, 1), \text{ then } c(\neg p) \in (0, 1).$$

Next, we further examine the relation among the believability-degrees of multiple flexible propositions.

Definition 24.2 Let flexible proposition p_i: $A_i(x)$, $i = 1, 2, ..., n$. If $A_1, A_2, ..., A_n$ be a group of complementary basic flexible linguistic values on measurement space U; then, we say propositions $p_1, p_2, ..., p_n$ are a group of complementary flexible propositions on the space U.

It can be seen that when $n = 2$, the group of complementary flexible propositions is a pair of relatively negative flexible propositions. From the probabilities of compound flexible events, it is not hard to derive that if $p_1, p_2, ..., p_n$ are a group of complementary flexible propositions on the space U, then

$$c(p_1) + c(p_2) + \cdots + c(p_n) = 1 \tag{24.11}$$

So, the believability-degrees of complementary flexible propositions still obey the **complement law of believability-degrees**.

24.2.2 Believability-Degree of an Implicational Flexible Proposition

Let propositions p: $A(x)$ and q: $B(y)$, and A and B be flexible linguistic values on measurement space U.

By the definition of believability-degree of a proposition, the believability-degree $c(p \rightarrow q)$ of implicational flexible proposition $p \rightarrow q$ is also the probability of proposition $p \rightarrow q$ being true, that is, $c(p \rightarrow q) = P(p \rightarrow q)$. While $P(p \rightarrow q) = P(e_A \rightarrow e_B)$ and $P(e_A \rightarrow e_B) = P(e_B \mid e_A)$, $P(p \rightarrow q) = P(e_B \mid e_A)$. Thus,

$$c(p \rightarrow q) = P(e_B \mid e_A) \tag{24.12}$$

That is to say, the believability-degree of implicational flexible proposition $p \rightarrow q$ is also the corresponding conditional probability $P(e_B \mid e_A)$.

Since flexible linguistic values A and B belong to the same space U, x and y as random variables are uniform distribution; from Eq. (23.15), it follows that

$$c(p \rightarrow q) = \text{cont}(A, B) \tag{24.13}$$

That is to say, the believability-degree of implicational flexible proposition $p \rightarrow q$ made up of flexible propositions p and q on the same space equals numerically to the inclusion-degree between corresponding sets, and vice versa.

The deriving from cont(A, B) to $c(p \rightarrow q)$ is as follows:

$$\text{cont}(A, B) \frac{|A \cap B|}{|A|} = \frac{|A \cap B|/|U|}{|A|/|U|} = \frac{P(e_A \wedge e_B)}{P(e_A)} = P(e_B|e_A) = c(p \rightarrow q)$$

Also by Definition 21.9, it follows that

$$\text{impl}(p, q) = \text{cont}(A, B)$$

Thus, we have

$$c(p \rightarrow q) = \text{impl}(p, q) \tag{24.14}$$

That is to say, the believability-degrees of some implicational propositions on the same space equal numerically to the implication-degrees of their premises to conclusions, and vice versa.

It should be mentioned that there are also scholars in China who presented the concept of the implicational degree, but there, the implicational degree was treated as the truth values of implicational compound propositions, so it means differently from the implication-degree in this book. Besides, there is literature that discussed the degree of rule $A \rightarrow B$ itself being true through the concept of "the strength of a rule." However, it seems not clear enough whether this degree of being true refers to the believability-degree or truth-degree of an implicational compound proposition.

Next, we further discuss the relation between the implication relation between propositions and the magnitude relation between the corresponding believability-degrees.

Lemma 24.1 *Let $U = [a, b]$, $A_1, A_2, ..., A_n$ be flexible linguistic values on U, and q_i: x is A_i, $x \in U$, $i \in \{1, 2, ..., n\}$. Then for arbitrary $i, j \in \{1, 2, ..., n\}$,*

$$q_i \Rightarrow q_i \text{ if and only if } A_i \subset A_j$$

Proof Suppose $A_i \subset A_j$. Then, by the definition of the inclusion of flexible sets, it follows that $m_{A_1}(x) \leq m_{A_2}(x)$; thus, when $m_{A_1}(x) > 0.5$, it also follows that $m_{A_2}(x) > 0.5$, that is, when q_i is roughly-true, q_j is also certainly roughly-true. Thus, we have $q_i, \Rightarrow q_j$.

Conversely, suppose $q_i, \Rightarrow q_j$. Then, it shows that when q_i is true, q_j is certainly true. That is, when $m_{A_1}(x) > 0.5$, it also follows that $m_{A_2}(x) > 0.5$; thus, it shows that $m_{A_1}(x) \leq m_{A_2}(x)$. Then, from the x being arbitrary, we have $A_i \subset A_j$.

From this lemma, further we have the following corollary.

Corollary 24.1 $q_1 \Rightarrow q_2 \Rightarrow \cdots \Rightarrow q_n$ if and only if $A_1 \subset A_2 \subset \cdots \subset A_n$.

Theorem 24.3 *Let* $\{\mathbf{Q}_n = \{q_i|q_i : x \ be \ A_i, x \in U, A_i \subset U, i \in \{1, 2, \ldots, n\}\}$. *Then,* *for arbitrary* $q_i, q_j \in \mathbf{Q}_n$, *if* $q_i \Rightarrow q_j$, *then* $c(q_i) < c(q_j)$.

Proof Suppose $q_i \Rightarrow q_j$. Then, by Lemma 24.1, it follows that $A_i \subset A$, whereas when $A_i \subset A_j$, shar(A_i) < shar(A_j), and by the definition of believability-degree, it follows that shar(A_i) = $c(q_i)$ and shar(A_j) = $c(q_j)$, and consequently, we have $c(q_i) < c(q_j)$.

This theorem means that when proposition q_i implies proposition q_j, then proposition q_j is more believable than proposition q_i. For instance, if Jack is very tall, then Jack is certainly tall. Therefore, Jack is tall is more believable than Jack is very tall.

Actually, the conclusion in Theorem 24.3 is also tenable for general propositions p and q with implication relation. That is, if proposition p implies proposition q, then q is more believable than p. Because when p implies q, p only is a sufficient condition for q.

By Theorem 24.3, we further have

Corollary 24.2 $q_1 \Rightarrow q_2 \Rightarrow \cdots \Rightarrow q_n$ then $c(q_1) < c(q_2) < \cdots < c(q_n)$.

24.3 Believability-Degrees of a Flexible Proposition with a Composite Linguistic Value and a Compound Flexible Proposition from Distinct Spaces

1. Believability-Degrees of a flexible proposition with a composite linguistic value

Let flexible propositions with a composite linguistic value be $p_{A_1 \wedge A_2 \wedge \cdots \wedge A_n}$: $A_1 \wedge A_2 \wedge \cdots \wedge A_n(x_{1_0}, x_{2_0}, \ldots x_{n_0})$, $p_{A_1 \vee A_2 \vee \cdots \vee A_n}$: $A_1 \vee A_2 \vee \cdots \vee A_n(x_{1_0}, x_{2_0}, \ldots, x_{n_0})$, and $p_{A_1 \oplus A_2 \oplus \cdots \oplus A_n}$: $A_1 \oplus A_2 \oplus \cdots \oplus A_n(x_{1_0}, x_{2_0}, \ldots, x_{n_0})$, where $A_1 \wedge A_2 \wedge \cdots \wedge A_n$, $A_1 \vee A_2 \vee \cdots \vee A_n$, and $A_1 \oplus A_2 \oplus \cdots \oplus A_n$ are flexible linguistic values on product space $U_1 \times U_2 \times \cdots \times U_n$. By the definition of believability-degree of a proposition, it follows that

$$c(p_{A_1 \wedge A_2 \wedge \cdots \wedge A_n}) = P(p_{A_1 \wedge A_2 \wedge \cdots \wedge A_n}) = P(e_{A_1 \wedge A_2 \wedge \cdots \wedge A_n}) \tag{24.15}$$

$$c(p_{A_1 \vee A_2 \vee \cdots \vee A_n}) = P(p_{A_1 \vee A_2 \vee \cdots \vee A_n}) = P(e_{A_1 \vee A_2 \vee \cdots \vee A_n}) \tag{24.16}$$

$$c(p_{A_1 \oplus A_2 \oplus \cdots \oplus A_n}) = P(p_{A_1 \oplus A_2 \oplus \cdots \oplus A_n}) = P(e_{A_1 \oplus A_2 \oplus \cdots \oplus A_N}) \tag{24.17}$$

That is, the believability-degrees of the flexible propositions with a composite linguistic value are reduced to the probabilities of the corresponding flexible events with a composite linguistic value.

2. Believability-Degree of a compound flexible proposition from distinct spaces

A compound flexible proposition from distinct spaces is the compound proposition in which the flexible linguistic values are on distinct measurement spaces, which is also the proposition that describes a compound flexible event from distinct spaces. Thus, the believability-degree of a compound flexible proposition from distinct spaces is reduced to the probability of a compound flexible event from distinct spaces.

Let propositions p_i: A_i (x_i), A_i be a flexible linguist value on the measurement space U_i $(i = 1, 2, ..., n)$, and $U_1, U_2, ..., U_n$ be different from each other. By the definition of believability-degree of a proposition and the probability computation formula of a compound flexible event from distinct spaces (see Sect. 23.3), it follows that

$$c(p_{A_i} \wedge p_{A_i} \wedge \cdots \wedge p_{A_i}) = P(p_{A_i} \wedge p_{A_i} \wedge \cdots \wedge p_{A_i}) = P(e_{A_1 \wedge A_2 \wedge \cdots \wedge A_n})$$
$$= c(p_{A_1 \wedge A_2 \wedge \cdots \wedge A_n})$$

Thus, we have

$$c(p_{A_i} \wedge p_{A_i} \wedge \cdots \wedge p_{A_i}) = c(p_{A_1 \wedge A_2 \wedge \cdots \wedge A_n}) \qquad (24.18)$$

In the same reason, we have

$$c(p_{A_i} \vee p_{A_i} \vee \cdots \vee p_{A_i}) = c(p_{A_1 \vee A_2 \vee \cdots \vee A_n}) = \sum_{i=1}^{n} c(p_{A_i}) - c(p_{A_1 \wedge A_2 \wedge \cdots \wedge A_n}) \quad (24.19)$$

While when $x_1, x_2, ..., x_n$ as random variables are all uniform distribution and mutually independent,

$$c(p_{A_i} \wedge p_{A_i} \wedge \cdots \wedge p_{A_i}) = c(p_{A_i})c(p_{A_2})...c(p_{A_n}) \qquad (24.20)$$

$$c(p_{A_i} \vee p_{A_i} \vee \cdots \vee p_{A_i}) = \sum_{i=1}^{n} c(p_{A_i}) - \sum_{1 \le i < j \le n} c(p_{A_i})c(p_{A_j})$$
$$+ \sum_{1 \le i < j < k \le n} c(p_{A_i})c(p_{A_j})c(p_{A_k}) - ... + (-1)^{n+1} c(p_{A_i})c(p_{A_2})...c(p_{A_n})$$

$$(24.21)$$

3. Believability-Degree of an implicational flexible proposition from distinct spaces

Let p: $A(x)$, q: $B(y)$, and A and B be separately flexible linguistic value on measurement spaces U and V, $U \ne V$. By the definition of believability-degree of a proposition, it follows that

$$c(p \rightarrow q) = P(p \rightarrow q) = P(e_B|e_A) \tag{24.22}$$

That is to say, the believability-degree of a implicational flexible proposition from distinct spaces is still the corresponding conditional probability $P(e_B|e_A)$.

Then, Generally, the believability-degrees of implicational flexible propositions with multiple conditions are as follows:

$$c(A_1(x_1) \wedge A_2(x_2) \wedge \cdots \wedge A_n(x_n) \rightarrow B(y)) = P(e_B|e_{A_1} \wedge e_{A_2} \wedge \cdots \wedge e_{A_n}) \tag{24.23}$$

$$c(A_1(x_1) \vee A_2(x_2) \vee \cdots \vee A_n(x_n) \rightarrow B(y)) = P(e_B|e_{A_1} \vee e_{A_2} \vee \cdots \vee e_{A_n}) \tag{24.24}$$

and those of implicational flexible propositions with a composite linguistic value are

$$c(A_1 \wedge A_2 \wedge \cdots \wedge A_n(x_1, x_2, \cdots x_n) \rightarrow B(y)) = P(e_B|e_{A_1 \wedge A_2 \wedge \cdots \wedge A_n}) \tag{24.25}$$

$$c(A_1 \vee A_2 \vee \cdots \vee A_n(x_1, x_2, \ldots x_n) \rightarrow B(y)) = P(e_B|e_{A_1 \vee A_2 \vee \cdots \vee A_n}) \tag{24.26}$$

$$c(A_1 \oplus A_2 \oplus \cdots \oplus A_n(x_1, x_2, \ldots x_n) \rightarrow B(y)) = P(e_B|e_{A_1 \oplus A_2 \oplus \cdots \oplus A_N}) \tag{24.27}$$

Since linguistic values A and B belong to distinct measurement spaces, when x and y as random variables are uniform distribution, by Eq. (23.62), it follows that

$$c(p \rightarrow q) = \mathrm{corr}(A, B) \tag{24.28}$$

That is to say, speaking from definition, the believability-degrees of some implicational flexible propositions from distinct spaces equal numerically to the correspondence-rates between corresponding flexible sets, and vice versa.

Also by Definition 21.9,

$$\mathrm{impl}(p, q) = \mathrm{corr}(A, B)$$

Thus, we have

$$c(p \rightarrow q) = \mathrm{impl}(p, q) \tag{24.29}$$

That is to say, speaking from definition, the believability-degrees of some implicational flexible propositions from distinct spaces equal numerically to the implication-degrees between corresponding flexible propositions, and vice versa.

Obviously, the above two Eqs. (24.28) and (24.29) also can be generalized to the implicational flexible propositions with multiconditions and implicational flexible propositions with a composite linguistic value.

24.4 Reasoning with Believability-Degrees

The usual logical inference is an inference for the truth of propositions. Believability-degree is a measure portraying the believability of propositions, so with believability-degree, the inference for the believability of propositions can be realized. We call the inference with the believability-degree of a proposition to be **reasoning with believability-degrees**.

The basic rules and schemes of reasoning with believability-degrees is the same as that of the usual logical inference, but the premise and conclusion of an argument are both with a belicvability-degree, and the believability-degree of conclusion is the computation result of that of premise. As thus, when conducting reasoning with believability-degrees, besides the symbolic deduction, the believability-degree computation can also be done. Therefore, the principle of reasoning with believability-degrees can be simply expressed as

Reasoning with Believability-Degrees = Symbolic Deducing + Believability-Degree Computing

Specifically speaking, the general scheme of reasoning with believability-degrees based on the *modus ponens* is

$$\frac{(p \to q,\ c(p \to q))}{(p_0,\ c(p_0))}{(q,\ c(q))} \tag{24.30}$$

Here, $c(p \to q)$, $c(p_0)$, and $c(q)$ are separately the believability-degrees of the major premise, minor premise, and conclusion.

When reasoning, the believability-degrees $c(p \to q)$ and $c(p_0)$ of premises already have been given, so it only needs to compute the believability-degree $c(q)$ of conclusion q. Here, we employ the multiplication operation, namely

$$c(q_0) = c(p_0) \cdot c(p \to q) \tag{24.31}$$

The theoretical basis of us employing the formula (24.31) is the total probability formula in probability theory. In fact, in practical problems:

(1) If there only is probably one case of $p \to q$, then it shows that there is no $\neg p \to q$; thus, $c(\neg p \to q) = 0$. Then, by total probability formula, it follows that

$$c(q_0) = c(p_0) \cdot c(p \to q) + c(\neg p_0) \cdot c(\neg p \to q) = c(p_0) \cdot c(p \to q)$$

(2) Besides that $p \to q$ is possible, if there also are probably $p_1 \to q, p_2 \to q, \ldots,$ $p_{n-1} \to q$ (of course, the events described by these uncertain propositions should be mutually exclusive), then, by total probability formula, adding up the corresponding believability-degrees $c(q_0)_1, c(q_0)_2, \ldots, c(q_0)_{n-1}$ and $c(q_0)$ together to be the final believability-degree $c(q_0)^*$ of conclusion q_0, that is

$$c(q_0)^* = c(q_0) + c(q_0)_1 + c(q_0)_2 \ldots + c(q_0)_{n-1}$$

Thus, computing the believability-degree of conclusion with the formula (24.31) in reasoning with believability-degrees based on the *modus ponens* is always correct.

Actually, here, major premise $p \rightarrow q$ is just a production rule. When the linguistic values therein are flexible linguistic values, it is also a flexible rule. That means the reasoning with believability degrees is also a kind of inference with rules or flexible rules.

Example 24.3 Suppose there is a modal rule: If the weather is sultry and almost windless, then it is probably going to be a storm (for modal rules, we will discuss in Chap. 25), and suppose the believability-degree of uncertain rule "if the weather is sultry and almost windless, then it is going to be a storm" described by the modal rule is 0.75. And there is a forecast: The believability-degree of "it will be sultry tomorrow" is 0.95, and the believability-degree of "almost windless" is 0.90. Find the believability-degree of "it will storm tomorrow" according to the given rule and facts.

Solution Let p_1: It is sultry, p_2: It is almost windless, and q: It is storming. Then, from the problem given, it follows that $c(p_1 \wedge p_2 \rightarrow q) = 0.75$, $c(p_1) = 0.95$, and $c(p_2) = 0.90$. Obviously, p_1 and p_2 are propositions on distinct spaces; assuming the corresponding two events are mutually independent, then by the Eq. (24.23), it follows that

$$c(p_1 \wedge p_2) = c(p_1)c(p_2) = 0.95 \times 0.90 \approx 0.86$$

Thus, we have the following reasoning with believability-degrees:

$$\frac{(p_1 \wedge p_2 \rightarrow q, 0.75)}{(p_{1_0} \wedge p_{2_0}, 0.86)}$$
$$\overline{(q_0, 0.65)}$$

where $0.65 \approx 0.86 \times 0.75 = c(q_0)$. Thus, the believability-degree of "it storms tomorrow" is about 0.65.

Since the believability-degrees between 0 and 1 and modal words "likely," "highly likely," and so forth can be converted mutually, so in reasoning with believability-degrees, the rules whose believability-degrees are between 0 and 1 (greater than 0 and less than 1) are also uncertain rules. Actually, reasoning with believability-degrees is mainly concerned with propositions and rules with believability-degrees between 0 and 1, whereas it is known from Theorem 24.1 that the truth values of propositions with believability-degrees between 0 and 1 cannot be determined. Therefore, the reasoning with believability-degrees belongs to usually called uncertain reasoning, while the definitions, computation formulas, and methods of believability-degrees we given above form a so-called uncertain reasoning model. Of course, the corresponding threshold values also can be set in practical problems.

Actually, expression (24.30) also can be seen as a rule of inference in the sense of believability-degree. Look from the form, it is the *modus ponens* with believability-degrees, so we call it to be the believability-degrees *modus ponens*. Besides this believability-degrees *modus ponens*, there also can be other rules of inference in reasoning believability-degrees. For instance,

$$\frac{\begin{array}{c}(p \rightarrow q, c(p \rightarrow q)) \\ (q \rightarrow r, c(q \rightarrow)) \end{array}}{\therefore (p \rightarrow r, c(p \rightarrow r))} \tag{24.32}$$

also is a rule of inference with believability-degrees, which can be called the believability-degree *hypothetical syllogism*.

The computation formula of believability-degree in this rule of inference with believability-degrees is $c(p \rightarrow r) = c(p \rightarrow q)c(q \rightarrow r)$. But using this formula is conditional, that is, the uncertain implication relation between the propositions p, q, and r needs to satisfy transitivity. In the following, we prove this formula as a theorem.

Theorem 24.4 *Let propositions p: x is A, q: y is B, and r: z is C, and A, B, and C are quantifiable linguistic values, and "if p then possibly q" and "if q then possibly r" are two uncertain implicational propositions (rules). If the implication relation between propositions p, q, and r satisfies transitivity, then*

$$c(p \rightarrow r) = c(p \rightarrow q)c(q \rightarrow r) \tag{24.33}$$

Proof

$c(p \rightarrow q) = c(q \rightarrow r) = P(e_B|e_A)P(e_C|e_A)$ (e_A, e_B, e_C

are flexible events described separately by propositions p,q and r)

$$= \frac{P(e_A e_B)}{P(e_A)} \frac{P(e_B e_C)}{P(e_B)}$$

$$= \frac{P(e_A e_B)P(e_B e_C)}{P(e_A)P(e_B)}$$

$$= \frac{P(e_A e_B e_B e_C)}{P(e_A)P(e_B)} \quad \text{(because } e_A e_B \text{ and } e_B e_C \text{ are independent mutually)}$$

$$= \frac{P(e_A e_B e_C)}{P(e_A)P(e_B)} = \frac{P(e_B e_A e_C)}{P(e_A)P(e_B)}$$

$$= \frac{P(e_B)P(e_A e_C|e_B)}{P(e_A)P(e_B)} \quad \text{(by multiplication formula)}$$

$$= \frac{P(e_A e_C|e_B)}{P(e_A)}$$

$$= \frac{P(e_A e_C)}{P(e_A)} \quad \text{(because } e_A e_C \text{ is independent from } e_B \text{ at the time)}$$

$$= P(e_C|e_A) = c(p \rightarrow r)$$

We call Eq. (24.33) to be the **transitive formula of believability-degrees**.

From the expression of definition of believability-degree, it can be seen that believability-degree includes implication-degree, and implication-degree in turn includes inclusion-degree and correspondence-rate. Therefore, this transitive formula of believability-degrees includes actually transitive formula of implication-degrees (in Sect. 21.6) and then also includes transitive formulas of inclusion-degrees and correspondence-rates (in Sect. 20.6).

Remark For the measure of uncertainty, people have used many different methods and have proposed many of uncertain reasoning model. For instance, in *certainty factor theory*, a measure of *certainty factor* (CF) as the degree of believability of a proposition is used, while the computation formulas of the degrees of believability of compound propositions are

$$CF(q \wedge p) = \min\{CF(p), CF(q)\}, CF(q \vee p) = \max\{CF(p), CF(q)\}$$

Besides *certainty factor theory*, there are also the *proof theory* and *subjective Bayesian approach*, as well as the probabilistic reasoning with Bayesian network, etc. Since uncertain reasoning is not the subject of research of this book, here, the principles of these methods are not discussed.

24.5 Dual Reasoning with B-D(T)

Reasoning with believability-degrees is a kind of inference concerned with the believability of propositions, which is a kind of uncertain reasoning, and solves the inference problems about uncertain information, while the reasoning with degrees (truth-degrees) given in Chap.15 is the inference concerned with the truth of propositions, which is a kind of imprecise reasoning, and solves the inference problems about imprecise information. However, there is still some of information being uncertain as well as imprecise, or imprecise as well as uncertain (see Sect. 1.3). Then, how to deal with the inference with this kind of information?

It is not difficult to see that if we combine the reasoning with believability-degrees and the reasoning with degrees (truth-degrees) to realize a kind of dual reasoning that has both reasoning with believability-degrees and reasoning with degrees (truth-degrees), the problem will be solved.

In fact, as long as the truth-degreed function of major premise (that is, rule) and the truth-degree of minor premise (that is, evidentiary fact) are given in the reasoning with believability-degrees, the dual reasoning with believability-degrees and truth-degrees just can be realized. The specific scheme is

$$\frac{(p \to q, c, (p \to q), f_t(t))}{(p_0, c(p_0), t_{p_0})} \over (q_0, c(q_0), t_{q_0})} \tag{24.34}$$

where $c(q_0) = c(p_0)c(p \to q)$, $t_{q_0} = f_t(t_{p_0})$, $t_{p_0} > 0.5$.

Now, also let p: $A(x)$ and q: $B(y)$, then (p_0, t_{p_0}) is equivalent to $(A(x_0), d_{A_0})$, and (q_0, t_{q_0}) is equivalent to $(B(y_0), d_{B_0})$. Correspondingly, truth-degreed function $f_t(t)$ is equivalent to degreed function $f_d(d)$. Thus, from expression (24.34), we can also have the following dual reasoning with believability-degrees and degrees:

$$\frac{(A \to B, c(A \to B), f_d(d))}{(A_0, c(A_0), d_{A_0})} \\ \overline{(B_0, c(B_0), d_{B_0})} \tag{24.35}$$

where $c(B_0) = c(A_0)c(A \to B)$, $d_{B_0} = f_d(d_{A_0})$, $d_{A_0} > 0.5$.

Note that the meaning of the reasoning outcome $(q_0, c(q_0), t_{q_0})$ in expression (24.34) above is as follows: The believability-degree of proposition q_0 is $c(q_0)$, and when q_0 is true indeed, the truth-degree of which is t_{q_0}; and the meaning of the reasoning outcome $((B_0, c(B_0)), d_{B_0})$ in expression (24.35) is as follows: The believability-degree of conclusion B_0 is $c(B_0)$, and when B_0 is true indeed, the degree of which is d_{B_0}. However, we cannot understand the two outcomes as: The believability-degree of q_0 whose truth-degree being t_{q_0} is $c(q_0)$ and the believability-degree of B_0 whose degree being d_{B_0} is $c(B_0)$, because the q_0 whose truth-degree being t_{q_0} is equivalent to the B_0 whose degree being d_{B_0}, and the latter is equal to a certain number (or vector) y_0 in measurement space V. And from probability theory, it can be known that $P(X = y_0) < P(X \in B_0 \subset V)$; especially, when V is a continuous set, $P(X = y_0) = 0$ (i.e., an infinitesimal).

Since the degree and truth-degree of a proposition are interchangeable, and adjoint degreed function and adjoint truth-degreed function of a rule are also interchangeable, expression (24.35) of the kind of dual reasoning with believability-degrees and degrees is essentially the same (also interchangeable) as expression (24.34) of the kind of dual reasoning with believability-degrees and truth-degrees. Therefore, we tie expressions (24.34) and (24.35) together to call the dual reasoning with believability-degrees and degrees (truth-degrees) and simply write as dual reasoning with B-D(T).

Note that in this dual reasoning, the reasoning with believability-degrees and the reasoning with degrees (truth-degrees) are mutually independent.

We say that the above dual reasoning with B-D(T) on the believability and truth at the same time also has practical significance and background. As a matter of fact, in many cases, our reasoning is for truth as well as for believability. For instance, reasoning is needed in disease diagnosis to figure out what disease a patient may have and the severity if the disease was suffered indeed. This kind of reasoning is just on the believability and truth at the same time, which involves believability-degree and degree (truth-degree). Therefore, this kind of reasoning is dual reasoning with B-D(T). For another example, the weather forecast also needs reasoning done to estimate the weather of a future certain day(s) and its intensity. This reasoning is also on the believability and truth at the same time, which also involves believability-degree and degree (truth-degree). Thus, this kind of reasoning is dual reasoning with B-D(T).

Example 24.4 Add the degreed function of rule and the degrees of evidentiary facts to the problem in Example 24.3 and then do dual reasoning with B-D to find the believability-degree of rainstorm tomorrow and its strength.

Solution Denote "sultry" as A_1, "almost windless" as A_2, and "storm" as B, suppose $d_{A_1} = 1.25$ and $d_{A_2} = 0.96$, and employ directly Eq. (14.12) in Sect. 14.3.1, that is,

$$d_B = \frac{\beta_R - 0.5}{\beta_\wedge - 0.5}(d_{A_1 \wedge A_2} \quad 0.5) + 0.5$$

where $\beta_\wedge = \min\{\beta_1, \beta_2\}$, $d_{A_1 \wedge A_2} = \min\{d_{A_1}, d_{A_2}\}$, $(d_{A_1}, d_{A_2}) \in (0.5, \beta_1] \times (0.5, \beta_2]$, as an expression of adjoint degreed function of the rule. And suppose also $\beta_1 = 2.5$, $\beta_2 = 1.5$, and $\beta_B = 3$, then $\beta_\wedge = \min\{\beta_1, \beta_2\} = \min\{2.5, 1.5\} = 1.5$, substitute 3 and 1.5 into the above expression, obtaining the adjoint degreed function of the rule is

$$f_d(d) = 2.5d - 0.75, d = \min\{d_{A_1}, d_{A_2}\}$$

Since $\min\{d_{A_1}, d_{A_2}\} = \min\{1.25, 0.96\} = 0.96 > 0.5$, thus, we have the following dual reasoning with B-D:

$$\frac{(A_1 \wedge A_2 \rightarrow B, 0.75, f_d(d))}{(A_{1_0} \wedge A_{2_0}, 0.86, 0.96)}{(B_0, 0.65, 1.65)}$$

The outcome is as follows: The believability-degree of rainstorm tomorrow is about 0.65; and when the conclusion is true, the strength of rainstorm is 1.65.

Example 24.5 Suppose there is a modal rule: If someone has a slight fever plus a stuffy nose slightly plus a light of aversion to wind, then highly likely he gets a slight cold, and suppose the believability-degree of uncertain rule "if someone has a slight fever plus a stuffy nose slightly plus a light of aversion to wind, then he gets a slight cold" described by the modal rule is 0.95. There are also the facts: Someone has a slight fever with degree of 1.06, a stuffy nose slightly with degree of 0.85, and a light of aversion to wind with degree of 0.97. And suppose the believability-degree of the fact is 1. According to given rule and facts, what is the believability-degree of this person catching slight cold? And if he/she catching slight cold, then what is the degree of the slight cold?

Solution It can be seen that this is a dual reasoning with B-D problem. Denote "a slight (fever)" by A_1, "(stuffy nose) slightly" by A_2, "a light (of aversion to wind)" by A_3, and "a slight (cold)" by B, and then, the original rule can be symbolized as

$$A_1 \oplus A_2 \oplus A_3 \rightarrow B$$

With reference to the Eq. (14.14) in Sect. 14.3.1, let the adjoint degreed function $f_d(d_A)$ of the rule be

$$d_B = \frac{\beta_B - 0.5}{\beta_\oplus - 0.5}(d_A - 0.5) + 0.5$$

where $\beta_\oplus = w_1\beta_{A_1} + w_2\beta_{A_2} + w_3\beta_{A_3}$, $d_A = \sum_{i=1}^{3} w_i d_{A_i}, 0 < d_{A_j} \leq \beta_{A_j} (j = 1, 2, 3)$.

Suppose the maximum of consistency-degrees of "a slight (fever)" is $\beta_{A_1} = 2.5$, the maximum of consistency-degrees of "(stuffy nose) slightly" is $\beta_{A_2} = 1.8$, and the maximum of consistency-degrees of "a light (of aversion to wind)" is $\beta_{A_3} = 2.0$, and suppose the weights $w_1 = 0.5$, $w_2 = 0.3$, $w_3 = 0.2$. Then $\beta_\oplus = \sum_{i=1}^{3} w_i\beta_{A_i} = 0.5 \times 2.5 + 0.3 \times 1.8 + 0.2 \times 2.0 \approx 2.2$. Suppose then the maximum of consistency-degrees of "slight (cold)" is $\beta_B = 2.0$. Now, substituting the two numbers into the above functional expression, we have an adjoint degreed function of the rule:

$$d_B = \frac{15}{17}d_A + \frac{1}{17}$$

where $d_A = 0.5d_{A_1} + 0.3d_{A_2} + 0.2d_{A_3}$, $0 < d_{A_j} \leq \beta_{A_j} (j = 1, 2, 3)$.

And from the facts given, it follows that $d_{A_1} = 1.06$, $d_{A_2} = 0.85$, and $d_{A_3} = 0.97$; then the overall degree of facts is

$$d_{A_1 \oplus A_2 \oplus A_3} = 0.5\, d_{A_1} + 0.3\, d_{A_2} + 0.2\, d_{A_3} = 0.5 \times 1.06. + 0.3 \times 0.85 + 0.2 \times 0.97$$
$$\approx 0.98 > 0.5$$

Additionally, $c(A_1 \oplus A_2 \oplus A_3 \to B) = 0.95$, $c(A_1 \oplus A_2 \oplus A_3) = 1$. Thus, we have the following dual reasoning with B-D:

$$\frac{(A_1 \oplus A_2 \oplus A_3 \to B, 0.95, f_d(d_A))}{(A_1 \oplus A_2 \oplus A_3, 1, 0.98)}$$
$$\overline{(B, 0.95, 0.92)}$$

The reasoning outcome says that the believability-degree of this person caught a slight cold is 0.95, and if this conclusion is true, then the degree of the slight cold is 0.92.

The dual reasoning with B-D(T) is also uncertain imprecise dual reasoning. From the scheme and examples of the dual reasoning above, the connection and relationship between uncertain reasoning and imprecise reasoning can be clearly seen. Actually, viewed from dual reasoning perspective, pure reasoning with degrees (truth-degree) and pure reasoning with believability-degrees are all special dual reasoning—the former is the dual reasoning in which all believability-degrees are fixedly 1, and the latter is the dual reasoning in which all truth-degrees are fixedly 1 or >0.5. More general, the pure uncertain reasoning is precise uncertain reasoning, and the pure imprecise reasoning is certain imprecise reasoning.

24.6 Partial Implication, Relational Inference with Implication-Degrees, and Reasoning with Believability-Degrees

In Sect. 21.6, we presented the concept of partial implication (and partial equivalence and partial possession). Since being "partial implication" but usual "complete implication," the truth of the implicational compound proposition $p \rightarrow q$ made up of two propositions p and q having partial implication cannot be determined; that is, the implicational proposition is a uncertain proposition. For uncertain propositions, the logical inference in usual sense cannot be done. In this chapter, we introduced the method of reasoning with believability-degrees to solve the inference with uncertain propositions. Thus, for uncertain propositions $p \rightarrow q$ and p_0 (here p: $A(x)$ and q: $B(y)$), using believability-degrees *modus ponens*, we have

$$
\begin{array}{c}
(p \rightarrow q, c(p \rightarrow q)) \\
\dfrac{(p_0, c(p_0))}{(q_0, c(q_0))}
\end{array}
\tag{24.36}
$$

where $c(q_0) = c(p_0)c(p \rightarrow q)$.

And by Eqs. (24.14) and (24.29), $c(p \rightarrow q) = \text{impl}\,(p \rightarrow q)$, and by Eq. 24.3, $c(p_0) = \text{poss}(A)$ and $c(q_0) = \text{poss}(B)$. Thus, the reasoning with believability-degrees shown by expression (24.36) can be translated into the following relational inference:

$$
\begin{array}{c}
(p \mathrel{..\!\!\rhd} q, \text{impl}(p, q)) \\
\dfrac{(p, \text{poss}(A))}{\therefore\ (q, \text{poss}(B))}
\end{array}
\tag{24.37}
$$

where $\text{poss}(B) = \text{poss}(A)\text{impl}(p, q)$.

Similarly, by believability-degree *hypothetical syllogism*, we have the relational inference below:

$$
\begin{array}{c}
(p \mathrel{..\!\!\rhd} q, \text{impl}(p, q)) \\
\dfrac{(q \mathrel{..\!\!\rhd} r, \text{impl}(q, r))}{\therefore\ (p \mathrel{..\!\!\rhd} r, \text{impl}(p, r))}
\end{array}
\tag{24.38}
$$

where $\text{impl}(p, r) = \text{impl}(p, q)\text{impl}(q, r)$.

Now, we consider conversely if there occurs firstly the relational inference shown by expressions (24.37) and (24.38) (actually, here, expression (24.38) is just the relational inference expression (22.3) in Sect. 22.2.1, and the relational inference expression (22.6) in Sect. 22.2.2 includes the expression (24.37) here), then by $c(p \rightarrow q) = \text{impl}(p \rightarrow q)$, and $c(p_0) = \text{poss}(A)$ and $c(q_0) = \text{poss}(B)$, we also can have the reasoning with believability-degrees shown by expressions (24.36) and

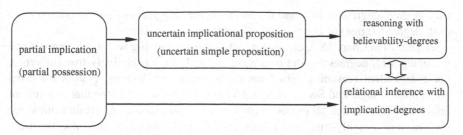

Fig. 24.2 Diagram of relation between partial implication, uncertain implicational proposition, relational inference with implication-degrees, and reasoning with believability-degrees

(24.32). That is to say, the relational inference with implication-degrees also can be translated into the reasoning with believability-degrees.

To sum up, the relational inference with implication-degrees and some reasoning with believability-degrees originate from partial implication relation between propositions [further tracing, which will originate from partial correspondence (inclusion)], and the two kinds of inference also can be translated to each other through "bridge"—$c(p \rightarrow q) = \text{impl}(p \rightarrow q)$. Thus, we have the following diagram (Fig. 24.2).

From Fig. 24.2, we can see intuitively the relationship between partial implication, uncertain propositions, relational inference with implication-degrees, and reasoning with believability-degrees. In consideration that implication-degree is to portray partial implication, i.e., to portray imprecision, the corresponding relational inference is to deal with imprecise information, while believability-degree is to portray uncertainty, and reasoning with believability-degrees is to deal with uncertain information; therefore, this diagram also reveals an origin of uncertain information and the correlation between imprecise-information (processing) and uncertain-information (processing); that is, some uncertain information originates from imprecise information, and imprecise-information (processing) and uncertain-information (processing) can be translated to each other in some conditions.

24.7 A Note on Reasoning with Truth-Degrees (Degrees)

Finally, we intend to make a discussion and explanation for reasoning with truth-degrees (degrees).

From the above dual reasoning, we see that reasoning with truth-degrees (degrees) is using truth-degrees (degrees) of fact and through the truth-degreed (degreed) function between antecedent and consequent of a rule to compute truth-degree (degree) of conclusion, while reasoning with believability-degrees is using believability-degrees of fact and believability-degree of a rule to compute believability-degree of conclusion. Then, whether reasoning with truth-degrees (degrees) can use the truth-degree (degree) of rule and the truth-degrees (degrees) of fact to compute the truth-degree (degree) of conclusion? The answer is negative.

In fact, from Chap. 15, it can be known that our reasoning with truth-degrees and reasoning with degrees are the reasoning in the frame of relatively-true inference, but independent reasoning based on truth-degrees or degrees. And from usual mathematical logic and Sects. 11.6 and 12.6 in the book, we know that any rule of inference demands that all premises must have simultaneously a certain truth value such as true, roughly-true, and relatively-true, and conclusion has also the truth value; On the other hand, in a reasoning, the corresponding rule is always appointed or supposed to be true, roughly-true, or relatively-true (otherwise, corresponding reasoning will is insignificant). That is to say, the real truth values of a rule do not play any role in reasoning; thus, there is no situation that rule and fact with different truth-degrees participate in one and the same reasoning; that is, there occurs no following reasoning with truth-degrees:

$$\frac{\begin{array}{l} p \rightarrow q, t_1 \\ p, t_2 \end{array}}{\therefore q, t_3} \tag{24.39}$$

where $t_1 \neq t_2$, $t_3 = f(t_1, t_2)$.

Similarly, there also occurs no following reasoning with truth-degrees following hypothetical syllogism (Transitivity):

$$\frac{\begin{array}{l} p \rightarrow q, t_1 \\ q \rightarrow r, t_2 \end{array}}{\therefore p \rightarrow r, t_3} \tag{24.40}$$

where $t_1 \neq t_2$, $t_3 = f(t_1, t_2)$.

As for reasoning with degrees, there is function relation between the degrees of the antecedent and the consequent of a rule; however, the rule as a whole has no something "degree"; therefore, there is no situation that a rule with a degree to participate in reasoning liking expressions (24.39) and (24.40).

24.8 Summary

In this chapter, we introduced a measure called believability-degree for the believability of propositions and on the basis of the theories of traditional probability and flexible event's probability presented the computation principles and methods of the believability-degrees of compound propositions on the same space and from distinct spaces, in particular, the computation principles and methods of propositions with a composite linguistic value and implicational propositions, thus founding the believability-degree theory of propositions. Then, we discussed the inference on the believability of propositions and presented an uncertain reasoning scheme based on believability-degrees—reasoning with believability-degrees and the corresponding reasoning model. And then, we discussed the dual reasoning about uncertainty and imprecision for the believability and truth of propositions simultaneously and presented a scheme and model of dual reasoning. Besides, we also discussed the correlation between partial implication, uncertain implicational proposition, relational inference with implication-degrees, and reasoning with believability-degrees.

This chapter has the following conclusions:

- A proposition is true if and only if it is completely believable; a proposition is false if and only if it is completely unbelievable; a proposition being true or false cannot be determined if and only if it is believable with some degree.
- The sum of the believability-degrees of complementary flexible propositions is 1.
- When proposition p implies proposition q, q is more believable than p.
- Relational inference with implication-degrees and some reasoning with believability-degrees originate from partial implication relation between propositions, and the two kinds of inference also can be translated to each other in some conditions.
- Some uncertain information originates from imprecise information, and imprecise-information (processing) and uncertain-information (processing) can be translated to each other in some conditions.

Reference

1. Lian S (2009) Principles of imprecise-information processing. Science Press, Beijing

Chapter 25
Some Discussions About Possibly-Type Modal Propositions and Rules

Abstract This chapter introduces firstly the concepts and terminologies of necessarily-type, possibly-type, and flexible possibly-type modal propositions and rules and presents a representation with believability-degree of (possibly-type) modal propositions and rules; next, it presents the definition and computation formulas of truth values of modal propositions; presents a reasoning method with modal propositions and rules; and derives a transitive formula of believability-degrees, and presents a model of relational inference with believability-degrees; then it analyzes and reveals the origin of possibly-type modal propositions, and then reveals also an origin of uncertain information; finally, it introduces flexibly-quantified propositions and their formal representation, and then reveals also a correlation between imprecise information and uncertain information.

Keywords Possibly-type modal propositions · Uncertain information · Imprecise information

First, we call such flexible linguistic values as "fairly possibly/probably/likely," "highly possibly/probably/likely," "not very possibly/probably/likely," and so forth to be the **flexible modal words**.

In daily information exchange, people always use modal or flexible modal words such as "possibly/probably/likely," "fairly possibly/probably/likely," "very possibly/probably/likely," and "not highly possibly/probably/likely" to describe uncertainty. The sentences using modal or flexible modal words to describe uncertainty themselves are a kind of propositions, and these words are all quantifiable linguistic values, so the type of propositions describing uncertainty themselves is also imprecise. Then, how are the truth values of the type of propositions to be determined? How is the corresponding reasoning to be done? These are the problems to be discussed in this chapter. In addition, we will analyze the origin of possibly-type modal propositions and reveal the origin of uncertain information as well as a relation between imprecise information and uncertain information.

© Springer Science+Business Media Singapore 2016
S. Lian, *Principles of Imprecise-Information Processing*,
DOI 10.1007/978-981-10-1549-6_25

25.1 (Possibly-Type) Modal Propositions and Rules and Their Representation with Believability-Degree

Definition 25.1

(1) We call the proposition containing modal word "necessarily" and "possibly/probably/likely" to be a modal proposition and call a modal (universal) implicational proposition to be the modal rule.

(2) We call the propositions and rules containing modal words "necessarily," "certainly," and so on to be the necessarily type modal propositions and rules and call the propositions and rules containing modal words "possibly/probably/likely" to be the possibly type modal propositions and rules. Expressly, we call the possibly-type propositions and rules containing flexible modal words "fairly possibly/probably/likely," "highly possibly/probably/likely," "not very possibly/probably/likely," and so forth to be the flexible possibly-type modal propositions and rules.

Observe these propositions below:

① It will certainly rain tomorrow.
② It will probably rain tomorrow.
③ If it is overcast then it will necessarily rain.
④ If it is overcast then it will probably rain.
⑤ It will fairly probably rain lightly tomorrow.
⑥ If it is heavily overcast, then it will very likely rain heavily.

It can be seen that all the above propositions are modal proposition, and ③, ④, and ⑥ of them are modal rules; further subdividing, then, of them, ① and ③ are necessarily-type modal propositions; ②, ④, ⑤, and ⑥ are possibly-type modal propositions; ③ is also a necessarily-type modal rule; ④ is also a possibly-type modal rule; ⑤ is also a flexible possibly-type modal proposition; and ⑥ is also a flexible possibly-type modal rule.

Note that the propositions and rules in the definitions above include also flexible propositions and flexible rules. Therefore, if to subdividing, then the corresponding various modal flexible propositions and flexible rules will follow. For example, the above ⑤ is just a modal flexible proposition and also a possibly-type modal flexible proposition and a flexible possibly-type modal flexible proposition; while ⑥ is a modal flexible rule, and also a possibly-type modal flexible rule and a flexible possibly- type modal flexible rule.

Actually, modal propositions (including flexible possibly-type modal propositions) are a kind of nested compound proposition, which can always be stated as the following form:

$$p \text{ is } A \tag{25.1}$$

where p is a proposition and A is a modal or flexible modal linguistic value. For example, the above modal proposition "It will probably rain tomorrow" can also be stated as "'it will rain tomorrow' is probable," and flexible possibly-type modal proposition "if it is overcast then it will very probably rain" can also be stated as "'if it is overcast then it will rain' is very possible." This is to say, viewed from the angle of syntax, a modal proposition is just a main-clause-structured compound sentence [1].

From the main-clause structure, it can be clearly seen that the linguistic value A in modal proposition "p is A" is to describe the possibility of subordinate clause (proposition) being true. From the relation between the possibility of a proposition being true and the believability of it, here linguistic value A is tantamount to linguistic value A' that describes the believability of subordinate clause p. Thus, proposition "p is A" is also tantamount to proposition "p is A'". For instance, "'it will rain tomorrow' is possible" is tantamount to "'it will rain tomorrow' is believable with some degree." Thus, a modal proposition can also be formally represented as

$$(p, \text{believability}, A') \quad \text{or} \quad (p, A') \tag{25.2}$$

For example, "it will probably rain tomorrow," that is, "'it will rain tomorrow' is possible," can be represented as

$$(\text{it will rain tomorrow, believability, believable with some degree})$$

or

$$(\text{it will rain tomorrow, believable with some degree})$$

And "if it is overcast then it will very probably rain," that is, "'if it is overcast then it will rain' is very possible," can be represented as

$$(\text{if it is overcast then it will rain, believability, very believable})$$

or

$$(\text{if it is overcast then it will rain, very believable})$$

Now, we see that modal propositions are actually a main clause in which the believability of subordinate clause is also expressed, or in other words, modal propositions are a kind of compound proposition that expresses the believability of its subordinate clause.

On the other hand, we know that the believability of a proposition can also be portrayed precisely by believability-degrees. Thus, converting linguistic value A' representing believability into a believability-degree, that is, a number, then modal proposition "p is A" can be represented as

$$(p, \text{believability}, c) \quad \text{or} \quad (p, c) \tag{25.3}$$

For example, "it will probably rain tomorrow" can also be represented as

(it will rain tomorrow, believability, 0.6)   or   (it will rain tomorrow, 0.6)

here, we convert modal word "probably" into 0.6. And "if it is overcast then it will very probably rain" can also be represented as

(If it is overcast then it will rain, believability, 0.8)

or

(If it is overcast then it will rain, 0.8)

Definition 25.2 We call tuples $(p, \text{believability}, c(p))$ or $(p, c(p))$ made up of proposition p and its believability-degree $c(p)$ to be a proposition with believability-degree. In particular, if p is a (universal) implicational proposition, then $(p, \text{believability}, c(p))$ or $(p, c(p))$ is called a rule with believability-degree.

Thus, the above expression (25.3) and examples are the proposition with believability-degree or rule with believability-degree. Thus, we can also say that a modal proposition or rule (mainly refers to the possibly-type modal propositions and rules) can be represented as a proposition with believability-degree or a rule with believability-degree.

It can be seen that the method of representing a modal proposition or rule as a proposition with believability-degree or a rule with believability-degree is: first convert corresponding modal word (linguistic value) into a numerical value, that is, a believability-degree, then replace modal word with the believability-degree, and then take the subordinate clause in original modal proposition or rule, that is, the "sub" proposition or rule described by original proposition or rule, together with the believability-degree to form a tuple. As for the conversion from modal or flexible modal linguistic values to believability-degrees, the method is the same as that from general linguistic values to numerical values in Sect. 7.3. For the definitions of corresponding flexible-modal linguistic values, see Fig. 24.1 in Sect. 24.1.

Note that there is another conversion approach, that is, we do not consider the modal word in a modal proposition or rule, but take its subordinate clause and its real believability-degree to form a proposition with believability-degree or rule with believability-degree. But since the real believability-degree of a subordinate clause, in general, is not equal to the believability-degree obtained by converting from corresponding modal word, the proposition with believability-degree or rule with believability-degree obtained by the approach is not certainly equivalent to the original modal proposition or rule. Then, to convert a modal proposition or rule to proposition with believability-degree or rule with believability-degree, which approach should we use? This would be decided according to the practical problems.

Of course, conversely, a proposition with believability-degree or rule with believability-degree can also be converted into a modal proposition or rule, and the method is also to convert believability-degree into the corresponding modal word.

25.2 Truth Values of Possibly-Type Modal Propositions and Reasoning with Possibly-Type Modal Propositions

1. Truth values of possibly-type modal propositions

The possibly-type modal propositions (including flexible possibly-type modal propositions) are also propositions; then, ① how do we determine the truth of this kind of proposition? ② How do we do reasoning with this kind of propositions? For example, how is the truth of possibly-type modal proposition "it will probably rain tomorrow" to be determined? For another example, can we do reasoning with rule "if it is overcast then it will probably rain" and fact "it will probably be overcast tomorrow"?

From the relation between believability-degree and linguistic values "believable with some degree," "very believable," "rather believable," and so on, as well as the definition of truth of flexible propositions (see Sect. 11.1), the truth-degree of modal proposition "p is A" is numerically equal to the consistency-degree $c_A(c(p))$ of the believability-degree $c(p)$ of subordinate clause p with the flexible linguistic value A, namely

$$t(p \text{ is } A) = c_A(c(p)) \tag{25.4}$$

Equation (25.4) is the computation formula of truth values of all flexible possibly-type modal propositions.

Note that here believability-degree $c(p)$ is the real believability-degree of proposition p, rather than believability-degree c in (p, c) converted from modal proposition "p is A." So consistency-degree $c_A(c)$ cannot be treated as truth-degree t $(p \text{ is } A)$. In fact, since number c is obtained by converting linguistic value A, always $c_A(c) > 0.5$; but $c_A(c(p))$ may be >0.5 and may also be =0.5 or <0.5.

If we define the consistency function of modal word "possibly" as

$$c_{\text{possibly}}(x) = \begin{cases} 1, & 0 < x < 1 \\ 0, & x = 0 \text{ or } x = 1 \end{cases} \tag{25.5}$$

then Eq. (25.4) is also the computation formula of truth values of all possibly-type modal propositions.

Thus, for all of possibly-type modal propositions, only when a modal word matches with the believability-degree of corresponding subordinate clause, the corresponding modal proposition is true. That is like a person's height can be described exactly by numerical value $y(m)$, but can also generally be described by linguistic values such as "tall" and "short." However, for one and the same person, the two kinds of descriptions must be well matched or compatible. For instance, for 1.90 m, we say it is a "tall" is correct of course, but if it is said to be a "short," which is then certainly wrong. Similarly, for 1.50 m, we say it is a "tall" is wrong, but saying it is a "short" is then correct.

If we define still the consistency functions of modal words "necessarily" and "impossibly" as

$$c_{necessarily}(x) = \begin{cases} 1, & x = 1 \\ 0, & x \neq 1 \end{cases} \tag{25.6}$$

$$c_{impossibly}(x) = \begin{cases} 1, & x = 0 \\ 0, & x \neq 1 \end{cases} \tag{25.7}$$

then Eq. (25.4) is the computation formula of truth values of all modal propositions.

Thus, for the truth of modal propositions in usual modal logic (here it refers to modal logic in the narrow sense, that is, the modal logic of truth theory), we have the following **judging rules**:

(1) A modal proposition with modal word "necessarily" is true if and only if the believability-degree of its subordinate clause is 1;
(2) A modal proposition with modal word "possibly" is true if and only if the believability-degree of its subordinate clause is bigger than 0 and smaller than 1;
(3) A modal proposition with modal word "impossibly" is true if and only if the believability-degree of its subordinate clause is 0.

If the notation of a modal proposition in modal logic is used, then the three judging rules are as follows:

$$t(\Box p) = 1 \Leftrightarrow c(p) = 1 \tag{25.8}$$

$$t(\Diamond p) = 1 \Leftrightarrow 0 < c(p) < 1 \tag{25.9}$$

$$t(\neg \Diamond p) = 1 \Leftrightarrow c(p) = 0 \tag{25.10}$$

Example 25.1 Let p: Sun rises from the east tomorrow. Obviously, $c(p) = 1$. Therefore, $t(\Box p) = 1$.

Let q: Sun rises from the west tomorrow. Obviously, $c(q) = 0$. Therefore, $t(\neg \Diamond q) = 1$.

Let r: Jack is probably a translator. Suppose according to the analysis and estimation, $0 < c(r)$, then $t(\Diamond p) = 1$.

Note that in daily language, the word "necessarily" in necessarily-type modal propositions is often omitted, which leads to a necessarily-type modal proposition (as "p is necessary") to be simplified as the subordinate clause (p) it describes. This, in turn, is equivalent to say, those usual propositions that are not modified by modal word "necessarily" are really defaulted as "being necessary." From the relation between the possibility of a proposition being true and the believability of the proposition (see Sect. 24.1), that a proposition (p) is necessary is equivalent to say that the proposition is completely believable (i.e., $c(p) = 1$); from the relation between believability and truth of a proposition, that a proposition is completely believable $(c(p) = 1)$ is equivalent to that the proposition is true $(t(p) = 1)$ (see Theorem 24.1 in Sect. 24.1). Thus, that a proposition is defaulted as to be believable is equivalent to that the proposition is defaulted as to be true. The latter just is the natural logical semantics of a proposition we say (see Sect. 12.4). That is to say, that usual a proposition implicates itself being true is coincident with that it is defaulted as to be believable. This shows from another angle that the natural logical semantics of propositions we discover and present is existent and correct.

2. The reasoning with possibly-type modal propositions

We observe the question: Suppose there are possibly-type modal rule "If it is overcast then it will probably rain" and possibly-type modal proposition "It is probably overcast tomorrow" as a fact, question is: will it rain tomorrow?

It can be seen that since the evidentiary fact "probably overcast" does not match totally with the premise condition "overcast" of rule, it is not hard to do reasoning by using directly given possibly-type modal proposition and rule. However, possibly-type modal propositions and rules can be expressed as the form with believability-degrees, that is, converting them into proposition with believability-degree and rule with believability-degree; for the proposition with believability-degree and rule with believability-degree, we can do reasoning with believability-degrees stated in Chap. 24. Of course, the outcome of reasoning with believability-degree is also a proposition with believability-degree, but as long as it is converted into a modal proposition, the modal proposition is the conclusion to which original modal proposition and rule correspond.

Thus, we represent the rule "if it is overcast then it will probably rain" as

$$(\text{it is overcast then it will rain}, c_1)$$

and the fact "it will probably be overcast tomorrow" as

$$(\text{it is will overcast tomorrow}, c_2)$$

Then, we have the following reasoning with believability-degrees:

$$\frac{((\text{If it is overcast then it will rain, } c_1)}{(\text{It is overcast tomorrow, } c_2)}$$
$$\therefore (\text{It rains tomorrow, } c_3)$$

where $c_3 = c_1\, c_2$. Now, we convert c_3 into a modal or flexible modal linguistic value. Suppose the flexibl-modal linguistic value converted by believability-degree c_3 is "fairly likely," then the conclusion obtained by the modal proposition and rule given above is as follows: It rains tomorrow fairly likely.

From the above example, we see that for possibly-type modal propositions and rules, we can convert them into the propositions and rules with believability-degree and do the corresponding reasoning with believability-degrees; then, if needed, we convert the obtained conclusion into a corresponding modal proposition again.

It can be seen that the above method is actually translating the reasoning with the possibly-type modal propositions into the reasoning with believability-degrees with the subordinate clause of them. That is to say, the reasoning with possibly-type modal propositions here is actually the reasoning with believability-degrees with respect to the subordinate clauses, but not the reasoning with degrees or truth-degrees with respect to the main clauses. As for the reasoning with degrees or truth-degrees with respect to the main clauses, the principles and methods, in principle, are the same as that in previous Chap.15.

One problem worthy of our attention here is although a modal rule itself has true or false, or truth-degree, the real truth values of the modal rules do not play any role in reasoning. In fact, ① if a modal rule is directly employed to do reasoning (if that can be done), then it always is treated as true or rough-true, because if not, the reasoning could not be done; ② if converting directly a corresponding modal rule into a rule with believability-degree to do reasoning, then since the believability-degree in the rule with believability-degree is from conversion of the modal or flexible modal linguistic value in original modal rule, but not the real truth-degree of subordinate clause of original modal rule, and the rule with believability-degree always is still treated as true or rough-true; therefore, the truth-degree of the rule that is obtained from the real believability-degree of the rule is still not used in reasoning; ③ if the rule with believability-degree in reasoning is made up of the subordinate clause of corresponding modal rule and its believability-degree, then not to mention that such a rule with believability-degree is unnecessarily equivalent to the original modal rule, even if they are equivalent, the rule with believability-degree is still treated as true or rough-true. In a word, the above three cases of reasoning are all not related to truth value of rule with believability-degree. This shows again that the truth values of implicational propositions can only be used for the judgment of the truth or validity of itself, but cannot be used in the corresponding reasoning.

25.3   Origin of Possibly-Type Modal Propositions

Let A be a subset or flexible subset of U and B be a subset or flexible subset of V. And let $x_0 \in U$, $A(x_0)$ be a simple proposition, $y_0 \in V$, and $B(y_0)$ be a simple proposition.

By Definition 21.3, it is known that

$$\mathrm{poss}(A) = \mathrm{shar}(A) \qquad (25.11)$$

Thus, when shar $(A) \in (0, 1)$, $\mathrm{poss}(A) \in (0, 1)$. And $\mathrm{shar}(A) \in (0, 1)$ is equivalent to saying that set A holds the partial share of universal set U, and poss $(A) \in (0, 1)$ is equivalent to saying that linguistic value A can only be possessed by the partial objects in U. Thus, for $x_0 \in U$, the possibly-type modal simple proposition would be occurred: $A(x_0)$ is possible.

That the logical relation between "partial share," "partial possession," and possibly-type modal simple proposition above is to say that from "partial share," the "partial possession" can be drawn forth; from "partial possession," the possibly-type modal simple propositions can be drawn forth.

By Definition 21.9, it is known that

$$\mathrm{impl}(A(x), B(y)) = \mathrm{corr}(A, B) | \mathrm{cont}(A, B) \qquad (25.12)$$

Thus, when $\mathrm{corr}(A, B) \in (0, 1)$, or $\mathrm{cont}(A, B) \in (0, 1)$, $\mathrm{impl}(A(x), B(y)) \in (0, 1)$. And $\mathrm{corr}(A, B) \in (0, 1)$ is equivalent to saying that set A is partly corresponded by set B, $\mathrm{cont}(A, B) \in (0, 1)$ is equivalent to saying that set A is partly contained by set B, and $\mathrm{impl}(A(x), B(y)) \in (0, 1)$ is equivalent to saying that proposition $A(x)$ implies partly proposition $B(y)$. Thus, for $x_0 \in U$ and $y_0 \in V$, the possibly-type modal implicational proposition would be occurred: $A(x_0) \rightarrow B(y_0)$ is possible.

That the logical relation between "partial correspondence" or "partial inclusion," "partial implication," and possibly-type modal implicational proposition above is to say that from "partial correspondence" or "partial inclusion," the "partial implication" can be drawn forth; from "partial implication," the possibly-type modal implicational propositions can be drawn forth.

To sum up, a "route map" drawing possibly-type modal propositions is shown in Fig. 25.1.

From this diagram, it can be seen intuitively that the possibly-type modal propositions originate from the feature of "partial share" and the relations of "partial correspondence" or "partial inclusion" of relevant sets.

For "partial share," the range of shares is the real interval $(0, 1)$, and for "partial correspondence" and "partial inclusion," the range of correspondence-rates and the range of inclusion-degrees are also the interval $(0, 1)$. Therefore, viewed abstractly, interval $(0, 1)$ is also the range of numerical values to which quantifiable linguistic value "partial" corresponds. Thus, the flexible linguistic values of "most," "majority," "overwhelming majority," "few," "tiny minority," and so on can be defined on the interval $(0, 1)$. These flexible linguistic values are all to describe quantities,

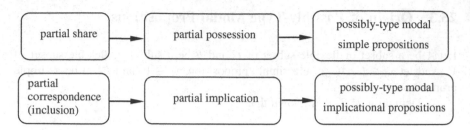

Fig. 25.1 A "route map" of possibly-type modal propositions being drawn (1)

so they can also be used to describe "partial possession" and "partial implication." For example,

```
Majority of trees have been sprouting.
    In most cases, if the deposit interest rates fall, the stock is
up.
```

Obviously, this sentence is also a flexible proposition. In the following, we introduce several terms and their notations and give the formal representation of this kind of flexible propositions.

Definition 25.3 We call collectively the words that represent imprecise number such as "majority," "vast majority," "overwhelming majority," "minority," and "tiny minority" to be the flexible quantifiers and denote them in order as Mx, BMx, $VBMx$, Ux, and VUx; the propositions quantified by these flexible quantifiers are called the flexibly-quantified propositions.

For instance, $MxA(x)$ is a flexibly-quantified proposition, which means there are a majority of $x \in U$, and x is A.

For another instance, $BMx \, \exists \, y(A(x) \rightarrow B(y))$ is also a flexibly-quantified proposition which means for great majority of $x \in U$, there exists $y \in V$, and if x is A, then y is B.

It can be seen that flexibly-quantified propositions are actually a kind of statements of specific "partial possession" and "partial implication." Therefore, flexibly-quantified propositions originate also from the "partial share" and the "partial correspondence" or "partial inclusion" of sets. On the other hand, it is not hard to see that from flexibly-quantified proposition, possibly-type modal propositions can also be drawn forth.

For example, given a flexibly-quantified proposition $MxA(x)$ $(x \in U)$, then by the semantics of $MxA(x)$, for $\forall x_0 \in U$, we have possibly-type modal proposition: A (x_0) is likely.

Similarly, from flexibly-quantified proposition $BMx \, \exists \, y(A(x) \rightarrow B(y))$, we can have flexible possibly-type modal proposition:

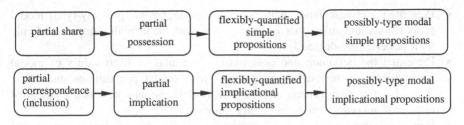

Fig. 25.2 A "route map" of possibly-type modal propositions being drawn (2)

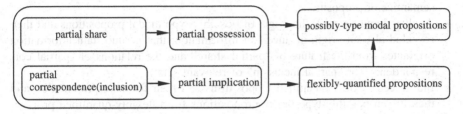

Fig. 25.3 A "route map" of possibly-type modal propositions being drawn (3)

If $A(x)$ then very likely $B(y)$, that is, $A(x) \rightarrow B(y)$ is very likely

Thus, we obtain another "route map" drawing possibly-type modal propositions (see Fig. 25.2).

Synthesizing Figs. 25.1 and 25.2, we have the following as in Fig. 25.3.

The diagram shows that it is just from the feature of "partial share" of relevant sets and the relations of "partial correspondence" or "partial inclusion" between relevant sets that possibly-type modal propositions are drawn. Since a possibly-type modal proposition is actually to describe its uncertain subordinate proposition(s), Fig. 25.3 shows and reveals also an origin of uncertain information. Also, since flexibly-quantified propositions express imprecise information, thus, the diagram shows and reveals also a correlation between imprecise information and uncertain information.

25.4 Summary

In this chapter, the work and results of us are as follows:

- Introduced the terminology of flexible modal words, subdivided modal propositions, and rules and presented the concepts and terminologies of necessarily-type, possibly-type, and flexible possibly-type modal propositions and rules.

- Presented a representation with believability-degree of (possibly-type) modal propositions and rules, that is, proposition with believability-degree and rule with believability-degree.
- Presented the definition and computation formulas of truth values of modal propositions; gave a reasoning method with modal propositions and rules; derived a transitive formula of believability-degrees; and presented a model of relational inference with believability-degrees.
- Revealed the relationship between the simplified representation of necessarily-type modal propositions and the natural logical semantics of propositions, thus showing the correctness of our view about the logical semantics of propositions.
- Analyzed and revealed the origin of possibly-type modal propositions and then revealed also an origin of uncertain information, that is, uncertain information originates from the feature of "partial share" and the relations of "partial correspondence" or "partial inclusion" of relevant sets.
- Introduced flexibly-quantified propositions and their formal representation and then drawn possibly-type modal propositions from flexibly-quantified propositions. Thus, also a correlation between imprecise information and uncertain information is revealed, that is, uncertain information can also be drawn from some of imprecise information.

Reference

1. Lian S (2009) Principles of Imprecise-Information Processing. Science Press, Beijing

Chapter 26
Several Further Research Directions and Topics

Abstract This chapter discusses further work, which includes application development and theoretical and technical explorations. In what follows, the author, from his perspective, presents directly 6 research directions and some research problems for references.

Keywords Imprecise information · Flexible linguistic values · Artificial intelligence · Anthropomorphic intelligent systems · Natural language processing

All previous chapters of this book have founded a theoretical and technological system of imprecise-information processing, which provides a technological platform for relevant applications and lays a theoretical foundation for further researches. In this chapter, we discuss further work, which includes application development and theoretical and technical explorations. But since the problems involving imprecise-information processing are very extensive, this chapter cannot exhaust all research directions and topics. In what follows, the author, from his perspective, presents directly 6 research directions and some topics for reference [1].

26.1 Development of Anthropomorphic Computer Application Systems and Intelligent Systems with Imprecise-Information Processing Ability

It is obvious that the anthropomorphic computer application systems and the intelligent systems with imprecise-information processing ability are more flexible, friendly, and more intelligent. The ability to deal with imprecise information is necessary for some systems, especially for the computer systems involving human's natural language.

This research direction puts the imprecise-information processing technology given in the book directly into practices to solve relevant practical problems and

© Springer Science+Business Media Singapore 2016   633
S. Lian, *Principles of Imprecise-Information Processing*,
DOI 10.1007/978-981-10-1549-6_26

develop corresponding computer application systems (such as the imprecise-problem solving systems about classifying, recognition, judging, controlling, diagnosis, forecasting, and planning) and intelligent systems (such as intelligent computers, intelligent robots, intelligent Internet/Web, intelligent mobile phones, and expert systems), especially anthropomorphic computer application systems and intelligent systems. Dividing from technology, this direction mainly has the following subjects:

1. Development of various N-L and L-N conversion interfaces

Almost all of the computer application systems and intelligent systems with imprecise-information processing ability require the conversion interfaces between numerical values and flexible linguistic values. And the anthropomorphic computer application systems and intelligent systems more require the input and output interfaces as human's perceiving and expressing. In addition to the related hardware devices (such as sensors), the key technologies inside these interfaces are also the N-L and L-N conversions as we said. Therefore, the development of the N-L and L-N conversion interface is an important project. Obviously, to build a common conversion interface is difficult and also not necessary; we only need to build the conversion interfaces with respect to different problems.

To do N-L and L-N conversions, there must be the corresponding flexible linguistic values. Thus, a conversion interface of N-N or L-N needs a flexible linguistic value base to support it. The key of a flexible linguistic value is its mathematical model, i.e., consistency function. It is relatively easy to establish a normal model of a flexible linguistic value, but the dynamic model is difficult. This is also a problem that needs to be further studied.

Of course, for a high-level anthropomorphic intelligence system, having only these conversion interfaces is not enough, but which also requires a support of the machine understanding and generation techniques of imprecise information. This involves natural language processing. About it, we will discuss specially in the following Sect. 26.3.

2. Development of problem-solving systems with imprecise knowledge

(1) **Development of problem-solving systems with membership or consistency function**

This subject is to develop the computer application systems which are based on flexible linguistic values or flexible sets (or more general quantifiable linguistic values or ordered sets) and can solve relevant practical problems or engineering problems. The main work this subject involves is the determination and acquirement of relevant membership or consistency function, and the designing of relevant computer software.

It should be pointed out that though pattern recognition and comprehensive judging are also important application areas of fuzzy technology, the flexible recognition and flexible judging given in the book are new technologies different from fuzzy recognition and fuzzy judging. Besides, the flexible programming proposed in the book is also a kind of new technology yet to be put into use. These new technologies and their applications are all problems for further research.

Information fusion is a research hot spot at present. The synthesizing of flexible linguistic values in this book has similarities with information fusion. Then, can the synthesizing of flexible linguistic values be applied to information fusion? That is also a research problem. Obviously, applying flexible linguistic-valued synthesis to information fusion will add new content for information fusion, and the research on this problem has important significance and application value for the build of intelligent robots.

(2) **Development of problem-solving systems with flexible linguistic rules or flexible linguistic functions**

This subject is to develop computer application systems which base on flexible linguistic rules or flexible linguistic functions and can solve relevant practical problems and engineering problems. The main work this subject involves is acquiring relevant flexible linguistic rules and its corresponding adjoint-degreed functions or measured functions, acquiring relevant flexible linguistic functions, and designing relevant computing and reasoning software systems.

What should be pointed out is that for the classical control problem of inverted pendulum control, though by using multiple methods such as traditional control, fuzzy control and humanoid control and so on, people have already done successfully experiments and implementations, we still need to experiment with the flexible control method presented in this book. In doing so, on the one hand, flexible control method can be tested and advanced; on the other hand, it can also be compared with other methods.

3. **Common development platforms for flexible linguistic rule/function-based systems**

In order to help develop intelligent systems with flexible linguistic rules and flexible linguistic functions, we can develop relevant common development platforms, of which the main work are the design and implementation of flexible linguistic rules base and its management system, inference and computing components, definition and computing components of flexible linguistic functions, N-L conversion and L-N conversion components, input and output interfaces, etc. Here, it involves the description language of flexible linguistic rules, the structures of rule bases and function bases, relevant function modules of management and maintenance, explanation modules, inference algorithms, imprecise-reasoning models, etc. In addition, it also involves some sub-problems such as dual reasoning modes and models for imprecision and uncertainty.

4. **Adaptive flexible linguistic rule/function-based systems**

In order to further improve the intelligent level and effect of flexible linguistic rule/function-based systems, adaptive flexible linguistic rule/function-based systems must be developed. This kind of adaptive system mainly includes two parts of the automatic acquisition and real-time optimization of flexible linguistic rules and flexible linguistic functions. That is to say, this kind of flexible linguistic rule/function-based system has self-learning ability, which can learn from the work environment to find flexible linguistic rules or flexible linguistic functions and make them optimized continuously, thus improving its intelligent level continuously. Here, it involves the dynamic expanding or contracting and the dynamic flexible clustering and flexible partition (even hierarchical flexible clustering and flexible partition) of a measurement space, self-organizing of flexible linguistic values, self-optimizing of rules and functions, etc.

26.2 Imprecise-Knowledge Discovery and Machine Learning with Imprecise Information

On this direction, there are at least the following research subjects:

1. **Flexible cluster analysis**

In Sect. 19.4, some algorithms on flexible cluster analysis were already presented, but these algorithms still need to be put into practice and get checked and improved in practical problems. In addition, this research subject also needs further exploration.

2. **Discovery of flexible linguistic rules**

In Sect. 19.5, some basic ideas and approaches for the discovery of flexible linguistic rules were already given. But there the method is actually tantamount to rigidening (hardening) the flexible linguistic values firstly and then processing it according to the usual decision tree learning method. Then, if with flexible classes directly, how is the decision tree learning conducted? Here, it would involve the computation of the corresponding information entropies and the problem of one and the same object possessing simultaneously multiple flexible linguistic values.

3. **Discovery of flexible linguistic (flexible-numbered) functions**

In Sects. 19.6 and 19.7, some ideas and methods for the discovery of flexible linguistic functions and flexible functions (i.e., flexible-numbered functions) were already given. However, the mathematical background of the practical problems may be various, so the discovery of flexible linguistic functions is not natural. For example, to a flexible subset of measurement space U there correspond many flexible subsets of measurement space V. Then, for this situation, the corresponding mathematical background needs to be analyzed or processed, such as by increasing the dimensions of the definition domain to decrease the number of the corresponding values. For another example, it is also a problem whether the local linguistic function formed by several pairs of flexible linguistic values (or rules with flexible linguistic values) can always be extended into a global flexible linguistic function. Flexible linguistic (flexible) function discovery is very important to the complex system modeling; then, can we design an algorithm of summarizing the corresponding flexible linguistic (flexible) function with the relevant sample data? This is a problem worth considering. Here, we can consider the hierarchical flexible clustering and flexible partition, thus designing a set of stepwise refinement (i.e., the size of flexible linguistic values is diminishing) flexible linguistic functions. In a word, on the discovery of flexible linguistic (flexible) functions, more research needs us to do.

4. Function learning with flexible numbers or flexible sets

Function learning is to acquire functions that satisfy certain requirements with some known sample data and by using certain machine learning methods. Usually, the sample data are all the ordered pairs consisted of ordinary numbers (real numbers) or vectors, and the corresponding data sets are also ordinary sets. Then, when the sample data are the ordered pairs consisted of flexible numbers or flexible vectors, or corresponding data sets are flexible sets, how is the corresponding function learning to be done?

Actually, now there already appeared some researches and results on this aspect. For example, people now introduce the idea of fuzzy set and membership-degree to support vector machine (SVM); thus, there occurs a new technique of fuzzy SVM. Thus, we can consider how to combine flexible numbers, flexible vectors, and flexible sets with SVM to realize "flexible support vector machine."

Also, on the basis of SVM, some scholars also proposed Core Vector Machines (CVM) [2] and Ball Vector Machines (BVM) [3]. Core vector machine can be said to be a kind of fast SVM suitable for large data sets. The idea of core vector machines is to transform the Kernel methods including SVMs into the Minimum Enclosing Ball (MEB) problem in computational geometry. We find that the minimum enclosing ball here has some similarities to the flexible ball proposed in this book. Therefore, we can further study the connection and relation between the two and then find the machine learning methods based on flexible balls and other geometrical flexible classes.

Of course, we can also combine flexible numbers, flexible vectors, and flexible sets with other machine learning techniques such as artificial neural network and intelligent algorithms. Thus, this research subject is also the problem of how to combine imprecise-information processing with traditional machine learning techniques.

26.3   Natural Language Understanding and Generation with Flexible Concepts

The words that represent flexible concepts can be found everywhere in our daily language, so the natural language understanding and generation with flexible concepts appear to be very necessary and important. Therefore, it naturally becomes a direction necessitating further research. On this direction, several tentative ideas have already been proposed in Sect. 22.5. Overall, on the one hand, the mathematical models of flexible concepts can be introduced into natural language processing, or in other words combining the theories of flexible linguistic values with the existing natural language processing theories and techniques, to study the natural language processing with flexible concepts. For instance, we can study the semantics theory with flexible linguistic values. For another instance, embedding the interconversion mechanism between numerical values and linguistic values into the process of the usual natural language processing is also a considerable approach. On the other hand, we can explore new theories and methods of natural language processing according to the characteristics of flexible concepts and the mathematical models given in this book. For example, we can combine the membership or consistency function of flexible concepts to build the knowledge bases or corpuses of flexible concepts to accomplish the understanding of flexible concepts. Besides, there is also the dynamic modeling problem of flexible concepts. In short, there still exits many problems on natural language generation and understanding with the flexible concepts to need us research. Of course, the natural language processing referred to here is only at the level of linguistics and computer programming, as to the natural language understanding and generation at the level of human brain is yet in need of the breakthrough of brain and cognitive science.

26.4   Flexible Logic Circuits and Flexible Computer Languages

Traditional logic circuits are all designed based on "rigid" logic (including two-valued logic and multivalued logic). Just the same, traditional programming languages (including machine language and advanced programming language) are all designed with "rigid" linguistic values. Now, with truth-degreed logic and flexible linguistic-truth-valued logic, we can also design corresponding logic circuits and computer languages based on the two logics. Further, these logic circuits can be used to form all kinds of special purpose chips and computers.

On the other hand, flexible computer languages with flexible linguistic values can also be developed. Specifically speaking, they include those from the machine flexible instructions of lowest level to flexible requirements description language in flexible engineering, flexible description language of software architectures, etc., and those from the general flexible machine languages and flexible advanced programming languages to flexible command languages directly used in decision and control. Certainly, the research in this area would also involve flexible formal language and automaton theory with flexible linguistic values.

With these computer hardware and software oriented to imprecise information, we can directly use them to process imprecise information. Obviously, this will be very significant and valuable to some instruments and meters, installations and facilities, sensors, effectors, etc., especially sensor networks and intelligent robots.

It can be seen that there are many subjects and subsubjects being worthy of study in this direction.

26.5 Exploring on Brain Model of Flexible Concepts and Qualitative Thinking Mechanism of Human Brain

Membership function and consistency function are the mathematical models of a flexible concept, but they are actually a kind of relation and transformation model from the physiological numerical information of human brain to its psychological linguistic information. Then, viewed from the angle of anatomy, what kinds of model and mechanism are the expression, relation, and transformation of these two kinds of information in human brain? Such is just the brain model of flexible concepts we speak. Closely connecting with the brain model of flexible concepts is the qualitative thinking mechanism of human brain. Thus, the brain model of flexible concepts and qualitative thinking mechanism of human brain will form another research direction. For this research direction, we only give the following tentative thoughts.

For the brain models of flexible concepts, what we might first think of is neural network. The classical neural network models have forward network, feedback network, self-organizing network, etc. Which is more suitable for the representation of flexible concepts?

The work process of forward BP network is the input-processing-output of information, which is actually a function and thus suitable for processing of information. Thus, BP network can be used to realize the membership functions and consistency functions of flexible concepts as the brain model of flexible concepts. The widely interconnected Hopfield network is good at expressing states, thus suitable to realize the memory and association of information. Then, we can also consider using Hopfield network to store flexible concepts. Self-organizing feature map network Kohonen is a kind of unsupervised competitive learning network, which is good at realizing clustering. Then, if the winner unit of this network represents the core of a flexible concept, then its neighborhood is just tantamount to the support set. Thus, in comparison, self-organizing map network seems more suitable to represent flexible concepts. However, neural networks generally need learning and training, while most flexible concepts, especially commonsense flexible concepts, generally do not need learning. Therefore, even if using traditional neural networks to represent flexible concept can be realized forcedly, there are still some limitations.

With the development of quantum information technology, people proposed the concept of Quantum neural computing. By researching, it is discovered that there occurs quantum effect in human brain, and there is a certain association between the quantum phenomenon and the consciousness of human brain, and the collapse of quantum wave function is very similar to the reconstructing of neural pattern in the memory process of human brain, and there is astonishing similarity between the dynamic equations of quantum process based on quantum potential and brain process based on neural potential, and a single quantum neuron has just the non-linear mapping ability that is tantamount to that of the traditional two-layered forward neural network. Quantum neural computing is now becoming a research hot spot, and a number of quantum neural network models have been proposed such as quantum parallel self-organizing mapping model (SOM) and quantum associative memory model. The weight update of SOM is completed through a series of synchronous operations, thus making the traditional repeating training process become a one-time learning process, which is more similar to the one-time learning and memory function of human brain. Quantum associative memory model can save 2^n patterns within the time of $O(mn)$ by using n quantum neurons and also recall one pattern in the time of $O(\sqrt{2^n})$. It can be seen that this kind of quantum neural network model has exponential improvement in memory capacity and recalling speed compared with traditional Hopfield network model. Therefore, if this kind of quantum neural network is used to store flexible concepts, then the performance is more superior.

Like other concepts, flexible concepts are a kind of feeling or sensation at the "macro"-psychological level of human brain. In fact, the thinking activities (such as our memory, association, reasoning, computing, and thinking) that human brain can be aware is carried out at the psychological level. The thinking processes at the psychological level can be explicitly expressed in linguistic symbols; thus, human brain's thinking process can be expressed in words and be modeled in logic. However, the thinking of human brain at the macro-psychological (or linguistic) level is closely connected to the group behavior at neuronic level—it is just the group behavior of neurons at the low micro-physiological level that forms the advanced thinking activities at the macro-psychological level. The relation between the group behavior of neurons and the advanced thinking activities of human brain is the "gush" or "emergence" phenomenon called in system science. Therefore, a flexible concept should also a kind of overall attribution gushed or emerged from a system formed by multiple basic units in human brain. If these basic units are neurons, then neuron networks can be used for modeling of flexible concepts. Further, if the whole attribute emerged from a neuron network is at the macro-physical level (such as electric level), then the traditional neuron networks can be used for modeling; if the whole attribute emerged from a neuron network is at the micro-physical level (such as electron, atom, or molecule), then the quantum neuron networks can be used for modeling. However, if these basic units supporting the thinking of human brain are not neurons, and neurons are rather only a kind of transporting and processing channels of information, then we need to further probe what on earth are these basic units. Some studies have shown that a new physics that connects quantum phenomenon and general relativity can explain some human brain activities such as understanding, perception, and awareness. Based on that, people transfer attention from studying the network structure formed by neurons to focusing on analyzing the supermicro-structure and molecular combination within a neural cell such as canaliculus, microfilament, and neurofilament inside a neuron.

Since the psychological activities and thinking of human brain are closely connected with language, studying the linguistic mechanism of human brain should be very helpful for revealing the mystery of these basic units. In addition, the formation principles of all kinds of senses (such as visual sense, hearing, taste, sense of touch, and hot and cold feelings) of human brain are also helpful for the revealing of this mystery of the basic units, while the physiological principles about language and senses need to be found in the layers and subareas of cerebral cortex.

On the other hand, whether the material basis which forms the psychological feelings and thinking of human brain is electrical or chemical is also a key problem needing us considering. Because viewed from the information transmission mechanism of a neural system, it is both electrical (showing electric potential and impulse) and also chemical (showing various kinds of neurotransmitters).

It can be seen that in this direction there still exists many challenging problems to need us further research. It goes without saying that as long as the brain model of flexible concepts is made a breakthrough, the qualitative thinking mechanism of human brain will also be revealed along with.

In addition, it is worth considering whether or not the reasoning and computing with degree proposed and realized in this book can be treated as a clue to help exploring human brain's logical reasoning mechanism and further revealing the general qualitative thinking mechanism.

26.6   Related Mathematical and Logic Theories

The flexible sets, flexible relations, flexible linguistic values, flexible linguistic functions, flexible numbers, flexible functions, etc. that are proposed in this book form a basis of "imprecise mathematics" or "flexible mathematics," while the truth-degreed logic and flexible-linguistic-truth-valued logic and so on in the book form a basis of "imprecise logic" or "flexible logic." From that, we can further research and find related mathematical theories and logic theories.

In the mathematics, we can consider these topics below:

- On the basis of flexible sets and flexible relations, research and found corresponding theories of matrixes, graphs, lattices, algebraic systems, measures and integrals, topology and space, and so forth.
- Viewed from the angle of geometry, the flexible points, flexible lines, flexible planes, flexible circles, and flexible squares in this book are a kind of flexible geometry. In contrast, the usual geometries are rigid geometries. Thus, the flexible geometries are the extension of rigid geometries, while rigid geometries are the contraction of flexible geometries. Then, based on these flexible geometries, we can explore and found the corresponding "flexible geometry" theory. Besides, the multiple kinds of geometrical flexible classes in this book are all connected with computational geometry. Therefore, the two can be combined for research.
- Flexible numbers can be viewed as the extension of usual real numbers. Then, can we found a kind of "flexible mathematics" based on flexible numbers?
- Flexible vectors are a kind of extension of usual vectors, and usual vectors can be viewed as the contraction of flexible vectors. Then, can we further research the operations on flexible vectors and then found the corresponding "flexible-vector algebra" and "flexible-vector space" theories?
- Flexible functions are a kind of extension of usual real functions, and usual real functions can be viewed as the contraction of flexible functions. Then, can we research and found the theories of "flexible calculus" and "flexible differential equations" based on flexible functions?
- Theory and technique of approximate computation based on flexible linguistic functions.
- Theories on probability and mathematical statistics and entropy based on flexible sets, flexible linguistic values and flexible numbers, etc.

In the logic, on the basis of the truth-degreed logic, flexible linguistic-truth-valued logic founded in this book, we can further research negation-type logic, opposite-type logic, multivalued logic, etc., and connect and compare these logics with the existing logic theories to make clear and straighten out their relations to found a more complete logic theory. On the other hand, new logic branches can be developed on the basis of the logic theories in this book such as flexible modal logic and numerical modal logic, flexible cognitive logic, and flexible command logic, and it can also be studied whether there exists mutually complementary logic and mutually exclusive logic. Logic operations and logic systems also involve algebraic theories, and there are also some problems necessitating further research. Besides, the existing fuzzy mathematics theories can be restudied by using the flexible set theory.

As can be seen, there is a wide range of subjects for further research and exploration in mathematics and logic, which indicates that in these directions one can accomplish a great deal.

References

1. Lian S (2009) Principles of imprecise-information processing. Science Press, Beijing
2. Tsang IW, Kwok JT, Cheung P-M (2005) Core vector machines: fast SVM training on very large data sets. J Mach Learn Res 6(2005):363–392
3. Tsang IW, Kocsor A, Kwok JT (2007) Simpler core vector machines with enclosing balls. In: ICML '07 proceedings of the 24th international conference on machine learning, pp 911–918

Index

© Springer Science+Business Media Singapore 2016
S. Lian, *Principles of Imprecise-Information Processing*,
DOI 10.1007/978-981-10-1549-6

Printed in the United States
By Bookmasters